MEDICINAL MARINE MICROBIOTA

MEDICINAL MARINE MICROBIOTA

Santhanam Ramesh, PhD
Ramasamy Santhanam, PhD
Arumugam Uma, PhD
Raju Mohanraju, PhD

AAP | APPLE ACADEMIC PRESS

First edition published 2025

Apple Academic Press Inc.
1265 Goldenrod Circle, NE,
Palm Bay, FL 32905 USA

760 Laurentian Drive, Unit 19,
Burlington, ON L7N 0A4, CANADA

CRC Press
2385 NW Executive Center Drive,
Suite 320, Boca Raton FL 33431

4 Park Square, Milton Park,
Abingdon, Oxon, OX14 4RN UK

Library and Archives Canada Cataloguing in Publication

CIP data on file with Canada Library and Archives

Library of Congress Cataloging-in-Publication Data

CIP data on file with US Library of Congress

ISBN: 978-1-77491-640-7 (hbk)
ISBN: 978-1-77491-641-4 (pbk)
ISBN: 978-1-00350-496-2 (ebk)

About the Authors

Santhanam Ramesh, PhD
Vice-Principal, Karuna College of Pharmacy, Kerala University of Health Sciences, Palakkad, Kerala, India

Santhanam Ramesh, PhD, is Vice-Principal at the Karuna College of Pharmacy, Kerala University of Health Sciences, Palakkad, Kerala, India. He has 15 years of teaching and research experience in pharmaceutical sciences, and his fields of specialization include natural products, pharmaceutical nanotechnology, biomaterials, and marine wastes of pharmaceutical importance. Dr. Ramesh is also a visiting professor in the Department of Pharmacology, North-Caucasian State Humanitarian and Technological Academy (North-Caucasian State University), Cherkessk, Russia. To his credit, he has 13 books published with internationally reputed publishers. He is a member of the International Society for Pharmacoeconomics and Outcome Research, the British Society for Nanomedicine, and the Academic Pharmacy Group of Royal Pharmaceutical Society, London. He earned his PhD from Jawaharlal Nehru Technological University, Hyderabad, India.

Ramasamy Santhanam, PhD
Former Dean, Fisheries College and Research Institute (FC &RI), (Tamil Nadu Dr. J. Jayalalithaa Fisheries University), Thoothukudi, India

Ramasamy Santhanam, PhD, is Ex-Dean, Fisheries College and Research Institute, Tamil Nadu Veterinary and Animal Sciences University, Thoothukudi, India. He has about 50 years of teaching and research experience in marine/fisheries sciences. He has been serving as a fisheries expert for various government and nongovernment organizations in India and abroad. To his credit, Dr. Santhanam has 40 books on various aspects of marine life. He was a member of the American Fisheries Society, United States; World Aquaculture Society, United States; Global Fisheries Ecosystem Management Network (GFEMN), United States; and the IUCN's Commission on Ecosystem Management, Switzerland.

Arumugam Uma, PhD
Professor, Dr. MGR Fisheries College and Research Institute (Tamil Nadu Dr. J Jayalalithaa Fisheries University), Ponneri, Chennai, India

Arumugam Uma, PhD, is presently serving as Professor at Dr. MGR Fisheries College and Research Institute (Tamil Nadu Dr. J. Jayalalithaa Fisheries University), Ponneri, Chennai, India. She has 25 years of teaching and research experience, and her research interests include fish immunology and therapeutics. She is the inventor of two patented technologies and recipient of the Tamil Nadu Young Women Scientist and Best Researcher Award. She is a member of the Asian Fisheries Society, Manila, and World Aquaculture Society, USA. To her credit, Dr. Uma has two books and 30 research papers published in various peer-reviewed journals of international repute.

Raju Mohanraju, PhD
Professor, Department of Ocean Studies & Marine Biology, Pondicherry Central University, Port Blair, Andaman's, India

Raju Mohanraju, PhD, is presently serving as Professor in the Department of Ocean Studies and Marine Biology at Pondicherry Central University, Port Blair, Andaman's, India. He has 25 years of teaching and research experience, and his research interests include marine microbiology and marine natural products. He has undergone overseas training at the University of Iowa as well as the Woods Hole Oceanographic Institution, USA. He has so far guided 15 PhD students. He received the Dr. Bernie-Davies award from the Woods Hole Oceanographic Institution, USA, and three consecutive best teacher awards from Pondicherry University for his outstanding work. He has also participated as a chief scientist in research cruises on both ocean and coastal international research vessels. To his credit, Dr. Mohanraju has 70 research papers published in various peer-reviewed journals of international repute.

Contents

Preface .... *ix*

1. **Introduction** **1**
2. **Toxins of Marine Microalgae and Their Potential Therapeutic Applications** **17**
3. **Medicinal Marine Microalgae** **29**
4. **Medicinal Bacteria Associated with Marine Plants and Animals: Their Compounds and Bioactivities** **163**
5. **Medicinal Fungi Associated with Marine Plants and Animals: Their Compounds and Bioactivities** **337**

Index ***521***

Preface

The marine biota viz. microbes, plants, and animals account for about 50% of the total world biodiversity. But their importance as a rich source of bioactive compounds and their applications in the development of new drugs has only recently been understood through several scientific advancements. In other words, the marine environment is a largely unexplored resource for understanding the diverse communities of medicinally important organisms. Unlike the terrestrial ecosystems, marine habitats provide unique environments for the growth of microbes. Further, the co-evolution of many marine invertebrates with these microorganisms has led to the development of symbiotic relationships between the host organisms and a number of species of marine bacteria and fungi.

Microbes such as bacteria and fungi have been reported to make a unique contribution to the health of people throughout the world. These microbes produce both primary metabolites such as amino acids, vitamins, and nucleotides; and secondary metabolites such as antibacterial, antiviral, and anticancer compounds which constitute more than half of the pharmaceuticals on the market today. Many of these bioactive compounds from marine microbes are largely due to the application of new screening technologies and utilizing the recent advances in the genetics of secondary metabolism in these microbes.

Marine microbiota viz. microalgae (cyanobacteria, dinoflagellates, diatoms etc.); bacteria and fungi of sea water, sediments; and bacteria and fungi associated with marine plants and animals serve as novel sources of bioactive compounds (secondary metabolites) that offer a vast scope as drug candidates for the treatment of various diseases in humans, such as cancer, diabetes, anemia, diarrhea, atopic dermatitis, obesity, Crohn's disease, etc. These metabolites are also promising sources of natural antioxidants, immunosuppressants, antibiotics, enzyme inhibitors, hypocholesterolemic agents, etc. As very limited scientific information is available on the bioactive potential of marine microbiota, intensive research and investigation are very much needed to explore the importance of marine microbiota as producers of novel drugs.

Though some books are presently available on marine natural products, a comprehensive book on the medicinally important marine microbiota viz. microalgae, bacteria, and fungi is hitherto wanting. Keeping this in consideration, this publication is being brought out for the first time by bringing together the experts of marine biopharmacy, marine microbiology, and marine biology disciplines. Aspects relating to the medicinally important marine microbiota such as medicinal compounds of marine microalgae (cyanobacteria, diatoms and dinoflagellates) and bacteria and fungi associated with marine plants and animals have been dealt with in this comprehensive volume. It is strongly hoped that this title will largely help the researchers and students of disciplines that include marine bio-pharmacy, marine biology, marine microbiology, marine biochemistry, and marine biotechnology as a standard reference for the libraries of colleges and universities and as a guide for the drug companies involved in the development of new drugs from marine microbiota.

I am highly indebted to Dr. K. Venkataramanujam, former Dean, Fisheries College and Research Institute, Tamil Nadu Dr. J. Jayalalithaa Fisheries University, India, for his valued comments and suggestions on the manuscript. I sincerely thank all my international friends who were very kind enough to share their images for the present publication. The services viz. photography and secretarial assistance rendered by Mrs.Albin Panimalar Ramesh are also gratefully acknowledged.

CHAPTER 1

Introduction

Modern lifestyle which is invariably associated with junk food, alcohol, and the use of antibiotics has been largely responsible for various diseases such as cancer, cardiovascular disease, and aging among humans. While microorganisms are resistant to drugs, medications possess side effects especially in the cancer treatment. This calls for novel drug candidates, especially from the microbiota of the aquatic biotopes viz freshwater, seas, and oceans. Among the different components of marine microbiota viz. microalgae or phytoplankton and microbes (bacteria and fungi), the microalgae which comprise cyanobacteria, the Gram-negative oxygenic photoautotrophs (blue-green algae), diatoms, and dinoflagellates assume greater importance as they form the world's largest biomass with about 30% of the world's photosynthesis.

MARINE MICROALGAE

The marine microalgae are presently attracting considerable interest worldwide due to their potential industrial applications. These microalgae have been reported to serve as colorant or supplement for feed and food industries, bioremediators in the waste water treatment, raw materials in the production of renewable fuels, and pharmaceuticals and cosmetics (Wu et al., 2021).

AQUAFEED

Several species of marine microalgae serve as feed in the aquaculture industries. They serve as direct or indirect feed sources in the fish or shrimp hatchery. While the smaller sized (46 m) microalgae ease the digestibility of the early juveniles, the high polyunsaturated fatty acids contents viz.

Medicinal Marine Microbiota. Santhanam Ramesh, Ramasamy Santhanam, Arumugam Uma & Raju Mohanraju (Authors)

docosahexaenoic acid (DHA) and eicosapentaenoic acid (EPA) make these microalgae as a nutritional and well-balanced diet. Further, species like Isochyrisis sp. serves as a living culture product and freeze-dried Isochyrisis Tahitianstrain T-ISO serves as a feed for animal plankton diets such as rotifers and copepods (Camacho et al., 2019).

COLORANT OR SUPPLEMENT FOR FEED AND FOOD INDUSTRIES

The β-carotene, astaxanthin, and phycocyanin pigments and polyunsaturated fatty acids (PUFAs) viz. EPA and DHA of the marine microalgae serve as colorants or food supplements for human food and livestock feed industries. Table 1.1. indicates the common species of marine microalgae and their applications.

TABLE 1.1 Marine Microalgae and Their Food and Feed Applications.

Nutritional Component	Source Species	Applications
β-carotene (carotenoid pigment)	*Dunaliella salina*	Food colorant; additive for feed and food supplement (as Provitamin A)
Astaxanthin (carotenoid pigment)	*Haematococcus pluvialis*	Additive for feed and food supplement
Phycocyanin (Phycobili-protein)	*Porphyridium* sp. and Arthrospira sp.	Food pigment reagents (fluorescent markers)
DHA	*Schizochytrium* sp. and *Cryptocodinium* sp.	Food supplement
EPA	*Nitzschia* sp., *Nannochloropsis* sp. and *Phaeodactylum* sp.	Food supplement

Abbreviations: DHA, docosahexaenoic acid; EPA, eicosapentaenoic acid.

Source: Enzing et al. (2014).

RENEWABLE FUELS

Several species of marine microalgae could serve as feedstock in the production of renewable fuels such as hydrogen, ethanol, biodiesel, and methane. Biodiesel produced from microalgae can serve as petroleum diesel and has no sulfur content. Further, in the use of algae-based diesel, there are reduced emissions of hydrocarbons, SOX, particulate matter, and CO (https://www.sciencedirect.com/topics/social-sciences/biodieseldia).

BIOREMEDIATION

Many species of marine microalgae have been reported to serve as bioremediators as they are capable of eliminating 100% ammonia nitrogen, 85% nitrate nitrogen, 62% nitrite nitrogen, and 78% phosphorus from wastewaters. Apart from treating wastewaters, these microalgae produce biomass materials simultaneously (Zheng et al., https://www.scientific.net/paper-keyword/isochrysis-zhanjiangensis).

COSMETICS

The secondary metabolites isolated from the marine microalgae may be of great use in cosmetic and cosmeceutical formulations so as to achieve health benefits, including skin function and structure. The cyanobacteria and other components of marine microalgae offer vast scope in thalassotherapy, which involves the use of seawater and marine organisms (Mourelle et al., 2017).

PHARMACEUTICALS

The marine microalgae are considered to be cell factories owing to their bioactive compounds of therapeutic value such as polysaccharides or PUFAs and pigments. Table 1.2. gives the unique and potential pharmaceutical applications of the carotenoid pigments of the marine microalgae.

The pharmaceutical applications of microalgae are also depicted in Table 1.3.

MARINE *CYANOBACTERIA*

The cyanobacteria (Phylum: Cyanophyta) are Gram-negative and oxygenic photosynthetic bacteria possessing a variety of habitats which are from hot springs to ice caps of Antarctica (Whitton and Potts, 2000). These organisms were mislabeled as "blue-green algae" due to their slightly bluish-green photosynthetic pigments in their cytoplasm and acting like plants with the production of their own "food" (polysaccharides) and oxygen. However, due to their prokaryotic characteristics unlike other eukaryotic algae, they are now classified as bacteria. The species of marine cyanobacteria are known for their rich bioactive compounds (about 450), such as lipopeptides, depsipeptides, peptides,

TABLE 1.2 Bioactive Carotenoids of Marine Microalgae with Potential Pharmaceutical Applications

Carotenoid Type	Source Species	Activity
Asthaxanthin	*Chlorella vulgaris, Haematococcus pluvialis*	Antioxidant
Cantaxanthin	*Chlorella vulgaris, Haematococcus pluvialis*	Antioxidant
Lutein	*Haematococcus pluvialis, Scenedesmus obliquus, Chlorella pyrenoidosa*	Antioxidant
Violaxanthin	*Chlorella pyrenoidosa*	Antioxidant
β-carotene	*Dunaliella salina*	Antioxidant
Asthaxanthin	*Haematococcus pluvialis*	Apoptosis induction; antiproliferative activity of HCT-116 (colon cancer) HepG2 (hepatic cancer) MCF-7 (breast cancer)cells
Violaxanthin	*Chlorella ellipsoidea*	Antiproliferative activity of AGS (stomach) MCF-7 (breast) HeLa (cervical) DLD (colon) Hep-G2 (liver)cells
Violaxanthin + antheraxanthin + zeaxanthin	—	Apoptosis induction; antiproliferative activity of HCT116 (Colon)cells
Diadinochrome A, B, Diatoxanthin/cynthiaxanthin	*Peridinium bipes*	Cytotoxic effect on HeLa cells
Fucoxanthin	*Isochrisys* aff. *galbana, Chaetoseros* sp., *Odontella aurita,*	Antiproliferative activity of A549 (broncho-pulmonar) NSCLC-N6 (bronchopulmonar) SRA 01/04 (epithelial) cells
β-Carotene	*Dunaliella salina*	Inhibition of tumor; Hepatic cancer
α-Carotene	*Dunaliella salina*	Antiproliferative activity of lung, skin, liver and colon neuroblastoma cells

Source: Amaro, https://repositorio-aberto.up.pt/bitstream/10216/102689/2/181481.pdf

TABLE 1.3 Pharmaceutical applications of microalgae

Application	Activity
Antimicrobial	Antibacterial and antiviral
Anti-inflammatory and anticancer	Antiproliferative and apoptotic
Recombinant proteins	Subunit vaccines and monoclonal antibodies
Drug delivery	Diatom-based nanocarriers in cancer therapy

Source: Khavari et al., (2021)

alkaloids, macrolides/lactones, polyketides, terpenes, polysaccharides, and lipids; and associated therapeutic activities like their anticancer, antiviral, antimicrobial, antiprotozoal, antimalarial, antituberculosis, immunosuppressant, anti-inflammatory, anticoagulant, anti-HIV (human immunodeficiency virus), and anti-coronavirus (SARS-CoV-2) and are therefore aptly called as "Blue-green Pharmacy" and "Cyanotherapeutics" (El-fayoumy et al., 2021). Further, among the marine cyanobacteria, the species of *Lyngbya* and Oscillatoria have been reported to possess 93% of the total 450 bioactive compounds including toxin metabolites with potential anticancer properties. Furthermore, the marine cyanobacteria live invariably with host organisms such as fungi, protists, sponges, corals, and ascidians and their association with sponges, however, attracted significant research interest with biotechnological applications in the last decades (Konstantinou et al., 2020; Khalifa et al., 2021). Figure 1.1. gives the major bioactive compounds and their activities of the marine cyanobacteria.

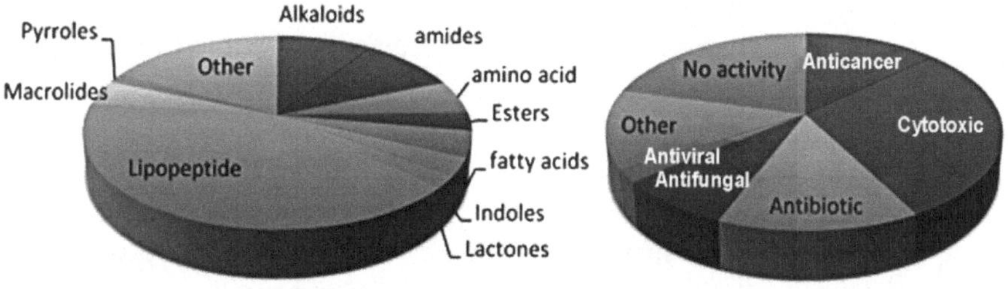

FIGURE 1.1 Compounds and biological activities of marine cyanobacteria.

Image credit: Telma Encarnação (Reproduced with permission)

MARINE DIATOMS

Diatoms, a major component of phytoplankton (unicellular microalga), are found in both marine and fresh water environments. The biomedical importance of marine diatoms is rather limited. However, recent research findings state that their frustules, which possess a distinct 3D architecture of silica cell walls with a highly ordered nano-/micropore structure and pattern (Lomora et al., 2019) are of great value. The biogenic silica of the frustules of diatoms has been reported to show potential biotechnological applications in the fields of biomedical/pharmaceutical sciences (Rabiee et al., 2021).

MARINE DIATOMS AND THEIR DIAGNOSTIC AND THERAPEUTIC APPLICATIONS

i) Marine Diatoms in Diagnostic Applications

- **a) Bioimaging:** The Diatom frustules of the marine diatom species *Thalassiosira weissflogii* has yielded a photoluminescence material when they were conjugated with thiophenebenzothiadiazole-thiophene-based fluorophore. This biohybrid luminescent nanoparticle has been reported to be useful in bioimaging which is nothing but visualizing functional images of living systems.
- **b) Optical biosensors:** The marine diatoms' frustules have also been reported to serve as optical biosensors for which monoclonal antibodies are attached to these frustules to specifically bind their antigens.
- **c) Biological scaffold:** Marine diatoms also serve as ideal biological scaffold. In this process, the diatom frustules become an ideal ultrasensitive biosensor when their surface is coated with gold nanoparticles (Au NPs). The Au NPs are capable of amplifying the signal in surface-enhanced Raman spectroscopy (SERS), and it is believed to be the most sensitive technology in the identification of biomarkers from the complex clinical samples.
- **d) Drug release:** The marine diatoms have also been reported to be useful in the drug delivery, which is made possible by encapsulating magnetic nanoparticles and small molecule drugs inside diatoms of 10 μm in diameter, with 500 nm pores. The external magnetic field of such diatoms is helpful in targeting drug release at the tumor site (Lomora et al., 2019).

ii) Marine Diatoms in Therapeutic Applications

- **a) Drug delivery:** Surface modifications of the diatom frustules have been reported to be useful in controlling the delivery of both hydrophobic and hydrophilic drugs

b) **Enhancement of the properties of the diatom scaffold:** The photocatalytic ability of Titanium dioxide (TiO2) in the depletion of carbon is well known. This facility could be made possible with the marine-centric diatom species *Thalassiosira weissflogii* added with TiBALDH (Titanium (IV) bis-(ammonium lactato)-dihydroxide. Further, the addition of TiO2 induced the breakage of oxidative DNA strand and depletion of cancer cells. Further, this sort of attachment helps in the formation of a multi-faceted platform to target and deliver the encapsulated chemotherapeutic toxic drugs to tumor cells only (Lomora et al., 2019). The potential therapeutic application of the marine phytoplankton is shown in Figure 1.2.

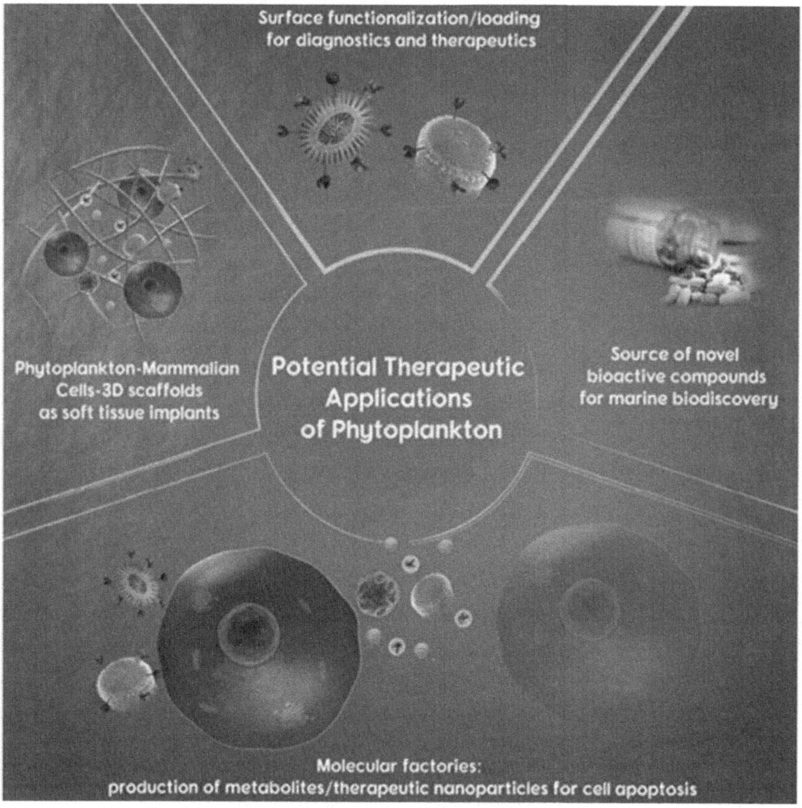

FIGURE 1.2 Overview of potential applications of phytoplankton with focus on diatoms and coccolithophores/coccoliths.

Image credit: Lomora et al. (Reproduced with permission)

MARINE COCCOLITHOPHORES

Coccolithophores, a unicellular species of marine phytoplankton produce tightly regulated, intricate exoskeleton structures that are made of calcite (calcium carbonate). The formation of biogenic calcium carbonate is either extracellularly forming tiled-like' plates called as coccoliths or intracellularly forming wheel-like' calcite plates. The formed coccoliths are transformed into coccosphere which serves as a protective wall (Walsh et al., 2018).

a) **Therapeutic carrier:** Like the biosilica of marine diatoms, the calcium carbonate (calcite) of the marine coccolithophore *Emiliania huxleyi* has also been reported to possess the large surface area to serve as therapeutic carriers besides showing extended drug release with water-insoluble drugs. In vitro-modified coccoliths with enzymes have shown a surface area five times higher than the synthetic calcium carbonate (Lomora et al., 2019).
b) **Drug delivery:** Due to the pH-sensitive characteristics, the CaCO3 nanoparticles of coccolithophores have been reported to serve as an effective alternative for delivery and release of anticancer drugs for the depletion of HeLa cells (Walsh et al., 2018).
c) **Bone repair:** Many marine-based products are presently available as clinically important bone graft products. The commercial products ProOsteon and Algipore are manufactured from the coral and mineralized microalgae, respectively. Recent research findings have shown that the coccolithophore could be successfully converted into hydroxyapatite, a bone graft material. The smaller particle size of the coccolithophores is believed to be the major constraint for their direct use in the development of a suitable bone substitute. On the other hand, these coccoliths could serve as an additive to composite materials usable in the bone repair owing to their unique calcite surface topographies. Walsh et al. (2018) have also reported that coccolithophore biomineralization could be employed in bone repair through protein-assisted self-assembly. Coccolithophore proteins could therefore offer vast scope for the creation of novel "smart interface" biomaterials for the repair of bones.

MARINE PHYCOTOXINS

The toxic secondary metabolites of the marine microalgae such as cyano-bacteria, diatoms, and dinoflagellates belong to different chemical classes such as (i) macrolides (amphidinolides and pectenotoxins); (ii) polycyclic polyketides (yessotoxins, ciguatoxins, brevetoxins, maitotoxins, and palytoxins); (iii) linear polyketides (dinophysistoxins and okadaic acid); (iv) alkaloids (saxitoxin, neosaxitoxin, and gonyautoxin); and (v) cyclic imines (spirolides, gymnodimines, and pinnatoxins). Several marine microalgal blooms are the causative agents for the production of these harmful toxins which can be hazardous not only to their ecosystems but also to human health. However, recent research findings have reported that these phycotoxins may contribute to a number of biomedical applications owing to their therapeutic functions like anticancer, cytotoxic, antifungal, anticholesterol, analgesic, anesthetic, immune-suppressive, etc (Camacho-Muñoz et al., 2021). Pradhan and Ki (2022) reported that the marine phycotoxins have been designated as potent pharmacophores as they possess significant abilities to prevent many stress-associated diseases in humans. The phycotoxins associated with the harmful marine algal blooms are known to modulate different diseases in humans (vide Chapter. 2).

MARINE MICROBES

Nowadays, one of the most important health problems is the increase, emergence, and spread of antimicrobial resistance among the different microorganisms (bacteria, fungi, virus, and parasites). In spite of the remarkable impact on health that the antimicrobials have achieved in the 1960s and 1970s, 40 years later infectious diseases remain the second-leading cause of death worldwide. In the case of bacteria, resistance to antibiotics is increasing in both community and hospital settings in association with an increase in mortality and morbidity. Microorganisms (microalgae, bacteria, protozoa, fungi, and viruses), which are a rich source of clinically useful natural products, have exerted a significant effect on medicinal science. Many of their metabolites possess bioactivities such as antibacterial, antiviral, antiparasitic, antifungal, anticoagulant, antimutagenic, antihyperglycemic, antitumoral, ant-iinflammatory, anticarcinogenic, antioxidant, immunomodulator, etc. (Adnan et al., 2018). Many of their metabolites possess bioactivities such as antimicrobial, antiviral, antitumoral, antiparasitic, antihyperglycemic, antimutagenic, ant-iinflammatory, anticarcinogenic, antioxidant, anticoagulant, immunomodulator, etc. (Adnan et al., 2018). However, the bioactive compounds possessing antimicrobial properties are particularly commonly used nowadays as antibiotics, which are found to be potential drugs for the treatment of diseases like HIV and neuropsychiatric sequelae. The percentage occurrence of microorganisms in different marine sources is given in Table 1.4.

TABLE 1.4 Marine Microorganisms and Their Sources

Source	%
Invertebrates	32
Macroalgae (Seaweeds)	7
Mangrove plants	8
Sea water	39
Mangrove sediment	6

Source: Raimundo et al. (2018).

ADVANTAGES OF MARINE MICROBES

Unfortunately, secondary metabolites of macroorganisms are available in smaller quantities, which is the major hurdle in using such metabolites. Therefore, the only alternative is the extraction of drug sources from marine microbiota like marine microalgae, bacteria, and fungi, which are the less investigated components of marine metabolic diversity (Debbab et al., 2010). Utilization of microorganisms for this purpose has several advantages viz. isolation of large-scale bioactive compounds through culture using advanced biotechnological methods like fermentations and bioprocessing (Adnan et al., 2018).

MARINE INVERTEBRATE-ASSOCIATED MICROBES

The yield of natural products from marine microbes such as bacteria and fungi is on an increasing trend compared to that of marine invertebrates. For example, 50% of natural marine products were from marine microorganisms and 40% from marine invertebrates in 2016. Further, nearly half of the marine bacteria and fungi from which novel bioactive compounds were obtained in 2013 were derived from host organisms, viz. marine invertebrates, especially octocorals and sponges. Marine invertebrate-associated microbes are therefore considered as the major source of novel metabolites. Among the different invertebrate phyla, Porifera and Bryozoa have been known to harbor rich microbes possessing therapeutically important secondary metabollites. Therefore, new technologies that focus on symbiont-driven production of invertebrate-derived bioactive molecules have the potential to overcome current hindrances in the commercialization of these drugs, contributing to a more sustainable blue economy (Raimundo et al., 2018). The percentage occurrence of microorganisms in the different groups of invertebrates is given in Table 1.5.

TABLE 1.5 Marine Microorganisms from the Different Groups of Invertebrates

Group	%
Sponges	45
Octocorals	22
Echinoderms	15
Molluscs	13
Tunicates[a]	3
Crustaceans	2

[a]*Included under this category as they are without backbones*

Source: Raimundo et al. (2018).

MARINE BACTERIA: BIOACTIVE COMPOUNDS AND THEIR ACTIVITIES

Among the chief constituents of marine microbes, viz. bacteria and fungi, marine bacteria which are predominantly gram-negative and motile occur in low concentrations in water and higher ones in bottom sediments. It is also known that Actinobacteria with its largest genus, Streptomyces, is one of the dominant bacterial phyla and are the source of almost 50% of all known bioactive microbial metabolites. Marine actinomycete bacteria have been reported as a vital resource for the discovery of drugs. For example, the therapeutically important compounds viz. macrolactin A–F were produced by a deep-sea gram-positive bacterium. Among these compounds, macrolactin A inhibits B16F10 murine melanoma cells. Similarly, the metabolite derived from Alteromonas spp. has shown anti-HIV activity. Majority of the antibiotic producers have been equated with Bacillus and Micrococcus, which are usually regarded as terrestrial organisms rather than representative of the true marine microfauna. The marine bacterium of Salinospora, sp. yielded a highly cytotoxic proteasome inhibitor salinosporamide A which is considered to be a promising source of cancer chemotherapeutics. Similarly, marine biooxins like tetrodotoxin, ciguatoxins, brevetoxins, and saxitoxin, produced by marine microalgae and microbes are considered to be valuable sodium channel blockers. These biotoxins have been reported to be very useful in the neuropharmacological and neurophysiological studies (Fenical and Jensen 2006).

Several species of marine bacteria, particularly invertebrates, are found as symbionts in the different phyla of marine macrobiota. For example, the compound loloatin B (cyclic decapeptide antibiotic) has been derived from the species of Bacillus associated with the marine worm. This antibiotic inhibited vancomycin-resistant Enterococcus and methicillin-resistant *Staphylococcus aureus*. Similarly, another antibiotic viz. Thiomarinol has been derived from the symbiotic bacterium, *Alteromonas rava*. Other antibiotics derived include sesbanimides from Bacillus; agrochelin from Agrobacterium; and pelagiomicins and pyrones from Pseudomonas (Javed et al., 2011).

Microbial symbionts of marine invertebrates provide opportunities for microbial biotechnology. For example, the biosynthesis of antitumor polyketide was made possible with a naturally collected bacterial symbiont from the marine sponge of the species, *T. swinhoei*. Joseph (2017) reported that the secondary metabolites from the symbiotic bacteria could largely help in the large-scale production of new drugs. The chemical classes and bioactivities of some selected marine bacteria are given in Table 1.6.

MARINE FUNGI

Fungi are widely distributed in marine environments from the deep sea to polar ice covers. They occur in sediments and are found in all kinds of living and dead organic matter. Their numbers in ocean waters are quite low compared to bacteria, and most of the investigations related to the fungi are restricted to marine sediments or certain organisms. Among the marine biota, the sponges receive greater attention due to their association with therapeutically important bacterial genera like *Alternaria, Acremonium, Aspergillus, Fusarium, Penicillium, Phoma,* and *Trichoderma*. Among these genera, *Aspergillus* dominates with more than 180 species, and the classes of bioactive compounds of its different species include alkaloids, polyketides terpenoids, xanthones, etc, with several therapeutic activities including cytotoxic, anti-HIV, and antimicobial (Raimundo et al., 2018).

BIOACTIVE COMPOUNDS OF MARINE FUNGI

It is estimated that more than 70% of the new fungal bioactive compounds have been derived from the marine animals and plants and the sediments

TABLE 1.6 Chemical Classes and Activities of Some Selected Marine Bacteria

Microorganism	Source	Chemical Class	Bioactivity
Actinomycete	Coastal sediment	Lactone	—
	Coelenterate	Depsipeptide	Antiinflammatory
	Deep-sea sediment	Bromonaphtoquinone	Antibiotic
	Sediment (undefined)	Lactonized FA	—
	Shallow water sediment	bromonaphtoquinone and lactone	antibiotic
	Shallow water sediment	Sesquiterpene	—
	Undefined	Glycosylated macrolide	
Alteromonas	—	Indole	Antifungal
	Open sea	Cyclic peptide	—
	Sponge	Macrolactame and amide ester	Cytotoxic
	Undefined	Dipyrrole	Antibiotic
	Undefined	Guanidine	Toxic
	Undefined	Amide ester	Antimicrobial
	Undefined	Cyclic aromatic FA	Bronchodilatator
Bacillus	Deep waters	Aminoglycoside	Antimicrobial
	Mollusk	Despsipeptide	Cytotoxic
	Polychaete	Cyclic peptide	Antimicrobial
	Sediment	N-isocoumarine and cyclic peptide	Antitumor
	Sponge	Cyclic despsipeptide	—
Bacteria Gram +	Deep sea sediment	C-237 Ac macrolide	Antiviral
	Undefined	Cyclic lysine	Cytotoxic
Pseudomonas	Fish skin	Guanidine	Toxic
	Polychaete	Cyclic peptide	Antimicrobial

TABLE 1.6 *(Continued)*

Microorganism	Source	Chemical Class	Bioactivity
	Red alga	Cyclic peptide	Antimicrobia
	Sponge	C50 carotene	—
		Diketopiperazine and phenazine	—
	Tunicate	Amide	—
	Undefined	Indole and quinolinol	Antimicrobial and antibiotic
		Guanidine diketopiperazine	Chitinase inhibitor
Streptomyces	Estuarine sediment	N-glycosylated flavonoid	—
	Fish	Peptide	—
	Gorgonian	FA lactone	Cytotoxic
	Mollusc	Macrolactame	Superoxide inhibitor
	Shallow water sediment	Phenazine	Antimicrobial
	Shallow water sediment	Lactone	—
	Sediment	Diketopiperazine, pyrrole	Enzyme inhibition
	Sponge	Phenazine and lactone amide	Antimicrobial and antibiotic
	Undefined	Naphtoquinone	
Vibrio	Fish	Indole	—
	fish pathogen	Amide	—
	Sponge	Bromo diphenyl ether	Antimicrobial
	Sponge	Indole and lactame	Antimicrobial
	Undefined	Guanidine and lactame	Toxic FA- Fatty acid

Source: Kelecom (2002).

contribute the rest. Though the bioactive compounds of the marine fungi have been reported to possess clinical applications, they are yet to be included in the drug pipelines, and unfortunately, no fungus-based drug is on the market. A cyclic dipeptide analogue of halimide viz. plinabulin which is derived from the marine fungus is presently in the phase II clinical trial for treating lung cancer (Jin et al., 2016). The percentage of occurrence of fungi in the non-living and living components of the marine environment, as well as the chemical classes of bioactive compounds derived from marine fungi and their bioactivities, are given in the following tables.

TABLE 1.7 Percentage Occurrence of Marine Fungi[a]

Habitat/Biota	%
Marine plants	42.5
Mangroves	25.5
Macroalgae (Seaweeds)	14.4
Marine animals	30.1
Sponge	9.2
Sediment	22.9
Sea water	4.6
Others	2.5

[a]*Data based on new bioactive compounds isolated from marine-derived fungi during 2014–2016.*

Source: Jin et al. (2016).

TABLE 1.8 Percentage Occurrence of Bioactive Compounds Derived from Marine Fungi[a]

Chemical Class	%
Alkaloids	27.0
Polyketides	25.7
Peptides	13.8
Terpenes	9.9
Lactone	3.9
Steroids	3.3
Others	16.6

[a]*Data based on new bioactive compounds isolated from marine-derived fungi during 2014–2016.*

Source: Jin et al. (2016).

TABLE 1.9 Bioactivities of Compounds Derived from Marine Fungi (data based on 2014–2016)

Biological Activity	%
Cytotoxicity	37.5
Antibacterial	18.4
Antifungal	7.9
Antiviral	7.2
Antioxidant	5.3
Lipid lowering	5.3
Others	18.4

Source: Jin et al. (2016).

SYMBIOTIC BACTERIA PRODUCED MARINE DRUGS IN CLINICAL PIPELINE

Ecteinascidin 743 (ET-743, Yondelis) is a clinically approved chemotherapeutic compound, which is derived from the gammaproteobacterial species viz. *Endoecteinascida frumentensis* found in the tunicate *Ecteinascidia turbinata*. This is widely used for cancer treatment. Furthermore, polyether compounds such as Nigericin, Tetronamycin, and K 41A are also suspected to have chances of microbial creation (Schofield et al., 2015).

FUTURE PERSPECTIVES

In spite of several successes in drug discovery from microbiota, marine microbiota have received very little attention. For instance, the present drawback in the large-scale production of pharmaceutically important secondary metabolites from marine microorganisms such as bacteria and fungi is largely due to the noncultivation of the majority of such microorganisms. It has often proven extremely difficult to get sufficient amounts of many of the bioactive compounds from these marine microbiota due to the limited amounts produced by the organism itself. Further, difficulties still exist in the isolation of target bioactive compounds from marine microbial symbionts. Therefore, extensive research and development efforts are required to meet the demands of the target compound of pharmaceutical importance.

CHAPTER 2

Toxins of Marine Microalgae and Their Potential Therapeutic Applications

Among the marine biota, the hepato and neurotoxin-bearing, microalgae (phytoplankton) such as cyanobacteria (blue-green algae), diatoms, and dinoflagellates are believed to be responsible for human deaths through their harmful algal blooms (HABs) and the marine phycotoxins of these algae have been reported to cause about 60,000 human intoxications per year. Among the toxic components of marine microalgae, the dinoflagellates possess about 75% of the toxic species that are responsible for HABs (Camacho-Muñoz et al., 2021).

MARINE PHYCOTOXINS AND THEIR TYPES

Depending on the chemical structure, there are various marine phycotoxins viz. (i) okadaic acid (OA) and dinophysistoxins (DTXs) (macrocyclic and linear polyethers; (ii) ciguatoxins (CTXs) and brevetoxins (BTXs) (ladder-frame polyethers); (iii) saxitoxins (STXs) and gonyautoxins (GTXs) (tetrahydropurines); (iv) spirolides (SPXs) and gymnodimines (GYMs) (macrocyclic imines); and (v) domoic acid (DA) (toxic secondary amines). However, according to the symptoms generated in humans after the consumption of HAB-contaminated seafood, the poisoning of these phycotoxins is classified as (i) PSP, paralytic shellfish poisoning due to saxitoxins (STX) and their analogs; (ii) CFP, ciguatera fish poisoning due to ciguatoxin (CTX) and maitotoxin (MTX); (iii) ASP, amnesic shellfish poisoning due to domoic acid (DA); (iv) DSP,diarrhetic shellfish poisoning due to okadaic acid (OA), dinophysistoxins (DTXs), pectenotoxins (PTXs) and yessotoxins (YTXs); NSP, neurotoxic shellfish poisoning due to brevetoxins (BTXs); and AZP, azaspiracid shellfish poisoning due to azaspiracids (AZAs) (Camacho-Muñoz et al., 2021).

Medicinal Marine Microbiota. Santhanam Ramesh, Ramasamy Santhanam, Arumugam Uma & Raju Mohanraju (Authors)

CHEMISTRY OF MARINE PHYCOTOXINS

The toxic secondary metabolites of the marine microalgae belong to different chemical classes such as (i) macrolides (amphidinolides and pectenotoxins); (ii) alkaloids (gonyautoxin, saxitoxin, and neosaxitoxin); (iii) polycyclic polyketides (yessotoxins, ciguatoxins, brevetoxins, maitotoxins and palytoxins); (iv) linear polyketides (dinophysistoxins and okadaic acid); and (v) cyclic imines (spirolides, gymnodimines and pinnatoxins).

POTENTIAL THERAPEUTIC APPLICATIONS OF MARINE PHYCOTOXINS

Several marine microalgal blooms are known to be responsible for the production of harmful toxins, which can be hazardous not only to their ecosystems but also to human health. However, these phycotoxins could contribute to a number of biomedical applications owing to their therapeutic functions like anticancer, cytotoxic, antifungal, anticholesterol, analgesic, anesthetic, immune-suppressive, etc (Camacho-Muñoz et al., 2021). Pradhan and Ki (2022) reported that the marine phycotoxins have been designated as potent pharmacophores as they possess significant abilities to prevent many stress-associated diseases in humans. The phytoplankton toxins, which can modulate different diseases in humans, are shown in Figure 2.1.

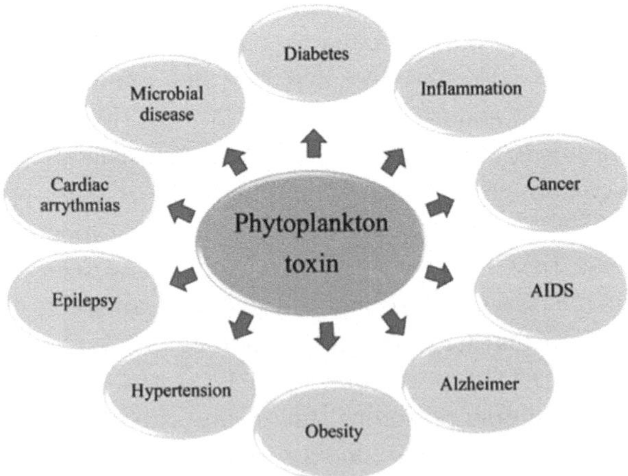

FIGURE 2.1 Phytoplankton toxin modulates different diseases in human

Source: Reprinted with permission from Pradhan and Ki (2022). https://creativecommons.org/licenses/by/4.0/

The most common and important toxins of the marine cyanobacteria, diatoms, and dinoflagellates and their potential therapeutic applications are given in Tables 2.1. and 2.2.

TABLE 2.1 Marine Cyanobacterial Toxins and Their Therapeutic Applications

Source	Toxin	Application
Lyngbya majuscula	Kalkitoxin	HCT-116 colon cell line
Lyngbya sp.	Oscillatoxin and its analogs, oscillatoxin E and 30-methyloscillatoxin D	Chinese hamster ovary (CHO) cell line
Nodularia spumigena	Nodularin	Breast carcinoma cell line MCF-7
Anabaena, Microcystis, Planktothrix, and *Aphanizomenon*	Microcystins	Breast carcinoma cell line MCF-7
Anabaena, Lyngbya, Planktothrix, Aphanizomenon, and *Cylindrospermopsis*	STX groups	Block sodium channels in nerve axons

Source: Zanchett and Oliveira-Filho, 2013; Pradhan and Ki, 2022

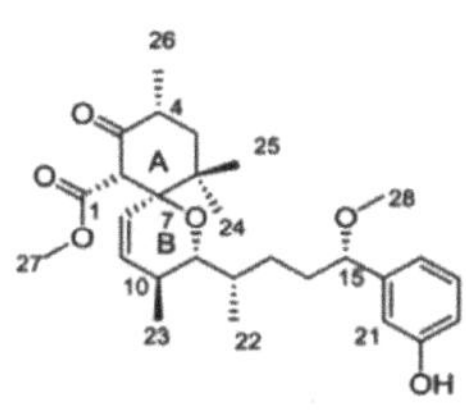

Oscillatoxin E

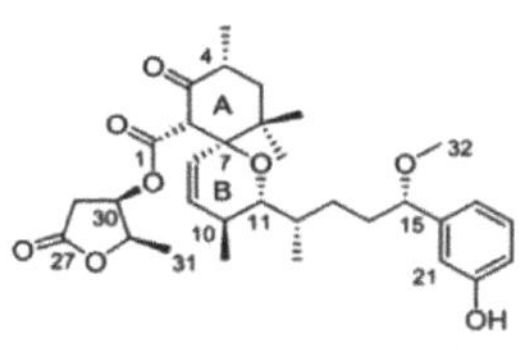

30-methyloscillatoxin D

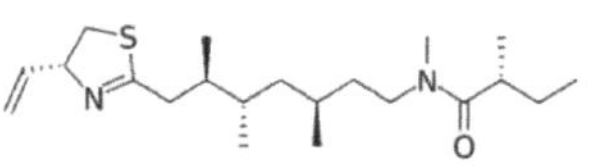

Kalkitoxin

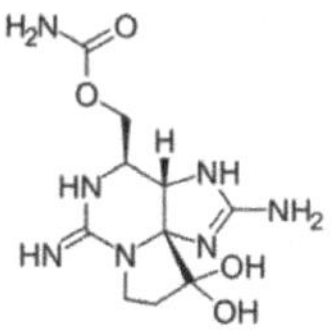

Saxitoxin

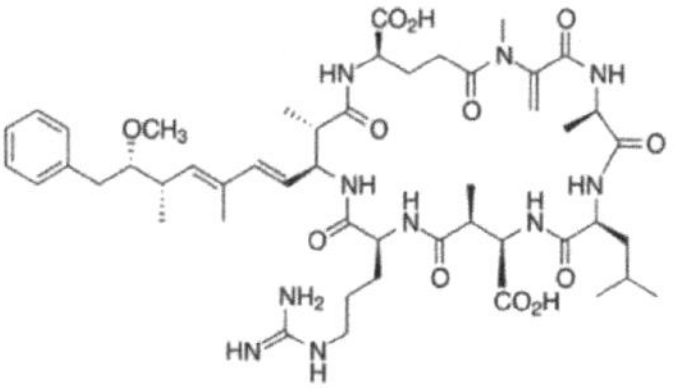

Microcystin

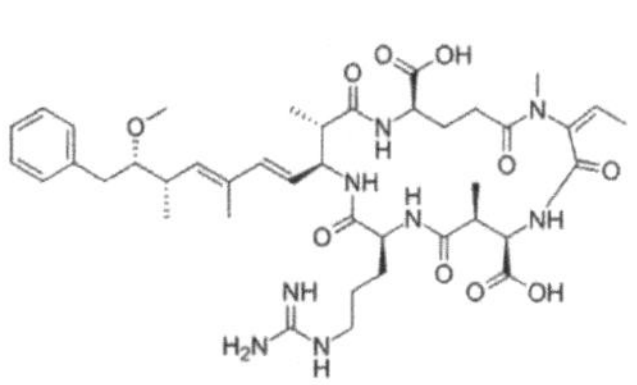

Nodularin

TABLE 2.2 Toxins of Marine Diatoms and Their Potential Therapeutic Applications

Toxin	Source	Applications[a]
Domoic acid	*Pseudo-nitzschia, Amphora*	Disease preventive actions against several deleterious diseases, such as cancer

Source: Pradhan and Ki, 2022.

[a]*Further research would confirm this.*

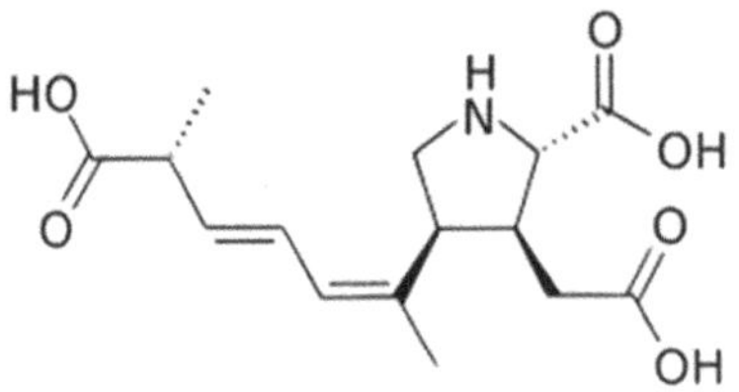

Domoic acid

TOXINS OF MARINE DINOFLAGELLATES AND THEIR POTENTIAL THERAPEUTIC APPLICATIONS

TABLE 2.3 Marine Dinoflagellate Toxins and their Potential Applications for Cancer

Source (Genus/Genera)	Toxin	Application
Prorocentrum and *Dinophysis*	Okadaic acid	Breast, intestine, blood, brain, lungs and hepatic, cancer cell lines, and neuroblastoma cell line
Gambierdiscus	Ciguatoxin	Gastrointestinal cell line
Dinophysis	Pectenotoxin and Pectenotoxin-2	Human breast, lung, and colon cancer cell lines
Ostreopsis	Palytoxin	Lymphoblastic cell line; neuroblastoma and intestinal cell lines
Karenia brevis	Brevetoxin	Jurkat E6-1 cell line
Protoceratium, Lingulodinium, and *Gonyaulax*	Yessotoxins	Epithelial cancer cell line
Ostreopsis	Ostreocin-D	Neuroblastoma and intestinal and cell lines
Protoceratium, Lingulodinium, and *Gonyaulax*	Yessotoxin and its analogues	Melanoma tumor cell line; and neuroblastoma cell line

TABLE 2.3 *(Continued)*

Source (Genus/Genera)	Toxin	Application
Protoceratium	Protoceratins I, II, III, and IV	Colon cancer cell line
Amphidinium	Caribenolide I	P388 murine leukemia cell line; and HCT 116/VM 46
Amphidinium	Amphidinolides and Colopsinols	HCT116 cell line; and epidermoid carcinoma KB cell line
Alexandrium	Gymnocin	Neuroblastoma cell line
Alexandrium and *Dinophysis*	Gymnocin and Okadaic acid	Several cancer cell lines
Alexandrium and *Gymnodinium*	Gymnocin-A	P388 murine leukemia cell line
Azadinium and *Amphidoma*	Azaspiracid	T-lymphocyte cell line

Source: Pradhan and Ki, 2022.

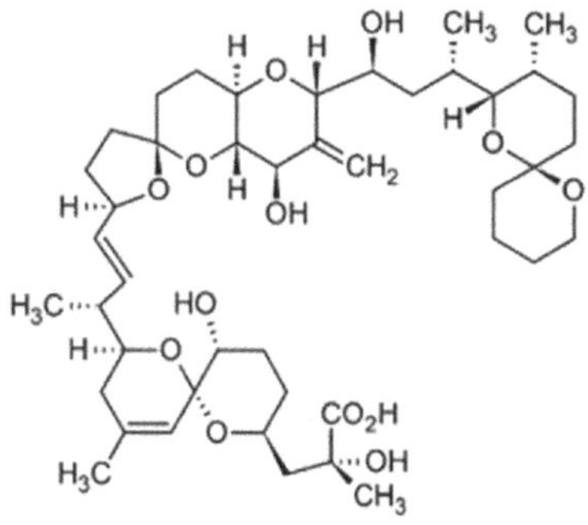

Okadaic acid

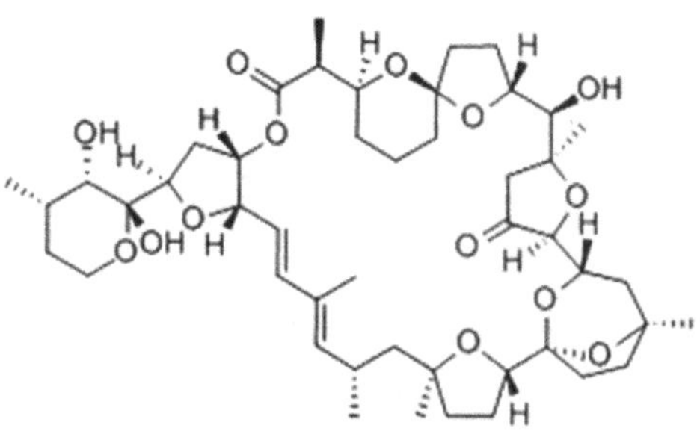

Pectenotoxin

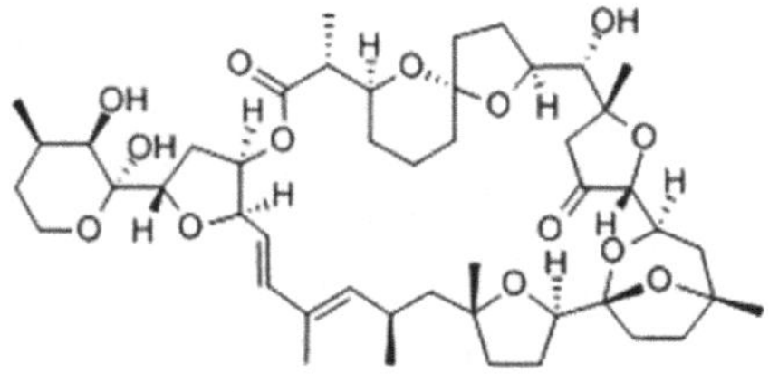

Pectenotoxin 2

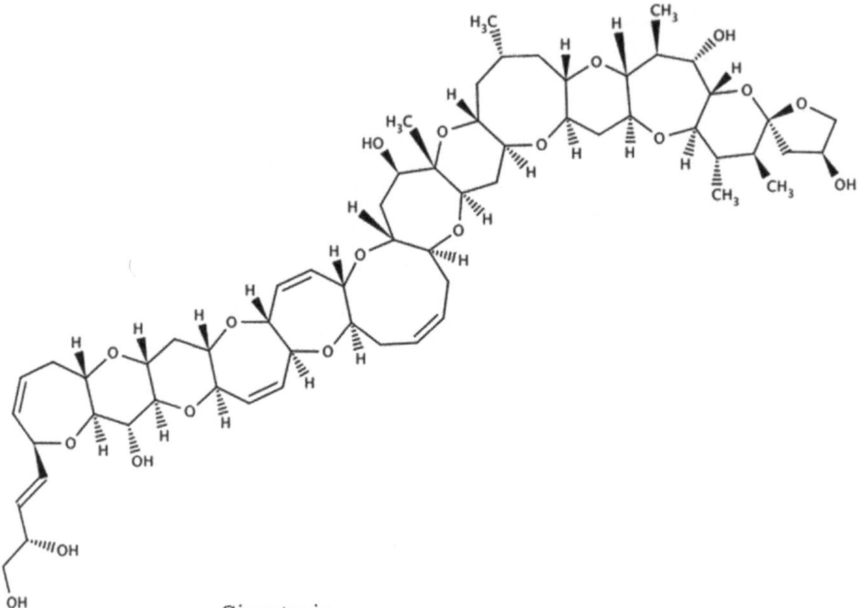

Ciguatoxin

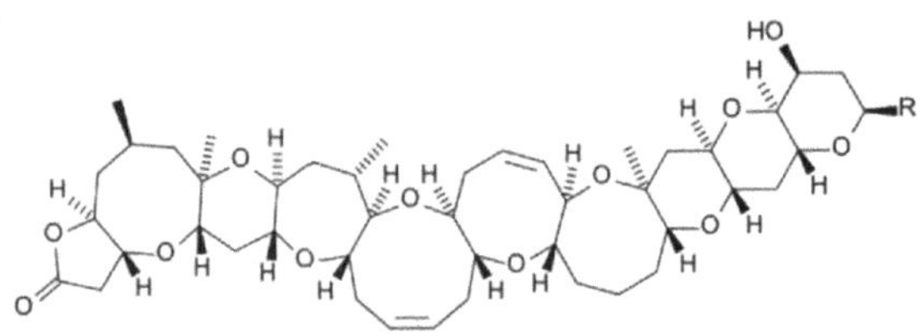

Brevetoxin

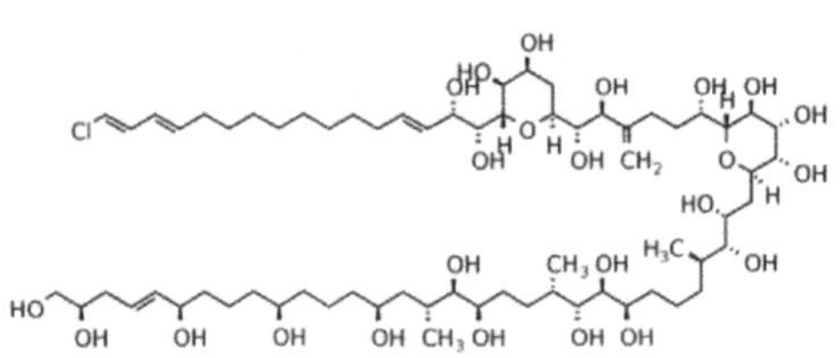

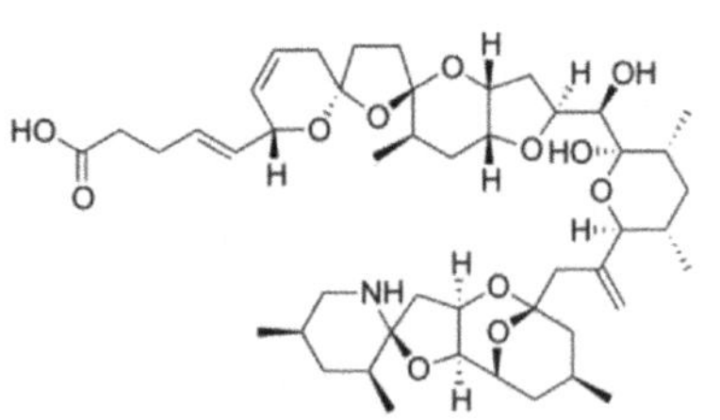

Karlotoxin

Azaspiracid

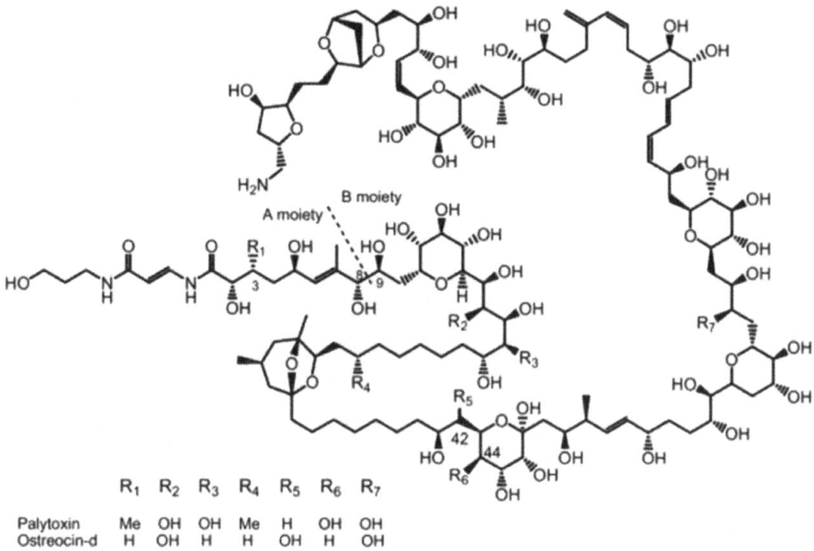

Palytoxin & Ostreocin-D

Image credit: Suzuki et al.

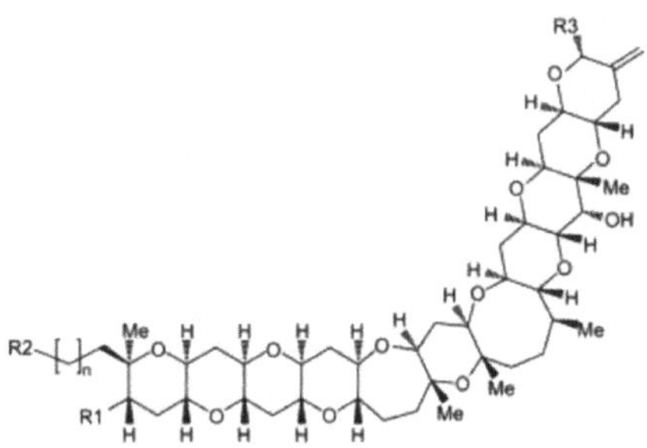

Yessotoxin

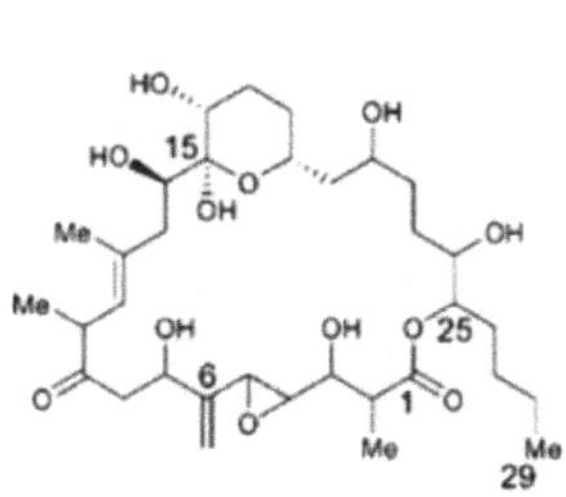

Amphidinolide

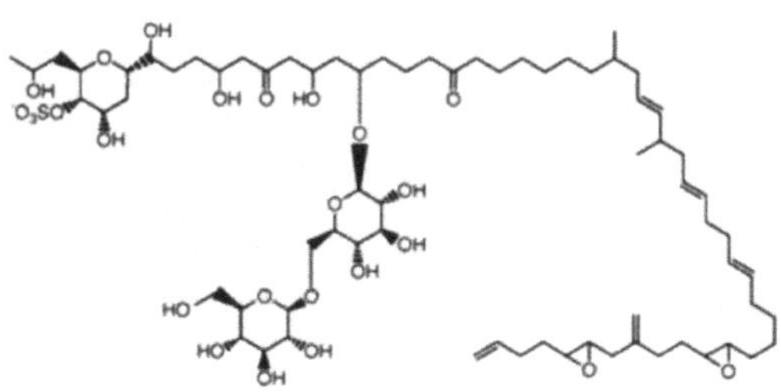

Colopsinol

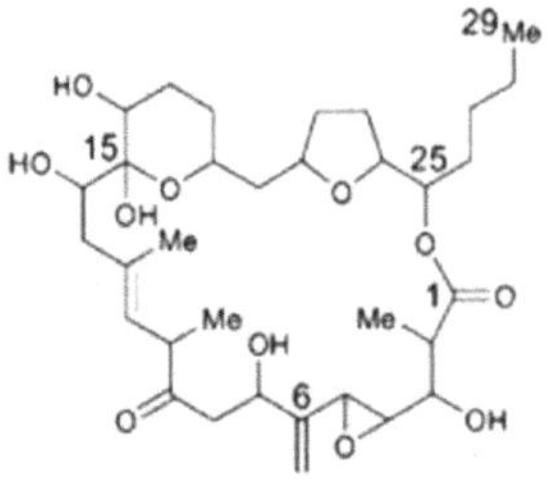

Caribenolide I

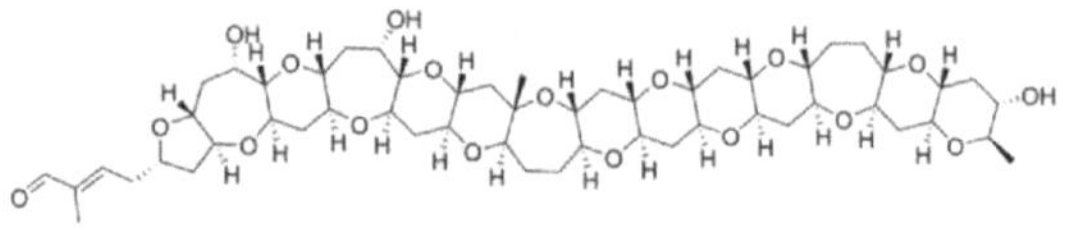

Gymnocin-A

TABLE 2.4 Marine Dinoflagellate Toxins s and Their Potential Applications for Certain Diseases

Disease	Toxin	Source (Genus/Genera)	Application
Alzheimer	Gambierol	*Gambierdiscus*	Inhibition of the level of t- and β-amyloid
	Gymnocin and Spirolides	*Alexandrium*	Inhibition of the level of t- and β-amyloid
Pain	Gonyautoxins	*Amphidinium*	—
	Gonyautoxins 2,3 and Tetrodotoxin	*Amphidinium, Alexandrium*	—
Schizophrenia and AIDS	Okadaic acid	*Dinophysis, Prorocentrum*	—
Diabetes	Okadaic acid	*Dinophysis, Prorocentrum*	—
	Gambierol	*Gambierdiscus*	—
Brain injury and rheumatoid arthritis	Gambierol	*Gambierdiscus*	—
Fungal disease	Okadaic acid	*Dinophysis, Prorocentrum*	Suppression of Candida albicans growth
	Karlotoxin	*Karlodinium*	Suppression of Candida albicans growth
Coronary heart disease	Karlotoxin	*Karlodinium*	—
Allergy and Asthma	Yessotoxin and its analogs	*Protoceratium, Lingulodinium,* and *Gonyaulax*	—
Brain disorder	Brevetoxin-2	*Karenia brevis*	—
Osteoporosis	Symbioimine	*Symbiodinium*	For postmenopausal women
Inflammatory disorders	Symbioimine	*Symbiodinium*	For cyclooxygenase-2-related disorder viz. inflammogenesis of cancer

TABLE 2.4 *(Continued)*

Disease	Toxin	Source (Genus/Genera)	Application
Pain	Gonyautoxins	*Amphidinium*	—
Fungal, bacterial, and protozoal diseases	Saxitoxin	*Alexandrium, Gymnodinium, Pyrodinium*	—

Source: Pradhan and Ki (2022).

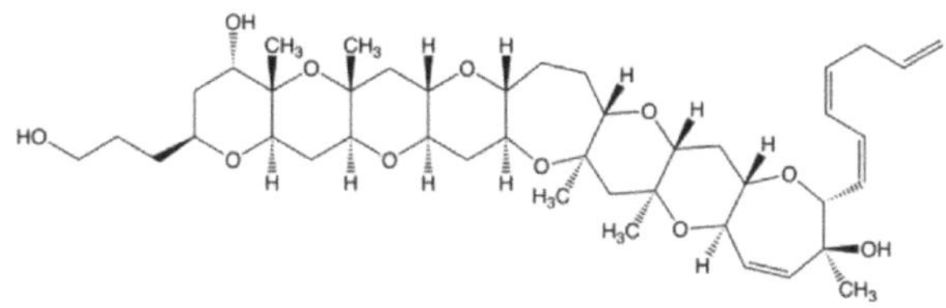

Gambierol

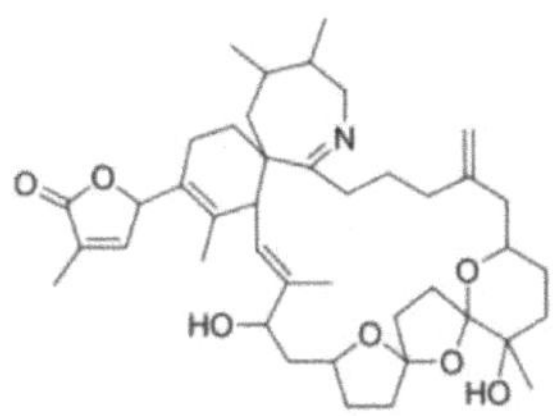

Spirolide (13dmC)

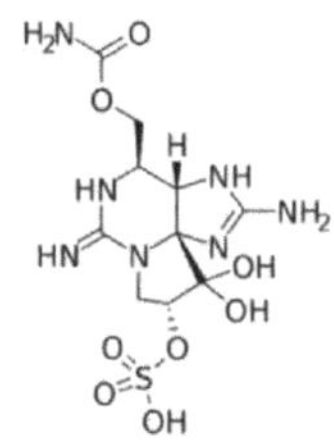

Gonyautoxin

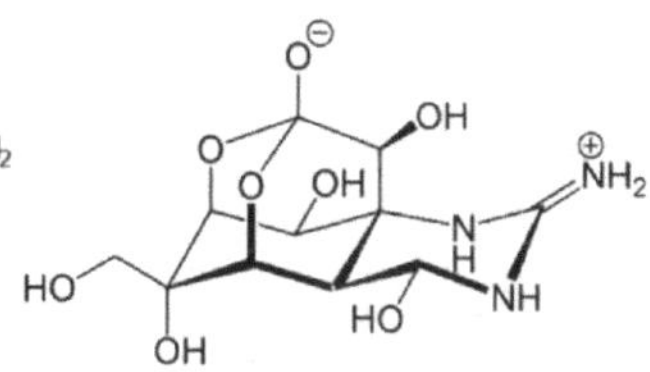

Tetrodotoxin

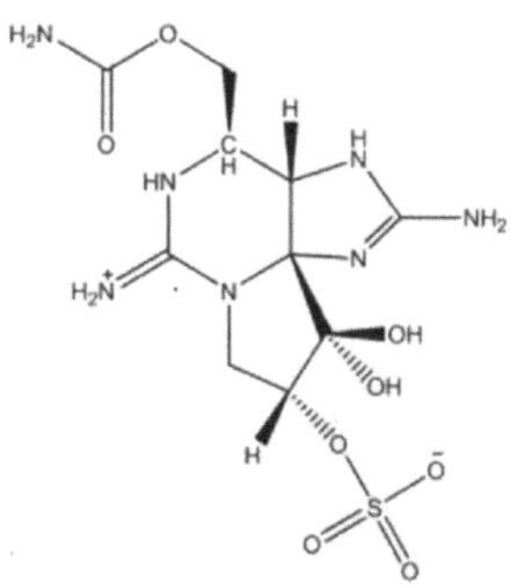

Gonyautoxin 2

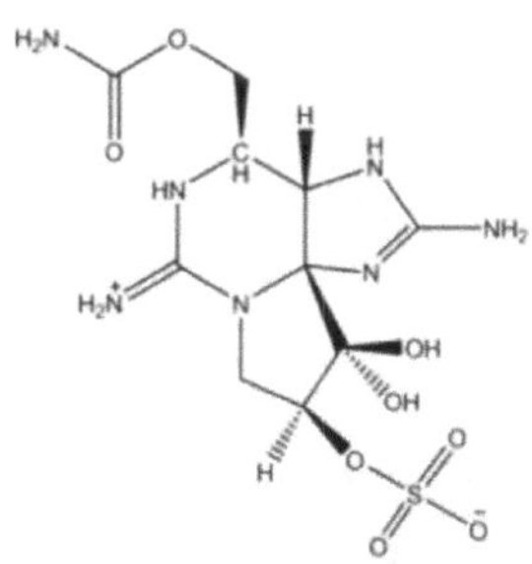

Gonyautoxin 3

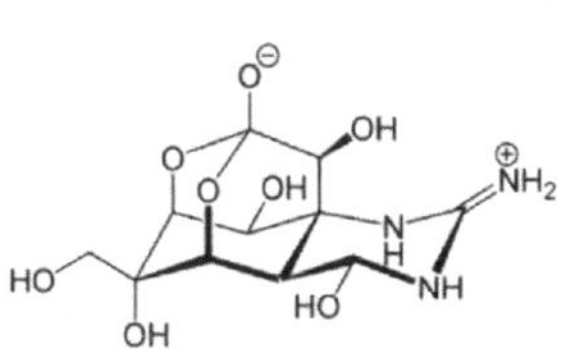

Tetrodotoxin

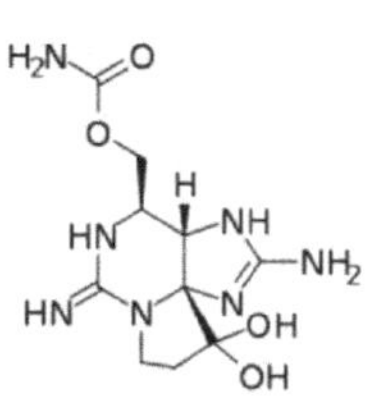

Saxitoxin

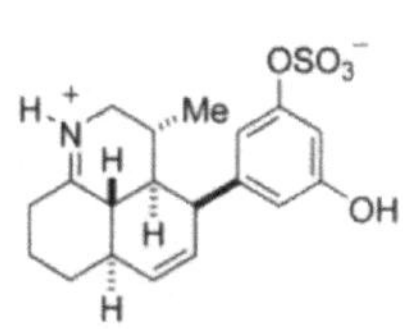

Symbiomine

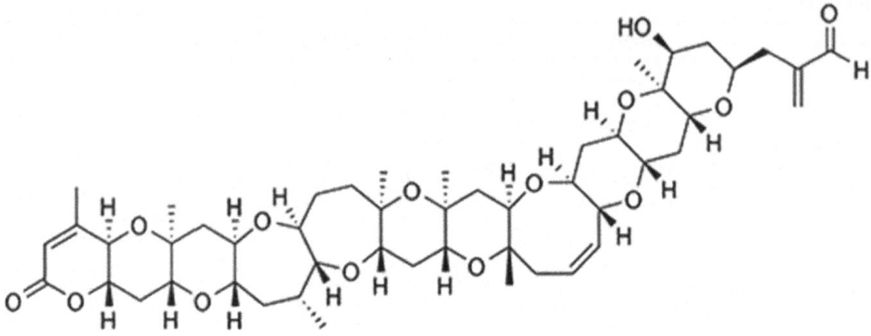

Brevetoxin-2

CONSTRAINTS

The availability of adequate quantities of marine phycotoxins has restricted research relating to their potential therapeutic functions. For instance, the chemical synthesis of the toxin, okadaic acid yielded only a poor yield due to the lengthy steps and its molecule complexity. Further, it is unreliable and impractical to make use of the naturally occurring phytoplankton species for the production of needed phycotoxins. In order to get rid of these hurdles, large-scale culture of the required species of marine phytoplankton with specialist facilities is the need of the hour (Camacho-Muñoz et al., 2021).

FUTURE PROSPECTS OF MARINE PHYCOTOXINS

Pradhan and Ki (2022) reported that the therapeutic values of marine phytoplankton toxins have been underestimated and are yet to be thoroughly analyzed. This calls for adequate expertise and research facilities. As marine phytoplankton toxins offer vast scope for use in marine bio-pharmacology and in the development of new phytoplankton toxin-based new drugs in the near future, adequate funding and awareness on the importance of marine phycotoxins are the need of the hour. A phytoplankton toxin-based drug molecule proposed is shown in Figure 2.2.

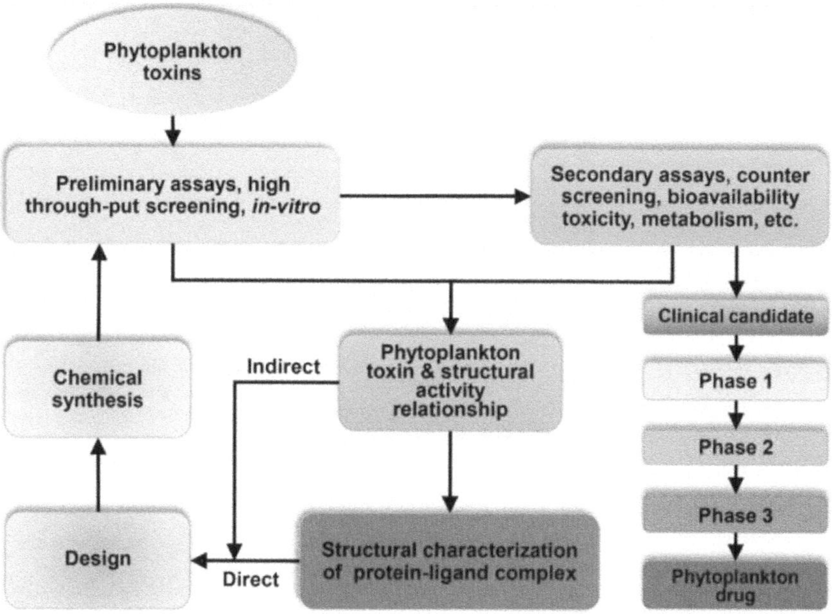

FIGURE 2.2 A proposed model for phytoplankton toxin as a future drug molecule.

Source: Reprinted with permission from Pradhan and Ki (2022). https://creativecommons.org/licenses/by/4.0/

CHAPTER 3

Medicinal Marine Microalgae

GREEN ALGAE

Phylum: Chlorophyta Class: Chlorophyceae Order: Volvoclaes
Family: Dunaliellaceae

Dunaliella salina

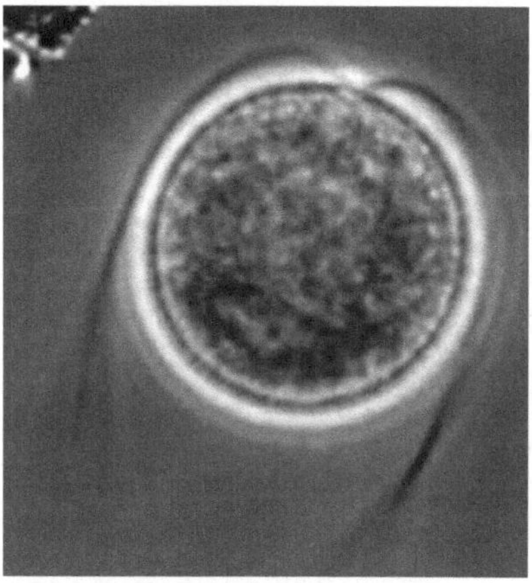

Image credit: Photo by Mike Dyall-Smith. https://creativecommons.org/licenses/by-sa/4.0/

Global distribution: Europe and North America.

Ecology: Hypersaline environments like salt lakes and salt pans

Medicinal compounds and their bioactivities:

Antibacterial activity: Extracts of this species showed antibacterial activity on *Pseudomonas aeruginosa*, *Staphylococcus aureus*, *Escherichia coli*, and *Klebsiella* spp.(Pane et al., 2015)

Medicinal Marine Microbiota. Santhanam Ramesh, Ramasamy Santhanam, Arumugam Uma & Raju Mohanraju (Authors)

Anti-inflammatory and immunomodulatory activities: Its omega-3 fatty acid concentrate (Ds-ω-3 FA) showed anti-inflammatory and immunomodulatory activities by inhibiting tumour necrosis factor α (TNF-α) and interleukin 6 (IL-6). Further, this concentrate served as a promising blocker of matrix metalloproteinase (MMP-2 and MMP-9) and cyclooxygenase (COX-2) expression (Chitranjali et al. 2015).

Luo et al. (2015) reported on the occurrence of phytosterols viz. ergosterol and 7-dehydroporiferasterol in these species. These are known to block the absorption of cholesterol.

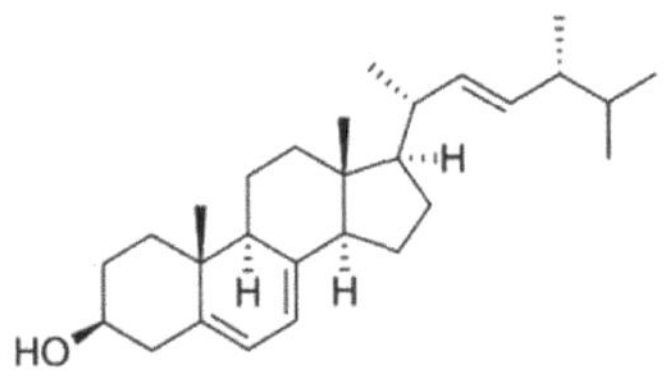

Ergosterol

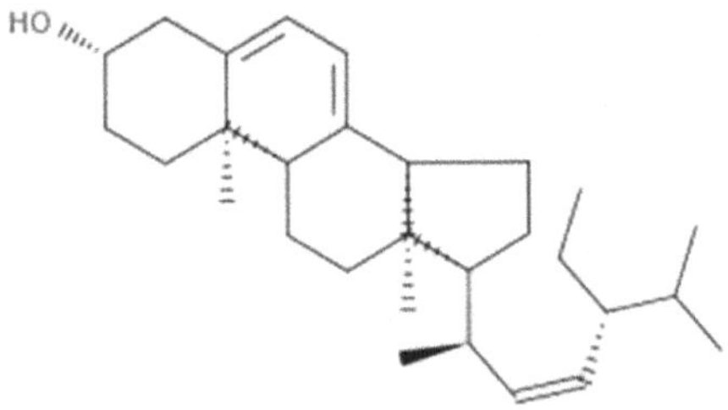

7-Dehydroporiferasterol

Dunaliella tertiolecta

Global distribution: It is an arctoboreal species and is common in UK

Ecology: Salt water ditches, salt lakes, and salt pans

Medicinal compounds and their bioactivities:

Antiproliferative activity: A glycoglycerolipid compound, glycerol 1-(9Z,12Z,15Z-octadecatrienoate)-2-(4Z,7Z,10Z,13Z-hexadecatetraenoate)-3-O-β-D-galactopyranoside isolated from the extracts of this species showed antiproliferative activity on A2058 cells with 35 and 44 % activity at 100 and 50 μg/mL concentrations, respectively (Martínez et al. 2022).

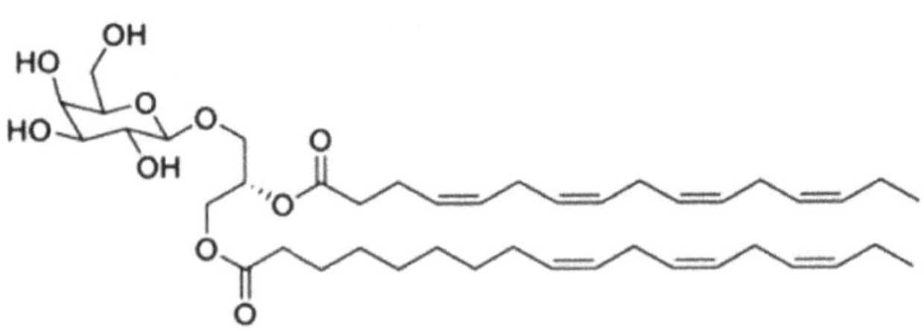

Glycerol 1-(9Z,12Z,15Z-octadecatrienoate) 2-(4Z,7Z,10Z,13Z-hexadecatetraenoate)-3-O-β-D-galactopyranoside

Immunomodulatory, neuromodulatory, and anti-inflammatory activities: Its ergosterol,7-dehydroporiferasterol, showed Immunomodulatory and anti-inflammatory activities. Further, this compound showed neuromodulatory activity in the brain areas of experimental rats (Luo et al. 2015).

Order: Volvocales Family: Haematococcaceae

Hematococcus pluvialis *(= Sphaerella lacustris)*

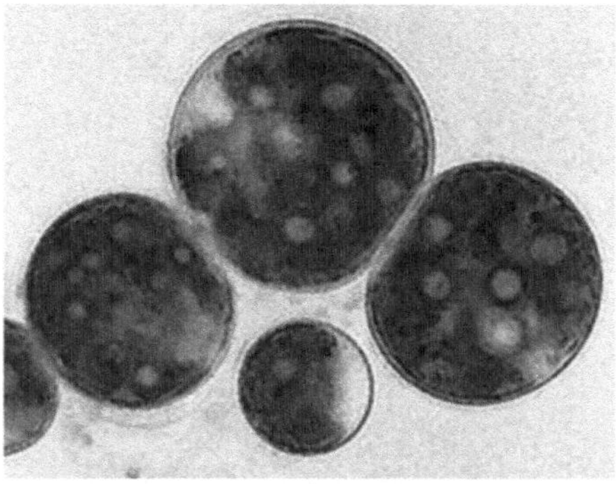

Image credit: Photo by Andrea.valentino. https://creativecommons.org/licenses/by-sa/4.0/

Global distribution: Widely distributed in most parts of the world

Ecology: Temporary water bodies like concrete basins and birdbaths.

Medicinal compounds and their bioactivities:

Antioxidant and anti-inflammatory activities: It contains the carotenoid pigment viz. astaxanthin which possesses both antioxidant and anti-inflammatory activities. Its antioxidant activity is by blocking the production of NF-κB activation and subsequent suppression of IkB kinase activity and IkB-α degradation. Further, its anti-inflammatory activity is due to its inhibition of inflammatory mediators by blocking iNOS and COX-2 activation. Its astaxanthin finds application in the human's functional food production (https://www.sciencedirect.com/topics/agricultural-and-biological-sciences/haematococcus-pluvialis)

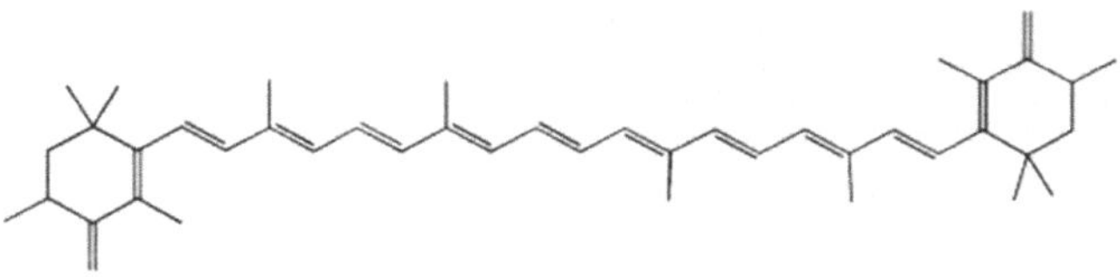

Astaxanthin

Class: Chlorodendrophyceae Order: Chlorodendrales
Family: Chlorodendraceae

Tetraselmis chui

Global distribution: It is globally distributed

Ecology: It occurs in a diverse water environments.

Medicinal compounds and their bioactivities:

Antioxidant activity: The gluten-free bread with the addition of this alga at 4% has been reported to possess rich phenol content with antioxidant capacity (Nunes et al. 2020)

Antibacterial activity: The extracts of this alga showed antibacterial activity against Vibrio strains (Kokou et al. 2012)

Immunomodulatory properties: The extracts of this species have been reported to possess immunomodulatory properties by inhibiting the production of interleukin 6 (IL-6). Further, oral administration of this lyophilized alga has been reported to possess hemolytic complement activity (Riccio and Lauritano 2020). presence of eicosapentaenoic acid (EPA) and docosahexaenoic acid (DHA) in this microalga in considerable quantities make this species suitable for in nutraceutical production.

Tetraselmis suecica

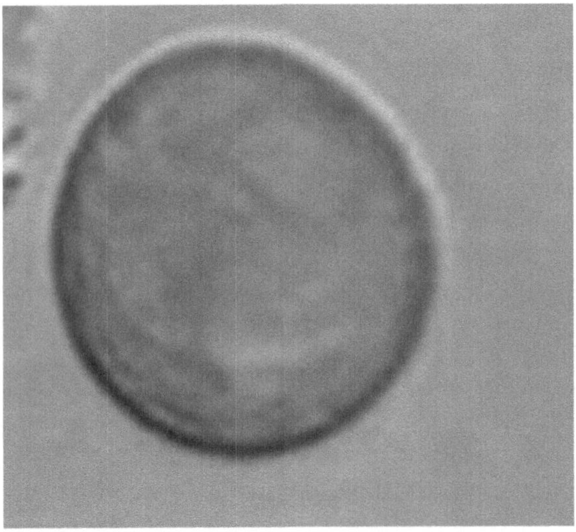

Image credit: Photo by Nobody16w. https://creativecommons.org/licenses/by-sa/4.0/

Global distribution: Temperate seas

Ecology: It occurs in a diverse water environments

Medicinal compounds and their bioactivities:

Cytotoxic and antioxidant effects: The extracts of this species have shown rich α-tocopherol content and other vitamins and have shown cytotoxic effects on MCF-7, HL-60, and NCI-H460 tumor cells and antioxidant activity. It is known to serve as a potential natural nutraceutical for the pharmaceutical industry (https://en.wikipedia.org/wiki/Tetraselmis_suecica)

Antibacterial activity: The peptides from one of the strains (AQ-1766) of this species have shown antibacterial activity on Gram-negative bacteria, like *E. coli*, *P. aeruginosa*, and as well as against the Gram-positive bacterial strains, like methicillin-resistant *S. aureus*, *S. typhimurium*, *B. cereus*, *M. luteus*, and *L. monocytogenes*; and Gram-negative bacteria, like *S. typhimurium*, *E. coli*, and *P. aeruginosa* (Wu et al. 2021; https://commons.wikimedia.org/wiki/File:Tetraselmis_suecica_03.jpgpeptide)

Antioxidant activity: The ethanol/water extract of this species contains pigments such as antheraxanthin, loroxanthin esters, lutein, violaxanthin and neoxanthin, which have shown antioxidant activity by reducing the activity of radicals. Further, this extract has demonstrated strong repairing activity in the A549 (human lung cancer) cell line (Sansone *et al.* 2017).

Tetraselmis sp.

Medicinal compounds and their bioactivities:

Antimicrobial activity: The extracts of this species have shown antibacterial activity on *Pseudomonas aeruginosa*, *Bacillus* sp., *S. aureus*, *Salmonella* sp., *Klebsiella* sp. and *Enterococcus* sp., and *Escherichia coli*; and antifungal activity on *Anidubans* sp., *Rhizopus* sp., and *Fusarium* sp. (Rajendran et al. 2014).

Phylum: Ochrophyta   Class: Eustigmatophyceae
Order: Eustigmatales   Family: Monodopsidaceae

Nannochloropsis granulata

Global distribution: China

Ecology: Brackish waters

Medicinal compounds and their bioactivities:

Anti-inflammatory activity: The lipid extract of one strain (RAW264.7) of this species containing pigments such as zeaxanthin, astaxanthin, and canthaxanthin and mono- and di-galactosyldiacylglycerols has shown anti-inflammatory activity (Wu et al. 2021).

Nannochloropsis oceanica

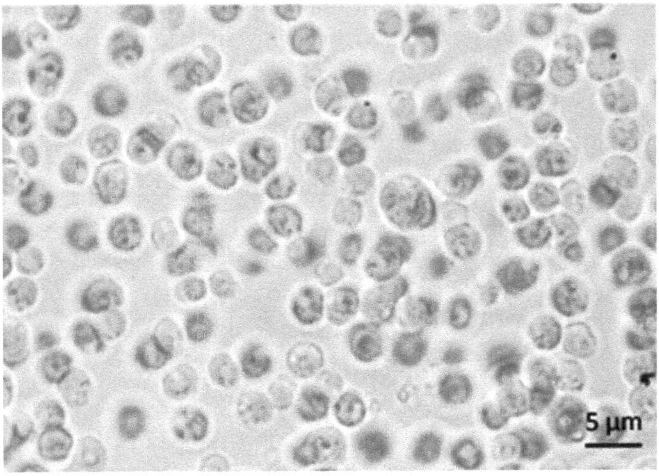

Image credit: Al-Hoqani et al. https://creativecommons.org/licenses/by-sa/4.0/

Global distribution: It occurs all oceans around the world including Southern Hemisphere.

Ecology: Coastal waters

Medicinal compounds and their bioactivities:

Antibacterial activity: The extracts of this species have shown antibacterial activity on *B. subtilis*, *S. aureus*, and *S. pyogenes* (Sushanth and Rajashekhar 2015).

Antioxidant activity: Its omega-3 fatty acid, eicosapentaenoic acid, has been reported to show antioxidant activity by countering the abeta-induced oxidative stress in the Neuro-2A cells. This alga may therefore serve as a functional food to protect against Alzheimer's disease (Lai 2015).

Nannochloropsis sp.

Medicinal compounds and their bioactivities:

Extracts of this unidentified species have shown antibacterial activity against six *Vibrio* strains (Kokou et al. 2012)

Phylum: Chlorophyta Class: Trebouxiophyceae

Order: Chlorellales Family: Chlorellaceae

Chlorella stigmatophora

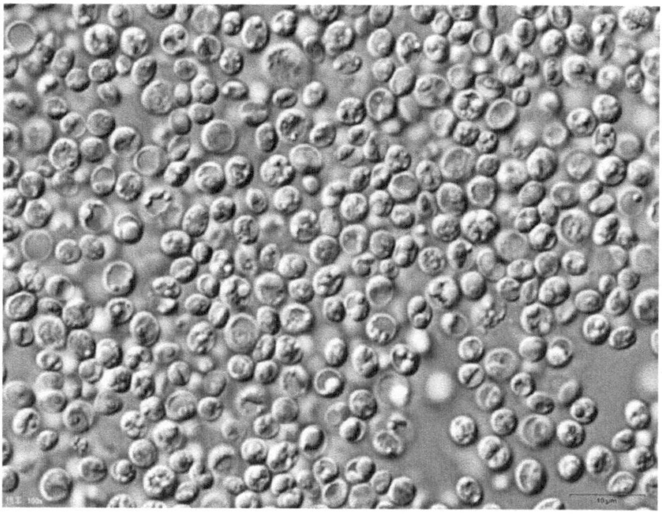

Image credit: N. Rybalka. https://creativecommons.org/licenses/by-sa/4.0/

Chlorella sp.

Medicinal compounds and their bioactivities:

Antiproliferative activity and cytotoxicity: The peptides isolated from the different marine species of *Chlorella* viz. *Chlorella spaerckii*, *C. stigmatophora*, *C. salina*, *C. marina*, and *C. ovalis* have been reported to possess antiproliferative activity, and they also showed activity against cytotoxicity induced by UVB; decreased cleaved PARP-1; prevented DNA damage and fragmentation; and reduced pFADD expression (Wu et al., 2021)

Phylum: Cryptophyta Class: Cryptophyceae
Order: Cryptomonadales; Family: Hemiselmidaceae

Hemiselmis andersenii

Global distribution: Caribbean Sea, Baltic Sea, and Isle of Man

Ecology: Coastal waters and oceanic areas

Medicinal compounds and their bioactivities:

Cholesterol-lowering properties: Its aliphatic alcohols have been reported to possess cholesterol-lowering properties. This species is also potential source of ω3 fatty acids which is usable in human nutraceuticals (Fernandez *et al.* 2020).

BLUE-GREEN ALGAE

Phylum: Cyanophyta Class: Cyanophyceae Order: Chroococcales Family: Chroococcaceae

Microcystis aeruginosa

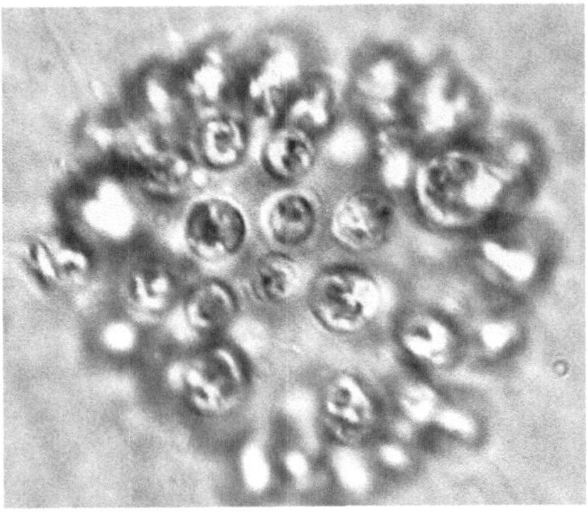

Image credit: Kristian Peters, https://creativecommons.org/licenses/by-sa/3.0/

Global distribution: It is found distributed throughout the world

Ecology: It occurs both in freshwater and brackish water environments.

Medicinal compounds and their bioactivities:

Phytochemical inhibiting activity: The compound, phycocyanobilin, isolated from this species showed phytochemical inhibiting activity against SARS-CoV-2 Mpro and PLpro proteases with IC50 values of 71 and 62 μM, respectively (Khalifa et al. 2021). The main protease (Mpro) and papain-like protease (PLpro) are believed to be important targets towards the inhibition of SARS-CoV-2 virus replication

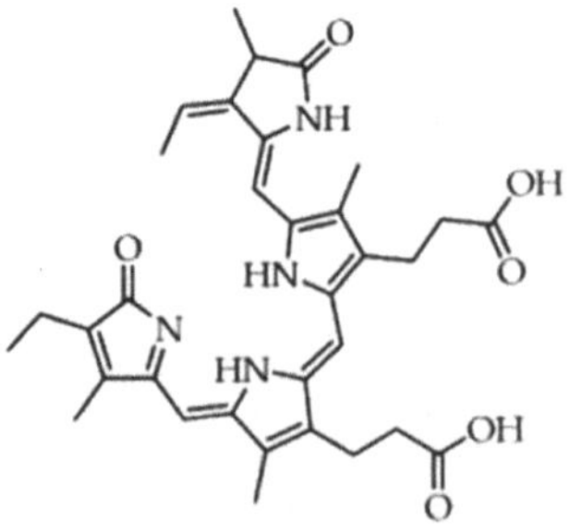

Phycocyanobilin

Phosphorylation of proteins and apoptosis: The compound, Microcystin-LR, of this species has been reported to enhance the phosphorylation of proteins in liver cells by inhibiting the activities of protein phosphatase type 1 (PP1) and type 2A (PP2A) (Khan et al. 2018). This phosphorylation of proteins is capable of altering the enzymatic activity of proteins. Pradhan and Ki (2022) reported that the microcystins of this species induced apoptosis in breast carcinoma cell line MCF-7 and human embryonic kidney HEK 293, cells by reducing the activity of PP1 and PP2A.

Anticancer activity: The analogues of microcystin of this species have shown anticancer activity by killing cancer cells which express organic anion transporting polypeptide (Pradhan and Ki 2022)

Order: Nostocales   Family: Nostocaceae

Anabaena circinalis *(= Dolichospermum sigmoideum)*

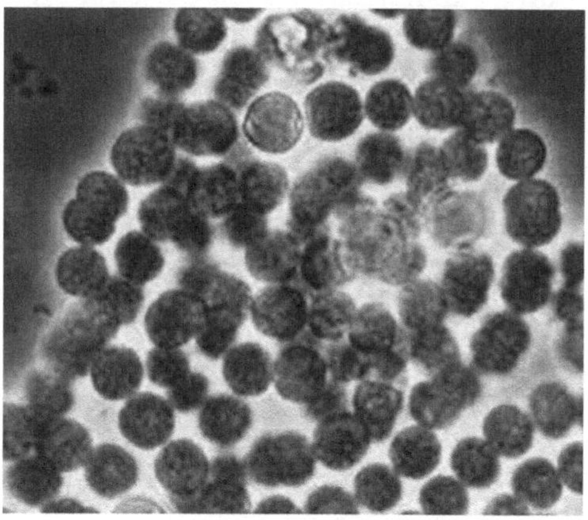

Image credit: Dr. Imre Oldal, https://creativecommons.org/licenses/by-sa/4.0/

Global distribution: It is seen worldwide.

Ecology: This dwells both in marine and freshwater biotopes

Medicinal compounds and their bioactivities:

As nicotinic agonist: The compound, Anatoxin-a, isolated from this species has been reported to serve as a nicotinic agonist which is a drug mimicking the action of acetylcholine (ACh) at nicotinic acetylcholine receptors (nAChRs). The nAChR plays an important role in neuronal communication (Wonnacott and Gallagher 2006; Khalifa et al. 2021).

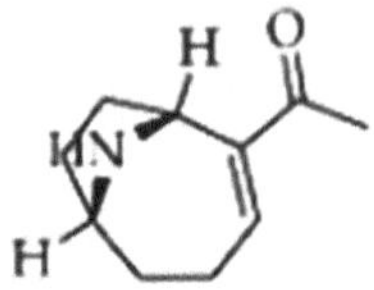

Anatoxin A

***Anabaena* sp.**

Medicinal compounds and their bioactivities:

Antioxidant activity: The methanolic extracts of this unidentified species of Anabaena have shown antioxidant activity (Khalifa et al. 2021)

Anti-anginal activity: Pradhan and Ki (2022) reported that the saxitoxin groups of the species of *Anabaena, Cylindrospermopsis, Lyngbya, Aphanizomenon,* and *Planktothrix* block sodium channels in nerve axons and these toxins could be potential antianginal drug candidates which are used for the treatment of chronic angina (a symptom of coronary artery disease). Khan et al. (2018) stated that the extracts of Anabaena sp. could induce apoptosis of a leukemia myeloid cell line

Anticancer activity: The extracts of Anabaena sp. M44, M30, and M27 strains triggered apoptosis and these extracts could serve as VGSC (voltage-gated Na+ channel) inhibiting drugs with activities against cancer migration, invasion and proliferation (Pradhan and Ki 2022)

Nodularia spumigena

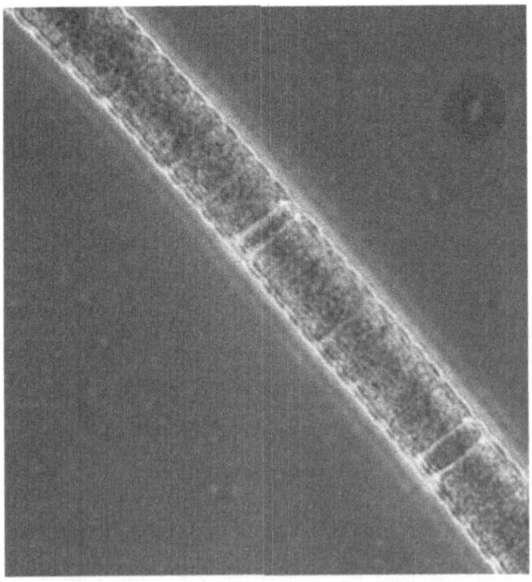

Image credit: Mireia Bertos Fortis, https://creativecommons.org/licenses/by-sa/4.0/

Global distribution: Australia and Baltic Sea

Ecology: It occurs mainly in brackish water environments

Medicinal compounds and their bioactivities:

Anticancer activity: The cyclic pentapeptide nodularin of this species has been reported to induce apoptosis in breast carcinoma cell line MCF-7 and human embryonic kidney HEK 293, cells by reducing the activity of PP1 and PP2A (Pradhan and Ki 2022)

***Nodularia* sp.** *LEGE 06071*

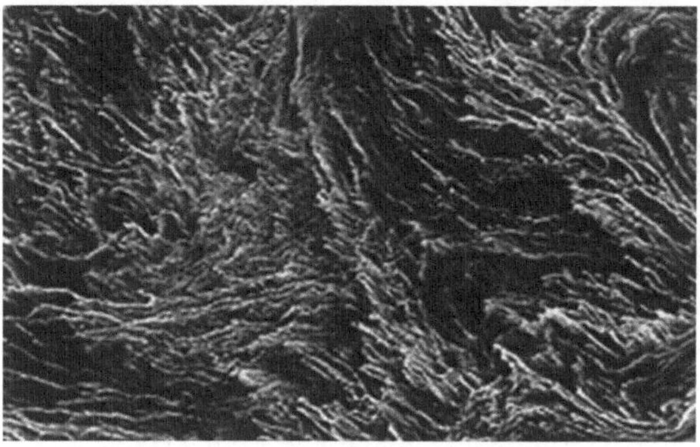

Image credit: NASA Earth Observatory/USGS. Public domain...

Global distribution: Baltic Sea. And Peel-Harvey Estuary in Western Australia

Ecology: It occurs both in salinic and brackish waters

Medicinal compounds and their bioactivities:

Antiproliferative effects: The compound Nocuolin A isolated from the strain Nodularia sp. LEGE 06071 has been reported to show antiproliferative activity against human hTERT-RPE-1 epithelial cells and HCT116 colon cancer cells with IC50 values of 1.2 and 1.35 μM, respectively (Sousa et al. 2019).

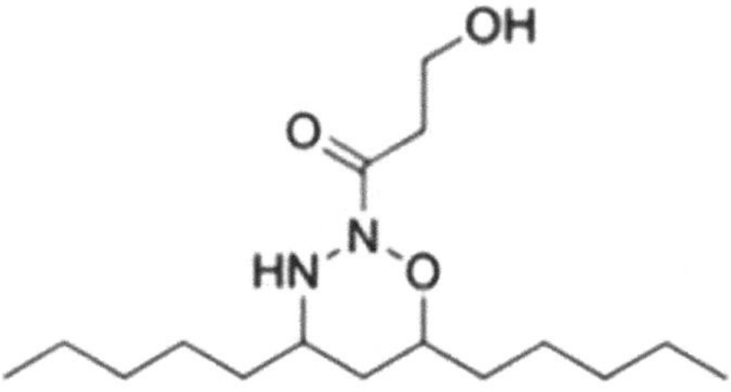

Nocuolin A

Nostoc commune

Image credit: Yamamaya, https://creativecommons.org/licenses/by-sa/3.0/

Global distribution: It occurs in several countries around the world

Ecology: It dwells in both freshwater and brackish water environments

Medicinal compounds and their bioactivities (Li and Quo 2018)

Antimicrobial activity:

i) The ethyl acetate and acetone extracts of this species displayed potent antibacterial activity against *P. aeruginosa*
ii) The methanol and ethanol extracts displayed potent antibacterial activity against *Alternaria altemata, Trichethecium roseum,* and *Peniicillium* sp.
iii) The methanol extracts inhibited the growth of *Staphylococcus aureus, Escherichia coli, Bacillus subtilis, Monilia albican* and *Serratia marcescens* with MIC alues of 1.5–1.0 mg/mL
iv) The fat-soluble compositions showed inhibitory effects against *Bacillus subtilis, Escherichia coli, Staphylococcus aureus, Pseudomonas stutzeri., Aspergillus nigere* and *Saccharomyces cerevisiae.* In the case of *Escherichia coli*, an MIC value of 64 µg/mL was recorded
v) A diterpenoid compound, Noscomin, isolated from this species showed antibacterial activity against *Bacillus cereus* (MIC 32 ppm), *Escherichia coli* (MIC 128 ppm and *Staphylococcus epidermidis* (MIC 8 ppm)

vi) Its compounds comnostins A–E showed moderate antibacterial activity against *Bacillus cereus*; comnostins A–D showed activity against *S. epidermidis;* comnostins A, B, and D were against *E. coli;* Comnostin C showed significant activity against *Escherichia coli*; and comnostin E showed potent activity against *Staphylococcus epidermidis*
vii) Its polysaccharide fraction showed significant antifungal activity against *Aspergillus nigere* and *Candida albicans* with an MIC value of 100 mg/mL.
viii) Its total flavonoids displayed dose- dependent antifungal activity against *Candida albicans* and *Aspergillus niger*
ix) Its lipopeptide compound, nostofungicidine showed potent antifungal activity against *Aspergillus candidus* with an MIC: value of 1.6 μg/mL.

Anti-inflammatory activity

i) Its heteroglycan compound viz. Nc-5-s showed anti-inflammatory response through the Akt/PI3K pathway and ERK1/2 pathway
ii) Its extract showed NO production in murine macrophage RAW264 cells
iii) Its b-ionone derivatives viz. nostocionone and its derivatives displayed NO production through decreased iNOS in murine macrophage RAW264 cells

Antioxidative properties

i) Its two polysaccharides showed O2·− scavenging capacity with 54.8% and 103.7%, respectively,
ii) The total flavonoids of tis algal extracts showed the higher antioxidative ability
iii) Its glycosylated mycosporine-like amino acids showed total radical scavenging activity

Anti-carcinogenic activity

i) The crude extracts of this species showed significant activity against A549 and human lung epithelium adenocarcinoma cells with IC50 vlaues of 24.8 and 51.3 μg/mL, respectively.
ii) Its reduced scytonemin compound yielded induction of autophagy on Human T-lymphoid cell line Jurkat cells by ROS formation

iii) The synthesized nostocionone derivative of this species showed activity against Human T-lymphoid cell line Jurkat cells. This derivative inhibited cell apoptosis through mitochondria via the release of Endo-G.
iv) The water stress proteins of this species inhibited cell proliferation on DLD1, HCT-116, HT-29, and SW480, human colon carcinoma cells with IC50 values of of 0.19, 0.21, 0.39, and 0.41 µg/mL, respectively.
v) The recombinant water stress proteins 1 of this species suppressed the cell growth of SW480, human colon carcinoma cells significantly.
vi) The extracelluar polysaccharides of this species inhibited cell proliferation in MCF-7, human breast cancer and DLD1, human colon carcinoma cells with IC50 values of 67 and 110 µg/mL, respectively.
vii) The extracelluar polysaccharides- of this species suppressed migration of NCI-H446 and NCI-H1688, human small cell lung cells, through blocking the EMT.

Immunomodulating property

The polysaccharide-rich extract of this species inhibited the growth of human peripheral blood-mononuclear cells and U937, human leukemia by downregulating IL-6 and IL-17.

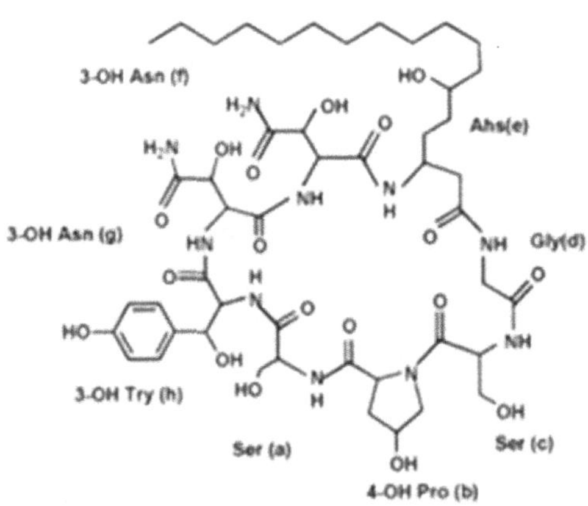

Nostofungicidine

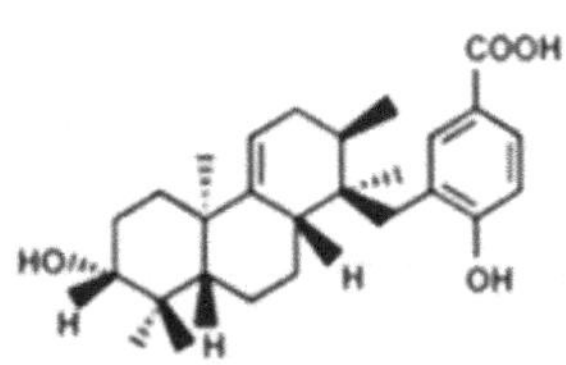

Noscomin

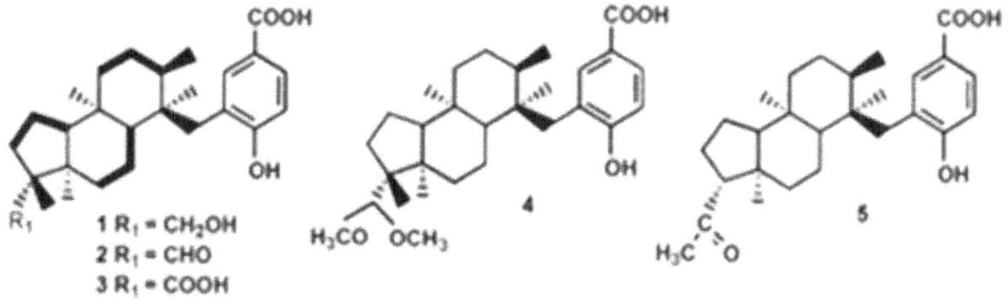

Comnostins A-E

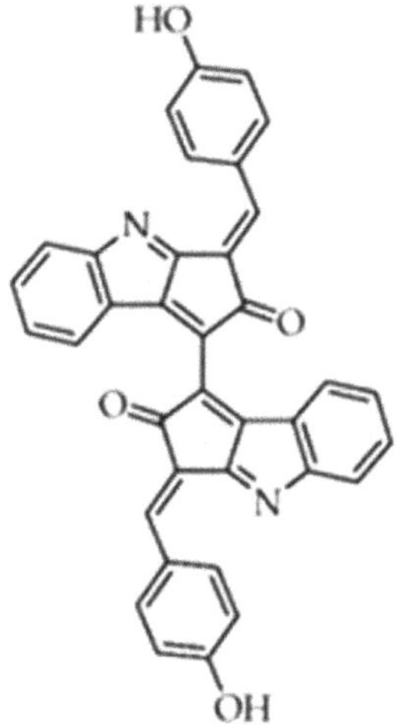

Scytonemin

Nostoc ellipsosporum

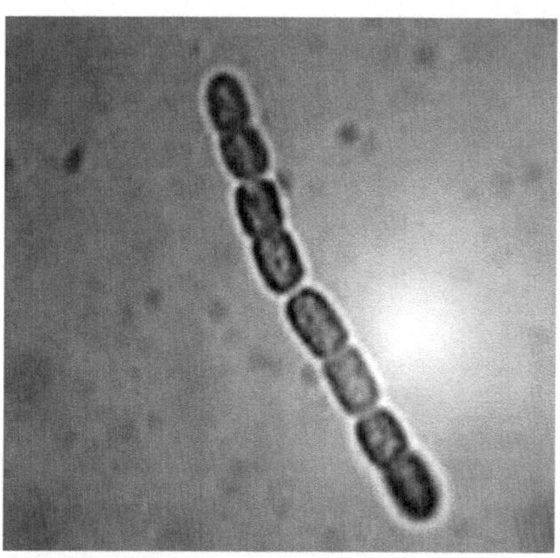

Image courtesy: Michele Greque de Morais (Reproduced with his kind permission)

Global distribution: India

Ecology: It occurs rarely in marine habitats

Medicinal compounds and their bioactivities:

Antiviral activity: Its cyanovirin-N which is a virucidal protein has been reported to possess antiviral activity inhibiting simian immunodeficiency virus and strains of HIV-1, HIV-2 (Khalifa et al. 2021; https://www.sciencedirect.com/topics/agricultural-and-biological-sciences/nostoc-ellipsosporum)

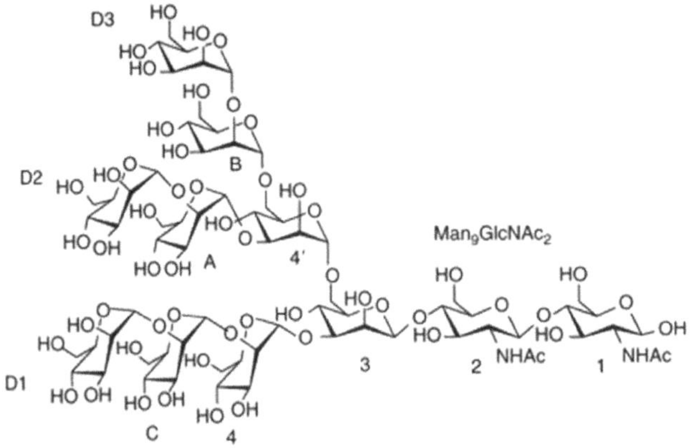

Cyanovirin-N

Nostoc linckia

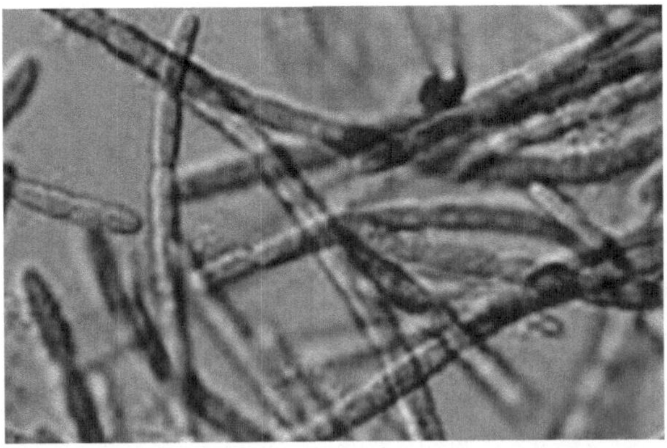

Image credit: Institute of Hydrobiology, Chinese Academy of Sciences (Reproduced with its kind permission)

Global distribution: China, Mediterranean

Ecology: It is hardy species living in a variety of environments like coastal water and freshwater. It is a sessile prokaryote species growing mainly in the canyon with l rock surfaces. It is capable of tolerating extreme fluctuations of desiccation, solar radiation and temperature.

Medicinal compounds and their bioactivities: (El-fayoumy et al., 2021)

Antitumor activity: The polyketide borophycin isolated from this species displayed antitumor activity against LoVo and KB (MIC with MIC values of 0.066 and 3.3 μg/mL, respectively (Khalifa et al. 2021)

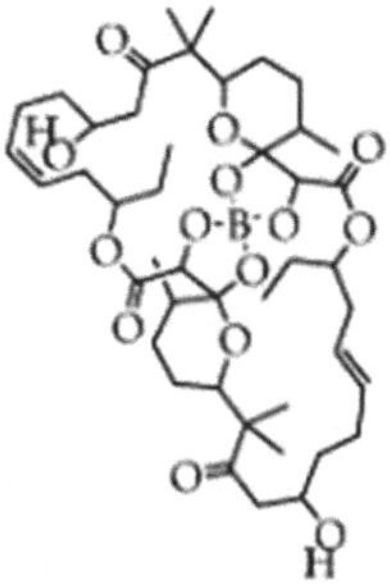

Borophycin

Antioxidant activity: The methylene chloride:methanol, (1:1) extracts of this species showed antioxidant activity against 2,2′-azino-bis(ethylbenzthiazoline-6-sulfonic acid) (ABTS·+) and 2,2-diphenyl-1-picrylhydrazyl (DPPH) radical assays with percentage values of 84% and 88 % respectively (El-fayoumy et al. 2021)

Anticancer activity: The above extracts also showed anticancer activity against HCT 116 cell line with an IC50 value of 155 μg/ml.

Antibacterial activity: The extracts showed antibacterial activity against *Streptococcus mutans* and *Staphylococcus aureus* with inhibition zone values of 19 mm and 21 mm, respectively.

Antiviral activity: The extracts showed antiviral activity against H5N1 virus with percentage inhibition values of 50% and 63.6% at the concentration of 7 and 28 μg/ml, respectively,

***Nostoc* sp.**

Medicinal compounds and their bioactivities:

Cytotoxic/anticancer activity: The peptolide cryptophycin isolated from this unidentified species showed cytotoxic activity. Further, this compound was found to be more active against human colorectal cancer cells. It has also been reported that this compound was several hundred to a thousand times more active than the conventional drugs viz. vinblastine or taxol (Khalifa et al. 2021).

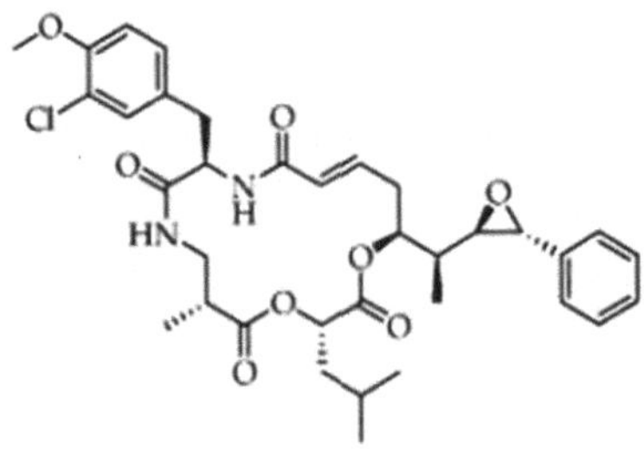

Cryptophycin

***Nostoc* sp.** Strain ATCC 53789

Medicinal compounds and their bioactivities:

The compound isolated form this strain showed inhibitory activities against the enzyme, papain-like protease (PLpro) which regulated the coronavirus, SARS-CoV-2 (Khalifa et al. 2021)

Order: Nostocales Family: Rivulariaceae

***Calothrix* sp.**

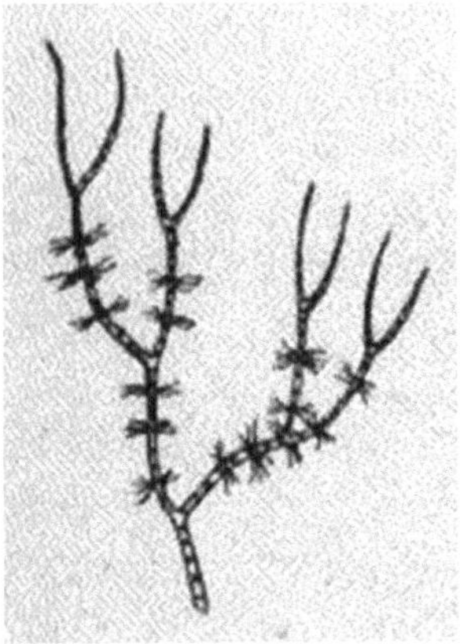

Emile Wuitner, Public domain.

Ecology: it inhabits marine, fresh and terrestrial environments

Medicinal compounds and their bioactivities:

Antioxidant activity: The indole alkaloid compound, scytonemin (SCY, 1), isolated from this species is not only therapeutic to the disorders of inflammation and proliferation but also showed strong antioxidant activity (Khalifa et al. 2021)

Antimalarial activity: The pentacyclic alkaloid compounds Calothrixins A and B of this species displayed promising activity against the malarial parasite *Plasmodium falciparum* with IC50 values of 58 nM and 180 nM, respectively (Khalifa et al. 2021)

Anticancer activity: Calothrixins A and B of this species also showed promising anticancer activity against human HeLa cancer cells with IC50 values of 40 and 350 nM, respectively. https://accedacris.ulpgc.es/bitstream/10553/74363/2/0767079_00000_0000.pdf

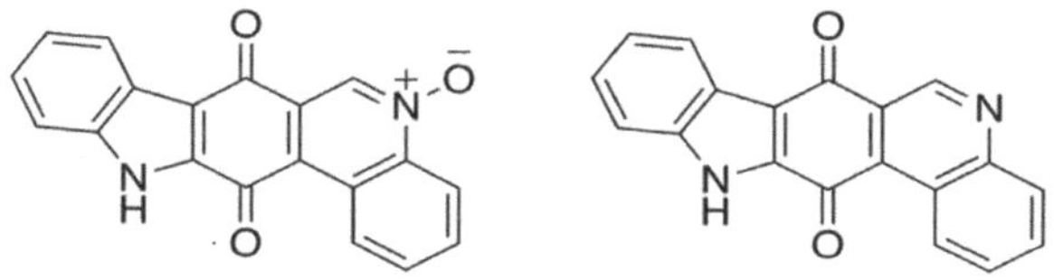

Calothrixins A,B

***Rivularia* sp.**

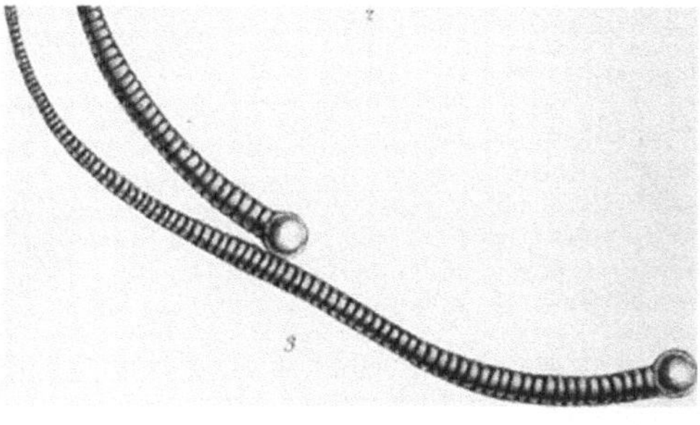

Image credit: Harvey, William, Public domain..

Global distribution: It has worldwide distribution

Ecology: It is seen in marine, freshwater, and terrestrial habitats where moist rocks and submerged stones are present

Medicinal compounds and their bioactivities:

Antioxidant activity: Its indole alkaloid compound, scytonemin (SCY, 1), exhibited antioxidant activity (Khalifa et al. 2021).

Order: Oscillatoriales Family: Coleofasciculaceae

***Coleofasciculus* sp.**

Image not available

Global distribution: Northern Line Islands,west-central Pacific Ocean

Ecology: It is common in island/atoll waters

Medicinal compounds and their bioactivities:

Cytotoxicity: The compound, tutuilamide C, isolated from this species showed cytotoxicity in H-460 lung cancer cell (Keller et al. 2020).

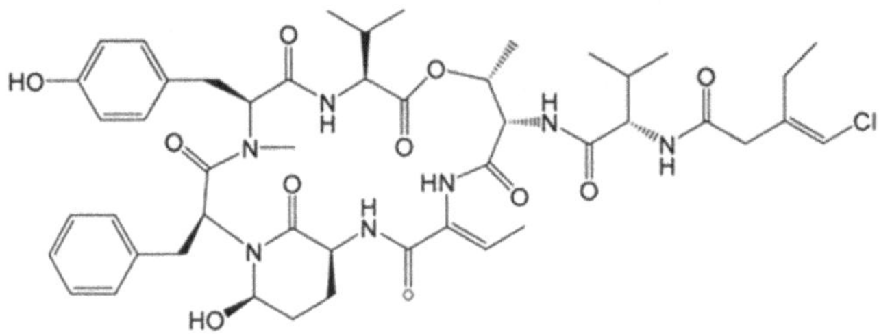

Tutuilamide C

Order: Oscillatoriales Family: Microcoleaceae

Arthrospira platensis *(= Spirulina platensis)*

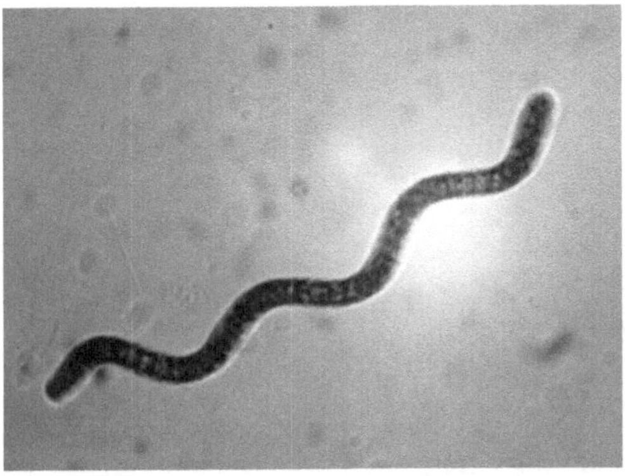

Image courtesy: Michele Greque de Morais (Reproduced with his kind permission)

Global distribution: It has a worldwide distribution

Ecology: It occurs both in marine and freshwater environments

Medicinal compounds and their bioactivities:

Antitumor activity: The polysaccharides isolated form this species showed antitumor activity (Wu et al., 2021).

Antiviral activity: Calcium spirulan (Ca-SP), a sulfated polysaccharide derived from this species has displayed a broad-spectrum antiviral activity against human viruses such as HIV-1, HSV-1, HCMV and HHV-6 (Khalifa et al. 2021). Wu et al. (2021) reported that its polysaccharides exhibited antiviral effect on the Ectromelia virus or Vaccinia virus, in vitro or in vivo.

Anti-inflammatory properties: The extracts of this species have shown anti-inflammatory activity by blocking the release of histamine from mast cells (https://www.sciencedirect.com/topics/agricultural-and-biological-sciences/arthrospira-platensis)

Antioxidant activity: The phenolic compounds such as caffeic acid, chlorogenic acid, transcinnamic acid, synaptic acid and salicylic acid present in this species have been shown to possess antioxidant activity and help in reducing the body weight (https://www.sciencedirect.com/topics/agricultural-and-biological-sciences/arthrospira-platensis). The β-carotene and xanthophylls viz. diadinoxanthin and diatoxanthin have shown the highest scavenging activity (Khalifa et al. 2021). It is also reported that its tetraspyrole chromophore, Phycocyanobilin (2) possessed antioxidant activity by efficiently quenching the different oxygen derivatives.

Immune activity: The dietary supplementation of this species for 12 weeks has been reported to improve immunosenescence by elevating indoleamine 2,3-dioxygenase enzyme activity and complete cell count in older people (Khalifa et al. 2021).

Antimicrobial activity: The methanolic extracts of this species have shown strong antimicrobial, antioxidant and cytotoxic activities (Khalifa et al. 2021)

Lipid- reducing activity: The chlorophyll derivative, 132 -hydroxy-pheophytin a isolated from its organic crude extract, displayed excellent lipid-reducing activity in the experimental fish with an EC50 value of 8.9 μM. Hfa may therefore be a potential drug for lipid reduction (Silva et al., 2020). Freitas et al. (2019) reported this compound also acted against neutral lipid accumulation in murine preadipocytes (Freitas et al. 2019)

Dietary supplement: As it contains proteins, polysaccharides, phycobiliproteins, polyunsaturated fatty acids, carotenoids, https://www.sciencedirect.com/topics/earth-and-planetary-sciences/polysaccharide vitamins (especially vitamin B12), and minerals, this species serves as a dietary supplement. Dietary supplementation of this species for 12 weeks significantly reduced inflammatory cytokine (TNF-α and IL-6) levels and plasma malondialdehyde level in Korean patients with type 2 diabetes (Lee et al. 2008). Oral administration of this species without any processing has been reported to be useful to human health especially in enhancing antioxidant, anticancer, and antiviral and immune system activities. Further, it is known to regulate the hyperlipidemia and cholesterol levels thus providing cell defence against a variety of conditions such as hepatotoxicity, immunomodulation, inflammation, obesity, allergies, asthma and diabetes. It is worth mentioning here that this species was used by NASA as a dietary supplement for astronauts in space (Khalifa et al. 2021).

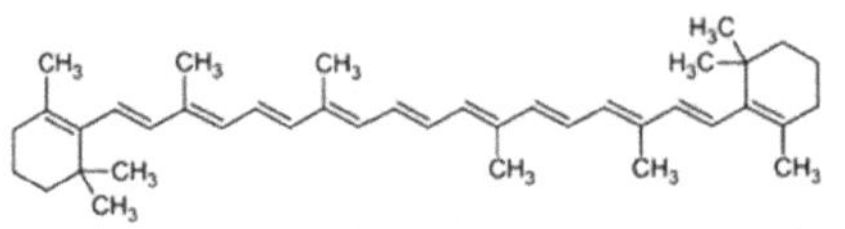

β-carotene

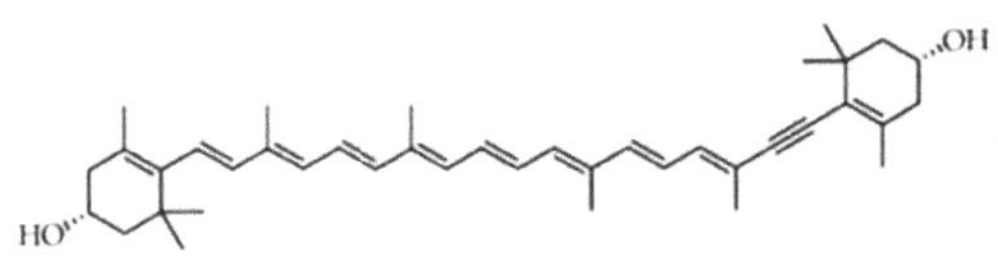

Diatoxanthin

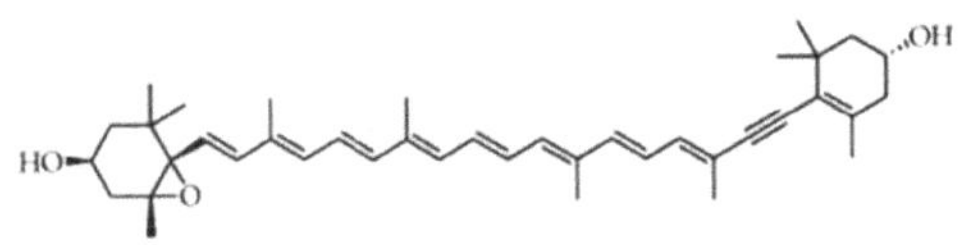

Diadinoxanthin

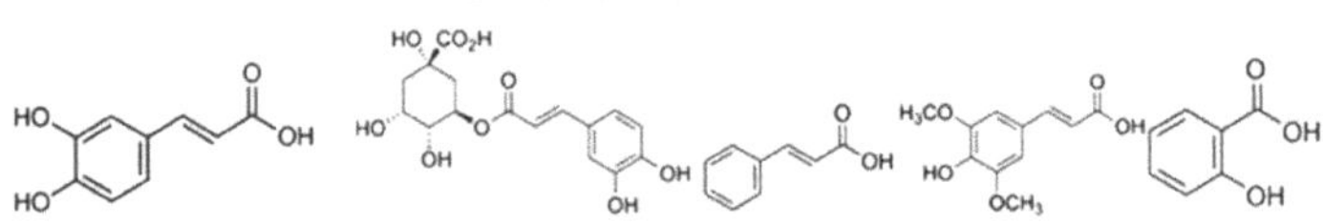

1 2 3 4 5

1: Caffeic acid 2: Chlorogenic acid 3: Trans-Cinnamic acid 4:Sinaptic acid 5:Salicylic acid

Order: Oscillatoriales Family: Oscillatoriaceae

Lyngbya ectocarpi *(= Phormidium ectocarpi)*

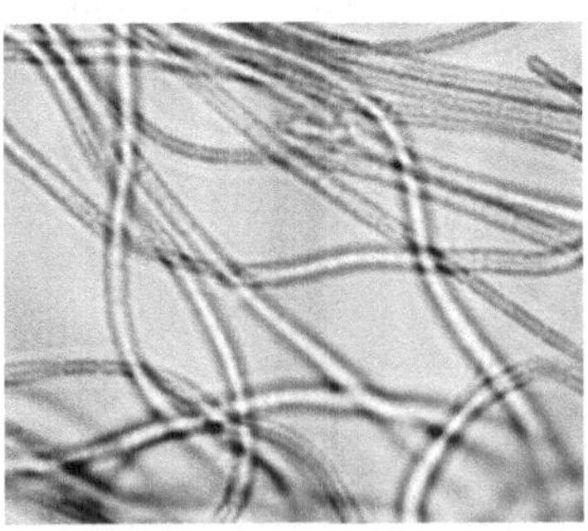

Image credit: PLOS ONE https://creativecommons.org/licenses/by/4.0/

Global distribution: Singapore

Ecology: It is a coastal, benthic species

Medicinal compounds and their bioactivities:

Antiplasmodial activity: The compound, hierridin B, isolated from the strain SAG 60.90 showed antiplasmodial activity (Leão et al. 2013)

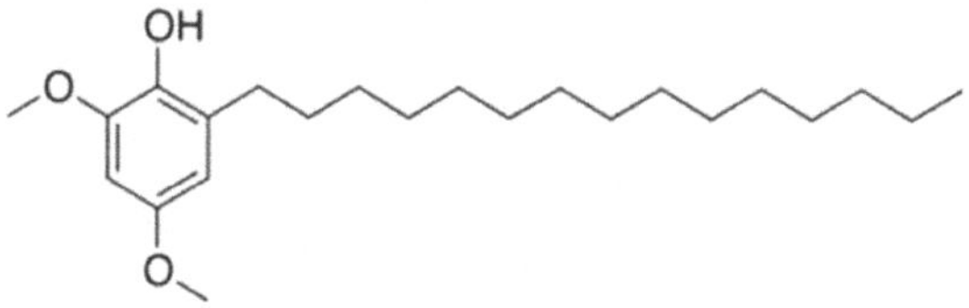

Hierridin B

Lyngbya majuscula

Image credit: Philippe Bourjon, https://creativecommons.org/licenses/by-sa/4.0/

Global distribution: It is a tropical and subtropical species distributed worldwide.

Ecology: This benthic species is found in coastal and estuarine environments where it is usually found associated with the sediment or loosely attached to rocks, sand, seaweeds, seagrass or corals. It may also form large mats with gas content.

Medicinal compounds and their bioactivities:

Antiproliferative and cytotoxic activities: The compounds malyngamide C and 8-O-acetyl-8-epi-malyngamide C isolated from this species have shown antiproliferating effects against heLa cell lines with EC50 values of 0.12 μM and 0.24 μM, respectively. Further, these compounds have shown cytotoxicity against HT29 cancer cell line, with IC50 values 5.2 and 15.4 μM, respectively. Furthermore, a cyclic depsipeptide compound viz. desmethoxymajusculamide C of this species showed cytotoxicity against H-460 human large cell lung carcinoma, neuro-2A murine neuroblastoma, HCT-116 human colon carcinoma, and MDA-MB-435 human carcinoma. The alkynoic lipopeptide, dragonamide B showed cytotoxicity against Vero cells with an IC50 value of 9.8 μM (Khalifa et al. 2021).

Cytotoxic activity: The compound kalkitoixn of this species showed cytotoxicity on HCT-116 colon cell line. Further, this compound has been reported to reduce the hypoxia-induced initiation of HIF-1 in the T47D breast tumor cell line with an IC50 value of 5.6 nM (Pradhan and Ki, 2022)

Anticancer activity: The peptide drug, glembatumumab vedotin (GV) of this species (in phase II study) is to be considered in the treatment of melanoma and breast cancer at a dose of 1.0–1.88 mg/kg. Further, another peptide drug, Brentuximab vedotin (Adcetris™) of this species has been approved by European Medicines Evaluation Agency (EMEA) and U.S. Food and Drug Administration (FDA) for the treatment of cancer (Khalifa et al., 2021)

Antimalarial activity: Among the alkynoic lipopeptides viz. carmabin A, dragomabin, dragonamide A, and dragonamide B isolated from this species, the first three compounds displayed antimalarial activity with IC50 values of 4.3, 6.0, and 7.7 μM, respectively (Khalifa et al. 2021)

Antileishmanial activity: The compounds, dragonamide A and E of this species showed antileishmanial activity against *Leishmania donovani* with IC50 values of 6.5 and 5.1 μM, respectively (Liu and Rein, 2010). Khalifa et al. (2021) reported that the cytotoxic lipopeptides viz. almiramids A–C of this species showed activity against *L. donovani*, *L. infantum*, and *L. chagasi*.

Antiviral activity: The compounds Apratoxin A and Cryptophycin 52 (an analogue of cryptophycin 1) showed antiviral activity against SARS-CoV-2 Mpro (Khalifa et al., 2021)

Antibacterial activity: The compound, lyngbyoic acid (a cyclopropane-containing fatty acid) of this species served as a *Pseudomonas aeruginosa* quorum sensing (QS) inhibitor (Kwan et al., 2011).

Antibiotic activity: At concentrations 3.6 – 57 μM, the antibiotic malyngolide of this species served as a QS inhibitor with an IC50 value of 12.2 μM (Dobretsov et al., 2010)

Immunosuppressive activity: The compound, microcolin-A, of this species showed Immunosuppressive activity (Khan et al. 2018)

Antifungal and cytotoxic activities: The chlorinated lipopeptide compound, hectochlorin of this species, showed both antifungal and cytotoxic activities. This compound has been reported to inhibit human cell lines through hyper-polymerization of actin (https://accedacris.ulpgc.es/bitstream/10553/74363/2/0767079_00000_0000.pdf)

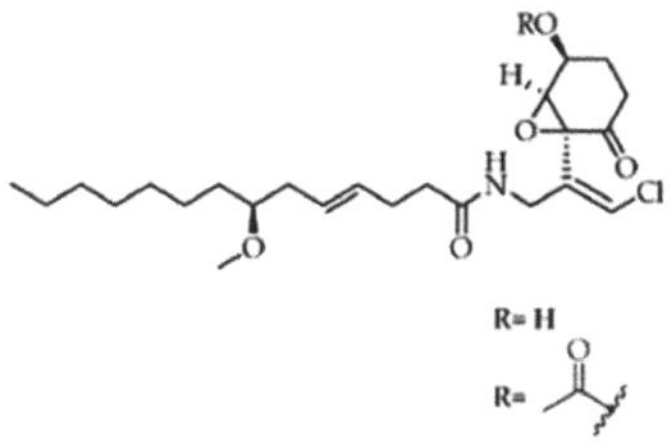

Malyngamide C and 8-O-acetyl-8-epi-malyngamide C

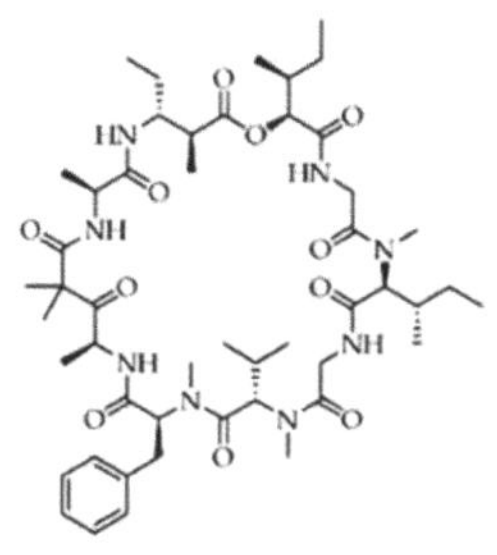

Desmethoxymajusculamide C

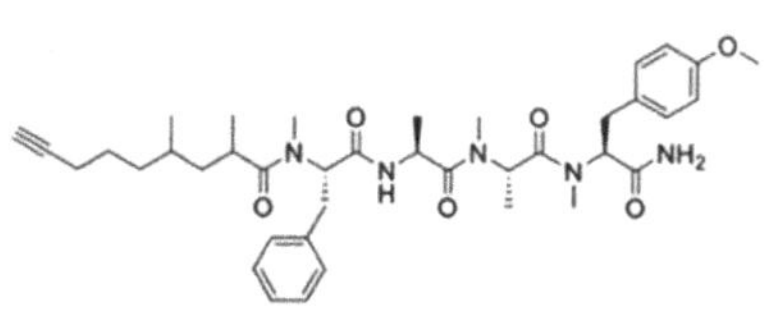

Carmabin A

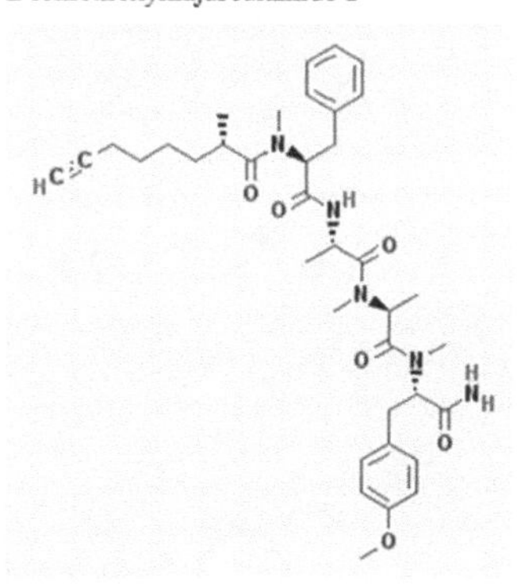

Dragomabin

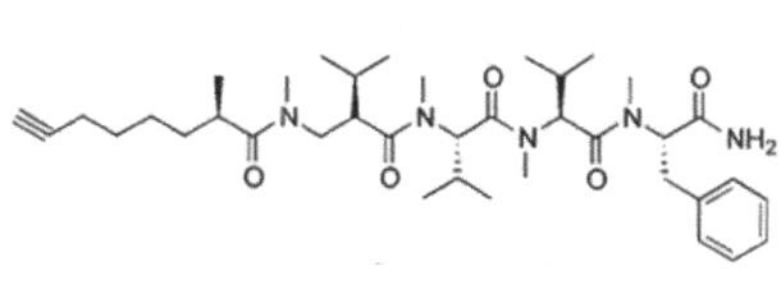

Dragonamide A

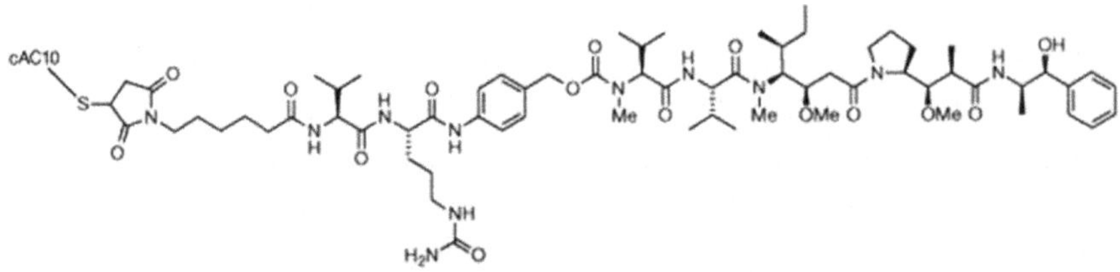

Brentuximab vedotin

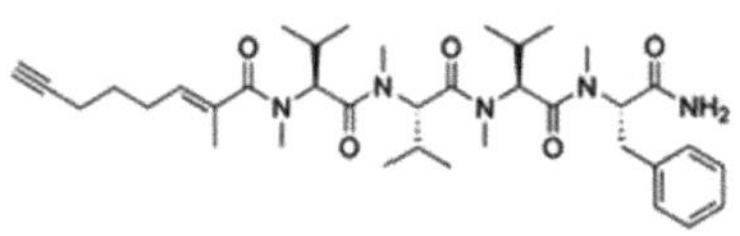

Dragonamide E

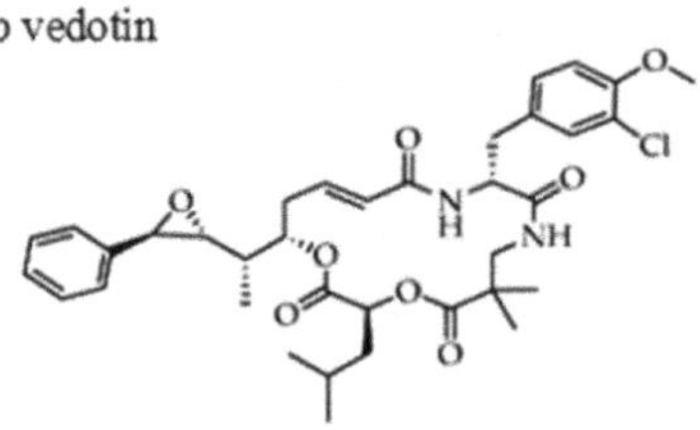

Cryptophycin 52

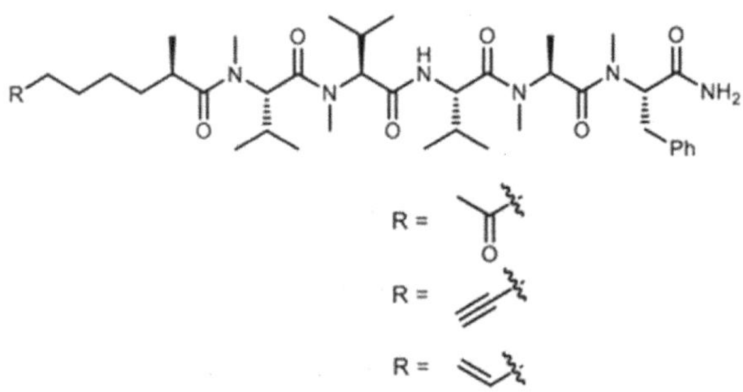

Almiramids A–C

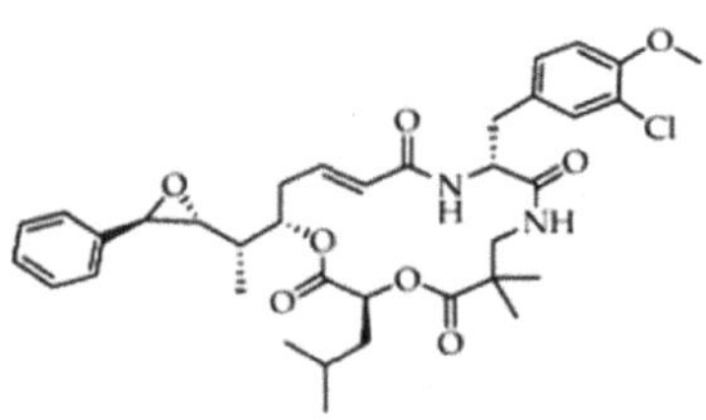

Cryptophycin 52

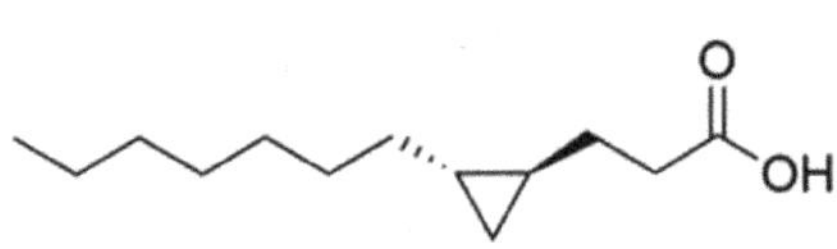

Lyngbyoic acid

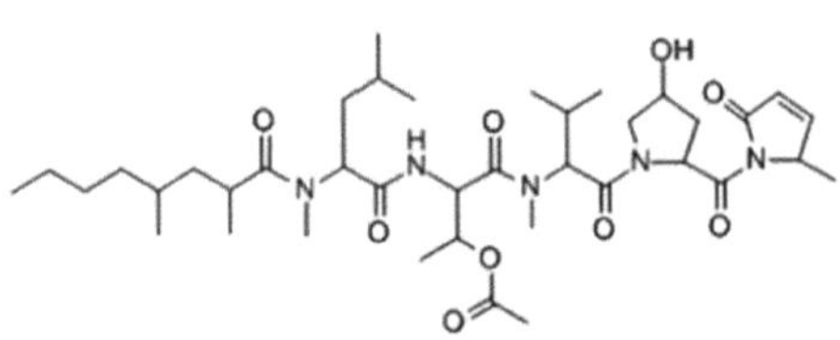

Microcolin-A

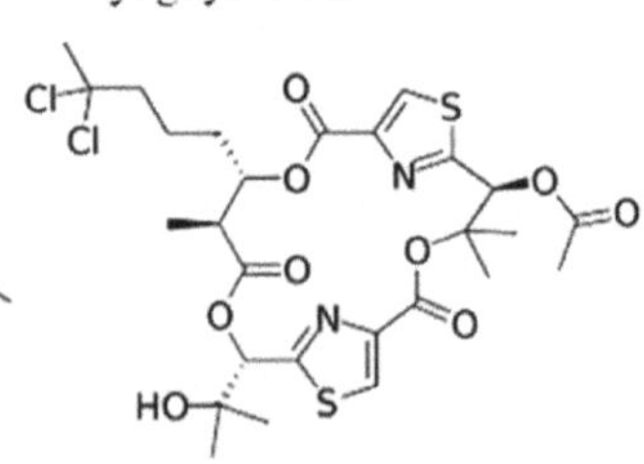

Hectochlorin

Lyngbya polychroa

Global distribution: Caribbean Sea

Ecology: It normally inhabits fore reef areas of the neritic waters

Medicinal compounds and their bioactivities:

Antimalarial activity: The compounds carmabin A, dragomabin, and dragonamide A of this species displayed significant antimalarial activity with IC50 values of 4.3, 6.0, and 7.7 μM, respectively (Liu and Rein 2010)

Antileishmanial activity: The compounds dragonamide A, E of this species showed antileishmanial activity against *Leishmania donovani* with IC50 values of 6.5 and 5.1 μM, respectively (Liu and Rein, 2010)

Anticancer activity: The compounds dragonamide C, D of this species showed weak anticancer activity against IMR-32 neuroblastoma cells with growth inhibition (GI50) values of 49 and 51 μM, respectively; against U2OS osteosarcoma cells with GI50 values of 56 and 59 μM, respectively; and against HT29 colon adenocarcinoma cells with GI50 values of 22 and 32 μM respectively (Liu and Rein 2010)

Cytotoxic activity: The cyclodepsipeptide carriebowmide isolated from the lipophilic extract of this species showed cytotoxicity (Gunasekera et al. 2008)

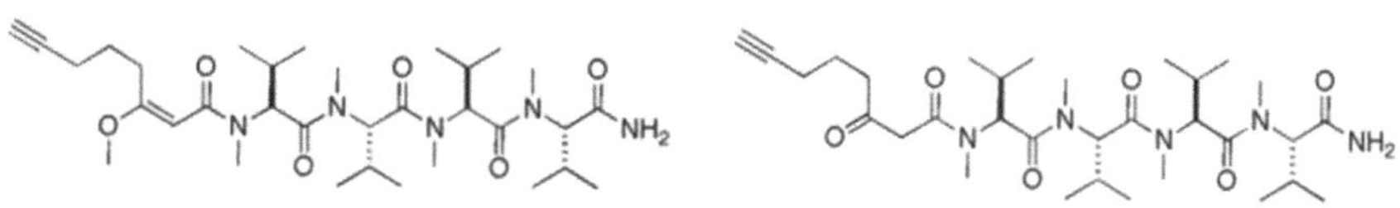

Dragonamides C, D

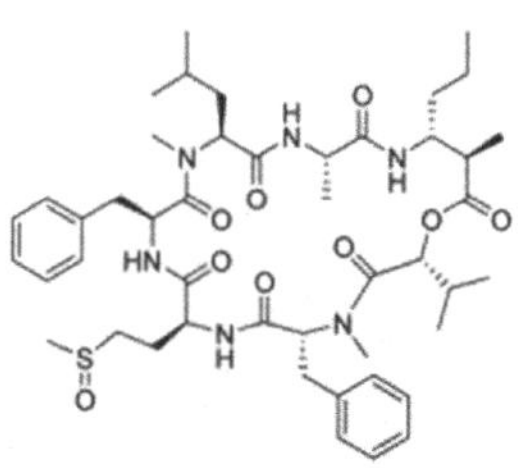

Carriebowmide

***Lyngbya* sp.**

Medicinal compounds and their bioactivities:

Cytotoxicity: The compounds grassypeptolide D, E of this unidentified species showed cytotoxicity against HeLa cell line with IC50 values of 335 and 192 nM, respectively; and mouse neuro-2a blastoma cells with IC50 values of 599 and 407 nM, respectively (Thornburg et al. 2011).

Others: The compounds oscillatoxin E and 30-methyloscillatoxin D of this species inhibited Kv1.5 expression in Chinese hamster ovary (CHO) cells with IC50 values of 0.8 and 1.5 μM, respectively (Pradhan and Ki 2022). Further, the glycoside, biselyngbyaside, isolated from this species showed apoptosis in mature osteoclasts (Khan et al. 2018)

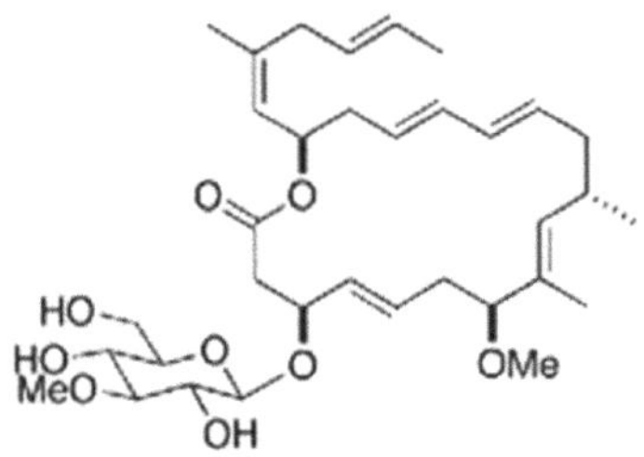

Biselyngbyaside

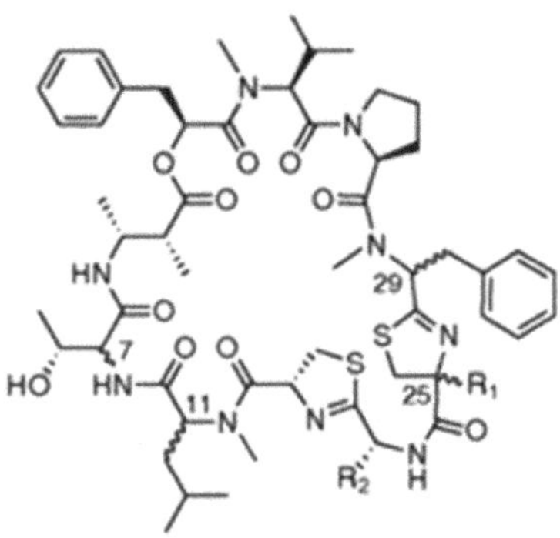

R1= CH3, R2= OH

Grassypeptolides D, E

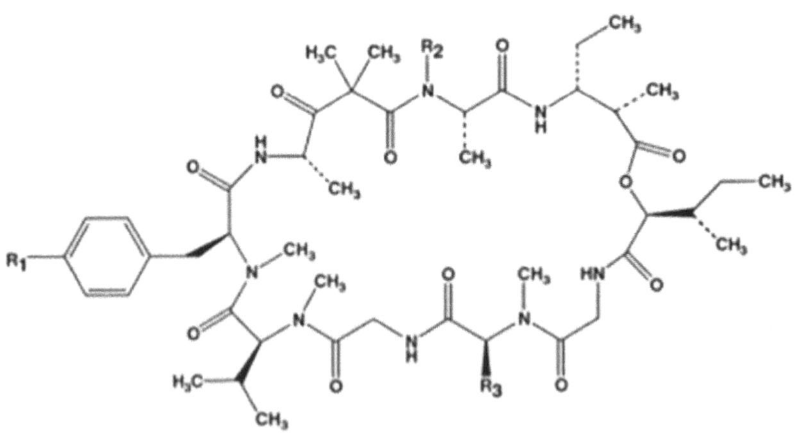

R1- H, R2= CH3, R3=(CH3)2CHCH2-

Dolastatin 12

Moorea bouillonii *(= Lyngbya boulloni)*

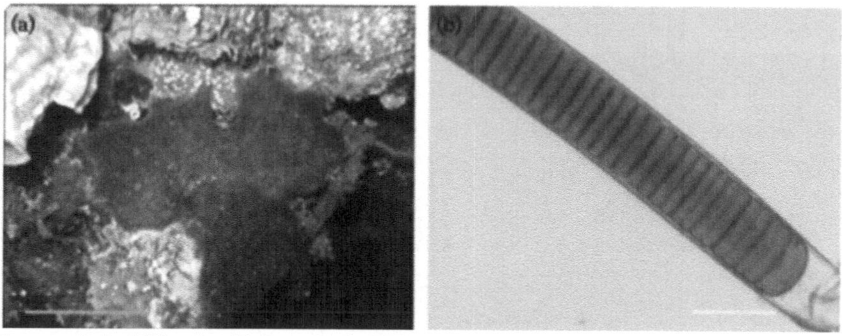

Image credit: Dr. William Gerwick (Reproduced with his kind permission)

Global distribution: It is a pan-tropical cosmopolitan species

Ecology: It inhabits neritic environments such as sandy beaches, coral reefs, and mangroves where it is strongly attached to any nearby substrate. Colonies of this species are often found associated with snapping shrimp, *Alpheus frontalis*.

Cytotoxicity: Compounds such as bouillonamide, ulongamide A, lyngbouilloside and apratoxin A have been isolated from this species. Among them, the cyclic depsipeptide, bouillonamide showed mild toxicity against the neuron 2a mouse neuroblastoma cells with an IC50 vlaue of 6 μM (Tan et al. 2013). Khalifa et al. (2021) reported that the cyclic depsipeptides Apratoxin A of this species showed cytotoxicity against adenocarcinoma cells.

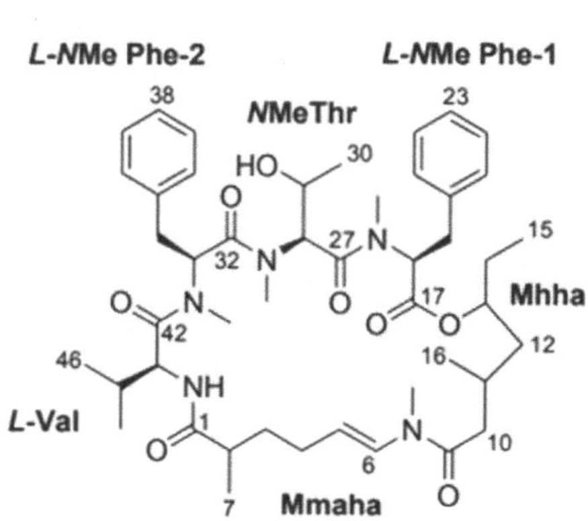

Bouillonamide

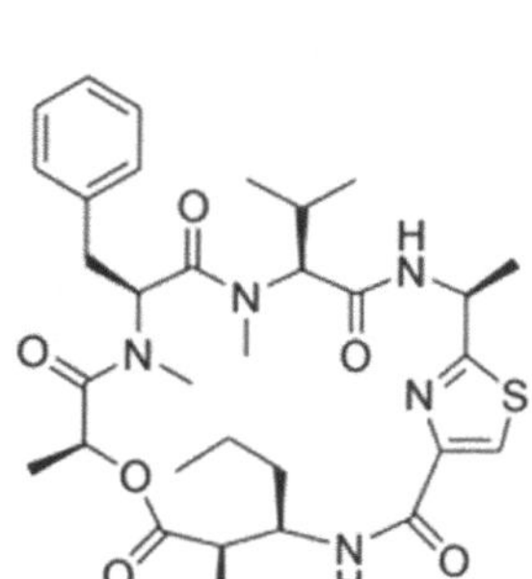

Ulongamide A

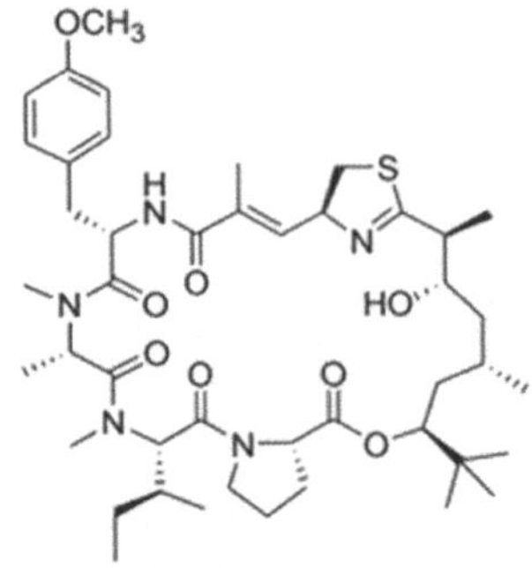

Apratoxin A

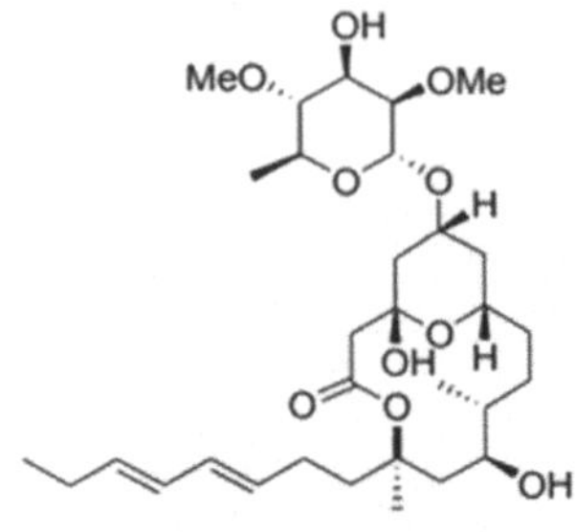

Lyngbouilloside

Moorea producens

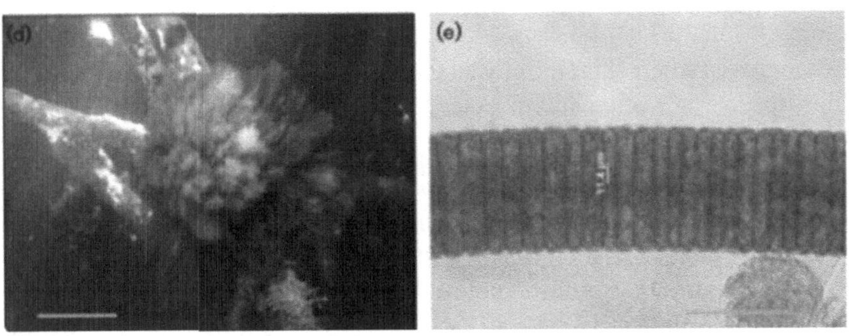

Image credit: Dr. William Gerwick (Reproduced with his kind permission)

Global distribution: It is a pantropical, cosmopolitan species

Ecology: it occurs in shallow-water marine marine habitats like sandy beaches, coral reefs, and mangroves.

Medicinal compounds and their bioactivities:

Cytotoxic/ antitumor activity: Lipopeptides, microcolins E–M of this species have shown strong cytotoxic activity against lung carcinoma cells (Khalifa et al. 2021).The compound palmyramide A of this species showed antitumor activity against Neuro-2a cells by blocking the sodium channel and cytotoxic with H-460 human lung carcinoma cells (https://accedacris.ulpgc.es/bitstream/10553/74363/2/0767079_00000_0000.pdf). Thornburg et al. (2013) reported that among its compounds viz. apratoxin H, apratoxin A sulfoxide, apratoxins A–C, lyngbyabellin B, and hectochlorin, apratoxin H, and apratoxin A sulfoxide displayed cytotoxicity to human NCI-H460 lung cancer cells with IC50 values of 3.4 and 89.9 nM, respectively.

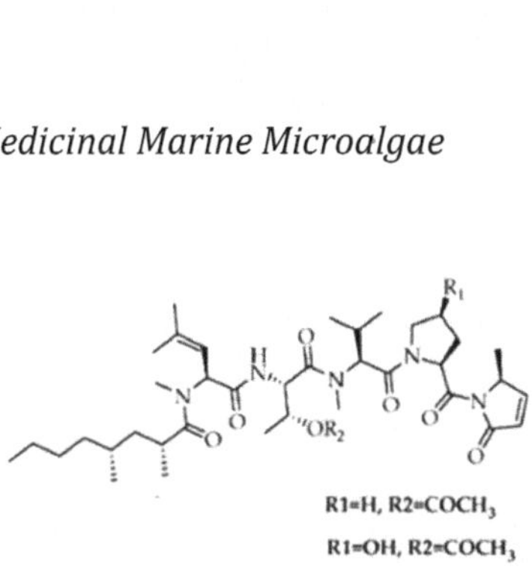

Microcolins E,F

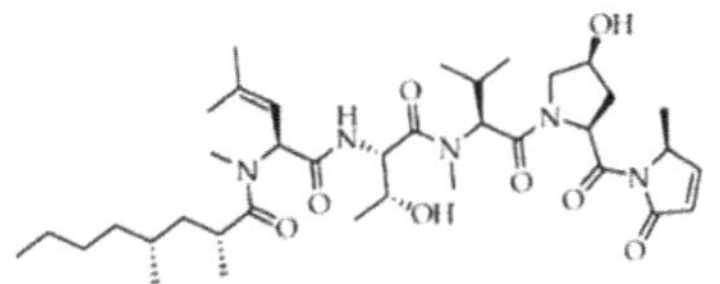

Microcolin G

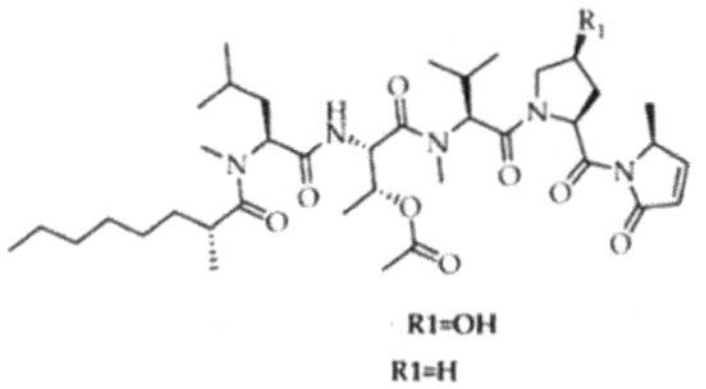

Microcolins H,I

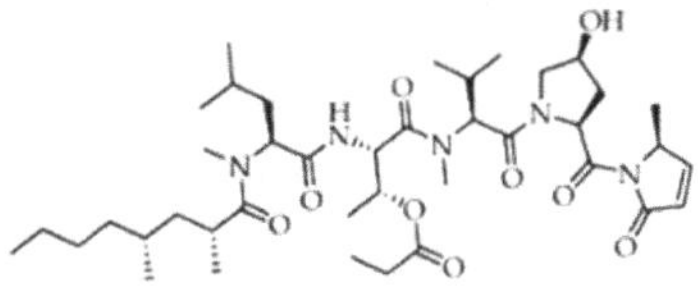

Microcolin J

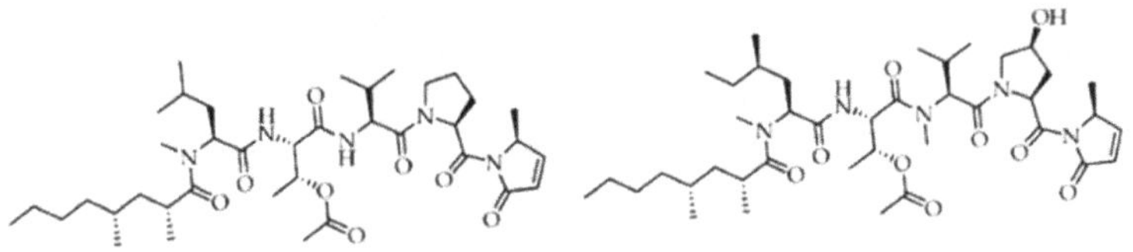

Microcolins K,L

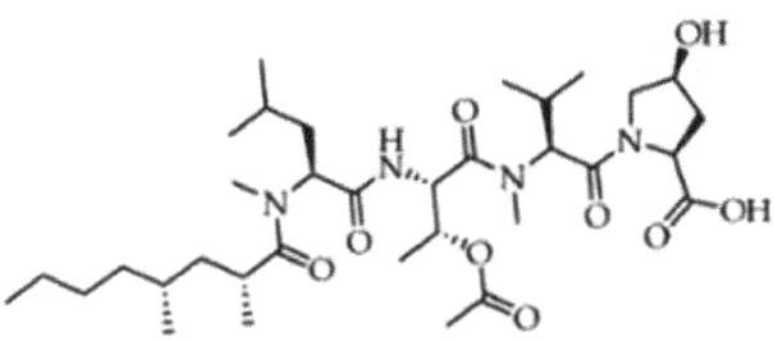

Microcolin M

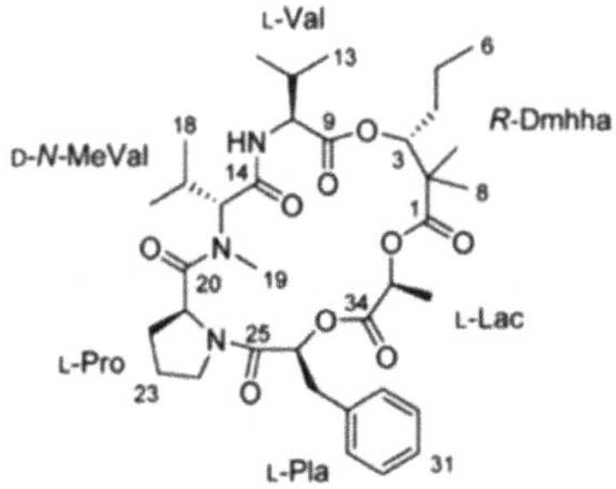

Palmyramide A

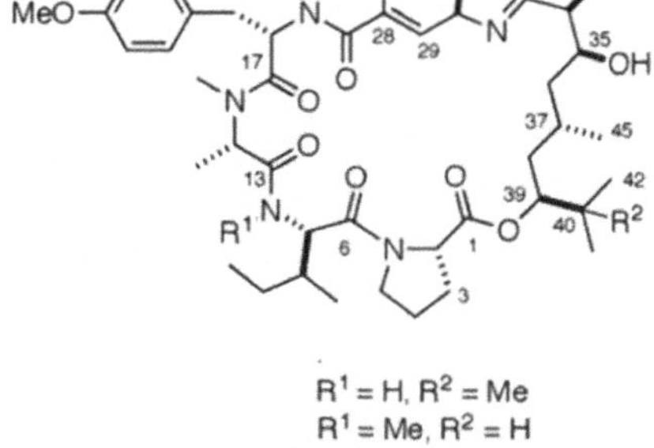

Apratoxin B,C

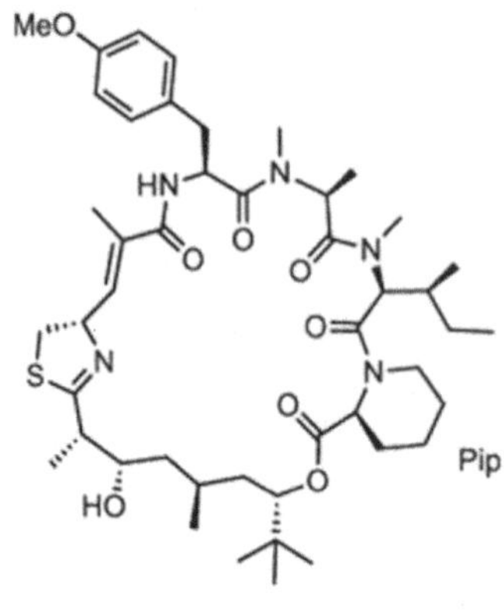

Apratoxin H

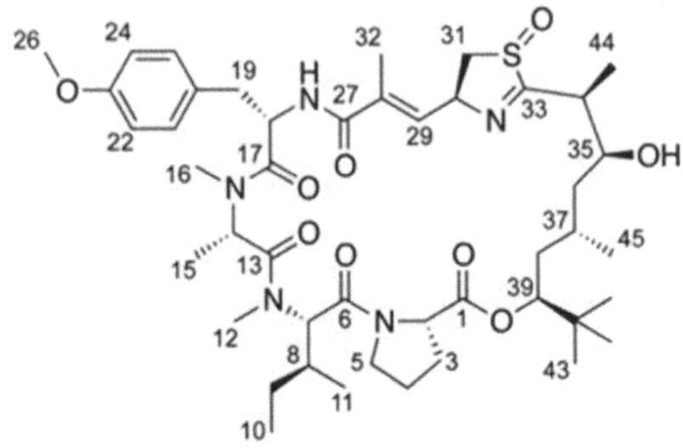

Apratoxin A sulfoxide

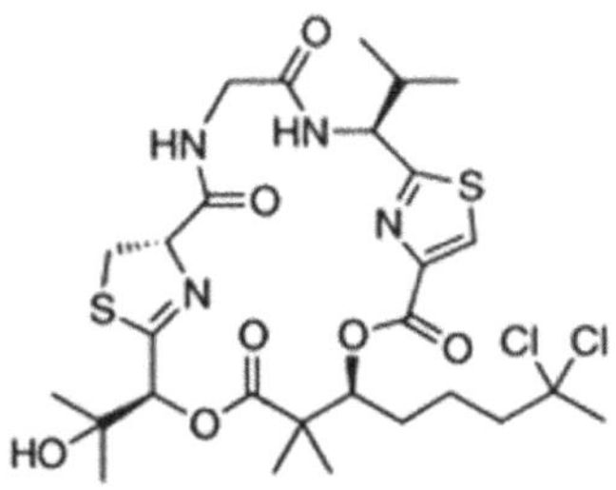

Lyngbyabellin B

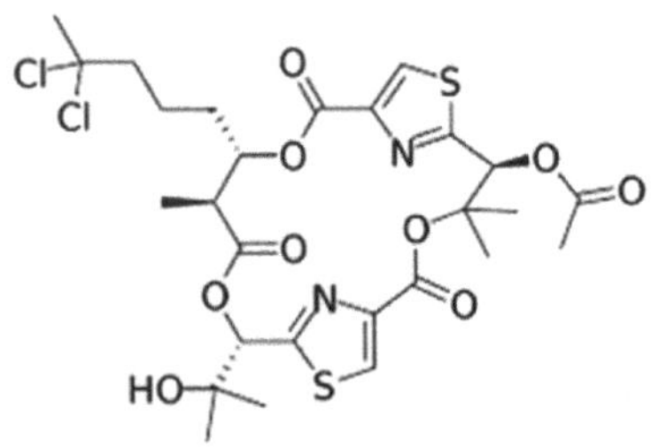

Hectochlorin

Okeania hirsuta

Global distribution: This globally distributed species are commonly found in tropical and subtropical seas

Ecology: It is commonly seen in the shallow-water benthos of neritic waters

Medicinal compounds and their bioactivities:

Anticancer activity: The lipopeptide viz. minnamide A derived from this species has shown growth-inhibitory activity toward HeLa cells with an IC50 value of 0.17 μM (Sumimoto et al. 2019).

Cytotoxicity and antimalarial activity: A polyhydroxy macrolide, bastimolide A, synthesized by this species displayed moderate cytotoxicity and significant antimalarial activity against MDR strains of *Plasmodium falciparum* with IC50 values ranging from 80 to 270 nM (Nweze et al. 2020).

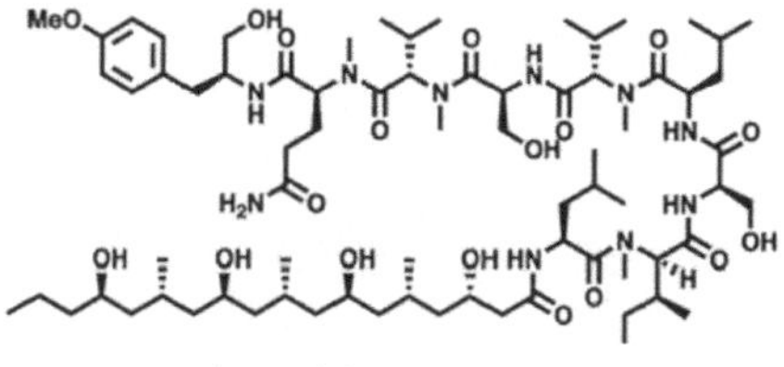

Minnamide A

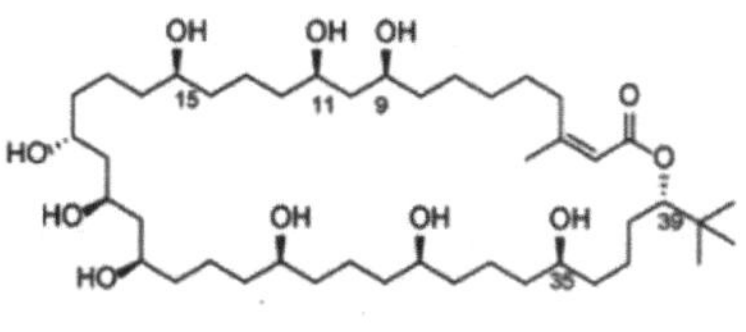

Bastimolide A

***Okeania* sp.**

Medicinal compounds and their bioactivities:

Cytotoxicity: The cyclodepsipeptide, odoamide, isolated from this species has shown promising cytotoxicity on HeLa S3 human cervical cancer cells with an IC50 value of 26.3 nM (Sueyoshi et al. 2016). Petitbois *et al.* (2017) reported that its compounds, serinolamides C and D; lyngbyabellins F, G, O and P; and dolastatin 16 showed cytotoxicity.

Antimalarial activity: A lipopeptide, mabuniamide, of this species displayed antimalarial activity against *Plasmodium falciparum* 3D7 clone in in vitro with an IC50 value of 1.4 µM (Khalifa et al. 2021). Iwasaki et al. (2020) reported on the antimalarial activity of its lipopeptide, ikoamide with an IC50 value of 0.14 µM.

Others: The macrolide glycosides, irijimasides A–E of this species have been reported to inhibit osteoclast formation (Yamano et al. 2020).

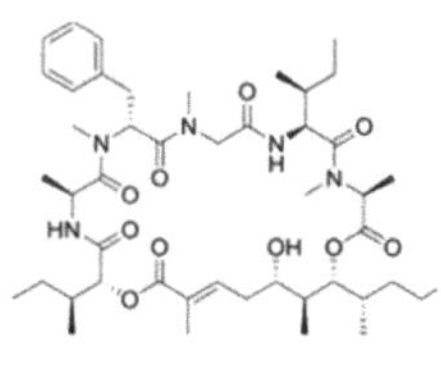

Mabuniamide

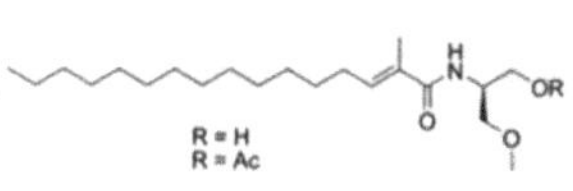

Serinolamide C, D

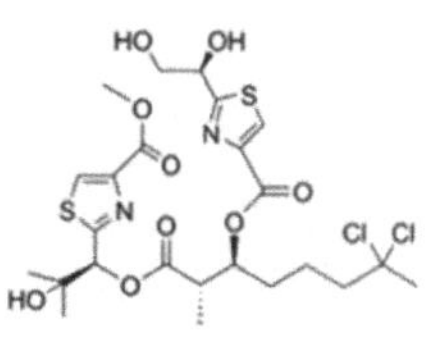

Lyngbyabellin O

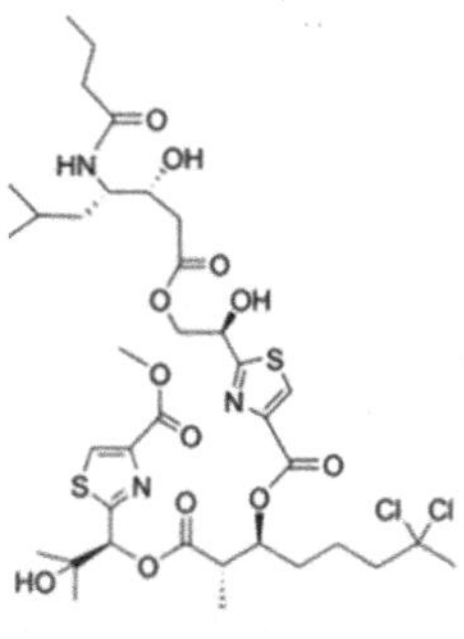

Lyngbyabellin P

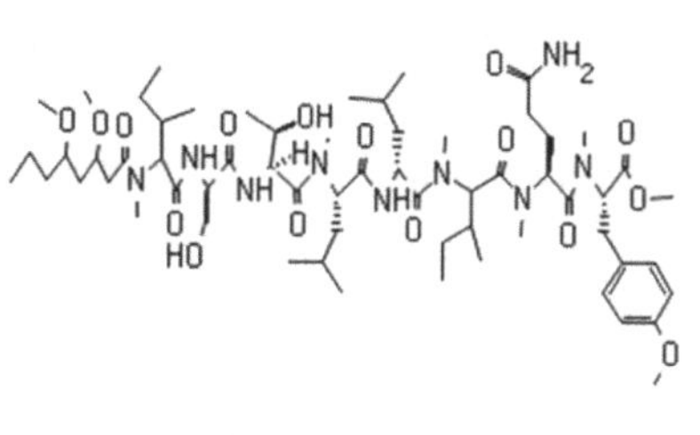

Ikoamide

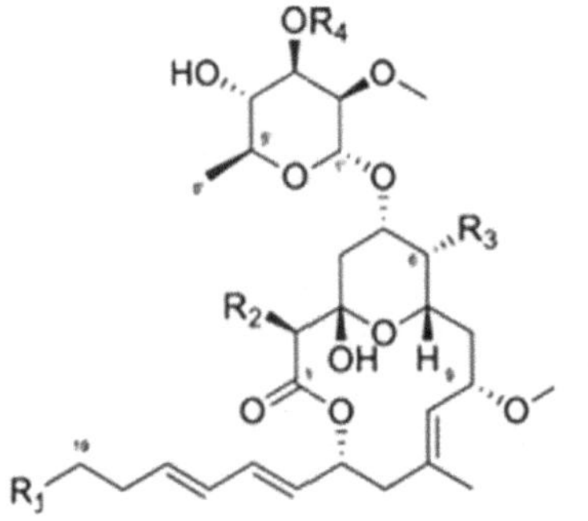

	R_1	R_2	R_3	R_4
1	CH=CH$_2$	CH$_3$	CH$_3$	CH$_3$
2	CH=CH$_2$	CH$_3$	H	CH$_3$
3	CH=CH$_2$	H	CH$_3$	CH$_3$
4	CH=CH$_2$	CH$_3$	CH$_3$	H
5	CH$_2$C≡CH	CH$_3$	CH$_3$	CH$_3$

1-5: Irijimaside A,E

***Oscillatoria* sp.**

Image credit: Wiedehopf20, https://creativecommons.org/licenses/by-sa/4.0/

Global distribution: it is found distributed in the seas of tropical and subtropical latitudes

Ecology: it inhabits a wide variety of environments viz. freshwater to marine and the different species of this genus may lead either planktonic or benthic life

Medicinal compounds and their bioactivities:

Antioxidant activity: The methanolic extracts of this unidentified marine *Oscillatoria* sp. showed antioxidant activity (Khalifa et al. 2021)

***Phormidium* sp.**

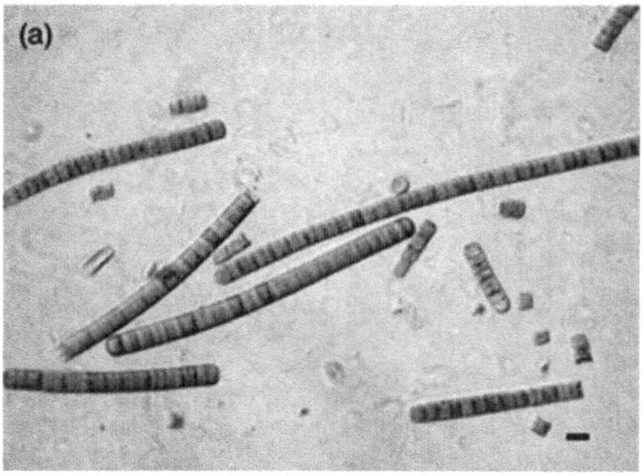

Image courtesy: Rhodes Lesley (Reproduced with his kind permission)

Global distribution: It is found distributed worldwide.

Ecology: Filaments of this species form mats on different substrates of marine littoral area, freshwater and even wet soil.

Medicinal compounds and their bioactivities:

Antioxidant activity: The methanolic extracts of this species showed antioxidant activity (Khalifa et al. 2021)

Cytotoxicity: The compounds, portoamides A and B of the strain, LEGE 05929 of this unidentified species showed cytotoxicity of the cancer cell line, HCT116 by disrupting the function of mitochondria (Sousa et al. 2020).

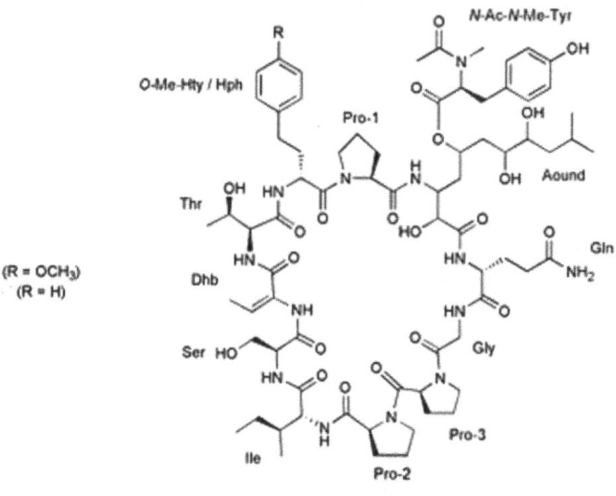

Portoamides A, B

Order: Oscillatoriales Family: Phormidiaceae

Symploca hydnoides

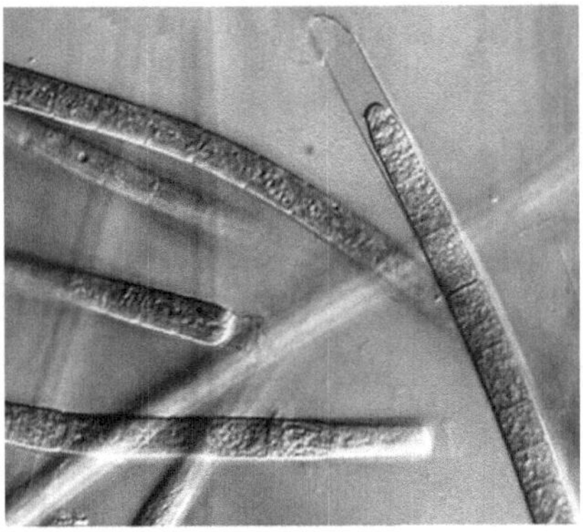

Image courtesy: Christopher S Lobban (Reproduced with permission)

Global distribution: It is commonly distributed in Pacific Mexico and Panama

Ecology: It occurs both in Marine and freshwater environments

Medicinal compounds and their bioactivities:

Anticancer activity: Several anticancer compounds have been isolated from this species. Its malevamide D was found active against colon carcinoma, lung carcinoma and melanoma cancer cells; its symplostatin 1 was anticancerous against epidermoid carcinoma, breast cancer, and ovarian carcinoma cells; its veraguamide A was against lung cancer, colon, and cervical carcinoma cells; and veraguamide B, C, D, E was acting against all cervical and colon carcinoma cells (https://www.sciencedirect.com/topics/agricultural-and-biological-sciences/symploca). The glembatumumab vedotin (GV) peptide of this species (a phase II study) is to be used in the treatment of melanoma and breast cancer at a dose of .0–1.9 mg/kg. Further, its peptide drug, .0–1.88 mg/kg has been approved for cancer treatment by the European Medicines Evaluation Agency (EMEA) and U.S. Food and Drug Administration (FDA) (Khalifa et al. 2021)

Cytotoxicity: A cyclic decapeptide, trikoramide A isolated from this species, has shown cytotoxicity against the AML2 and MOLT-4 cancer cell lines with IC50 values of 8.2 and 4.8 μM, respectively (Phyo et al. 2019; Ma et al. 2020).

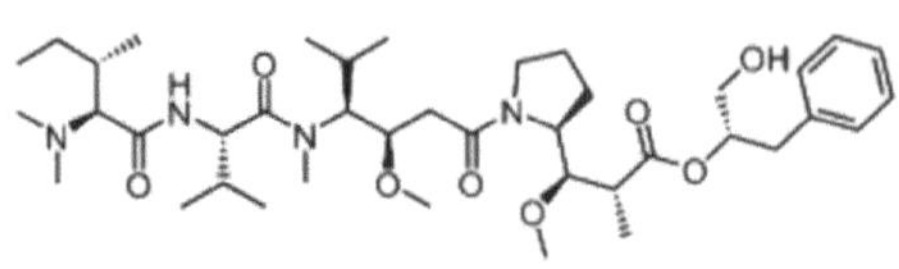

Malevamide D

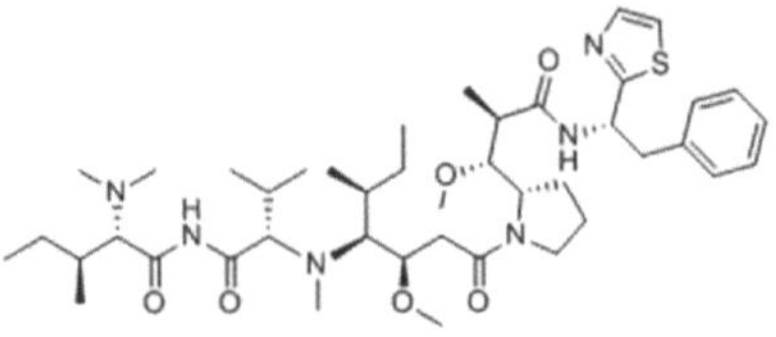

Symplostatin 1

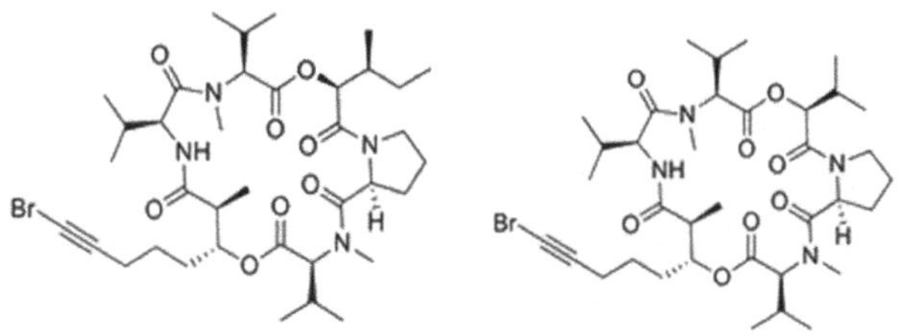

Veraguamides A,B

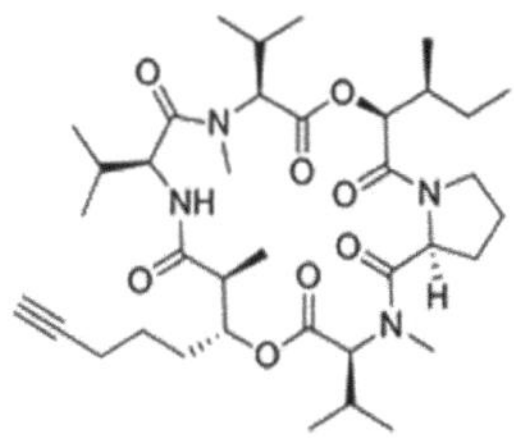

Veraguamide C

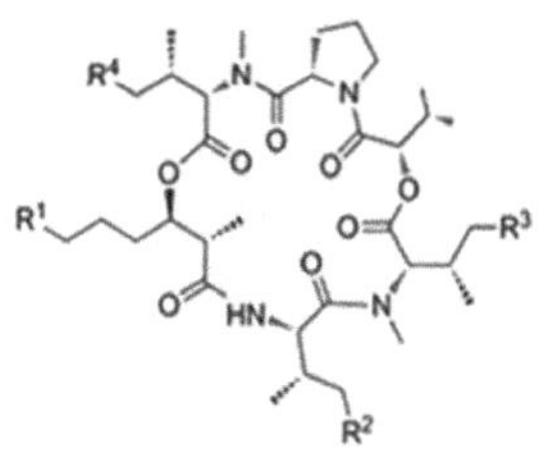

Veraguamide D = R2, H; R3, H; R4, CH3
Veraguamide E= R2, CH3; R3, Ch3; R4,H

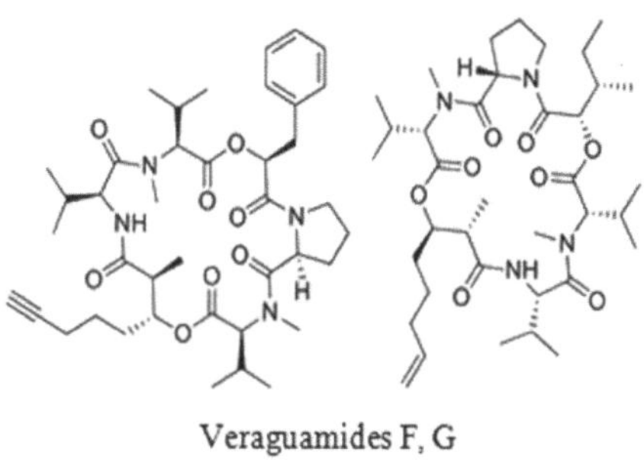

Veraguamides F, G

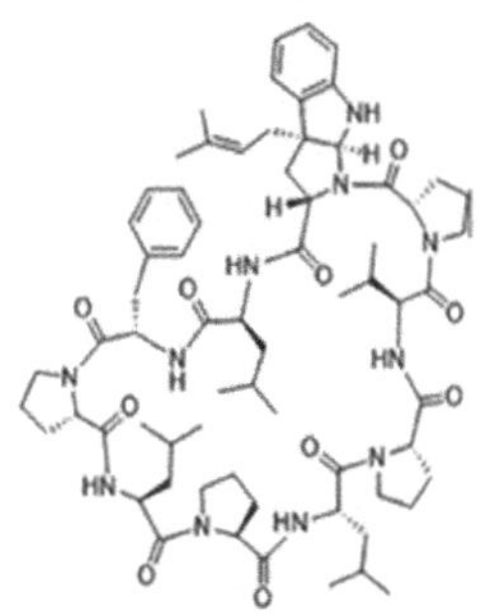

Trikoramide A

Symploca sp.

Medicinal compounds and their bioactivities:

Antiproliferative/ anticancer activity: Its dolastin 10 showed antiproliferative activity against A549 (lung cancer) by serving as a significant microtubule inhibitor, which arrests mitotic cell division. As it was discontinued in Phase II trial due to the development of peripheral neuropathy in 40% of the patients, its analog viz. dolastatin 10, TZT-1027 was found to be effective. Further, another linear peptide dolastatin 15 isolated from this species showed anticancer activity. Another analog, ILX-651 has also successfully passed the Phase II trial (https://accedacris.ulpgc.es/bitstream/10553/74363/2/0767079_00000_0000.pdf)

Its other compounds viz belamide A of this species inhibited HCT-116 (colon cancer); carmaphycin A,B acted against HCT-116 (colorectal carcinoma) and H-460 (lung adenocarcinoma); largazole was inhibiting HCT-116 (colorectal carcinoma), A549 (lung cancer), IMR-32 (neuroblastoma), and HT29 (colon cancer); symplocamide A was found to inhibit neuro-2a (neuroblastoma) and H460 (lung cancer); and tasiamide A and B acted against KB (oral epidermoid cancer. Among these compounds, the cyclic dipeptide largazole and its analogs have been reported to have the potential to serve as novel anticancer therapeutics (https://accedacris.ulpgc.es/bitstream/10553/74363/2/0767079_00000_0000.pdf).

Cytotoxicity: The depsipeptide compounds, tasipeptins A–B from *Symploca* sp. displayed cytotoxic activity against KB cells (human epithelial carcinoma cells) with IC50 values of 0.9 and 0.8 μM, respectively (Williams et al. 2003)

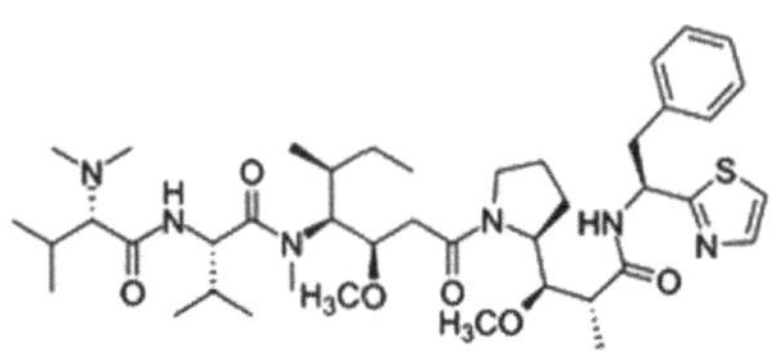

Dolastatin 10

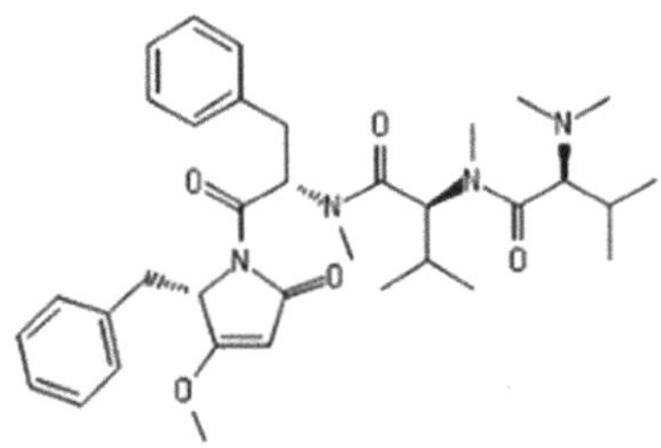

Belamide A

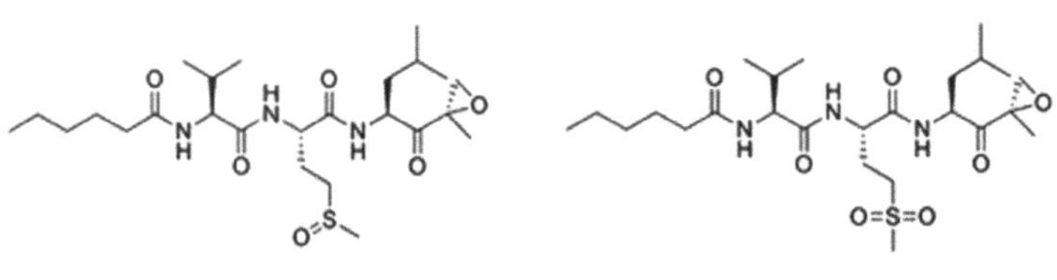

Carmaphycins A,B

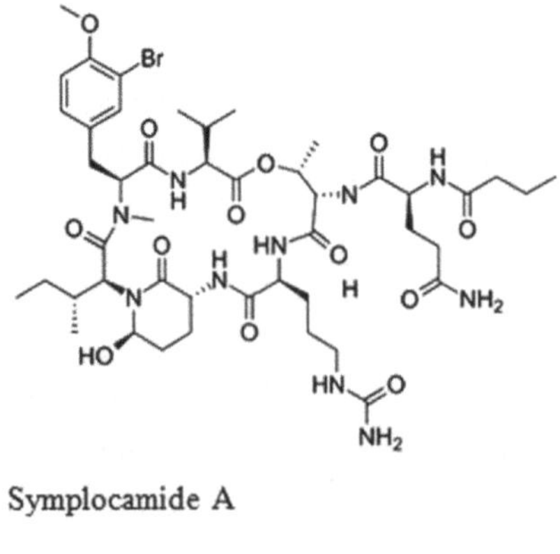

Symplocamide A

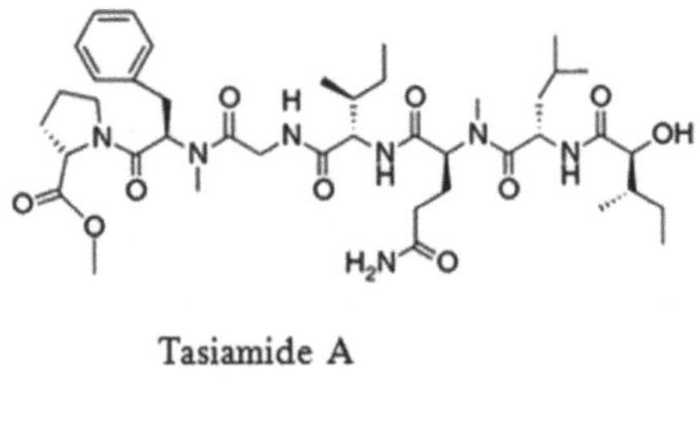

Tasiamide A

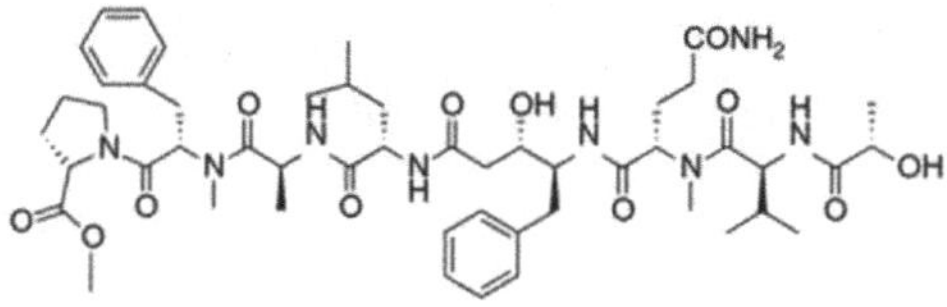

Tasiamide B

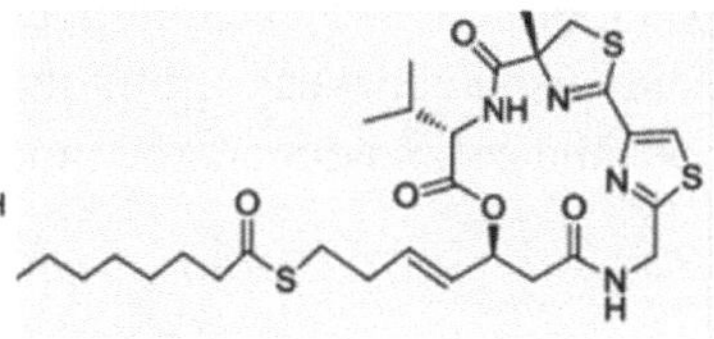

Largazole

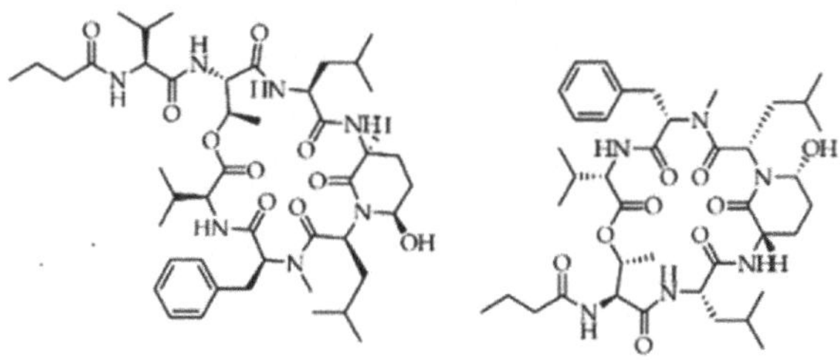

Tasipeptins A,B

Order: Oscillatoriales Family: Schizotrichaceae

Schizothrix calcicola

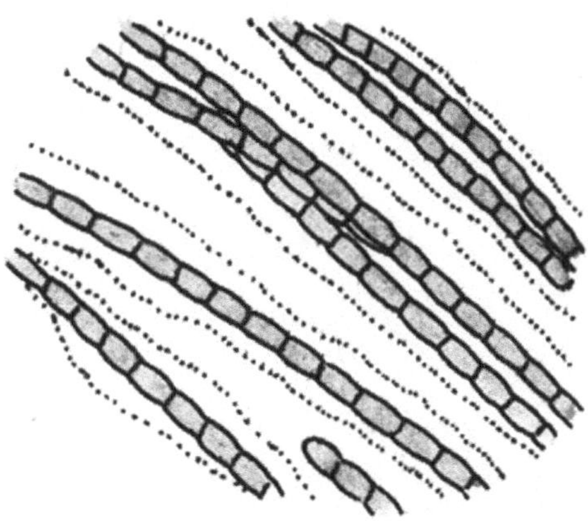

Image credit: Pentecost, Allan. https://creativecommons.org/licenses/by-sa/3.0/

Global distribution: USA

Ecology: Marine; It occurs in subtidal, limestone areas as submerged organisms.

Salient features: In this endolithic alga, the filaments which are dark blue-green and intertwined are thin and slender within a sheath. Cells are cylindrical and are longer than wide.

Medicinal compounds and their bioactivities:

Antitumor activity: The compound lyngbyastatin 1 from the extracts of this species displayed weak antitumor activity against mammary adenocarcinoma #16/C and colon adenocarcinoma #38 (Harrigan et al., 1998)

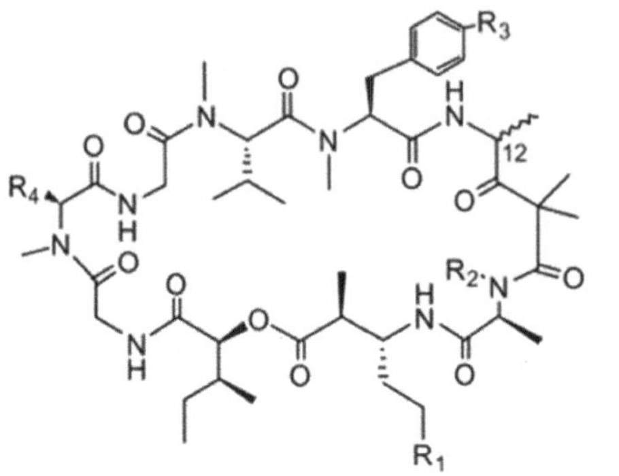

R1=H, R=Me, R3= OMe, R4= i-Bu

Lyngbyastatin 1

***Schizothrix* sp.**

Medicinal compounds and their bioactivities:

Cytotoxicity: The cyclic peptides, tutuilamide A, B of this species showed moderate cytotoxicity in H-460 lung cancer cell with IC50 values of 0.5 μM and 1.3 μM, respectively (Keller et al. 2020)

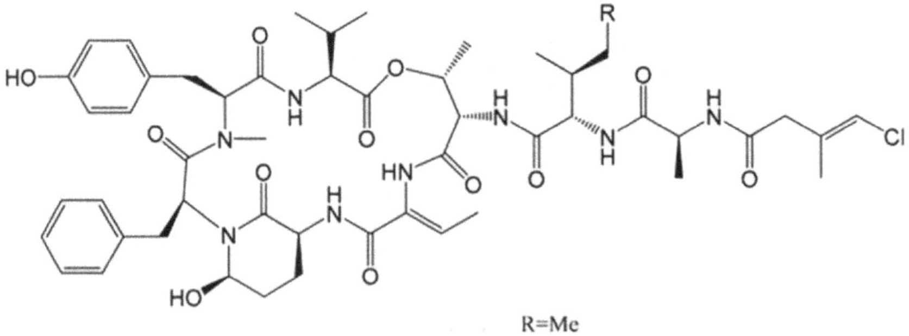

R=Me
R=H

Tutuilamide A, B

Order: Oscillatoriales; Family: Not reported

Dapis pleousa

Image not available

Global distribution: Subtropical and tropical marine habitats.

Ecology: It is largely seen in soft bottom habitats of shallow waters as benthic and planktonic forms. It may be from extensive blooms.

Salient features: The presence of gas vesicles, in connection with a mesh-like network of its filaments is a characteristic feature of this species. These gas vesicles which trap oxygen produced by photosynthesis help in the dispersal of this species in coastal waters.

Medicinal compounds and their bioactivities:

Antimalarial activity: The cyclodepside, malyngolide, isolated from this species has shown moderate antimalarial activity against chloroquine-resistant *Plasmodium falciparum* strain W2 with an IC50 value of 19 μM (https://www.sciencedirect.com/topics/pharmacology-toxicology-and-pharmaceutical-science/malyngolide)

Antibacterial activity: Dobretsov et al. (2010) reported that malyngolide possesses antibiotic activity by serving as a quorum sensing (QS) inhibitor at concentrations of 3.6– 57 μM with EC50 value of 12.2 μM)

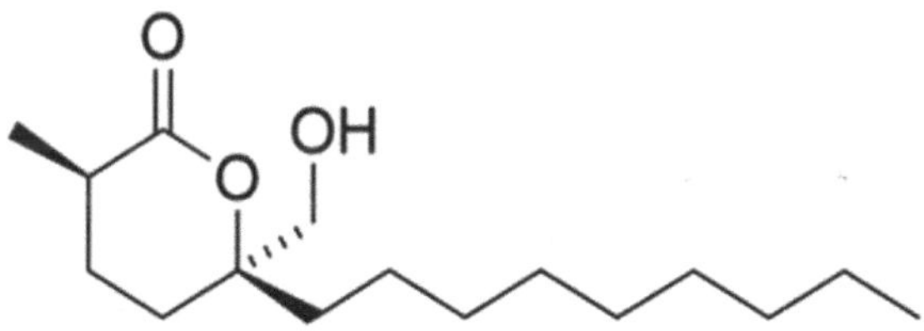

Malyngolide

Dapis pnigousa

Image not available

Global distribution: Subtropical and tropical marine habitats

Ecology: It is commonly seen in coral reefs below depths of about 10 m.

Salient features: No information

Medicinal compounds and their bioactivities:

Antibacterial activity: The compound, lyngbic acid, isolated from this species has been reported to strongly inhibit quorum sensing (QS) in the *Vibrio harveyi* QS reporters (Meyer et al. 2016; Engene et al. 2018).

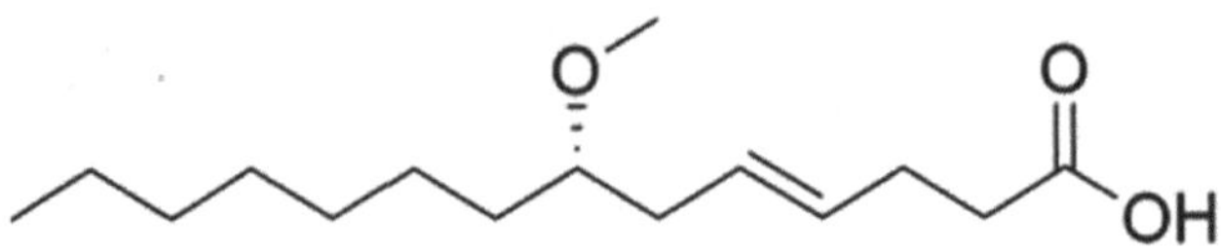

Lyngbic acid

***Dapis* sp.**

Medicinal compounds and their bioactivities:

Antitrypanosomal activity: Among the linear peptides, iheyamides A–C isolated from this species, Iheyamide A displayed weak antitrypanosomal activity against *Trypanosoma brucei brucei* and *Trypanosoma brucei rhodesiense* with an IC50 value of 1.5 μM. On the other hand, the other two compounds did not fair well as their IC50 values were about 20 μM (Kurisawa et al., 2020).

Cytotoxicity: Its iheyamides A sowed very weak cytotoxicity against human cancer cells, WI-38 with an IC50 value of 18 μM (Kurisawa et al. 2020; Khalifa et al. 2021).

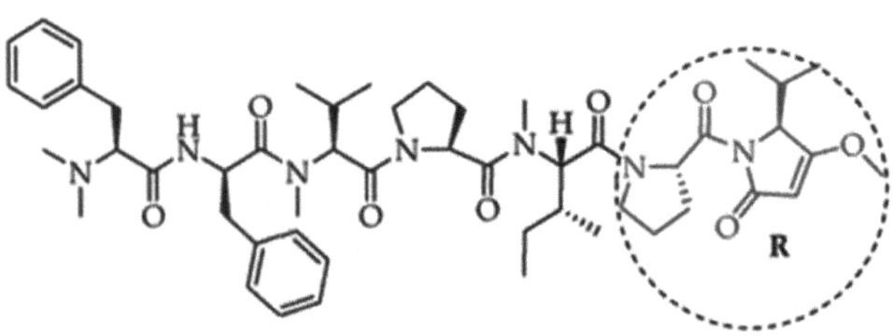

Iheyamide A

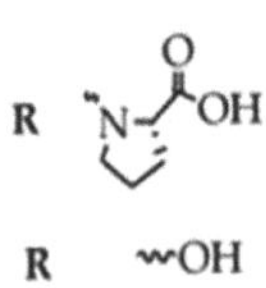

Iheyamides B,C

Order: Pleurocapsales Family: Xenococcaceae

***Chroococcidiopsis* sp.**

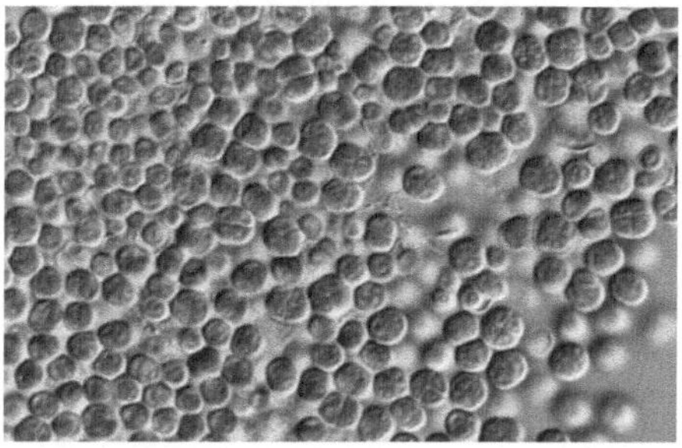

Image credit: T. Darienko, https://creativecommons.org/licenses/by-sa/4.0/

Global distribution: Not reported

Ecology: Aquatic, marine, sea water sample, coastal, surf zone. It can tolerate extreme conditions of radiation, temperatures, pH, desiccation, nutrient deprivation etc.

Medicinal compounds and their bioactivities:

Obesity-related comorbidities: Its biomass has been reported to possess valuable compounds which may be of great use as human food, animal feed, cosmetic and pharmaceuticals. Costa et al. (2019) reported on the use of its extracts in the treatment of comorbidity-associated obesity.

Antioxidant activity: Its high content of scytonemin, polysaccharides, phenolic compounds, and phycobiliproteins showed high antioxidant capacity (Assunção et al. 2021a).

Order: Pseudanabaenales Family: Prochlorotrichaceae

***Nodosilinea nodulosa* LEGE 06102**

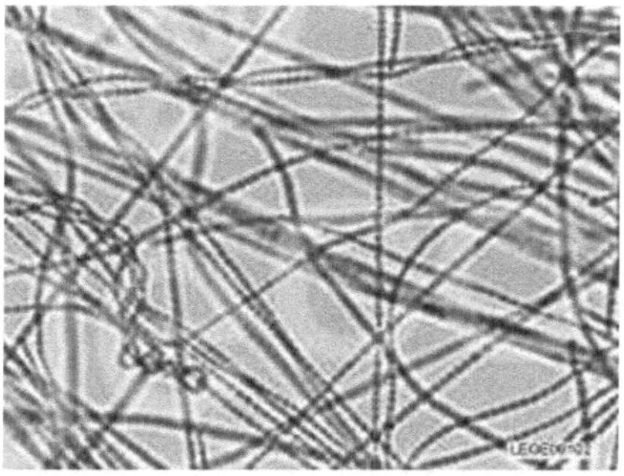

Image credit: Blue Biotechnology and Ecotoxicology Culture Collection (LEGE-CC) (Reproduced with its kind permission)

Global distribution: Subtropical seas worldwide

Ecology: This species is found in rock surface scrapings, tide puddles etc.

Medicinal compounds and their bioactivities:

Proteasome inhibitor: The polyunsaturated fatty acids and unidentified halogenated molecules isolated from this species showed activities against proteasomes which possess three types of active sites viz. trypsin-like, caspase-like and chymotrypsin-like. The proteasomes have been reported to degrade most proteins in human cells and are the known targets of anti-cancer drugs (Martins et al. 2020).

***Nodosilinea* sp.** LEGE 06001

Medicinal compounds and their bioactivities:

Lipid reducing activity: The chlorophyll derivative, 132-hydroxy-pheo-farnesin a (Hfa) isolated from its organic crude extract displayed excellent lipid-reducing activity in the experimental fish with an EC50 value of 15.5 μM. Hfa may therefore be a potential drug for lipid reduction (Silva et al., 2020). Freitas et al. (2019) reported this compound also acted against neutral lipid accumulation in murine preadipocytes

Family: Pseudanabaenaceae

Pseudanabaena curta

Global distribution: It has been reported form the Svalbard archipelago, High Arctic

Ecology: This benthic species dwells in tide puddle and on submerged stones in marine environments.

Medicinal compounds and their bioactivities:

Antimicrobial activity: The methanolic extract of this species displayed anti-microbial activity against *Salmonella* sp., *Staphylococcus aureus*, *Escherichia coli*, and *Candida albicans* (Oufdou et al. 2001).

Anticancer activity: The extracts of this strain LEGE 07160 showed activity against cancer cell lines HT-29, PC-3, RKO, and MG-63. Further another strain of this species LEGE 07169 showed similar activity against T47D, HT-29, PC-3, HepG2, MG-63, and SK-BR-3 cancer cell lines (Costa et al. 2014).

Anti-obesity activity: The extracts of this species have been reported to be useful in the treatment of comorbidities - associated obesity (Costa et al. 2019).

***Pseudanabaena* sp.** LEGE 06144

Medicinal compounds and their bioactivities:

Anticancer activity: The extracts of this strain showed activity against SK-BR-3, T47D, and HepG2 cancer cell lines (Costa et al. 2014)

Order: Pseudanabaenales Family: Romeriaceae;

Romeria gracilis

Global distribution: it is found commonly distributed in seas around Greece

Ecology: This benthic species is commonly found in brackish waters and mesotidal zone

Medicinal compounds and their bioactivities:

Anti-obesity: The extracts of this species may be used in the treatment of comorbidity -associated obesity (Costa et al., 2019).

Anticancer activity: The extracts of the strain LEGE 07310 of this species showed activity against SK-BR-3, T47D, HepG2, MG-63 and SH-SY5Y cancer cell lines (Costa et al. 2014).

***Romeria* sp.** LEGE 06013

Medicinal compounds and their bioactivities:

Anticancer activity: The extracts of the strain LEGE 06013 of this unidentified species showed activity against SH-SY5 and RKO cancer cell lines (Cost aet al., 2014).

Order: Synechococcales Family: Leptolyngbyaceae

Leptolyngbya crossbyana

Global distribution: It has been reported from the Honaunau reef off the Hawai'i island.

Ecology: Dense colonies have been reported to overgrow the coral branches of the genus *Porites* as extensive mats.

Medicinal compounds and their bioactivities:

Antibacterial activity: The heptabrominated polyphenolic ether compounds, crossbyanols A–D, have been isolated form this species. Among these compounds, Crossbyanol B has been reported to display antibacterial activity against MRSA (methicillin-resistant *Staphylococcus aureus*) with an MIC value of 2.0–3.9 μg/mL (Choi et al. 2010).

Anti-inflammatory activity: Among its compounds, honaucin A−C, honaucin A displayed anti-inflammatory activity through activation of Nrf2-ARE (nuclear erythroid 2-related factor 2 -antioxidant response element (Choi et al. 2012; Mascuch et al. 2018)

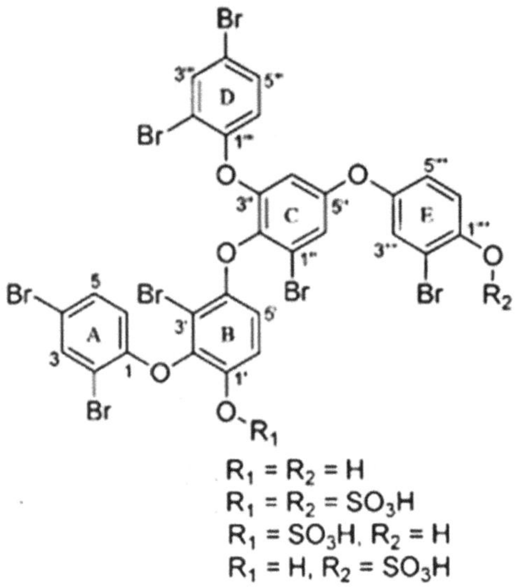

Crossbyanols A-D

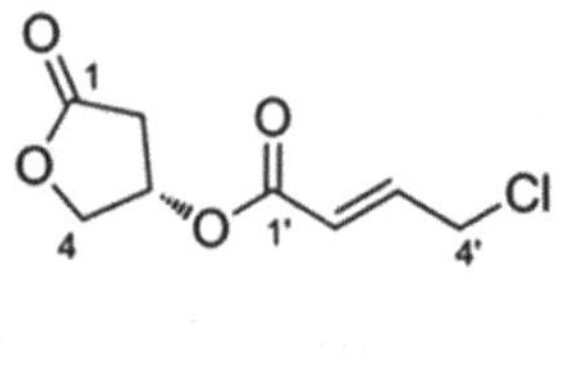

Honaucin A

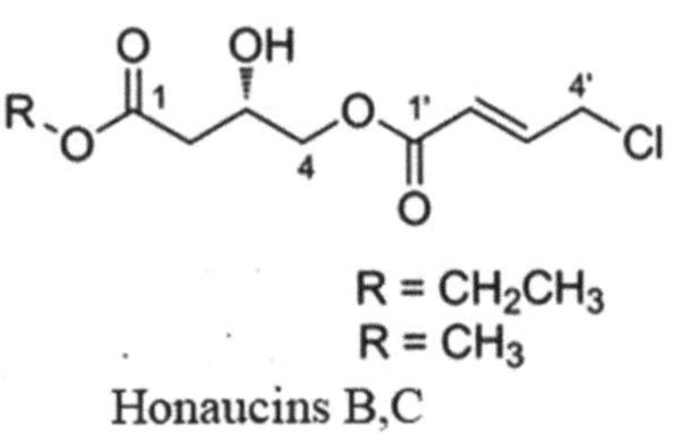

Honaucins B,C

Leptolyngbya halophila

Global distribution: It has world-wide distribuiton

Ecology: Marine biotopes

Medicinal compounds and their bioactivities:

Anticancer activity: The extracts of this species showed anticancer activity against SH-SY5Y, PC-3, MG-63, HT-29, HepG2, and RKO cancer cell lines (Costa et al., 2014)

Leptolyngbya mycoidea LEGE 06108, LEGE 06118, LEGE 06009

Global distribution: Portuguese Atlantic

Ecology: Coastal waters

Medicinal compounds and their bioactivities:

Anticancer activity: The extracts of the different strains of this species have been reportd to show anticancer activity. Its strain LEGE 061 was active against T47D cancer cell; the strain LEGE 06009 against SH-SY5Y, PC-3, T47D, and HT-29 cell lines; and strain LEGE 06118 was against HT-29, SH-SY5Y, MG-63, T47D, HepG2, and RKO cancer cell lines (Costa et al., 2014)

Leptolyngbya fragilis

Global distribution: it is a temperate species and is found distributed in the North Sea of Northeast Atlantic

Ecology: It is an epiphytic species commonly seen in salt marshes

Medicinal compounds and their bioactivities:

Anticancer activity: The extracts of this species displayed anticancer activity against several cancer cell lines such as T47D, HT-29, SH-SY5Y, PC-3, MG-63, SK-BR, HepG2 and RKO (Costa et al. 2014)

***Leptolyngbya* sp.**

Image credit: Juráň Josef (Reproduced with his kind permission)

Global distribution: The species of this genus have a cosmopolitan distribution.

Ecology: The species of this genus are basophilic and euryhaline and are found in a variety of aquatic environments such as freshwater, marine, and brackish waters. These species are epilithic, growing mainly on the surface of rocks.

Medicinal compounds and their bioactivities:

Antioxidant activity: The methanolic extracts of this unidentified species showed antioxidant activity (Khalifa et al., 2021)

Anticancer activity: The dipeptide coibamide A isolated from a strain of this genus showed anticancer activity against NCIH460 lung and mouse neuro-2a cell lines with an LC50 value of more than 23 nM. Further, this compound has been reported to inhibit VEGFR2/ VEGFA expression and suppress the tumor growth in glioblastona senografts (https://accedacris.ulpgc.es/bitstream/10553/74363/2/0767079_00000_0000.pdf; Khalifa et al. 2021)

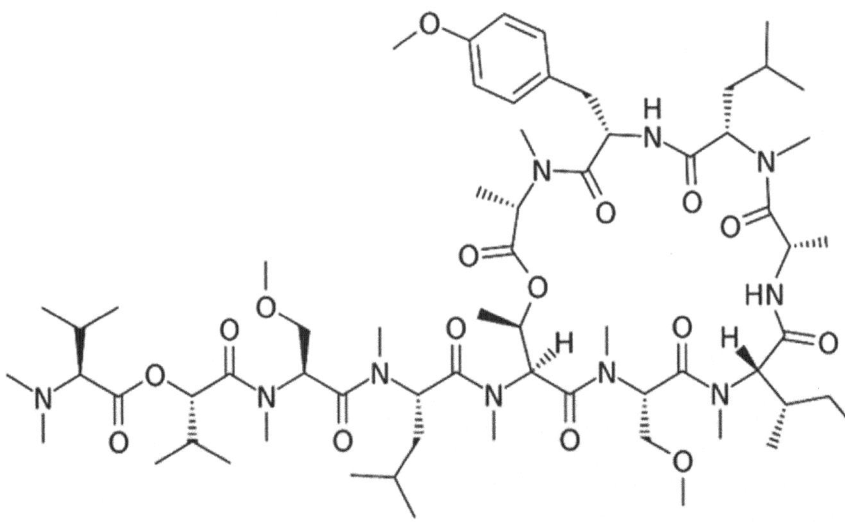

Coibamide A

Order: Synechococcales Family: Merismopediaceae

Nodosilinea nodulosa

Global distribution: North Atlantic

Ecology: It is commonly found in tide pools and on submerged stones

Medicinal compounds and their bioactivities:

Anticancer activity: The extracts of this species showed anticancer activity against SH-SY5Y, PC-3 and HT-29 cell lines (Costa et al., 2014)

Others: The chlorinated aromatic glycolipids, bartolosides A-D isolated from this species have not shown strong bioactive properties and further research is needed in this regard (da Silva 2018)

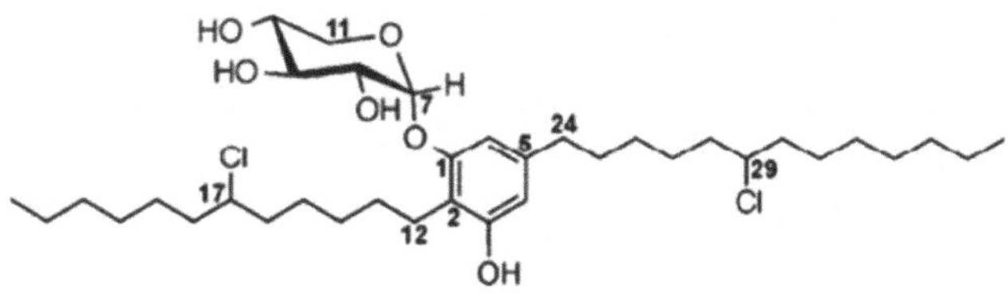

Bartoloside A

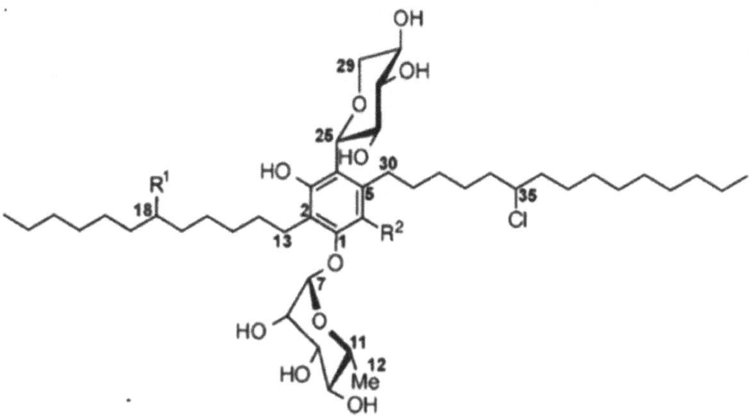

(B: R1=Cl,R2=H; C: R1=H, R2=H; D: R1=Cl, R2=Cl)

Bartolosides B-D

Nodosilinea sp. (LEGE 06102 and 06001

Medicinal compounds and their bioactivities:

Lipid-reducing activity/ anti-obesity activity: The molecule isolated form the strain LEGE 06001 of this species was more similar to hydroxy-pheophytin (with a farnesyl group), which has been reported to reduce the neutral lipids in an experimental fish with an IC50 value of 18.6 μM (Leão et al. 2015; da Silva 2018)

Synechocystis salina

Global distribution: Not reported

Ecology: It dwells largely in tide pool and rock surface

Medicinal compounds and their bioactivities:

Anti-obesity: The extracts of this species have been reported to be useful in the treatment of obesity-related comorbidities (Costa et al. 2019)

Antioxidant capacity: The aqueous extracts of this species have shown the highest antioxidant activity for for ABTS•+ and NO•−; and its acetonic and ethanolic extracts also displayed high antioxidant activity for DPPH• and O2 •−, respectively (Assunção et al. 2021)

Anticancer activity: The extracts of the different strains of this species showed anticancer activity as follows. The strain LEGE 06099 was found to be against SK-BR-3, SH-SY5Y, HepG2 and MG-63 cancer cell lines; the strain LEGE 07173 was against HT-29, SH-SY5Y, PC-3, HepG2, RKO, MG-63, SK-BR-3 and T47Dcell lines; and strain LEGE 06155 was against HT-29, SH-SY5, SK-BR-3, T47D, HepG2, RKO and MG-63 cell lines (Costa et al. 2014)

Others: Afonso et al. (2016) reported on the presence of the compounds bartolosides E-K from the strain LEGE 06099 of this species. However, the bioactivities of these compounds are yet to be known.

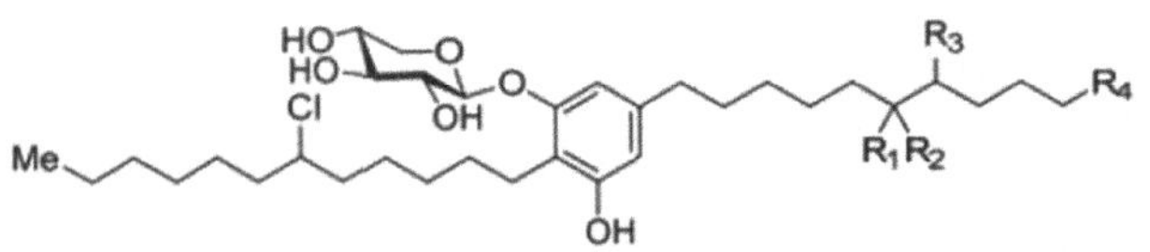

$R_1 = R_2 = H, R_3 = Cl, R_4 = CH_3$
$R_1 = Cl, R_2 = R_3 = H, R_4 = C_4H_9$
$R_1 = R_2 = R_3 = H, R_4 = C_3H_7$

Bartolosides E- G

$R_1 = Cl, R_2 = R_3 = H, R_4 = C_5H_{11}$
$R_1 = R_2 = Cl, R_3 = H, R_4 = C_3H_7$
$R_1 = R_2 = R_3 = H, R_4 = CH_3$
$R_1 = Cl, R_2 = R_3 = H, R_4 = C_2H_5$

Bartolosides H- K

***Synechocystis* sp.** PCC 6803

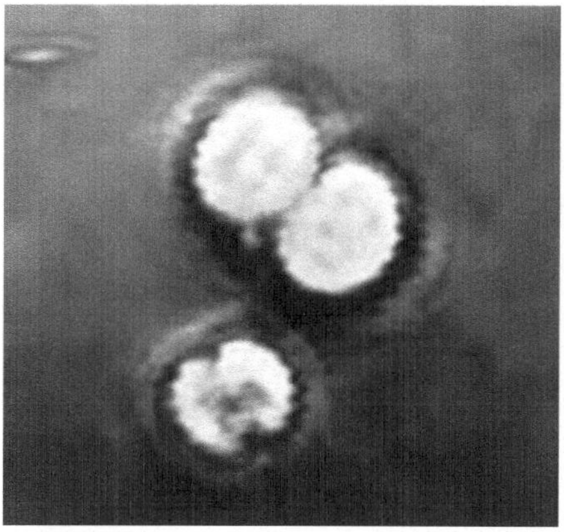

Image credit: Ankush Prasad, Michaela Sedlářová, Pavel Pospíšil. https://creativecommons.org/licenses/by-sa/4.0/

Medicinal compounds and their bioactivities:

Antioxidant activity: The methanolic extracts of *Synechocystis* sp. showed antioxidant actiivty (Khalifa et al. 2021)

Apoptosis-inducing activity: The extracts of this species have been reported to lead the HL-60 cells into apoptosis (Khan et al., 2018)

Antidiabetic activity: The intake of its stilbene compound, phytoalexin (resveratrol) has increased the release of insulin-dependent glucose transporter, GLUT4 in experimental rats with streptozotocin-induced diabetes. Further, this compound has been reported to induce the release the secretion of glucagon-like peptide 1 which helps releasing insulin from pancreas. Furthermore, this compound activates sirtuin 2 which regulates autophagy, a process of cleaning out damaged cells towards the regeration of new cells (Khalifa et al. 2021)

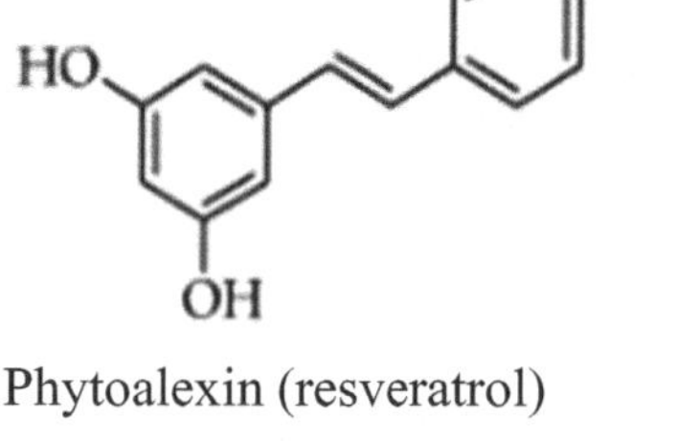

Phytoalexin (resveratrol)

Order: Synechococcales Family: Prochlorococcaceae

Cyanobium sp.

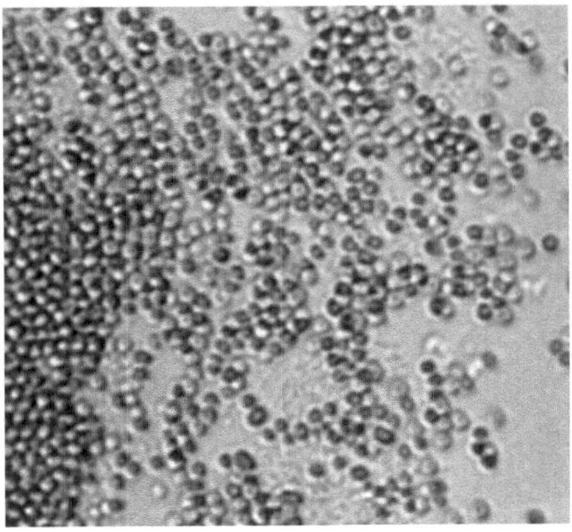

Image credit: Photo by Leao et al., 2013. https://creativecommons.org/licenses/by/4.0/

Global distribution: It has world-wide distribution

Ecology: I t is found in a variety of marine biotopes

Medicinal compounds and their bioactivities:

Cytotoxicity: The polyketide compound, hierridin B, isolated from the strain, LEGE 06113 displayed cytotoxicity against HT-29 colon adenocarcinoma cells and growth inhibition with an IC50 value of 100.2 μM (Lea˜o et al., 2013). Khalifa et al. (2021) reported similar activity with an IC50 value of 0.1 μM.

Anti-obesity: The extracts of this species have been reported for use in the treatment of obesity-related comorbidity (Costa et al. 2019)

Anticancer activity: The extracts of the different strains of this species viz. LEGE 06098, 06134, 07175, 07186, 06113, 06137, 06097, and 06139 have shown anticancer activity against cell lines viz. SH-SY5Y, MG-63, SK-BR-3; HT-29, MG-63; T47D, HepG2, MG-63; HT-29, PC-3, MG-63, T47D; SK-BR-3, SH-SY5Y, HepG2, MG-63; SH-SY5Y, RKO; T47D, HT-29, HepG2; and SH-SY5Y, RKO, T47D respectively (Costa et al. 2014)

Order: Synechococcales Family: Synechococcaceae

Synechococcus nidulans

Global distribution: Tropical and subtropical Indo-Pacific and Atlantic

Ecology: it is found in a variety of marine environments such as air- exposed rock surface tide puddle, and oceanic euphotic zone

Medicinal compounds and their bioactivities:

Antibacterial activity: The crude extract of this species displayed antibacterial activity against *Staphylococcus aureus* (Strieth et al. 2022)

Anticancer activity: The extracts of this species exhibited anticancer activity against SH-SY5Y, HepG2 and SK-BR-3 cancer cell lines (Costa et al. 2014)

Anti-obesity: The extracts of this species have been reported for use in the treatment of obesity-related comorbidity (Costa et al. 2019)

Synechococcus sp.

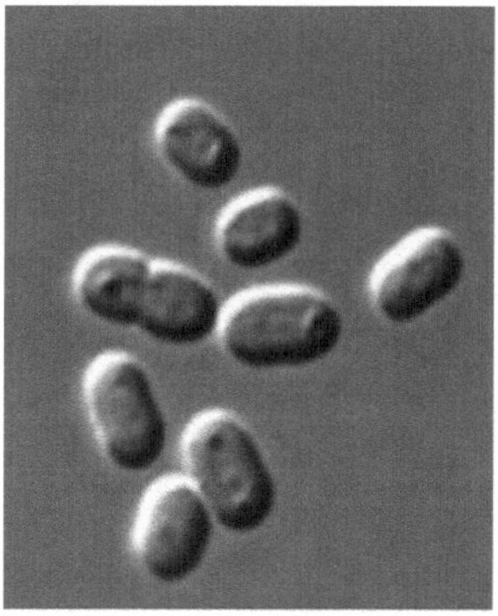

Image credit: Masur. Public domain.

Medicinal compounds and their bioactivities:

Apoptosis-inducing activity: Extracts of this species have shown apoptosis in HL-60 cells (Khan et al. 2018)

Anticancer activity: The extracts of the different strains of this species have shown anticancer activity. The strain LEGE 07172 was found to be against SH-SY5Y and RKO cancer cell lines; LEGE 06005 was against MG-63 cancer cell line; and LEGE 06026 showed activity against HT-29 and SK-BR-3 cancer cell lines (Cosa et al. 2014)

Leptothoe kymatousa, Leptothoe sithoniana and *Leptothoe spongobia*

Global distribution: Aegean Sea

Ecology: All these species are symbiotic with the marine sponges. The host sponge species of these *Leptothoe* spp. are *Chondrilla nucula (Leptothoe kymatousa), Petrosia ficiformis (Leptothoe sithoniana), Dysidea avara* and *Acanthella acuta* (*Leptothoe spongobia*)

Medicinal compounds and their bioactivities:

Antibacterial activity: The crude extracts of all these species displayed antibacterial activity against Gram- positive *Staphylococcus aureus* 9518 with an inhibition zone dia. of 10mm (Konstantinou et al. 2019)

Anticancer activity: The extracts of the strains viz. TAU-MAC 0915, *Leptothoe sithoniana;* TAU-MAC 1015, *Leptothoe spongobia;* TAU-MAC 1115; TAU-MAC 1215, *Leptothoe kymatousa* exhibited anticancer activity on MDA-MB-231, MCF-7, A-431 and HT-29 cancer cell lines. Among the strains, TAU-MAC 1015 and 1115 were reported to be the most potent ones with the IC50 (µg.mL-1) values of 26, 21, 14 and 18, 66, 35, 43 and 55, respectively (Konstantinou et al. 2019)

YELLOW-GREEN ALGAE

Phylum: Xanthophyta Class: Xanthophyceae
Order: Tribonematales Family: Tribonemataceae

Tribonema sp.

Global distribution: No information

Ecology: It is a brackish water species

Medicinal compounds and their bioactivities:

Immunomodulatory properties: Its sulfated polysaccharides have been reported to display immunomodulatory properties at 12.5–200 µg/mL via macrophage proliferation and improved expression of cytokines (Riccio and Lauritano 2020)

GOLDEN-BROWN, AND YELLOW ALGAE

Phylum: Haptophyta Class: Pavlovophyceae Order: Pavlovales
Family: Pavlovaceae

Diacronema lutheri *(= Pavlova lutheri; Monochrysis lutheri)*

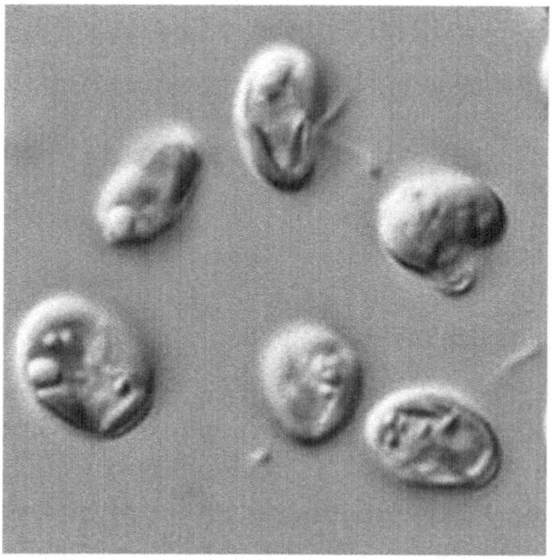

Image credit: CSIRO, https://creativecommons.org/licenses/by/3.0/

Global distribution: It has a world-wide distribution

Ecology: It occurs in all marine biotopes

Medicinal compounds and their bioactivities:

Anti-inflammatory and antithrombotic effects: The species of this genus have both eicosapentaenoic acid and docosahexaenoic acid. Further, they have been reported to synthesize betaine lipids viz. 1,2-diaclyglyceryl-3-O-2′-(hydroxmethyl)-(N, N, N-trimethyl)-β-alanine (DGTA), and 1,2-diacylglyceryl-3-O-carboxyhydroxymethylcholine (DGCC) and dihydroxylated sterols i.e. pavlovols (Kato et al. 1995; Hulatt et al. 2021). These lipids have been reported to possess anti-inflammatory effects besides serving as nutrient delivery vehicles to carry vitamins A, K, D and E throughout the body. Further, these lipids have antithrombotic activity against platelet-activating factor (PAF) which is said to play an important role in the formation of atherosclerosis (Lordan et al. 2018)

Antioxidant activity: A novel peptide obtained from this purified alga which was fermented by the yeast *Hansenula polymorpha* displayed antioxidant activity by scavenging DPPH, hydroxyl radicals and hydrogen peroxide (Oh et al. 2015). Zhong-Ji et al. (2012) also reported that the fermented extracts of this species displayed the highest antioxidant activity and inhibition of extracellular Myeloperoxidase (MPO) activity, which determines the oxidative stress in inflammatory diseases

Class: Prymnesiophyceae Order: Coccolithales
Family: Pleurochrysidaceae

Chrysotila carterae *(= Pleurochrysis carterae)*

Global distribution: UK, Norway

Ecology: it occurs in both marine and brackish waters

Medicinal compounds and their bioactivities:

Therapeutic carrier: Lomora et al. (2019) reported that the heterococcoliths of this species could serve as therapeutic carrier.

Source of vitamin B12: The vitamin B12 content of this lyophilized cells of this species has been reported to be 125 microg per 100 g dry weight. A 22-week-old B12-deficient rats which excreted a good amount of methylmalonic acid in its urine (a symptom of vitamin B12 deficiency) was fed with supplemented food containing these cells (at 10 g per kg diet) for 12 d, and it was found that there was considerable reduction in the excretion of methylmalonic acid and increase of hepatic vitamin B12 level in these experimental rats (Miyamoto et al. 2001)

Order: Isochrysidales Family: Noelaerhabdaceae

Emiliania huxleyi

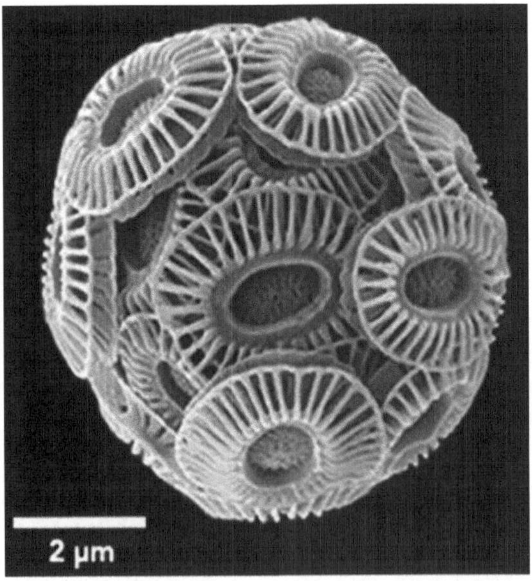

Image credit: Dr. Jeremy Young, https://creativecommons.org/licenses/by-sa/4.0/

Global distribution: found in almost all ocean ecosystems from the equator to sub-polar regions

Ecology: marine sediments; Its presence in plankton communities from the surface to 200m depth indicates a high tolerance for both fluctuating and low light conditions

Medicinal compounds and their bioactivities:

Therapeutic carrier: Lomora et al. (2019) reported that the heterococcoliths of this species could serve as therapeutic carrier.

Antioxidant activity: The cells of this species have been reported to produce NO extra- and intra-cellularly in lab culture. When these cells were treated with an NO scavenger, there was considerable reduction in viral production. This suggests that the NO of this species may have a cellular antioxidant activity (Schieler et al. 2019).

Order: Prymnesiales Family: Prymnesiaceae

Prymnesium parvum

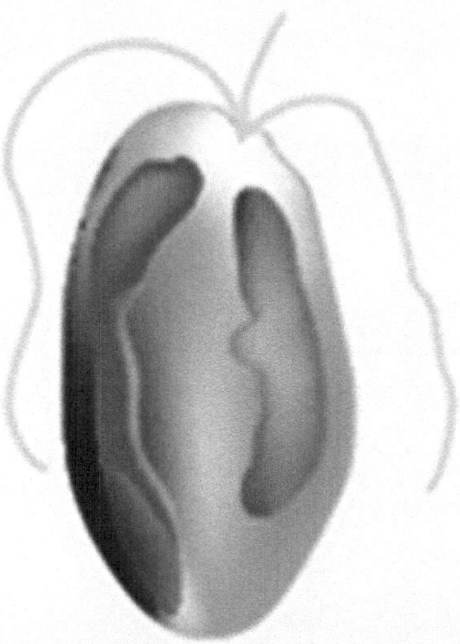

Image credit: Daniel Killerup Svenssen et al., https://creativecommons.org/licenses/by-sa/4.0/

Global distribution: It has worldwide distribution: the temperate zones of USA and Morocco, the Baltic Sea, Europe, China, and Australia

Ecology: It occurs in a variety of aquatic biotopes such as lakes, rivers, ponds, coastal waters, estuaries, fjords and brackish waters. This phytoflagellate species forms red tide especially in eutrophic waters

Medicinal compounds and their bioactivities:

Anticancer activity: The lipid-carbohydrate compounds (galactolipids, glycolipids or proteolipids) of this species are prymnesin 1,2. Among these compounds, prymnesin-2 has been reported to act against human cancer cell lines, Hela (https://en.wikipedia.org/wiki/Prymnesin-2)

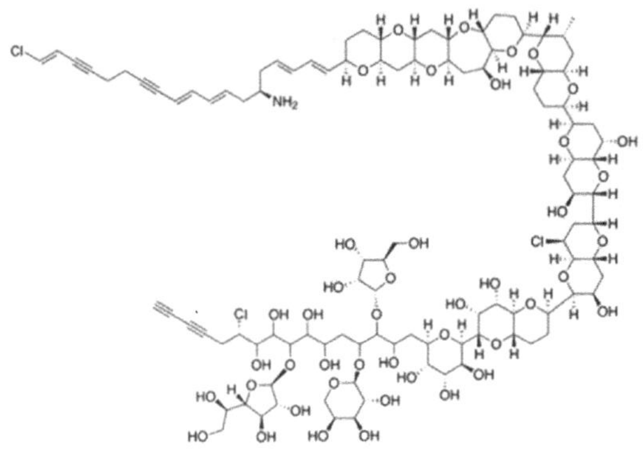

Prymnesin-1

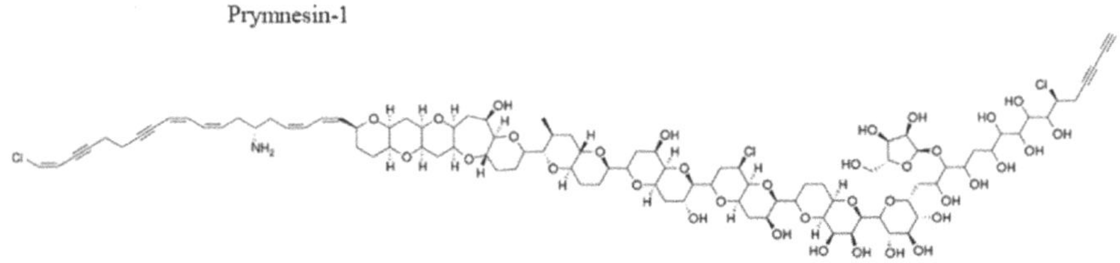

Prymnesin-2

Class: Prymnesiophyceae Order: Isochrysidales Family: Isochrysidaceae

Isochrysis galbana

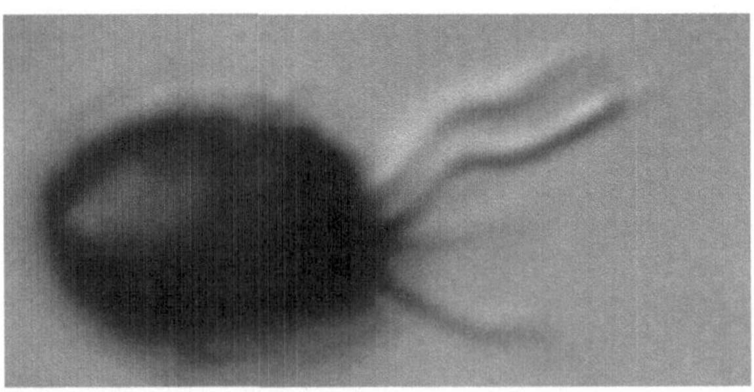

Image credit: F. Jouenne, https://creativecommons.org/licenses/by-sa/3.0/nl/

Global distribution: It is common in the Atlantic

Ecology: It dwells mostly in coastal waters

Medicinal compounds and their bioactivities:

Antimycobacterial activity (Antituberculosis): Its 24-Oxocholesterol acetate, Ergost-5-en-3β-ol, and Cholest-5-en-24-1,3-(acetyloxy)-3β-ol have been reported to possess antitubercular activity by acting against *Mycobacterium tuberculosis* with MIC values of 50–60 μg/mL (Luo et al., 2015)

Antioxidant activity: The extracts of this species have been reported to possess antioxidant activity due to its rich polyunsaturated fatty acid and the carotenoid pigment viz. fucoxanthin (18.2 mg/g dw) (https://en.wikipedia.org/wiki/Isochrysis_galbana). Matos et al. (2019) reported that the aqueous and ethanolic extracts of this species showed high phenolic content; and its freeze-dried biomass exhibited higher fucoxanthin content (6.1 mg per g dw), and β-carotene and Coenzyme Q10. The aqueous extracts of this species had significant antioxidant capacity (about 90% inhibition by ABTS method). Mishra and Mishra (2018) reported that the hexane and methane extracts of this species showed High DPPH, Hydroxyl, NO scavenging activity

Anticancer activity: The biomass and ethanolic extracts of this species displayed significant anticancer activity against human cervical cancer cells, HeLa, with IC50 values of 0.32 and 0.28 mg ml^{-1}, respectively (Matos et al. 2019). Mishra and Mishra (2018) reported that the acetone extract of this species showed anticancer activity by lowering the growth of human carcinoma Hep G-2 cells

Antitumour activity: The polysaccharide extract of this species showed promising antitumor activity by inhibiting the proliferation of U937 human leukemic monocyte lymphoma cell (Mishra and Mishra 2018)

Neuroprotective activity: The acetone extract of this species showed neuro-protective potential by increasing the chelate Fe2+, and inhibiting AChE (Mishra and Mishra 2018)

Hypoglycemic activity: The extract of this species has been reported to possess hypoglycemic activity by lowering the blood glucose level from 133 to 84 mg/dl in experimental rats (Mishra and Mishra 2018)

Hypolipidemic activity: The extract of this species has been reported to possess hypolipidemic activity by lowering the blood lipid level from 66 to 54 mg/dl in experimental rats (Mishra and Mishra 2018)

Hypocholesterolemic activity: The aqueous extract of this species has been reported to possess hypocholesterolemic activity by inducing a significant reduction in cholesterol absorption through the Caco-2 monolayers) (Matos et al. 2019)

Antibacterial activity: The extracts of this species containing EPA, DHA, oleic acid and linoleic acid displayed antibacterial activity against G+ bacteria (Wu et al., 2021). Mishra and Mishra (2018) reported that its axenic culture inhibited the growth of *Vibrio* sp.; and its extract possessed significant antibacterial activity against pathogens such as *Klebsiella pneumoniae, Escherichia coli, Proteus vulgaris, Streptococcus faecalis, Streptococcus aureus,* and *Salmonella typhii.*

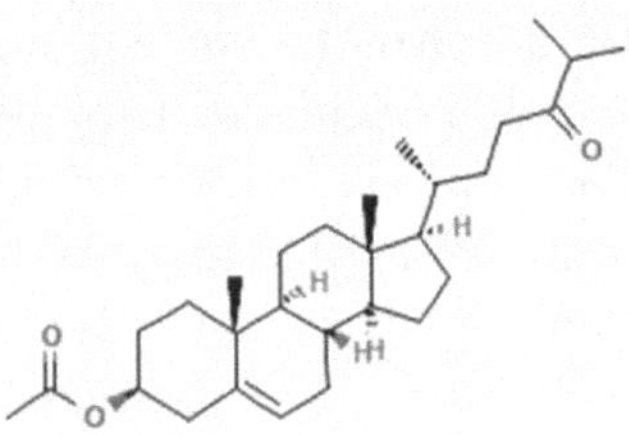

24-Oxocholesterol acetate

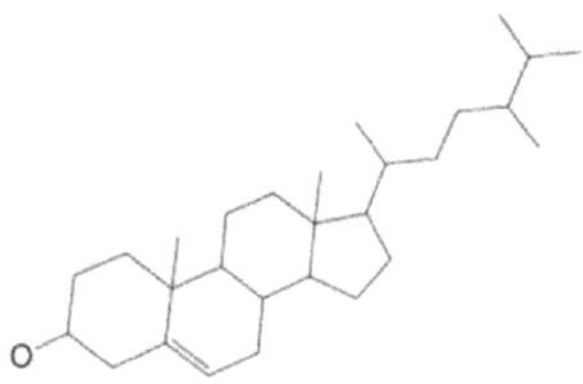

Ergost-5-en-3β-ol

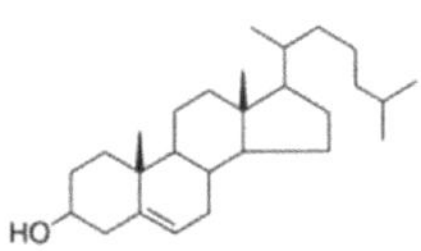

Cholest-5-en- 3β-ol

Isochrysis zhanjiangensis

Global distribution: It is found widely distributed in all tropical oceans

Ecology: It is a planktonic species

Medicinal compounds and their bioactivities:

This commercially important microalga is an excellent source of docosahexaenoic acid, is an important PUFA, and possesses several bioactivities like antiphotoaging and antihypertension.

Antioxidant activity: Its amino acid sequence viz. Asn–Asp–Ala–Glu–Tyr–Gly–Ile–Cys–Gly–Phe has been reported to display fairly high free radical scavenging ability by inhibiting ethanol-induced oxidative stress in the cancer cells, HepG2 (Vladkova et al., 2022). Chen et al. (2019) reported on the hepatoprotective effects of its purified peptide which acted against alcohol toxicity in HepG2 cells.

Anti-angiogenic and anti-inflammatory activities: The purified pentapeptides (AYP, Ala-Tyr-Ala-Pro-Glu) of this species have been reported to inhibit the invasion and migration of HUVECs (human umbilical vein endothelial cells) by controlling the activity of MMP-2/-9 (matrix metalloprotease-2/9) which are key tissue remodeling enzymes. At 100 μM AYP treatment, the ability of the pentapeptides to inhibit migration was found to be 64% for HUVECs and 68% for HT1080 cancer cells. Apart from this bioactivity, the pentapeptides also suppressed the activity of iNOS (nitric oxide synthase) and COX-2 (cyclooxygenase -2) by inhibiting the over secretion of VEGF in HT1080 cells and the activity of VEGFR-2 (Vascular endothelial growth factor receptor 2) in HUVECs. Owing to the capability of preventing inflammation and control tumor angiogenesis., these pentapeptides could be a promising drug source for these ailments (Tang et al., 2022)

Anti-atherosclerotic/anti-inflammatory activity: The nonapeptide, ETT (EMFGTSSET), of in vitro gastrointestinal digestion of this species has been reported to inhibit ox-LDL (oxidized low-density lipoprotein)-induced inflammation and atherosclerosis apoptosis. Therefore, it could be a promising drug candidate for the treatment of atherosclerosis (Pei et al. 2022).

Anti-photoaging properties: The heptapeptide, DAPTMGY isolated form this species has been reported to possess antiphotoaging activity by inhibiting Ultraviolet B (UVB) -induced apoptosis. The latter is mainly by its upregulation of Bcl-2 family proteins and downregulation of caspase-8, caspase-3 (family of protease enzymes) and Bax (member of the BCL-2 family of proteins) (Zheng et al. 2021).

Antiskin aging bioactivity: The pentapeptide, Ala-Tyr-Ala-Pro-Glu (AYAPE) of this species showed antiskin aging activity by acting against MMP-1 (matrix metalloproteinase-1) and UVB-induced photoaging in HaCaT cell line (He et al., 2022)

***Isochrysis* sp.**

Medicinal compounds and their bioactivities:

Extracts of this unidentified species showed antibacterial activity against six *Vibrio* strains (Kokou et al., 2012)

YELLOW-GREEN ALGAE

Phylum Ochrophyta Class: Bacillariophyceae Order: Achnanthales
Family: Cocconeidaceae

Cocconeis scutellum

Global distribution: It is found distributed from subpolar to tropical seas

Ecology: It occurs from marine to fresh waters

Medicinal compounds and their bioactivities:

Anticancer activity: Its diethyl ether extract has been reported to show anticancer activity against breast carcinoma (BT20) cells (Hussein and Abdullah, 2020)

Order: Bacillariales (Pennales) Family: Bacillariaceae

Cylindrotheca closterium

Global distribution: This cosmopolitanspecies has wide distribution in both high and low latitudes

Ecology: This benthic species occurs in coastal and brackish water regions

Medicinal compounds and their bioactivities: The fucoxanthin of this species has been reported to possess antioxidant, antiobesity, anticancer, antidiabetic, and anti-inflammatory properties (Khan et al., 2018)

Cylindrotheca fusiformis

Global distribution: It is found distributed in upwelling zones near the equator and polar oceans

Ecology: It is a coastal species

Medicinal compounds and their bioactivities: The frustules of this species have been reported to serve as orthopedic and dental implants (Lomora et al. 2019). Extracts of this species are believed to help as a skin moisturizing agent which is known to attract water to the upper layer of the skin. Its antioxidant properties are yet to be confirmed (https://whatsinmyjar.com/skincare-ingredients/cylindrotheca-fusiformis-extract)

Nitzschia frustulum *(= Synedra frustulum)*

Global distribution: it is commonly found in Spain and South America

Ecology: It is a marine benthic species

Medicinal compounds and their bioactivities: Qin et al. (2008) reported that the insertion of germanium in this species induced its nanocomb structure with blue colored photoluminescence. As such nanomaterials display optical properties, they are of great use in optoelectronics and semiconductors. Such nanomaterials with silica frustules of diatoms could be of great use in the study of antimicrobial activity.

Order: Chaetocerotales (Eupodiscales, Biddulphiales, Ccentrales)
Family: Attheyaceae

Attheya longicornis

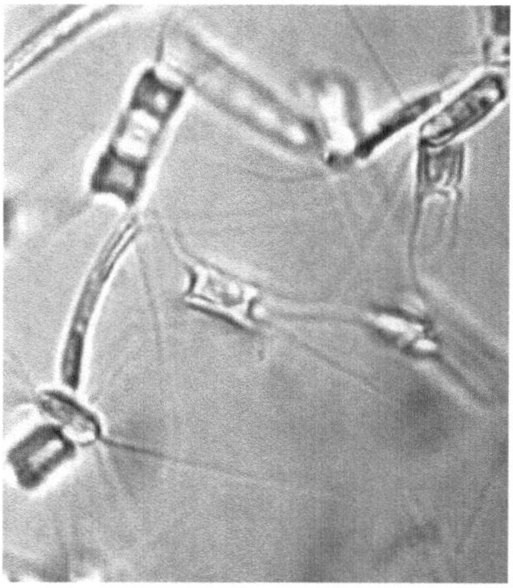

Image credit: Richard A. Ingebrigtsen. https://creativecommons.org/licenses/by-sa/4.0/

Global distribution: Northern Atlantic

Ecology: This marine species has been reported to be epiphytic on the diatom species of *Chaetoceros, Skeletonema* etc.

Medicinal compounds and their bioactivities:

Antibacterial activity: Extracts of this species showed significant antibacterial activity against methicillin resistant *Staphylococcus aureus* (MRSA) and *Staphylococcus aureus*; and weak activity was registered with *Escherichia faecalis* (Ingebrigtsen et al. 2016)

Order: Coscinodiscales Family: Coscinodiscaceae

Coscinodiscus concinnus

Global distribution: Mexico

Ecology: it is a pelagic, marine species i.e. member of marine phytoplankton

Medicinal compounds and their bioactivities: Due to the morphology and amorphous nature of this species, they are reported to be promising drug carriers candidates. Lomora et al. (2019) reported on the possible use of the valves of this species in oral drug delivery

Coscinodiscus wailesii

Global distribution: Atlantic Ocean and Pacific Ocean

Ecology: It is very common in estuaries also

Medicinal compounds and their bioactivities:

Biomedical applications: The valves of this species may have applications in the biosensing drug delivery (Lomora et al., 2019)

Phylum: Bacillariophyta Class: Mediophyceae Order: Bacillariales Family: bacillariaceae

Pseudo-nitzschia delicatissima

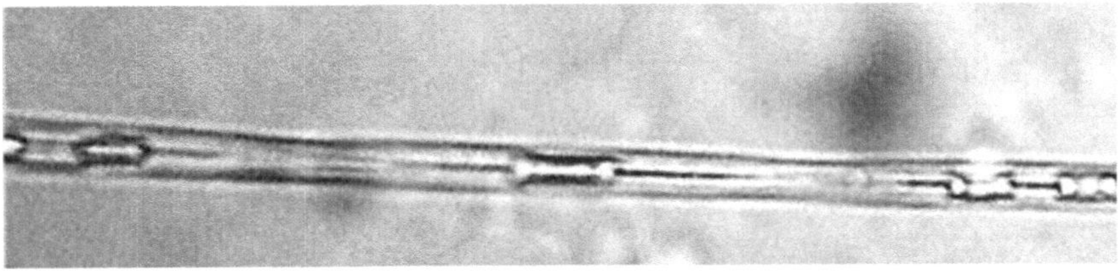

Image credit: Minami Himemiya, https://creativecommons.org/licenses/by-sa/3.0/

Global distribution: It has cosmopolitan distribution. In European waters, it is mainly found during summer

Ecology: It is a neritic species

Medicinal compounds and their bioactivities:

Anticancer activity: The unsaturated aldehydes present in the extracts of this species have shown anticancer activity against Caco-2 cells with IC50 values of 11 µg/mL (Hussein and Abdullah 2020)

Pseudo-nitzchia multistriata

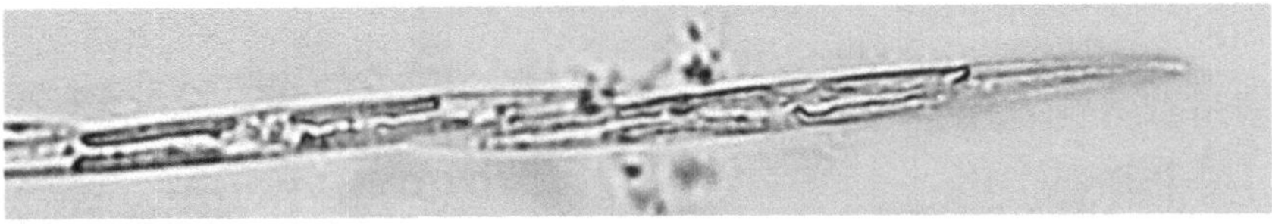

Image credit: Minami Himemiya. https://creativecommons.org/licenses/by-sa/3.0/

Global distribution: It is found distributed in all oceans, including the Arctic and Antarctic Oceans

Ecology: It may be found from coastal waters to open ocean

Medicinal compounds and their bioactivities:

Biomedical applications: The frustules of this species may be of great use in the chemical analysis of cell membranes, which regulate the movement of materials exiting and entering the cell (Lomora et al. 2019)

Order: Chaetocerotales Family: Chaetocerotaceae

Chaetoceros simplex* var. *calcitrans *(= Chaetoceros calcitrans)*

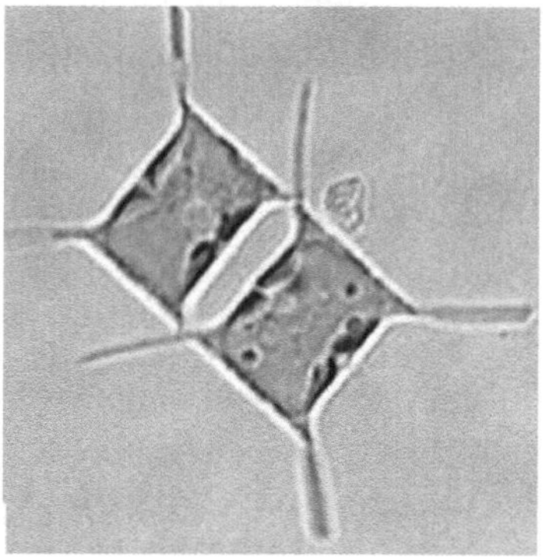

Image credit: Mona Hoppenrath. https://creativecommons.org/licenses/by-sa/2.5/

Global distribution: It is a common species of Japan waters

Ecology: This marine species has been reported to be abundant at a depth of 7.5 m

Medicinal compounds and their bioactivities:

Antibacterial activity: The n-hexane extracts of this species have shown antibacterial activity against *Staphylococcus aureus, Bacillus subtilis* and *Escherichia coli* with an MIC value of 250 μg/mL. (https://www.sciencedirect.com/topics/agricultural-and-biological-sciences/chaetoceros-calcitrans)

The extracts of this species have shown antibacterial activity against *B. subtilis, S. aureus* and *S. pyogenes* (Sushanth and Rajashekhar 2015)

Antioxidant activity: The presence of rich pigments (carotenoids and chlorophylls) and PUFAs in the extracts of this species have shown antioxidant activity by inhibiting 2,2-diphenyl-1-picrylhydrazyl (DPPH) free radical scavenging and nitric oxide (NO). Among the various carotenoid pigments, fucoxanthin of this species has shown outstanding antioxidative activity. Further, its chloroform and acetone extracts showed DPPH free radical scavenging activity and the percentages of inhibition were 35% and 43 %, respectively. Furthermore, the acetone, methanol, chloroform, and 70% ethanol extracts displayed significant NO inhibitory activity with IC50 values of 15.4, 3.8, 3.5, and 17.9 μg/mL, respectively (Azizan et al. 2018)

Anticancer activity: The EtOH extract of this species has shown anticancer activity against MCF-7 cancer cells with an IC50value of 3 μg/mL and its AcOEt extract was against Breast adenocarcinoma (MDA-MB-231) cells with an IC50value of 60 μg/mL(Hussein and Abdullah, 2020)

Immunomodulatory properties: The extracts of this species have been reported to activate IL-6 (interleukin,-6) /Human peripheral blood mononuclear cells (HPBMC), which are known to regulate immune responses (Riccio and Lauritano,2020)

Chaetoceros neogracile (= *Chaetoceros gracilis*)

Global distribution: It is commonly distributed in the northern and southern hemisphere

Ecology: it is an euryhaline species living in all marine environments with changes in salinity.

Medicinal compounds and their bioactivities:

Anti-inflammatory and antithrombotic effects: This species has been reported to synthesize 1,2-diacylglyceryl-3-O-carboxyhydroxymethylcholine (DGCC), which has been reported to possess anti-inflammatory effects besides serving as nutrient delivery vehicles to carry vitamins A, K, D and E, throughout the body. Further, this lipid has antithrombotic activity against platelet-activating factor (PAF) which is said to play an important role in the formation of atherosclerosis (Lordan et al., 2018; Cañavate et al., 2016)

Others: This species has been reported to possess rich Eicosapentaenoic acid (EPA) and fucoxanthin. While the EPA possesses both anti-inflammatory and anti-arteriosclerotic properties., the fucoxanthin shows anti-obesity and anti-oxidation effects on humans (Tachihana et al. 2020)

Chaetoceros muelleri

Global distribution: It is commonly distributed in the Devil's Lake, North Dakota

Ecology: It thrives in endorheic saline environments including saline lakes

Medicinal compounds and their bioactivities:

Antibacterial activity: The methanolic extracts of this species displayed antibacterial activity by inhibiting the growth of *Staphylococcus aureus* (Quijano-Scheggia1 et al. 2016)

Chaetoceros pseudocurvisetus

Global distribution: It is a subtropical or tropical species

Ecology: It is a marine, planktonic species commonly seen in neritic waters

Medicinal compounds and their bioactivities:

Antitubercular activity: Extracts of the different strains of this species grown in phosphate-starvation conditions showed anti-tubercular activity against *Mycobacterium bovis* and *Mycobacterium tuberculosis.* The extracts (800 µg/mL) of the starved strain FE85/3 showed activity against *Mycobacterium bovis* with 91% growth inhibition; and that of FE331/3 displayed 99% inhibition activity against *Mycobacterium tuberculosis*. Further, the organic extracts of the strain FE85 exhibited activity against *Mycobacterium tuberculosis* and *Mycobacterium bovis* at 200 and 800 µg/mL. On the other hand organic extracts of the strain FE85/3 showed activity against *Mycobacterium tuberculosis* and *Mycobacterium bovis* only at 800 µg/mL (Lauritano et al., 2018)

Chaetoceros socialis

Global distribution: It has cosmopolitan distribution. However, higher densities were reported from colder waters

Ecology: It is a very common benthic and neritic species

Medicinal compounds and their bioactivities:

Immunomodulatory properties: The extracts of this species have been reported to activate IL-6 (interleukin,-6) /Human peripheral blood mononuclear cells (HPBMC), which are known to regulate immune responses (Riccio and Lauritano 2020)

Order: Fragilariales Family: Fragilariaceae

Synedra acus

Global distribution: It has worldwide distribution.

Ecology: This attached species occurs in both marine and freshwater environments

Medicinal compounds and their bioactivities:

Anticancer activity: The compound, chrysolaminaran, of this species showed anticancer activity against human colon cancer cells (HT-29) with an IC50 value of 54.5 µg/mL; and on Colon cell line (DLD-1) with an IC50 value of 47.7 µg/mL (Hussein and Abdullah 2020)

Order: Naviculales Family: Naviculaceae

Haslea karadagensis

Global distribution: It is very common in the Black Sea

Ecology: It is a marine species

Medicinal compounds and their bioactivities:

Antiviral activity: The intracellular and extracellular blue-grey pigments (Marennine -like and polyphenolic) showed antiviral activity. The extracellular pigments were however more active than the intracellular form as the EC50 values were 23 µg mL^{-1} and 62 µg mL^{-1}, respectively (Gastineau et al. 2014).

Cytotoxicity: The intracellular pigments of this species also showed cytotoxicity against the fibroblastic Vero cells (Gastineau et al. 2014).

Navicula directa

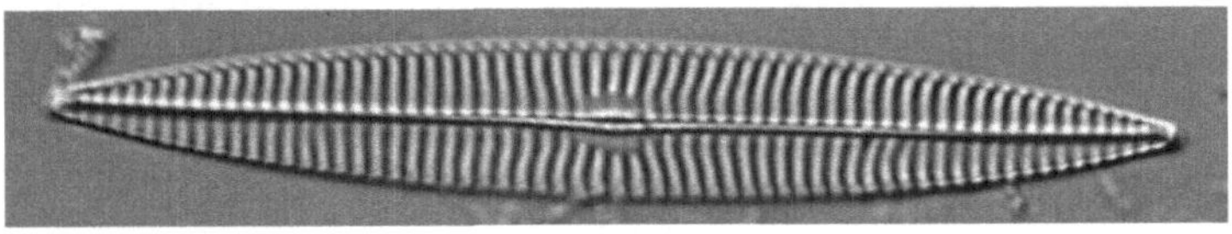

Image credit: Dr. Evelyn Gaiser & Mike Rugge (Florida Coastal Everglades Ecological Research) (Reproduced with his kind permission)

Global distribution: It has a cosmopolitan distribution

Ecology: It is commonly found in the neritic zone

Antiviral activity: The sulfated polysaccharide compound viz., naviculan isolated from this species displayed antiviral activity by inhibiting HSV-1 and HSV-2 (herpes simplex viruses type 1 and 2) with IC50 values of 7–14 µg mL^{-1} (Gastineau et al. 2014). Lee et al. (2006) reported that the above compound, naviculan displayed antiviral activity against HSV-1 and HSV-2 and influenza A virus with selectivity index (CC50/IC50) values of 270, 510 and 32, respectively. Further, this compound also exhibited inhibitory activity on cell-cell fusion between HIV gp160-expressing cells and CD4-expressing

Navicula salinicola *(= Navicula incerta)*

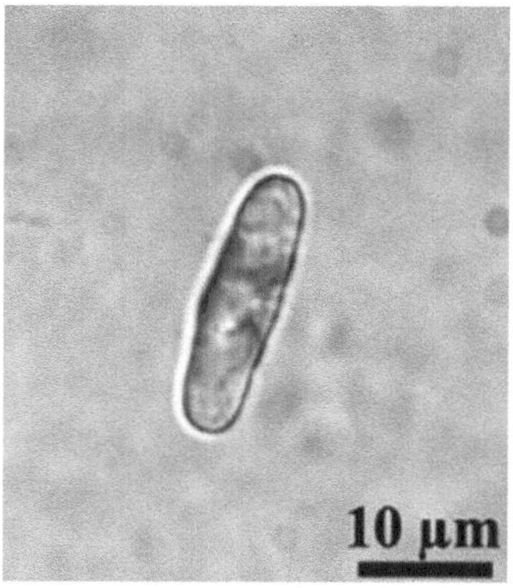

Image credit: G. Courtois de Vicose (Reproduced with his kind permission)

Global distribution: It is a common tropical species in Baja California, México,

Ecology: It mostly occurs in brackish waters

Medicinal compounds and their bioactivities:

Cytotoxicity: The compounds viz. stigmasterol and 5β-Hydroxysitostanol of this species displayed toxicity and induced apoptosis in HepG2 hepatocarcinoma cells. At the concentrations of 5, 10 and 20 µM, the percentage of toxicity was 40%, 43%, and 54%, respectively (Luo et al., 2015). Kim et al. (2014) reported that at 5, 10, and 20 µM concentrations, the stigmasterol induced significant DNA damage and morphological changes in HepG2 cells. Further, increased doses of stigmasterol reduced the the number of attached cells and made changes in the cell membrane.

Anticancer activity: The phytosterol, stigmasterol, of this species also displayed anticancer activity on liver hepatocellular carcinoma (HepG2) cells with an IC50 value of 8.25 µg/mL (Hussein and Abdullah 2020)

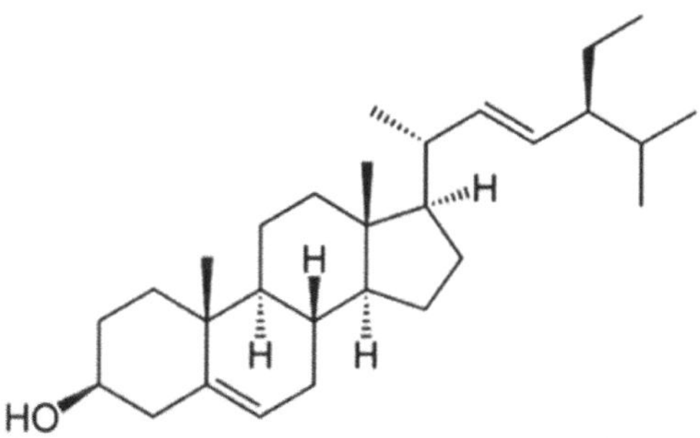

Stigmasterol

***Navicula* sp.**

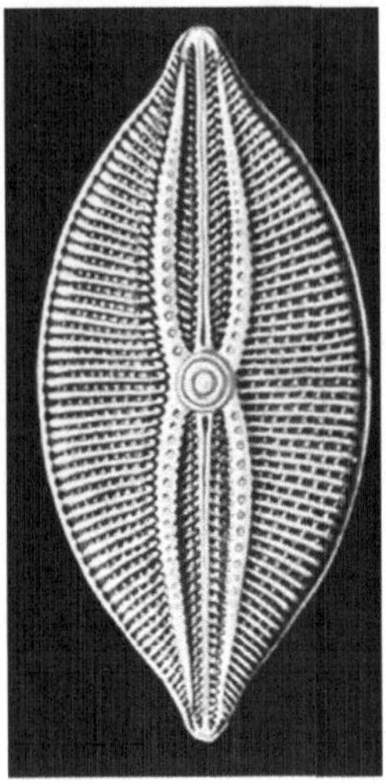

Image credit: Diatomeas-Haeckel. Public domain.

Medicinal compounds and their bioactivities:

Antioxidant activity: The sulfated polysaccharides of this unidentified species displayed DPPH scavenging activity which was found to be 49% (Fimbres-Olivarria et al. 2018)

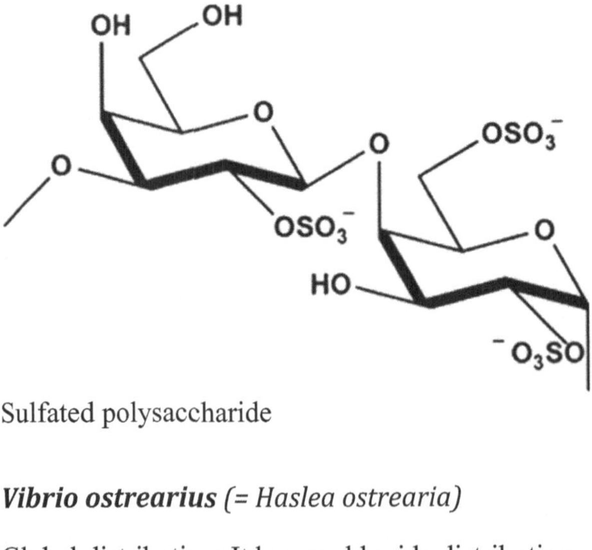

Sulfated polysaccharide

Vibrio ostrearius *(= Haslea ostrearia)*

Global distribution: It has world-wide distribution

Ecology: This marine, euryhaline, pennate species may lead a benthic, epiphytic or planktonic life. In shallow and nutrient-rich water, it has been reported to proliferate mainly during autumn/spring,

Medicinal compounds and their bioactivities:

Antibacterial activity: Its blue pigment viz. marennine has been reported to possess antibacterial activity against *Pseudoalteromonas elyakowii, Polaribacter irgensii* and *Vibrio aestuarianus* at concentration of 1 μg/m (Gastineau et al., 2014).

Antiviral activity: The marennine of both intracellular and extracellular forms of this species displayed antiviral activity against herpes virus, HSV-1 with IC50 values of 24.0 and 27.0 μg mL^{-1}, respectively (Gastineau et al. 2014).

Cytotoxicity: The marennine of both intracellular (IMn) and extracellular (EMn) forms of this species was cytotoxic against fibroblastic Vero cells (Gastineau et al. 2014).

Anticancer activity: Both IMn and EMn were found to possess anticancer activity by inhibiting the proliferation of several human cancer cell lines such as SKOV3 and SHIN3 ovarian cancer cells; M113 melanoma cells; R3111 kidney cancer cells; 1355 lung cancer cells; SW116 colon cancer cells; and MCF7 breast cancer cells) (Noujoud Gabed et al., 2022). Its compound hasla-6(17), 9, 13, 23- tetraene showed inhibition activity on human lung cancer (NSCLC-N6) cells with an IC50 value of 3.8 μg/mL. Further, its marennine showed activity against skin cancer (M96), lung cancer (NSCLC-N6), and kidney cancer (E39) cell lines with IC50 values of 30.2, 34.2, and 57.8 μg/mL, respectively (Hussein and Abdullah 2020).

Anticoagulant activity: The aqueous extracts of this species containing marennine showed anticoagulant property (Gabed et al. 2022)

Source of vitamin E: This species has been reported to be the natural source of vitamin E (Khan et al. 2018)

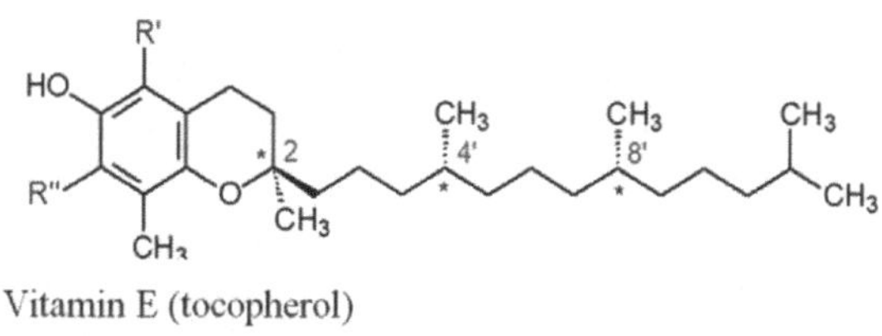

Vitamin E (tocopherol)

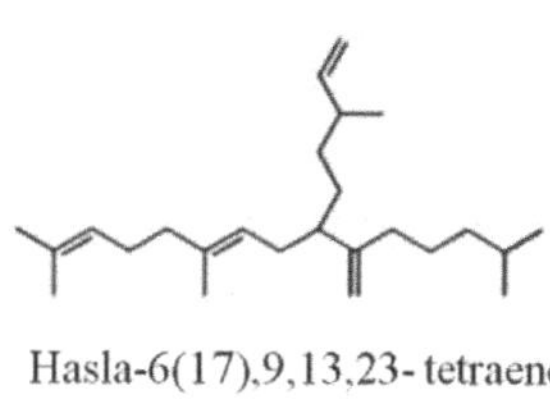

Hasla-6(17),9,13,23- tetraene

Order: Naviculales Family: Phaeodactylaceae

Phaeodactylum tricornutum

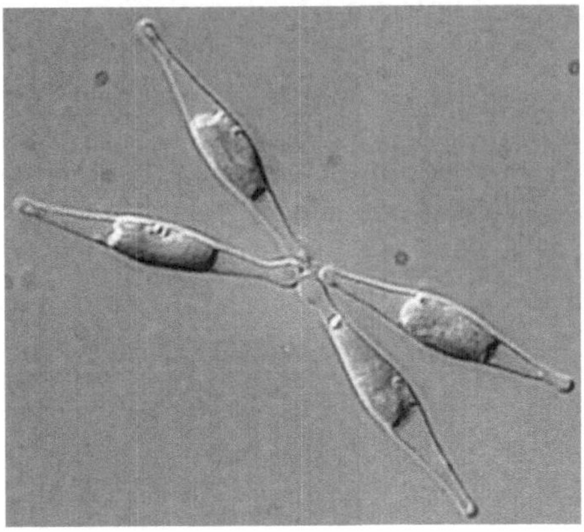

Image credit: Alessandra de Martino and Chris Bowler, Stazione Zoologica and Ecole Normale Supérieure. https://creativecommons.org/licenses/by/2.5/

Global distribution: it is commonly distributed in Atlantic and Pacific Oceans; Great Salt Lake

Ecology: it is abundant in coastal waters and inland saline waters

Medicinal compounds and their bioactivities: This species has been reported to be a cell factory for a spectrum of commercial products of industrial and pharmaceutical importance.

Anticancer activity: The fatty alcohol ester compound, nonyl 8-acetoxy-6-methyloctanoate isolated from this species displayed anticancer activity against Human promyelocytic leukemia (HL-60) and Human lung carcinoma (A549) with IC50 values of 22.3 µg/mL and 50 µg/mL, respectively. Further, its monogalactosyl diacylglycerols showed activity against Wild-types W2 and D3 with IC50 values of 52 µM and 64 µM, respectively. Furthermore, its fucoxanthin pigment was found to show anticancer activity against Hep G2, and HeLa (derived from cervical cancer cells) and Caco-2 (derived from a human colon adenocarcinoma) (Hussein and Abdullah, 2020)

Antibacterial activity: The synthesized eicosapentaenoic acid (EPA) of this species has been reported to be 25.65% (Tachihana et al., 2020) and this EPA showed in vitro antibacterial activity against *Vibrio harveyi, Vibrio anguillarum, Staphylococccus aureus* and *Listonella anguillarum* (Desbois et al., 2008; https://www.sciencedirect.com/topics/agricultural-and-biological-sciences/chaetoceros-calcitrans)

Antithrombic activity: Its 1,2-diaclyglyceryl-3-O-2′-(hydroxmethyl)-(N, N, N-trimethyl)-β-alanine (DGTA) showed antithrombotic activity against platelet-activating factor (PAF) which is said to play an important role in the formation of atherosclerosis (Lordan et al., 2018)

Antiaging activity: The active ingredient, megassane isolated from the cultivated marine microalgae of this species contained rich EPA and DHA and the above ingredient has been reported to possess antiaging activity by activating the proteasome. By this activity, megassane has been reported to help epidermis eliminate the oxidized proteins in human keratinocytes (Abida et al., 2013)

Others: The pigment, fucoxanthin, of this species has been reported to show anti-inflammatory, anti-cancer, anti-obesity, anti-diabetic, and anti-oxidant activities (Khan et al., 2018). Horrocks and Yeo (1990) reported that its docosahexaenoic acid (DHA) is largely helpful for the functional development and growth of the brain in infants and adults. Further, according to these authors, the DHA has the ability to reduce thrombosis by decreasing the level of triglycerides in the blood besides having a positive effect on diabetes mellitus, myocardial infarction, hypertension, arthritis, depression, and atherosclerosis. Furthermore, its synthesized betaine lipid, DGTA possessed anti-inflammatory effects besides serving as nutrient delivery vehicles to carry vitamins A, K, D, and E, throughout the body (Cañavate et al. 2016; Lordan et al. 2018)

Order: Thalassionematales Family: Thalassionemataceae

Thalassionema nitzschioides

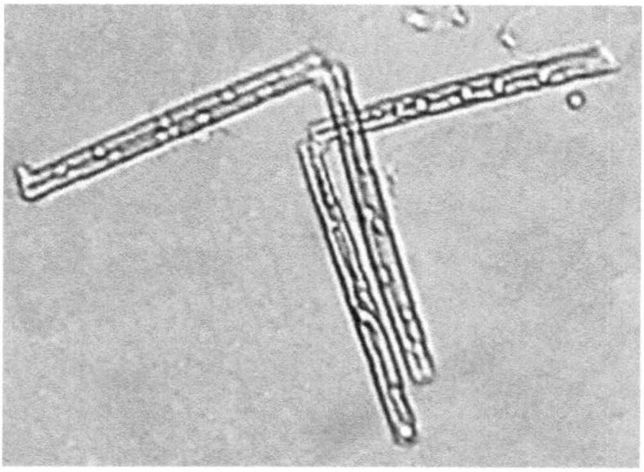

Image credit: Minami Himemiya. https://creativecommons.org/licenses/by-sa/3.0/

Global distribution: It has cosmopolitan distribution except in polar seas

Ecology: It is a neritic and pelagic species

Medicinal compounds and their bioactivities:

Anti-inflammatory effects: The phytosterols, 23-Methylcholesta-5,22E-dien-3β-ol and 23-Methyl-5α-cholest-22E-en-3β-ol of this species have shown anti-inflammatory effects in immune cells such as neutrophils and macrophages (https://lpi.oregonstate.edu/mic/dietary-factors/phytochemicals/phytosterols)

Hypocholesterolemic activity: The phytosterols of this species have been reported to show reductions in LDL cholesterol and total cholesterol concentrations when they are enriched in food (https://lpi.oregonstate.edu/mic/dietary-factors/phytochemicals/phytosterols). Luo et al. (2015) reported that the phytosterols of this species block cholesterol from being absorbed in human body. These authors also suggested that these phytosterols- enriched diet could largely improve serum lipid (cholesterol) profiles thereby reducing the risk of cardiovascular disease

Anticancer effects: Phytosterols of this species have the potential to induce apoptosis, inhibit proliferation, and reduce entry of cancer cells in culture (Luo et al. 2015)

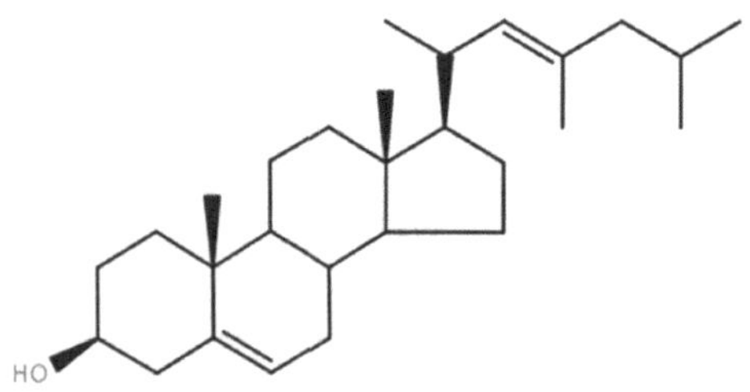

23-Methylcholesta-5,22E-dien-3β-ol

Order: Thalassiosirales Family: Skeletonemataceae

Skeletonema costatum

Image credit: "Judi" Li, NMFS/NOAA, https://creativecommons.org/licenses/by/2.0/

Global distribution: It has cosmopolitan distribution and is more commonly distributed in the coasts of Florida (USA), Hong Kong Island, China,, Sea of Japan, Uruguay, Brazil, Northern Queensland (Australia),

Ecology: It is the most common species in the neritic zone and brackish waters

Medicinal compounds and their bioactivities:

Antimicrobial activity: The methanol extracts of this species displayed antibacterial activity against pathogenic bacteria such as *Proteus vulgaris, Proteus aeruginosa, Proteus mirabilis, Staphylococcus aureus, Salmonella paratyphi,* and *Vibrio cholerae*. On the other hand, the ethanol extracts exhibited a weak activity against V*ibrio cholerae, Salmonella paratyphi, Proteus mirabilis,* and *Staphylococcus aureus* (https://www.sciencedirect.com/topics/agricultural-and-biological-sciences/chaetoceros-calcitrans). Naviner et al. (1999) reported that the lipophilic extracts of this species have shown activity against the Gram-negative marine bacterium, *Listonella anguillarum.* Sushanth and Rajashekhar (2015) stated that the organic extracts of this species showed antibacterial activity against *Staphylococcus aureus, Staphylococcus pyogenes* and *Bacillus subtilis*. These authors also reported on the antifungal activity of the extracts of this species on *Fusarium moniliforme*

Anticancer activity: The unsaturated aldehydes of this species have been reported to show anticancer activity against Caco-2 cells with an IC50 value of 11μg/mL (Hussein and Abdullah 2020)

Antitubercular activity: The extracts of the strain FE85 of this species cultured in phosphate-starvation conditions showed antitubercular activity against *Mycobacterium bovis* with 91% growth inhibition. Similarly, the organic extracts of this strain showed activity against both *Mycobacterium tuberculosis* and *Mycobacterium bovis* (Lauritano et al. 2018)

Skeletonema marinoi

Global distribution: It is a dominant species of temperate regions

Ecology: It is commonly found in coastal waters

Medicinal compounds and their bioactivities:

Antibacterial activity: The crude extracts of this species cultivated under phosphate and nitrogen starvation exhibited antibacterial activity against *Staphylococcus aureus* at a concentration of 50 μg/mL (https://www.sciencedirect.com/topics/agricultural-and-biological-sciences/chaetoceros-calcitrans). Lauritano et al. (2016) reported that the strains. FE60 and FE6 of this species showed antibacterial activity by inhibiting the growth of inhibited *Staphylococcus aureus* at 97 and 96%, respectively

Immunomodulatory properties: The extracts of this species showed immunomodulatory properties by activating IL (interleukin,) and PBMC (for human peripheral blood mononuclear cells) (Hussein and Abdullah 2020)

Cytotoxicity: The extract of FE60 strain of this species displayed cytotoxicity against human melanoma cells, A2058 (Hussein and Abdullah,2020)

Anticancer activity: The monoglycerides extracted from this species showed anticancer activity against hematological cancer cell line (U-937) and colon cancer cell line (HCT-116) with an IC50 value of 5um/mL/mL (Riccio and Lauritano 2020). These authors also reported that the polyunsaturated aldehyde, 2- trans,4-trans-decadienal showed activity against A549cells and COLO 205 cell line (epithelial cells isolated from ascitic fluid from a patient with colon cancer (Riccio and Lauritano 2020).

Order: Thalassiosirales Family: Thalassiosiraceae

Thalassiosira pseudonana

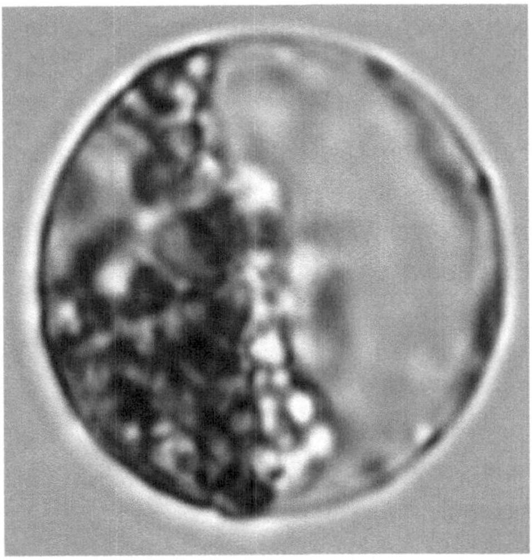

Image credit: Oregon State University. https://creativecommons.org/licenses/by-sa/2.0/

Global distribution: It is widely distributed in all the world's oceans

Ecology: it lives in brackish and coastal waters

Medicinal compounds and their bioactivities: This species has been reported to help in the targeted delivery of anticancer drugs to cancer cells (Lomora et al., 2019). As the valves of this species are capable of absorbing zinc and germanium to alter the pore size, this feature may be of great use in applica-tion s relating to photonic device (Sharma et al. 2021).

Thalassiosira rotula

Global distribution: It has a cosmopolitan distribution.

Ecology: It is a common neritic species

Medicinal compounds and their bioactivities:s

Anticancer activity: The unsaturated aldehydes of this species have shown anticancer activity against Caco-2 cells with an IC50 value of 11 μg/mL (Hussein and Abdullah 2020)

Thalassiosira weissflogii (= *Micropodiscus weissflogii*)

Global distribution: It is found distributed in many parts of the world including Atlantic and Pacific Oceans

Ecology: It is found in a variety of aquatic environments including coastal, brackish, and freshwaters

Medicinal compounds and their bioactivities:

Immunomodulatory properties: The extracts of this species showed immuno-modulatory properties by activating IL (interleukin) and PBMC (for human peripheral blood mononuclear cells) (Hussein and Abdullah 2020; Riccio and Lauritano 2020)

Drug carriers: Its morphology and amorphous nature make this species as promising drug carriers candidate (Sharma et al. 2021)

Antibacterial activity: The extracts of this species possess antibacterial activity (Lomora et al., 2019)

Others: Lomora et al. (2019) reported that this species may be of great use in the delivery of therapeutics, photocatalysis and in the functionalization of therapeutics.

PLANKTONIC FLAGELLATES

Phylum: Heterokontophyta Class: Raphidophyceae
Order: Chattonellales Family: Chattonellaceae

Heterosigma akashiwo

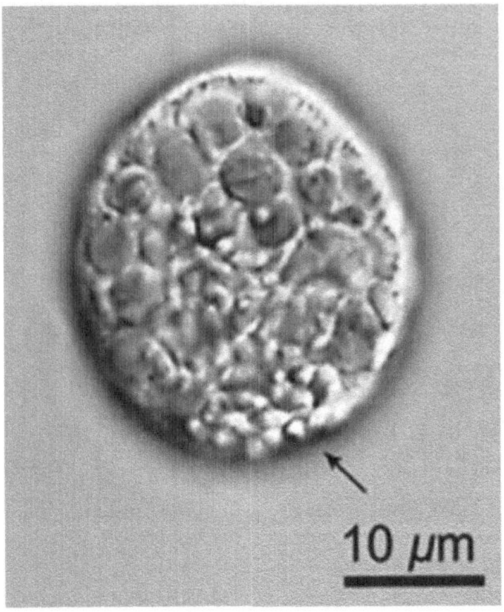

Image courtesy: Anette Engesmo (Reproduced with his kind permission)

Global distribution: This species has been recorded from many parts of the world like Ireland, Sweden, Norway, S.Korea, Hong Kong, Japan, Australia, New Zealand, Netherlands, Scotland, United States, Canada and Chile

Ecology: It is a common diatom species in shallow, neritic, surface waters of about 10 m deep. The formation of harmful algal blooms with this species may kill finfish.

Medicinal compounds and their bioactivities:

Antiviral activity: The cellular extracts (100 μg/mL concentration) of this species have shown antiviral activity by potentially inhibiting Chikungunya virus with an inhibition activity of about 80% (de Vera et al., 2018)

Anesthetic activity: The crude extracts the strain CCMP302 of this species displayed anaesthetic activity in the neuronal model neuro-2a by inhibiting the voltage-dependent sodium channel (Gallardo-Rodríguez et al. 2020)

DINOFLAGELLATES

Phylum: Pyrrophyta Class: Dinophyceae Order: Dinophysiales
Family: Dinophysiaceae

Dinophysis acuminata

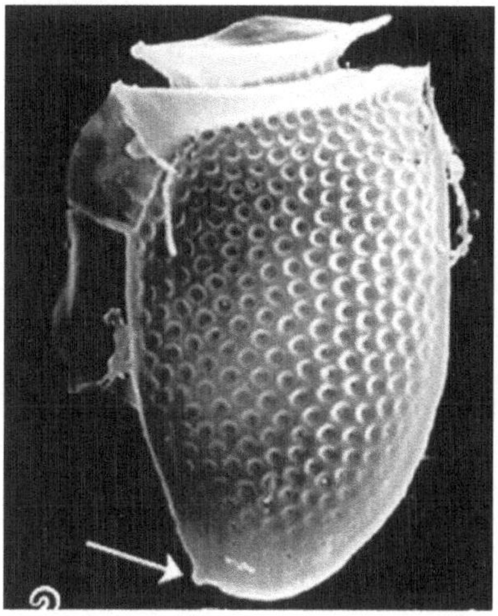

Image credit: Zingone et al., 1998. Public domain.

Global distribution: It is commonly distributed in the Pacific and north Atlantic oceans.

Ecology: It resides largely in coastal waters

Medicinal compounds and their bioactivities:

Cytotoxicity: The compound, okadaic acid of this species has shown cytotoxicity (Camacho et al. 2007). As a model potent neurotoxin, the okadaic acid is found to be very useful for examining the therapeutic effects of antipsychotic drugs which is used in the treatment of several neurodegenerative diseases

Chemotherapeutic agent: The pectenotoxins (PTXs) of the species of *Dinophysis* have been reported to be a promising cytotoxic agent against many human cancer cell lines. Further, the pectenotoxin-2 (PTX2) of this species which is an actin inhibitor is believed to be a strong chemotherapeutic agent against p53 gene - deficient tumors. The p53 gene haws been reported to be a novel tumor suppressor (Vogelstein et al. 2010)

Stimulation of actomyosin ATPase activity: Several compounds such as okadaic acid, 7-epi-PTX-2 seco acid, pectenotoxin II, dinophysistoxin I and acuminolide A have been isolated from this species. Among them, acuminolide A has been reported to stimulate actomyosin ATPase activity which is believed to help in muscle contraction (Su et al. 2014).

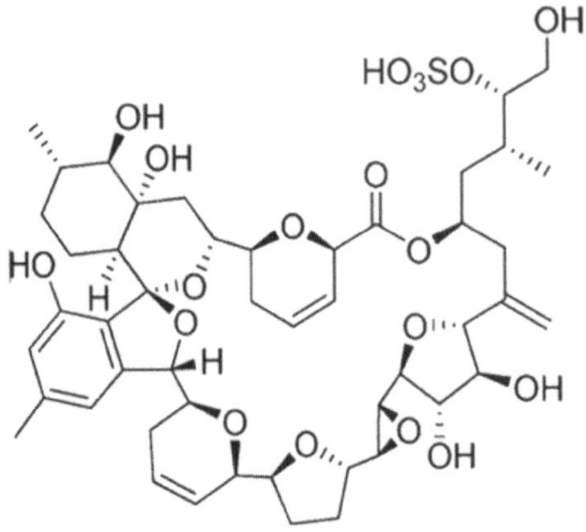

Acuminolide A

Dinophysis acuta

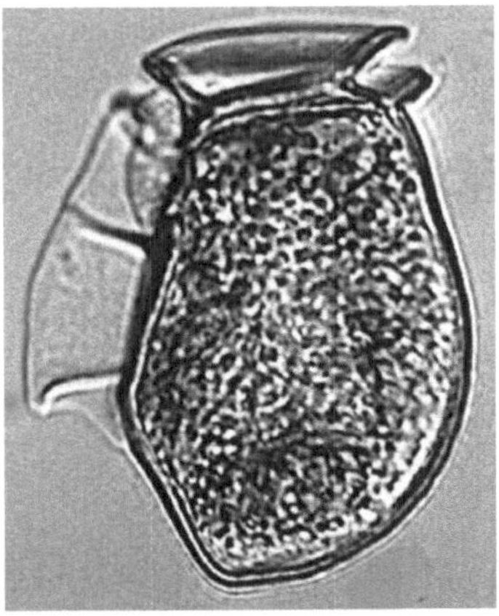

Image credit: Alexandra Kraberg. https://creativecommons.org/licenses/by-sa/3.0/

Global distribution: It is found distributed in temperate and cold waters world-wide

Ecology: It is a planktonic species found both in neritic and oceanic waters.

Medicinal compounds and their bioactivities:

The different strains of this species have been reported to contain okadaic acids (OAs) such as OA, DTX1, DTX2, and OA diol esters; and pectenotoxins such as PTX2, PTX11,PTX12, PTX 13 and PTX 14. PTX 13, and PTX 14 which are oxidized analogues of PTX-2 have been isolated from the extracts of this species. (Reguera et al., 2014; Miles et al., 2006). The lipophilic algal toxins viz. pectenotoxins have not yielded adverse acute health effects in humans. Espiña and Rubiolo (2008) reported that the pectenotoxins of this species induced apoptosis to a larger extent in p53-deficient tumor cells, which are often resistant to other chemotherapeutic agents

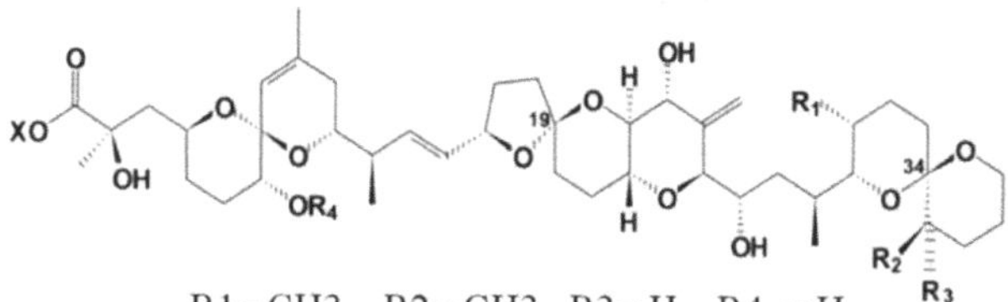

R1= CH3, R2= CH3, R3= H, R4 = H

R1= H, R2= H, R3= CH3, R4 = H

Dinophysis toxin 1,2

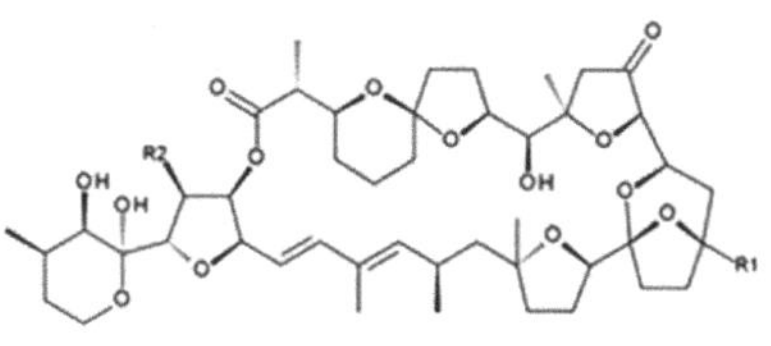

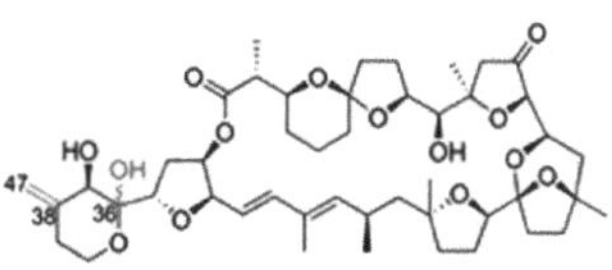

R1= CH3, R2=OH

Pectenotoxins 11

Pectenotoxin 12

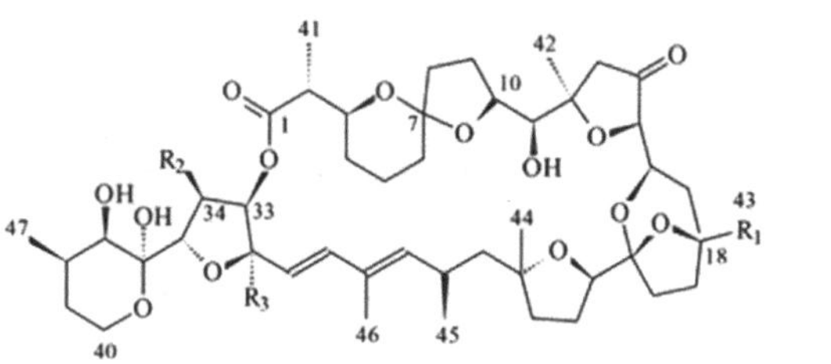

R1= CH3, R2= H, R3= OH, C-7= R

Pectenotoxin 13

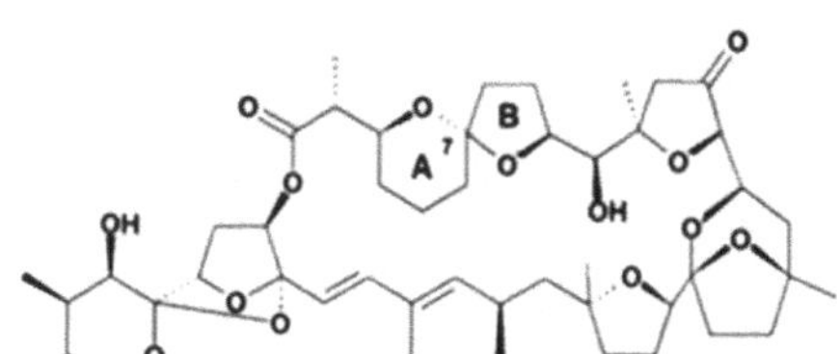

Pectenotoxin 14

Dinophysis caudata

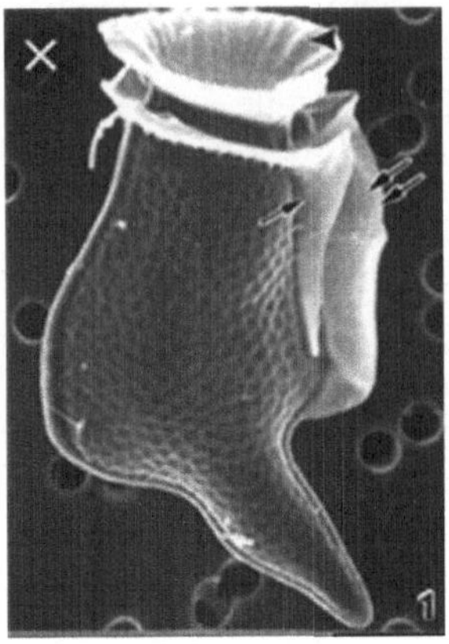

Image credit: Encyclopedia of Life. Public domain.

Global distribution: This cosmopolitan species is commonly distributed in temperate to tropical seas

Ecology: This planktonic species is found mainly in neritic waters

Medicinal compounds and their bioactivities:

Cytotoxicity: This species has been reported to possess toxic compounds like DTX 1, DTX 2, OA, PTX 2 (Reguera et al. 2014). Among them, the PTX 2 displayed strong cytotoxicity to several human cancer cell lines in which this toxin inhibited cytokinesis and mitotic separation by the depolymerization of actin filaments (Kim et al. 2011).

Dinophysis fortii

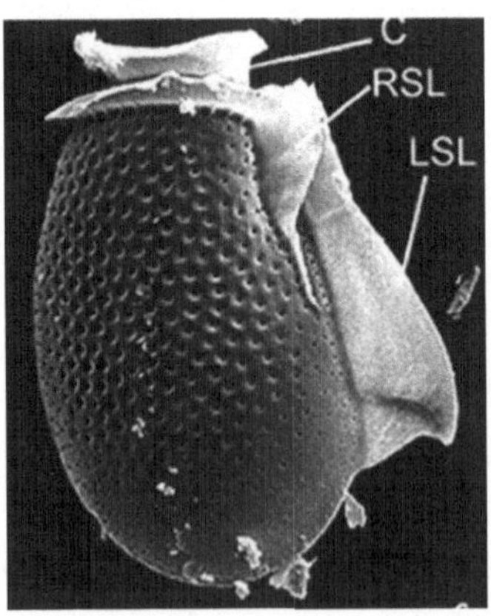

Image credit: Encyclopedia of Life Public domain.

Global distribution: It has world-wide distribution; subtropical to tropical, and cold temperate waters.

Ecology: This planktonic species dwells both in neritic and oceanic regions.. It is a common, noxious bloom-forming species Australia and Japan

Medicinal compounds and their bioactivities:

Cytotoxicity: This species has been reported to contain the phycotoxins viz. DTX 1 and PTX 2 (Reguera et al., 2014). Among these toxins, DTX-1 has shown significant cytotoxicity against MCF-7, Neuro-2a, and NG108-15 cell-lines (Soliño et al., 2015).

Antibacterial potency: The antibacterial potency of the polyether compounds isolated from this species was not found to be significant (Nagai et al. 1990).

Dinophysis miles

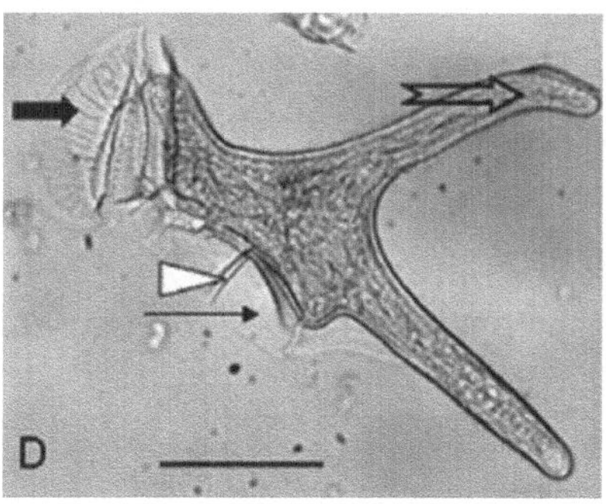

Image courtesy: Senjie Lin (Reproduced with his kind permission)

Global distribution: It is found distributed in tropical Western Indian Ocean including Iraq and Persian Gulf.

Ecology: It is marine, epiphytic species.

Medicinal compounds and their bioactivities:

This species has been reported to contain the phycotoxins viz. DTX 1 and OA (Reguera et al. 2014).

Dinophysis norvegica

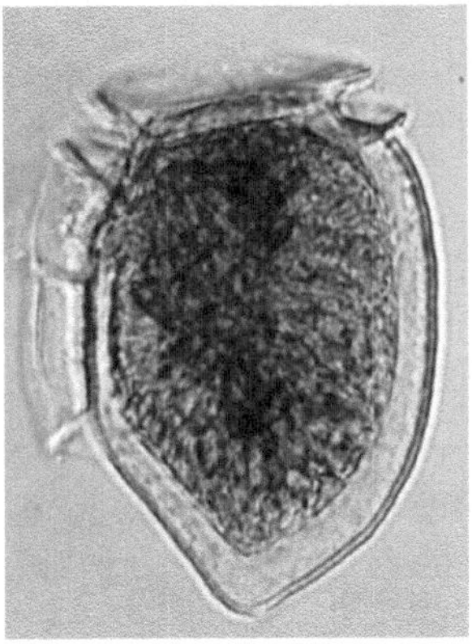

Image credit: Fisheries and Oceans Canada, Dalhousie University, Rajashree Gouda, https://creativecommons.org/licenses/by-sa/4.0/

Global distribution: it is commonly distributed in the northern cold, temperate waters

Ecology: It is a neritic species

Medicinal compounds and their bioactivities:

This species has been reported to contain the phycotoxins viz. DTX1, OA, PTX 2, and PTX 12 (Reguera et al. 2014). The medicinal properties of these compounds have already been dealt with.

Dinophysis rotundataum

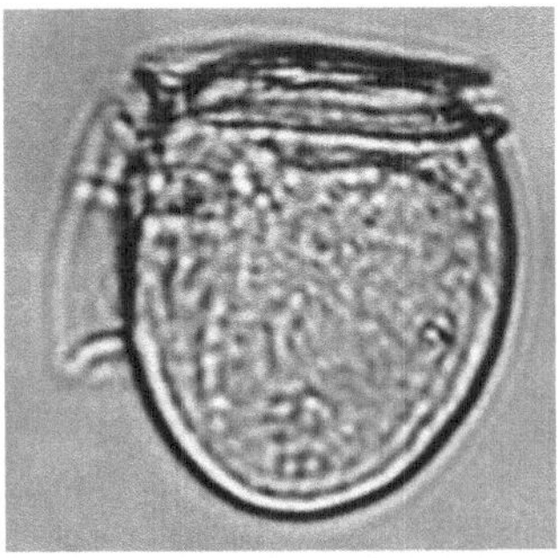

Image credit:Peter J. Bryant, University of California, Irvine (Reproduced with his kind permission)

Global distribution: It has world-wide distribution and is common in the subtropical and tropical waters.

Ecology: It is a marine species living in both neritic and oceanic areas.

Medicinal compounds and their bioactivities:

This species has been reported to contain the phycotoxin, DTX-1 (Lee et al. 1989). The medicinal properties of these compounds have already been dealt with.

Dinophysis sacculus

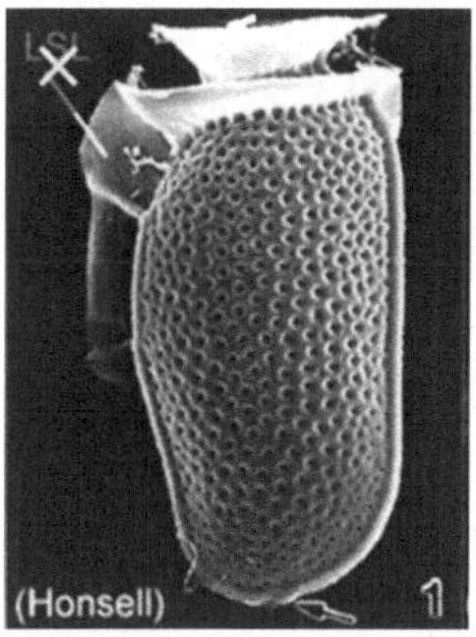

Image credit: Encyclopedia of Life (https://eol.org/pages/897518, CC-PD)

Global distribution: It is widely distributed in cold and temperate waters world- wide including Atlantic Ocean and the Mediterranean Sea

Ecology: It largely dwells in estuaries, lagoons and semi-enclosed basins

Medicinal compounds and their bioactivities:

This species has been reported to contain the phycotoxins viz. DTX-1 and PTX-2 (Reguera et al. 2014). The medicinal properties of these compounds have already been dealt with.

Dinophysis tripos

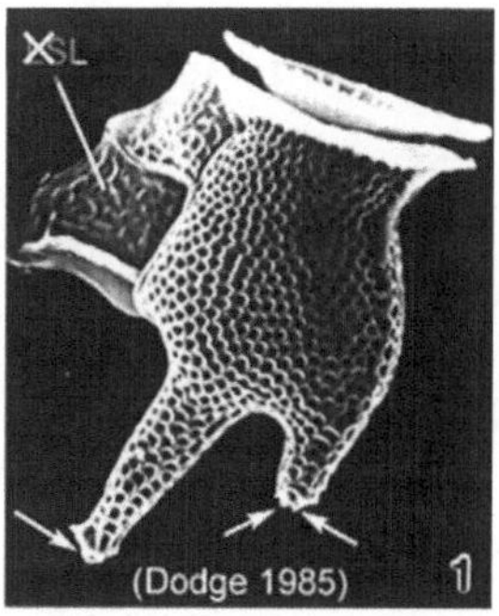

Image credit: Encyclopedia of Life. Public domain.

Global distribution: This species is commonly distributed intemperate and tropical waters, and also in colder regions occasionally

Ecology: It is commonly found in brackish estuaries and neritic and oceanic waters

Medicinal compounds and their bioactivities:

Cytotoxic activity: Its phycotoxins such as DTX-1, DTX-2, and OA have been reported to possess cytotoxicity. Among these compounds, DTX-1 was found to be potently toxic with an average EC50 value of 5.46 nM in Neuro-2a cells; 12.27 nM in MCF-7; and 5.66 nM in NG 108-15. On the other hand, DTX-2 was found to be less toxic than even OA (Soliño et al. 2015)

DTX-2 was the less toxic being the EC50 values approximately half than those exhibited by OA in Neuro-2a and NG 108-15 190 cells. In MCF-7 cells, toxicity of DTX-2 was lower than that of OA to 191 a less extent.

Anticancer activity: The anticancer activity of OA, DTX-1 and DTX-2 on colorectal adenocarcinoma cell line (Caco-2) and the intestinal muco-secreting cell line (HT29-MTX) showed that OA was more portent than DTX-1 and DTX-2 was the least potent compound. For Caco-2, and HT29-MTX, the IC50 values were found to be 49.7 nM and 75.3 nm, respectively, for OA; 22.5 nM and 22.0 n respectively M for DTX-1; and 106.0 and 213.3, respectively, for DTX-2 (Ferron et al., 2014)

Others: Pradhan and Ki, (2022) reported that the pectenotoxin-2 (PTX-2) of this species showed anticancer activity on human breast, colon and lung cancer cells. Further, the actin inhibiting PTX-2 of this species has been suggested to be a promising chemotherapeuticfor the treatment of p53- gene deficient malignancies.

Order: Dinophysiales Family: Dinophysiaceae

Phalacroma mitra *(= Dinophysis mitra)*

Global distribution: It has world-wide distribution in warm temperate to tropical waters

Ecology: This planktonic species dwells in oceanic and neritic waters.

Medicinal compounds and their bioactivities: Like other species of Dinophysis, this species also showed the presence of phycotoxins such as OA, DTX-1 and DTX-2 and their anticancer activity has already been dealt with (Reguera et al. 2014; Soliño et al. 2015; Ferron et al. 2014).

Phalacroma rotundatum *(=Dinophysis rotundata)*

Global distribution: North Atlantic ocean

Ecology: It is common in coastal waters

Medicinal compounds and their bioactivities:

Like other species of *Dinophysis*, this species also showed the presence of phycotoxins such as OA, DTX-1 and DTX-2 and their anticancer activity has already been dealtwith (Reguera et al. 2014; Soliño et al. 2015; Ferron et al. 2014).

Order: Dinophyceae incertae sedis Family: Amphidomataceae

Azadinium dexteroporum

Global distribution: It has been reported from the Mediterranean Sea

Ecology: It is a coastal water species.

Medicinal compounds and their bioactivities:

Cytotoxicity and cholesterol- lowering activity: The species of this genus have been reported to possess the phycotoxin polyethers viz. azaspiracid (AZA) and its derivatives (AZA-1,2,3) (Pradhan and Ki, 2022). Among these compounds, AZA-1,2 of *Azadinium dexteroporum, Azadinium poporum*, and *Azadinium spinosum* displayed cytotoxicity against several human cell types. Further, these compounds also lowered the cholesterol levels in T-lymphocyte cells (Pradhan and Ki 2022).

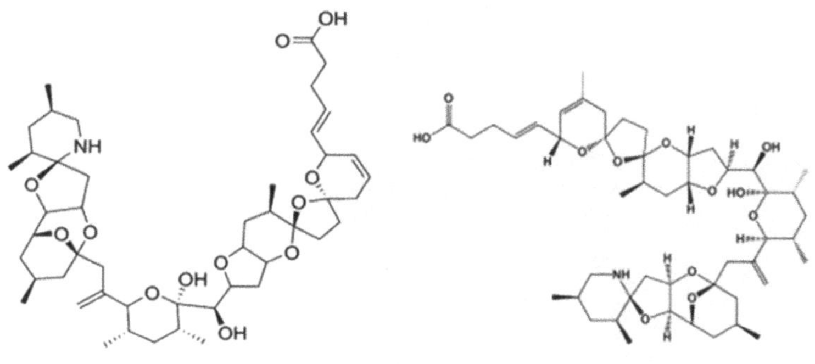

AZA- 1,2

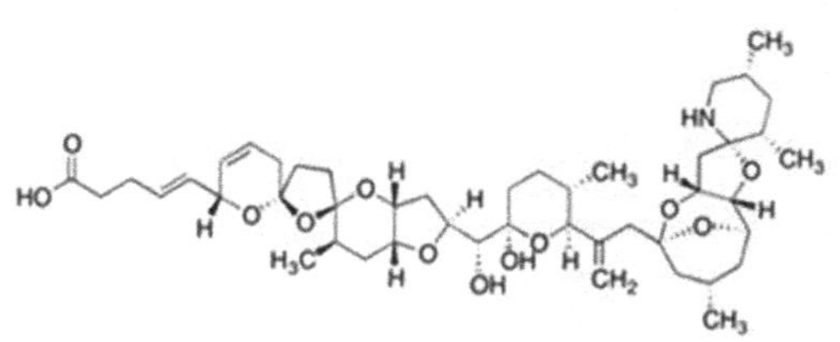

AZA-3

AZA-35, AZA-54 to AZA-58 and 3-epiAZA-7 (Rossi et al., 2017)

Azadinium poporum

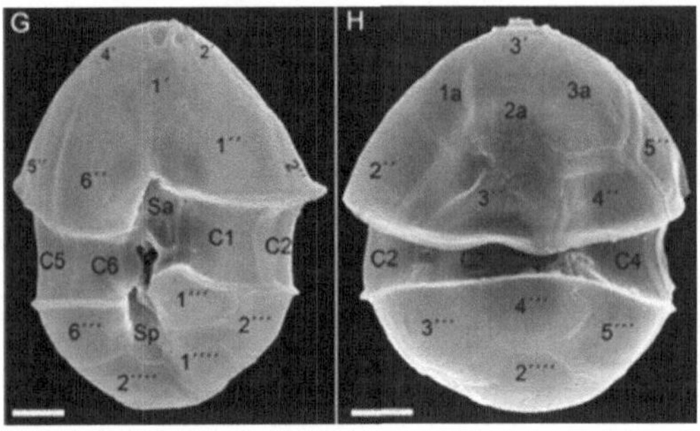

(G) ventral view, (H) dorsal view,

Image credit: Urban Tillmann (Reproduced with his kind permission)

Global distribution: It has been reported from the south-west Atlantic and southeast Pacific oceans

Ecology: It is a coastal species

Medicinal compounds and their bioactivities:

Cytotoxicity and cholesterol- lowering activity: vide *Azadinium dexteroporum*

Others: Tillmann et al. (2016) reported that this species has been reported to possess the compounds azaspiracid 2 and 11.

Azadinium spinosum

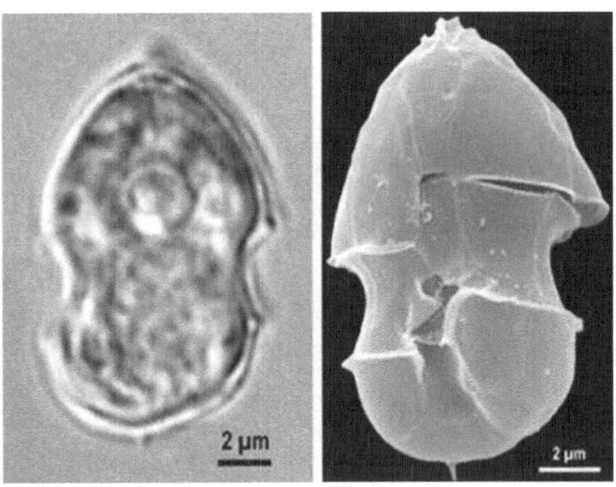

Image credit: Urban Tillmann

Global distribution: It is the most common species of NW Pacific ocean

Ecology: It is a coastal water species

Medicinal compounds and their bioactivities:

Cytotoxicity and cholesterol- lowering activity: vide *Azadinium dexteroporum*

Others: This species has been reported to possess azaspiracid (AZA) 1,2 and an isomer of azaspiracid 2 (https://en.wikipedia.org/wiki/Azadinium_spinosum). Rossi et al. (2017) reported that this species produced 7 azaspiracids viz. AZA-35, AZA-54 to AZA-58 and 3-epiAZA-7.. The occurrence of Azaspiracid 1 (AZA1) and its 11 analogues, AZA2–AZA12 has also been stated in this species (https://www.sciencedirect.com/topics/agricultural-and-biological-sciences/azadinium-spinosum). The bioactivities of these compounds are yet to be known.

Order: Gonyaulacales Family: Goniodomataceae

Alexandrium andersoni

Global distribution: it is commonly distributed in the Mediterranean Sea and south west coast of Ireland.

Ecology: It is a coastal water species

Medicinal compounds and their bioactivities:

Apoptotic Activity: The raw cell extracts of this species showed weak apoptotic activity of human lung carcinoma A549 and colorectal carcinoma HT29 cell lines (de Vera et al. 2018).

Cytotoxicity: Sansone et al. (2018) reported that the both its organic extract and related enriched extracts induced high cytotoxicity of the A549 lung cancer and HT29 colorectal cancer cell lines without affecting the normal cell viability.

Others: This species has been reported to possess phycotoxins of therapeutic importance such as

saxitoxin (STX) and neosaxitoxin (NeoSTX) (Ciminiello et al. 2000). The saxitoxin inhibits the entry of sodium ion into nerves and muscle by closing the voltage gated sodium channel. The neosaxitoxin apart from this activity may also serve as a long-acting pain blocker. As site-1 sodium channel blocker, these toxins undergo clinical trials as a local anesthetic with prolonged-duration (Templin et al. 2015)

Alexandrium catenella

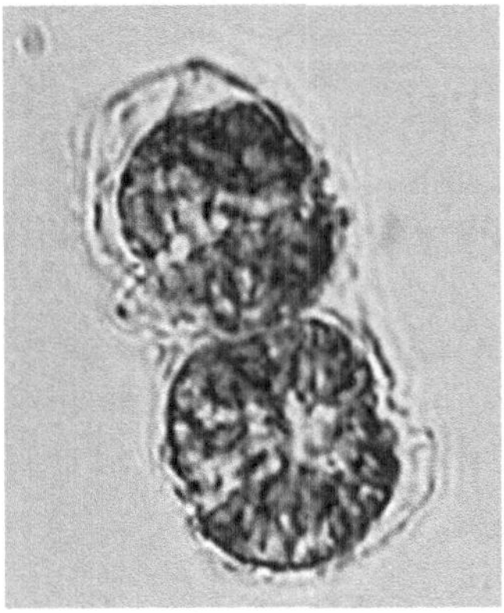

Image credit: Photo by Me.garneau, https://creativecommons.org/licenses/by-sa/3.0/

Global distribution: It is widely distributed in cold temperate regions of west coast of North America, Japan, Australia and Tasmania, western South Africa, Chile and Argentina

Ecology: it is a coastal water species.

Medicinal compounds and their bioactivities:

It contains saxitoxin (STX) which is a potential anesthetic agent. Pradhan and Ki (2022) reported that STXs possess significant antimicrobial and antiprotozoal activity in vitro. Further STXs are believed to block pain sensations

Alexandrium hiranoi

Global distribution: Atlantic and Pacific Oceans

Ecology: It is a neritic species and is commonly seen at a depth of 1 m.

Medicinal compounds and their bioactivities: vide *Alexandrium pseudogonyaulax*

Alexandrium monilatum

Global distribution: It is commonly distributed in the tropical to warm temperate seas of the America

Ecology: It is a coastal and estuarine species

Medicinal compounds and their bioactivities: vide *Alexandrium pseudogonyaulax*

Alexandrium pseudogonyaulax *(= Goniodoma pseudogoniaulax)*

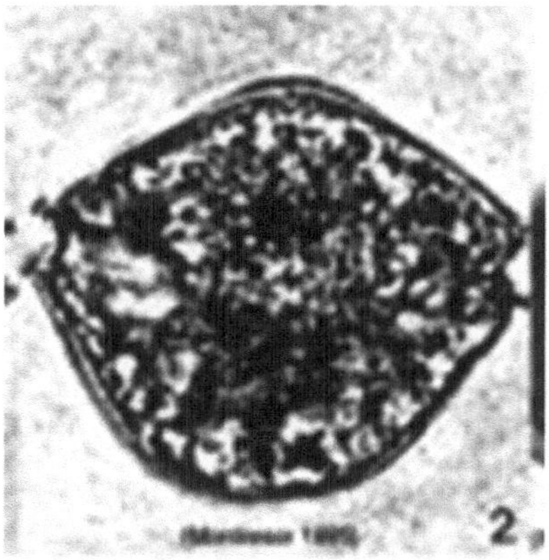

Image credit: Encyclopedia of Life. Public domain.

Global distribution: It is distributed in Europe and Pacific Ocean (British Columbia and Japan)

Ecology: It is a coastal and brackish waster species

Medicinal compounds and their bioactivities of *Alexandrium hiranoi, Alexandrium monilatum,* and *Alexandrium pseudogonyaulax*

These species have been reported to possess polyether macrolide compounds viz. goniodomin A,B. Among these compounds, goniodomin A has been reported to possess antifungal activity and cytotoxicity. Further, this compound is known to inhibit angiogenesis by reducing endothelial cell migration and tube formation which is caused by bFGF (basic fibroblast growth factor) (Pradhan and Ki 2022).

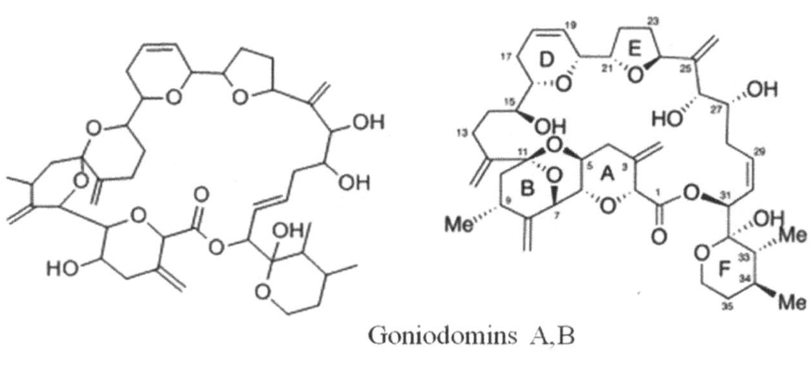

Goniodomins A,B

Alexandrium minutum

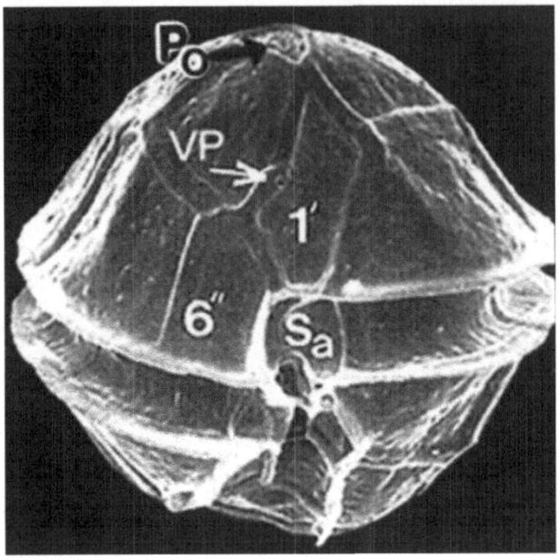

Image credit: Encyclopedia of Life. Public domain.

Global distribution: It is distributed in all temperate and warm coasts of the world. It has been reported to be abundant in the European north Atlantic, south- and south-east Asia, Australia, New Zealand Mediterranean Sea and Japan

Ecology: It dwells in coastal waters, lagoons and estuaries

Medicinal compounds and their bioactivities:

Antiviral activity: Cellular extracts of this species have shown antiviral activity against Chikungunya virus with a significant inhibition activity of about 80% at a concentration of 100 μg/mL (de Vera et al. 2018)

Apoptotic activity: Extracts of this species have been reported to possess the highest apoptotic potential on the human hepatocellular carcinoma cell line HepG2 (ATCC® HB-8065™) at a concentration of 100 μg/mL (de vera et al. 2018)

Others: Its saxitoxin (STX) has been reported to possess several therapeutic properties. It is believed to be a potential anesthetic agent besides minimizing pain sensations. Further, the extracts containing STXs possess antimicrobial and antiprotozoal activities in vitro (Pradhan and Ki 2022)

Alexandrium ostenfeldii

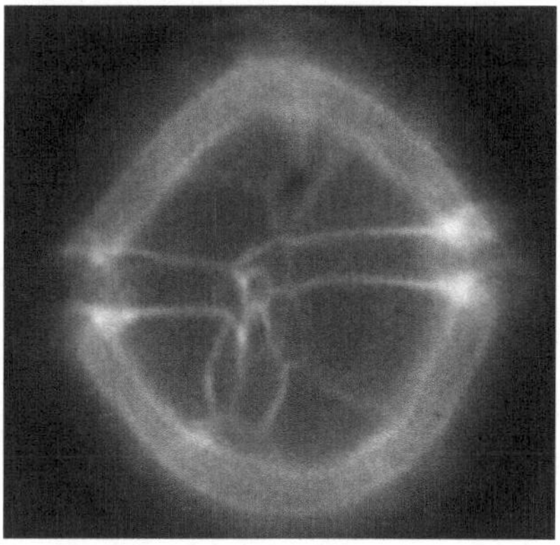

Image credit: Encyclopedia of Life. Public domain.

Global distribution: It is found distributed in the Baltic Sea and along the west coast of Europe.

Ecology: It is a cold-water coastal species

Medicinal compounds and their bioactivities:

Cytotoxicity: Among its compounds 13-desmethyl spirolide C, 20-methyl spirolide G, and spirolides A-C and H, only 13-desmethyl spirolide C has been reported to show cytotoxicity with different mammalian cancer cell lines (Munday et al. 2012).

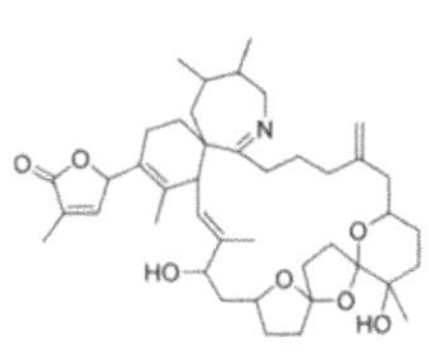

13-desmethyl spirolide C

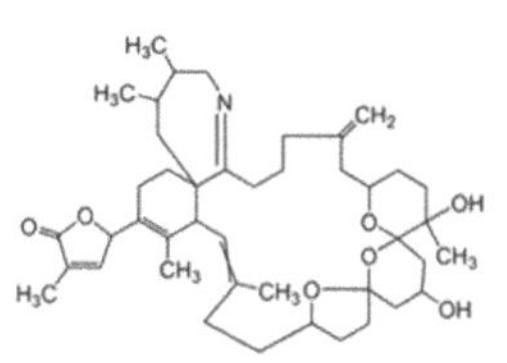

20-methyl spirolide G

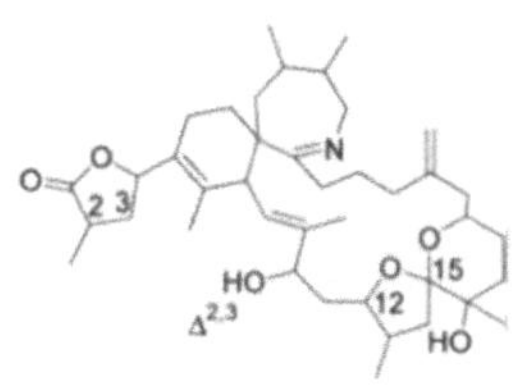

Spirolide H

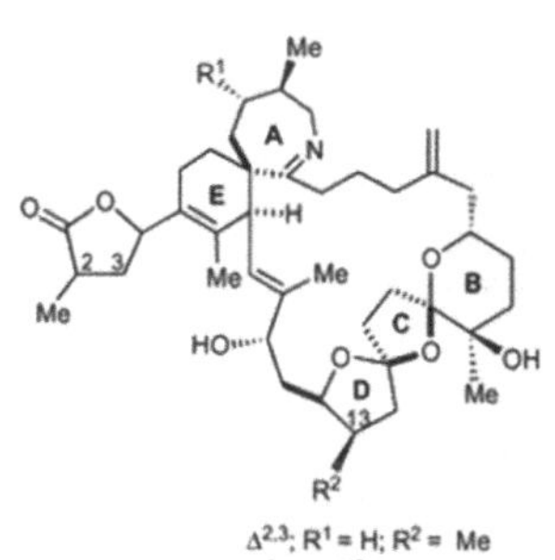

$\Delta^{2,3}$; R^1 = H; R^2 = Me
R^1 = H; R^2 = Me
$\Delta^{2,3}$; R^1 = Me; R^2 = Me

Spirolide A-C

Alexandrium peruvianum

Global distribution: North Carolina, USA

Ecology: It is an estuarine species

Medicinal compounds and their bioactivities:

Cytotoxicity: The compounds, 13-desmethy C and D spirolides have been isolated form this species (Tomas et al. 2012). Among them, 13-desmethy C has already been reported to show cytotoxicity with different mammalian cancer cell lines (Munday et al. 2012).

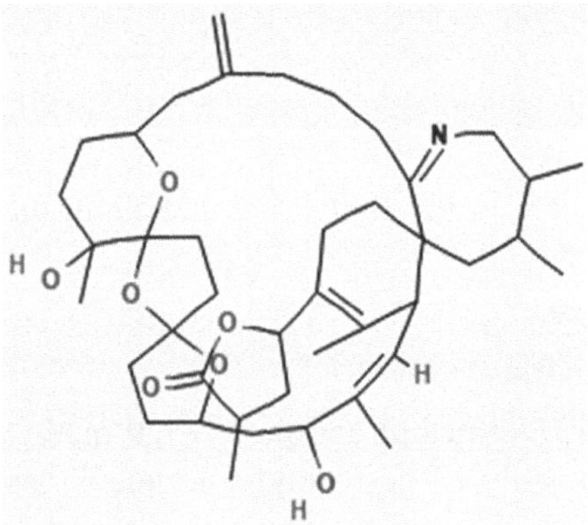

Alexandrium tamarense

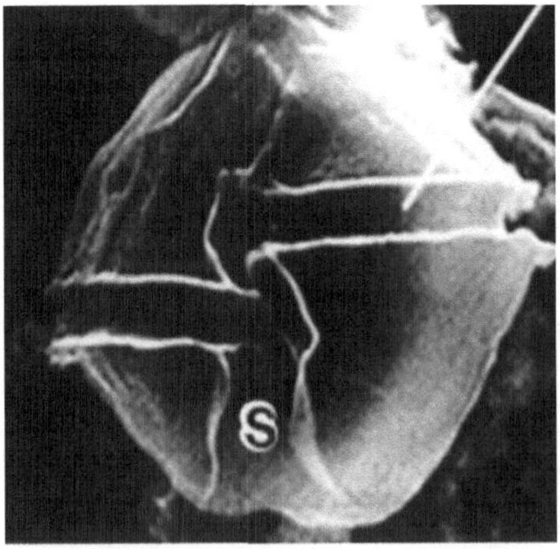

Image credit: Encyclopedia of Life, Public domain.

Global distribution: It has been reported from warmer waters world -wide; from cold to cold-temperate waters of North America, Japan and Europe and Japan; and from Gulf of Thailand, Australia and Venezuela

Ecology: It is a coastal and estuarine species

Medicinal compounds and their bioactivities:

Antiviral activity: The cellular extracts of this species have shown antiviral activity against Chikungunya virus with a significant inhibition activity of more than 82% at 100 μg/mL concentration (de Vera et al. 2018)

Apoptotic Activity: Extracts of this species have been reported to possess the highest apoptotic potential on the human hepatocellular carcinoma cell line HepG2 (ATCC® HB-8065™) at a concentration of 100 μg/mL (de vera et al. 2018)

Immunomodulatory properties: The total extracts of this species have been reported to display immunomodulatory properties by activating interleukin and human peripheral blood mononuclear cells (Riccio and Lauritano 2020)

Others: The tetrodotoxin (TTX) produced by this species has been reported to be a safe and effective and long-acting anesthetic. This toxin is believed to be effective in the treatment of heroin addiction. (Camacho et la., 2007). Further, administration of 5 and 10 μg TTX was found to be beneficial in relieving the symptoms of opium withdrawal with very minor side effects (Pradhan and Ki 2022). Pradhan and Ki (2022) reported that its saxitoxin (STX) is a potential anesthetic agent; capable of minimizing pain sensation; and is with antimicrobial and antiprotozoal activities.

***Alexandrium* sp.**

Medicinal compounds and their bioactivities:

The neurotoxin, gonyautoxin derived from certain species of *Alexandrium* served as a safe and effective painkiller. This activity of gonyautoxin has been demonstrated in patients with chronic tension-type headache for which 50 mg of goniautotoxins was infiltrated locally. Similar treatment also reduced pain during a surgery of arthroplasty in the knee (Ferrara 2020)

Pyrodinium bahamense

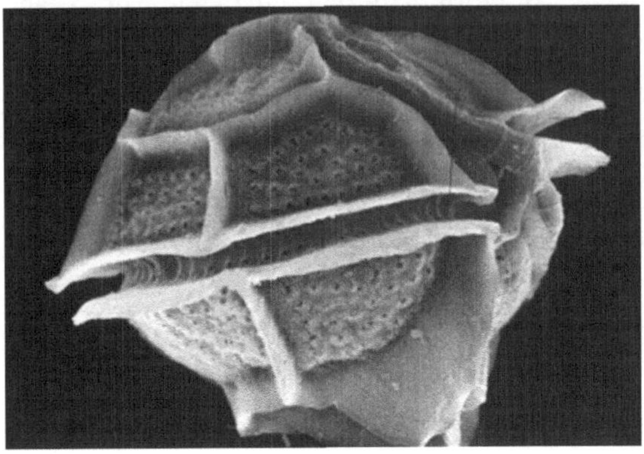

Image credit: CSIRO, https://creativecommons.org/licenses/by/3.0/

Global distribution: Pacific and Atlantic oceans.

Ecology: it is a coastal water species

Medicinal compounds and their bioactivities:

The saxitoxin (STX) of this species has been reported to be a promising anesthetic, antimicrobial, and antiprotozoal activities (Pradhan and Ki 2022)

Order: Gonyaulacales: Family: Gonyaulacaceae

Gonyaulax spinifera

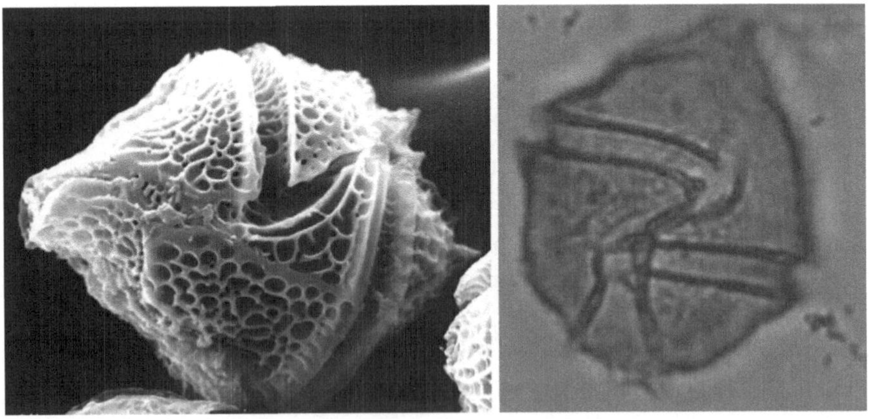

Image credit: Rossella Pistocchi (Reproduced with his kind permission)

Global distribution: It has world-wide distribuiton

Ecology: It lives in all neritic, estuarine and oceanic, estuarine environments

Medicinal compounds and their bioactivities:

The saxitoxin (STX) of this species has been reported to be a promising anesthetic, antimicrobial and antiprotozoal activities (Pradhan and Ki 2022). Riccardi et al.(2009) reported on the presence of sulphated polyether compounds viz. yessotoxins (YTX) in this species. Aslfonso et al. (2016) reported on the bioactivities of YTX. It displayed apoptotic effect in several cancer cell lines such as BE(2)-M17, HeLa S3, L6 and BC3H1 myoblasts, MCF-7, MDCK, HepG2, Bel7402, HL7702, and K-562. Further, this toxin has an immune-regulatory effect on T-lymphocyte EL-4 cells and is a promising therapy for Alzheimer's disease.

Lingulodinium polyedra

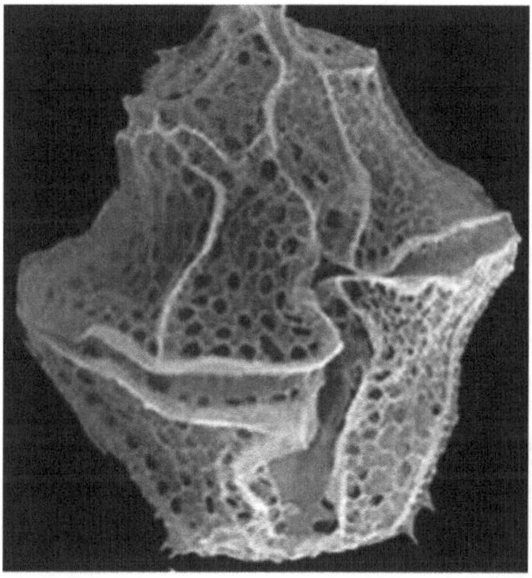

Image credit: Morseds, https://creativecommons.org/licenses/by-sa/4.0/

Global distribution: It is found distributed in subtropical and warm temperate waters

Ecology: It is a neritic water species

Medicinal compounds and their bioactivities:

Antibacterial activity: The aqueous, methanol, hexane and chloroform extracts of this species have shown significant antibacterial activity against *Vibrio vulnificus* and *Staphylococcus aureus* (Quijano-Scheggia et al. 2016).

Others; Yessotoxins (YTXs) have also been reported in this species. The therapeutic activities of this toxin have already been given earlier. Pradhan and Ki (2022) reported on the promising role of this toxin in the treatment of glucose and lipid metabolism-related disorders in glioma cells and on its minor role as a antiallergenic and antiasthmatic drug.

Protoceratium reticulatum

Global distribution: Atlantic coast of North America, North Sea, Baltic Sea, Gulf of Mexico South Africa, New Zealand, and South Africa and Russia

Ecology: It is a coastal water species

Medicinal compounds and their bioactivities

The therapeutic activities of the yessotoxins of this species have already been given (vide *Lingulodinium polyedra* and *Gonyaulax spinifera*)

Gambierdiscus australes

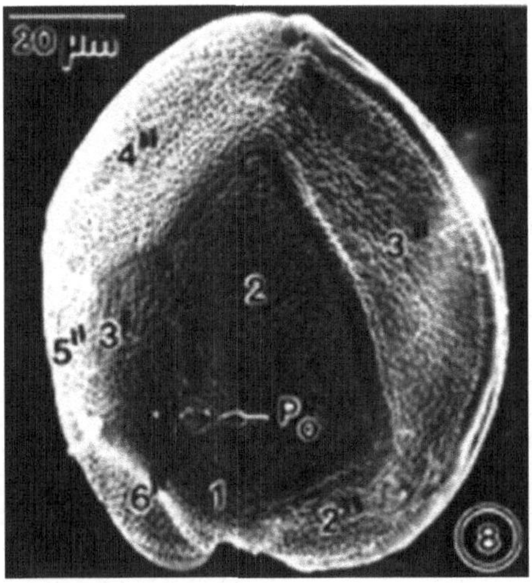

Image credit: M.A. Faust, courtesy of Smithsonian Institution (Reproduced with his kind permission)

Global distribution: It is commonly distributed in the Atlantic Ocean

Ecology: It is a benthic and epiphytic dinoflagellate sepcies

Medicinal compounds and their bioactivities:

Antiproliferative Activity: Extracts of this species have shown antiproliferative activity with more than 80% in all cell lines including MCF-7 cell line (de Vera et al. 2018).

Apoptotic activity: Extracts of this species have been reported to possess the highest apoptotic potential on the human hepatocellular carcinoma cell line HepG2 (ATCC® HB-8065™) at a concentration of 100 µg/mL (de vera et al. 2018)

Others: Pradhan and Ki (2022) reported on the occurrence of three different forms of maitotoxins viz. MTX-1, MTX-2, and MTX-3 in the species of *Gambierdiscus* such as *Gambierdiscus australes, Gambierdiscus pacificus,* and *Gambierdiscus toxicus.* According to these authors, the maitotoxins possess the ability to initiate intracellular cascades of events. Further, they are known to activate voltage-insensitive Ca2+ channels in C6-BU-1 glioma cells and stimulate the synthesis and secretion of nerve growth factor (Obara et al. 1999)

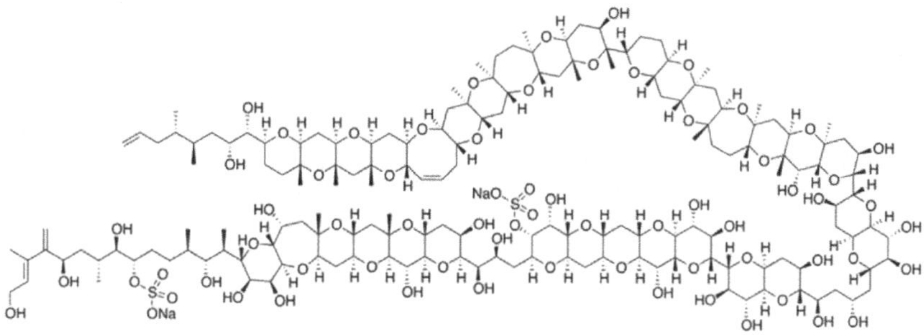

Maitotoxin

Gambierdiscus pacificus

Global distribution: This species is found distributed throughout tropical and subtropical regions world-wide

Ecology: It is a marine benthic and epiphytic species

Medicinal compounds and their bioactivities: This species possess the maitotoxins (MTXs) and ciguatoxins (CTXs) and the therapeutic properties of these toxins have already been given

Gambierdiscus toxicus

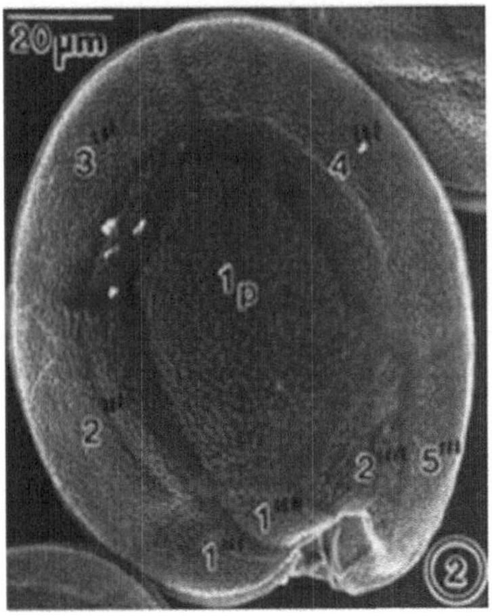

Image credit: Encyclopedia of Life, https://creativecommons.org/licenses/by-sa/4.0/

Global distribution: it is found distributed throughout the tropical and subtropical regions world-wide

Ecology: This marine, benthic species is frequently found as epiphytes on seaweeds and dead corals

Medicinal compounds and their bioactivities:

As in the case of *Gambierdiscus australes* and *Gambierdiscus pacificus*, this species is also possessing MTX-1, MTX-2, and MTX-3 (Pradhan and Ki 2022) and their therapeutic applications have already been given.

Apart from the above, this species contains several other compounds such as ciguatoxin (CTX), gambierol, and gambieric acids A- D have also been isolated form this species (Pradhan and Ki, 2022). The bioactivities of these compounds are given below

Ciguatoxin (CTX): This polyether compound has been reported to exhibit therapeutic properties through enhanced muscular contraction in cardiac tissue. Further, CTX is believed to be a useful tool for examining the bio-functions of a range of human diseases such as epilepsy, cancer, chronic pain and cardiac arrhythmias (Pradhan and Ki 2022).

Gambierol: This compound helps in neurite development, which is beneficial for the patients with brain injury; this compound serves as an mmunosuppressant in diseases which involve malfunctioning immune system such as rheumatoid arthritis, multiple sclerosis, and type 1 diabetes; two of its analogues viz. tetra and heptacyclic forms have been reported to reduce β-amyloid and tau hyperphosphorylation in Alzheimer's disease; and this compound has also served as a potent antagonist for the PbTx-2-induced cytotoxicity and Ca2+ influx (Pradhan and Ki, 2022).

Gambieric acids (GA): Gambieric acid A- D have been reported to possess antifungal activity against filamentous fungus. Further, GA-A and GA-B have have shown very significant activity against the fungus, *Aspergillus niger*. Furthermore, all these compounds have also shown cytotoxicity.

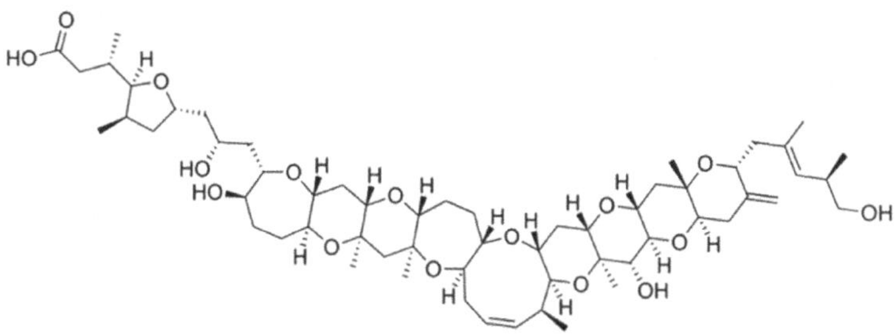

Gambieric acid A

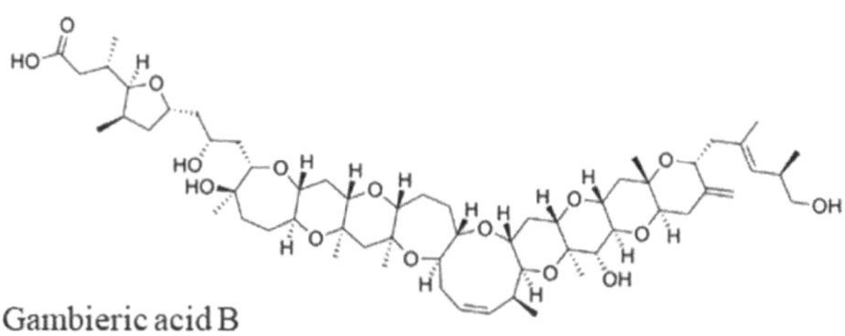

Gambieric acid B

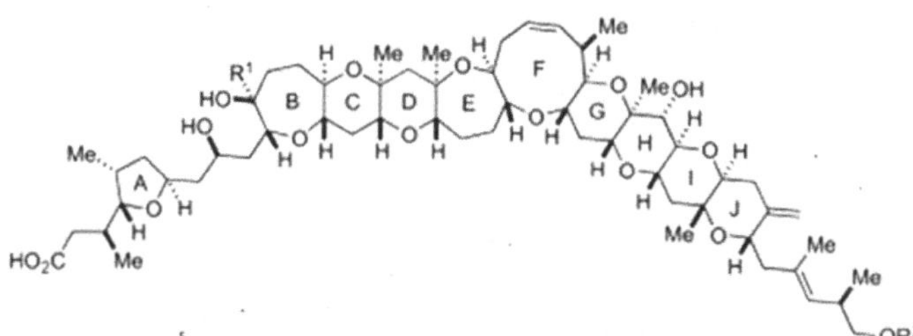

R1 = H, R2 = 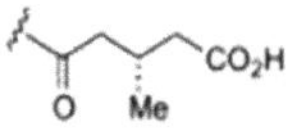

R1= Me, R 2= 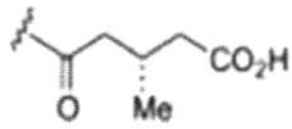

Gambieric acids C,D

Others: Nagai et al. (1990) reported that the polyether compounds derived from *Gambierdiscus toxicus, Prorocentrum lima* and *Dinophysis fortii* showed antifungal activity. Similarly, the okadaic acid and its two congeners, desulfated yessotoxin of these species, also showed potent antifungal activities

Gambierdiscus belizeanus

Global distribution: it is found distributed in tropical marine environments such as Caribbean Sea, Belize, South Water Cay and Carrie Bow Cay

Ecology: This benthic speciesis commonly found in coastal areas and lagoonal waters. It is an epiphytic species and is often associated with sea weeds coral rubble and colored sand in protected areas.

Medicinal compounds and their bioactivities:

The therapeutic properties of its ciguatoxin have already been dealt with

Gambierdiscus caribaeus

Global distribution: it has got circumtropical distribution

Ecology: Among the marine dinoflagellates, only he species of Gambierdiscus have been reported to be epiphytic and cooccur with diatoms such as *Nitzschia, Pleurosigma, Actinocyclus, Achnanthes* and *Amphora* and also with cnidarians like the colony of *Halecium* sp.

Medicinal compounds and their bioactivities:

The therapeutic properties of its ciguatoxin have already been dealt with

Gambierdiscus carolinianus

Global distribution: It has circumtropical distributionand is very common in North Carolina, USA.

Ecology: This seaweed- associated species is commonly found in a rocky tidal pools above 40 m off the coast

Medicinal compounds and their bioactivities:

The therapeutic properties of its ciguatoxin have already been dealt with

Gambierdiscus carpenteri

Global distribution: It has circumtropical distribution

Ecology: Like other species of this genus, it is also epibenthic.

Medicinal compounds and their bioactivities:

The therapeutic properties of its ciguatoxin have already been dealt with

Gambierdiscus excentricus

Global distribution: Northeast Atlantic Ocean
Ecology: This benthic species is epiphytic on seaweeds of tidal ponds
Medicinal compounds and their bioactivities:
The therapeutic properties of its ciguatoxin have already been dealt with
Order: Order: Gonyaulacales Family: Ostreopsidaceae

Ostreopsis fattorussoi

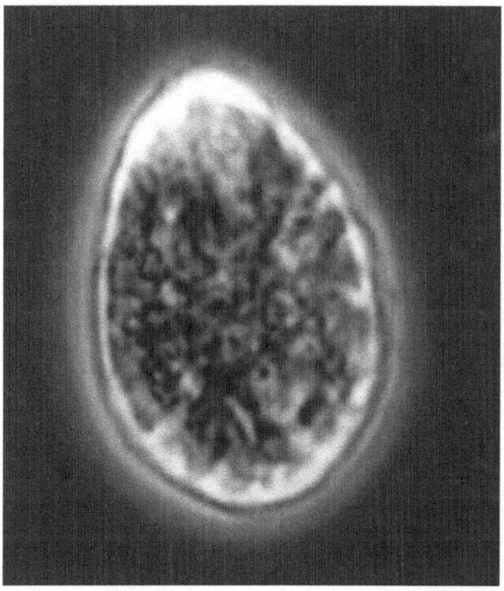

Ostreopsis sp.
Image credit: Cybergerac. https://creativecommons.org/licenses/by-sa/4.0/

Global distribution: It is commonly distributed in the coast of Lebanon and Cyprus and

Eastern Mediterranean Sea

Ecology: It is a benthic species.

Medicinal compounds and their bioactivities:

Pavaux et al. (2020) reported on the occurrence of ovatoxin, a, d, e, i, j1, j2, and k.

The polyether palytoxin (PLTX) of this species has been reported to modulate neurotransmitters such as norepinephrine (= noradrelaine) and acetylcholine by releasing prostaglandin-E2 and histamine (Pradhan and Ki, 2022). It is worth mentioning here that norepinephrine is both a hormone and a neurotransmitter, and it plays a key role in increasing and maintaining blood pressure in short-term serious health situations. Similarly, the acetylcholine which is a main neurotransmitter of the parasympathetic nervous system plays an important role in dilating blood vessels, and slowing heart rate besides contracting smooth muscle. PLTX may be a promising therapeutic agent in the prevention and treatment of leukemia (https://patents.google.com/patent/WO2015090591A1/en)

Ostreopsis lenticularis

Global distribution: It is commonly distributed in New Caledonia of the tropical Pacific Ocean

Ecology: It is found form shallow waters to offshore reefs. This benthic and epiphytic species is commonly associated with seaweeds and soft corals

Medicinal compounds and their bioactivities:

Pavaux et al. (2020) reported on the occurrence of ostreotoxin 1 and 3.

The therapeutic potential of its palytoxin are given elsewhere. Meunier et al. (1997) reported on the occurrence of ostreotoxins 1 and 3 in the cultures of this species. The therapeutic properties of these phycotoxins are however not known.

Ostreopsis mascarenensis

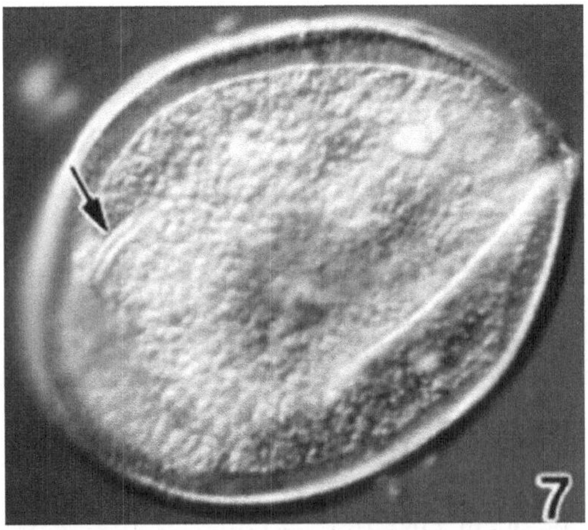

Image credit: Encyclopedia of Life. Public domain.

Global distribution: It is commonly distributed in the Caribbean Sea and SW Indian Ocean

Ecology: This benthic and epiphytic species is commonly found in shallow coral reef environments and lagoonal island. It is often associated with sediments, dead corals, and seaweeds like *Turbinaria* sp.

Medicinal compounds and their bioactivities:

Pavaux et al. (2020) reported on the occurrence of mascarenotoxin− A,B.

The therapeutic potential of its palytoxin is given elsewhere

Ostreopsis ovata

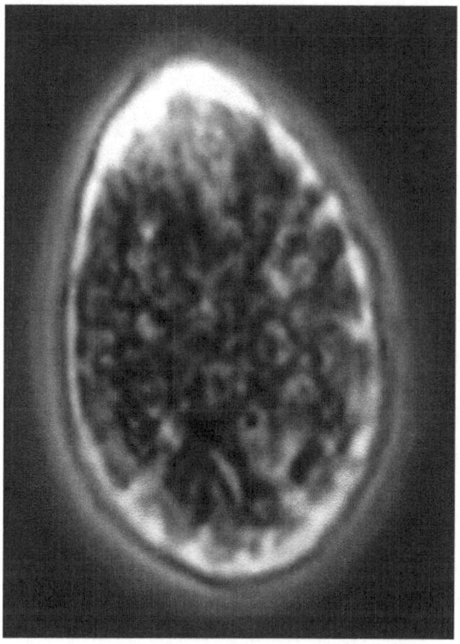

Image credit: Cybergerac-commons.wikimedia

Global distribution: It is very common species of the Mediterranean Sea

Ecology: This epiphytic species is found in shallow, costal environments where it is found settling on rocks or seaweeds.

Medicinal compounds and their bioactivities:

Pavaux et al. (2020) reported on the occurrence of ovatoxin a-h, ostreol A, mascarenotoxin-a, and mascarenotoxin-c in this species.

The therapeutic potential of its palytoxin is given elsewhere. This species has also been reported to possess the compound, Ostreol B which displayed medium cytotoxicity in Neuro-2a, HCT-116 and HepG2 cells (Hwang et al. 2018). Pelin et al. (2016) reported that among its 6 ovatoxins, (OVTX a-f), the OVTX a showed the weakest activity in reducing the viability of HaCaT cells.

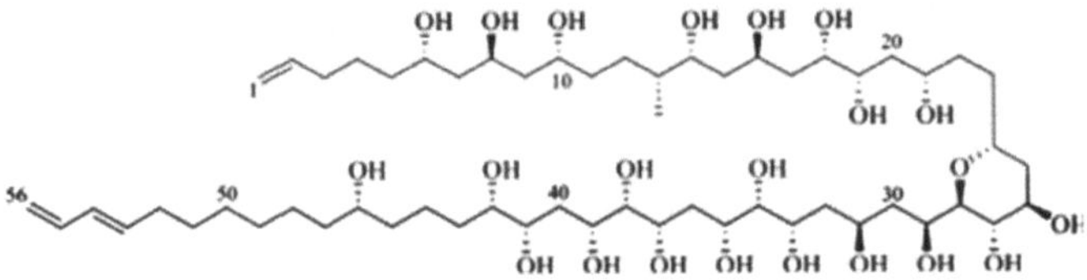

Ostreol B

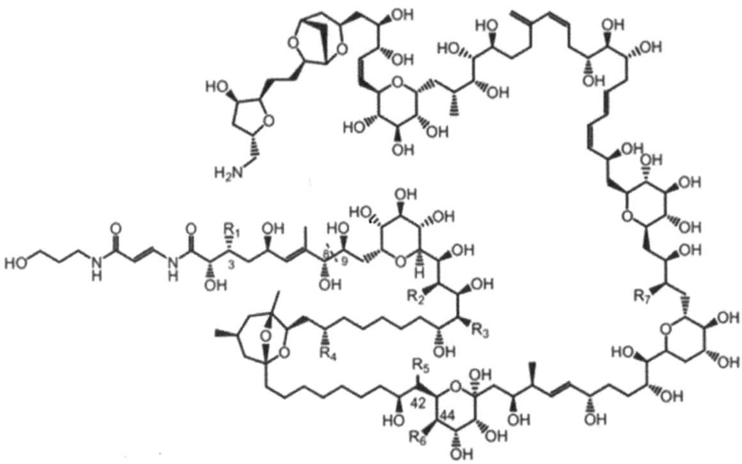

Ovatoxin A: R1= Me, R2= H, R3= OH, R4= Me, R5= OH, R6= H, R7 = H

Ostreopsis siamensis

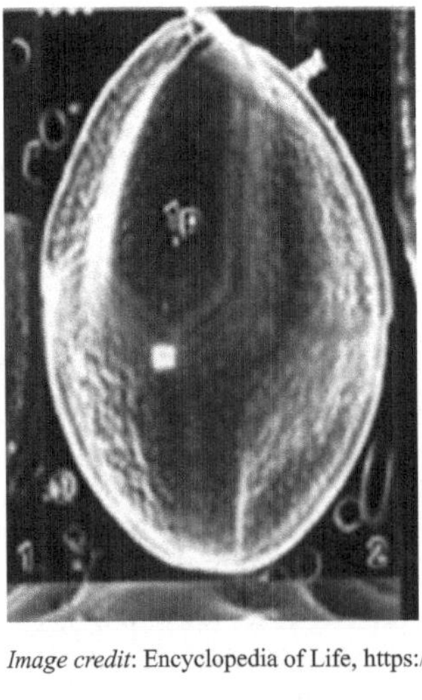

Image credit: Encyclopedia of Life, https://eol.org/media/9502345, Public domain.

Global distribution: It is commonly distributed in the tropical regions world-wide

Ecology: This benthic and epiphytic or tycoplanktonic species is often associated with sand and seaweeds

Medicinal compounds and their bioactivities:

Pavaux et al. (2020) reported on the occurrence of ostreocin-A,B,D and E1 in this species.

The therapeutic potential of its palytoxin and ostreocin D (https://www.sciencedirect.com/topics/pharmacology-toxicology-and-pharmaceutical-science/palytoxin) are given elsewhere.

Ostreopsis sp.

Medicinal compounds and their bioactivities:

PLTX of Ostreopsis spp. has been reported to be of great use in the prevention and treatment of cancers and other diseases such as parasitic diseases relating to *Plasmodium falciparum,* and *Leishmanii;* Ebola virus disease (EVD), human immunodeficiency virus (HIV) disease, skin diseases (Propionibacterium acnes, psoriasis, and nosocomial diseases (hospital-acquired infections) (https://patents.justia.com/patent/20200085783)

Order: Gymnodiniales Family: Gymnodiniaceae:

Amphidinium carterae

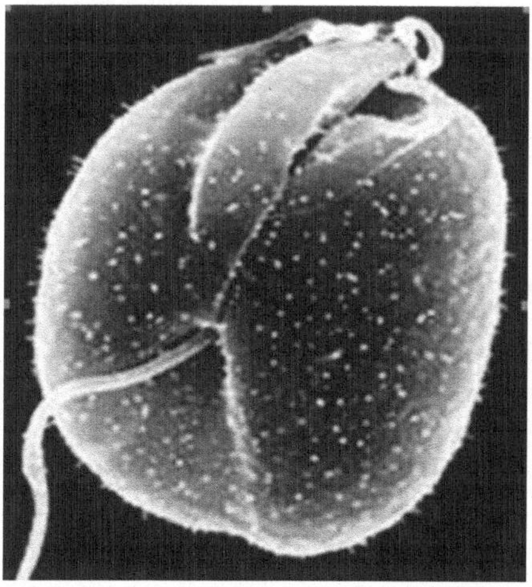

Image courtesy: Lydia Ferrara (Reproduced with his kind permission)

Global distribution: It occurs throughout North Atlantic Ocean and in Brazil and New Zealand

Ecology: It is largely found in estuaries and coastal bays where it is inhabited by the brown seaweed species *Sargassum wightii.*

Medicinal compounds and their bioactivities:

Antibacterial activity: The extracts of this species containing amphidinol type compounds AM-A, AM-B and AM-22 and dehydroAM-A exhibited antibacterial activity against *Staphylococcus aureus* and *Escherichia faecalis* with MIC values ranging between 16 and 256 μg/mL (Barone et al. 2021). Nuzzo et al. (2014) reported that its amphidinol 22 exhibited antibacterial activity against *Staphylococcus aureus MRSA* and *Staphylococcu aureus* MSSA.

Antifungal activity: Among its amphidinol analogs viz. amphidinol 9-13, amphidinol 2,4 showed antifungal activity (Echigoya et al. 2005). Cutignano et al. (2017) reported that among its amphidinol A, B, Amphidinol A displayed significant antifungal activity against *Candida albicans* with an MIC value of 19 μg/mL. Nuzzo et al. (2014) reported that the extracts of this species containing amphidinol 22 inhibited the growth of *Candida albicans* and *Aspergillus fumigatus* with an MIC value of 64 μg/mL; amphidinol 18 (AM18) showed activity against *Candida albicans* with an MIC value of 9 μg/mL; and the compounds AM2, AM4 and AM9 exhibited potent activity against *Aspergillus niger* with MIC values of 44.3, 58.2, and 32.9 μg/mL respectively. Nweze et al. (2020) reported that the methanol extract of this species containing the compound amphidinol 22 showed both antifungal and cytotoxic properties. This compound weakly inhibited the pathogenic fungus *Candida albicans* ATCC 64124 with an MIC value of 19 μg/mL) (Nweze et al. 2020)

Anticancer activity: Its amphidinol 22 has been reported to display anticancer activity against human skin melanoma A2058, hepatocyte carcinoma HepG2, breast adenocarcinoma MCF7, pancreas carcinoma MiaPaca-2 and human lung carcinoma A549 cell lines with IC50 values of 16.4, 6.8, 16.8, 8.6, and 8 μM, respectively (Martínez et al. 2019). Egler et al. (2016) reported on the anticancer activity of its L-asparaginase against leukemia.

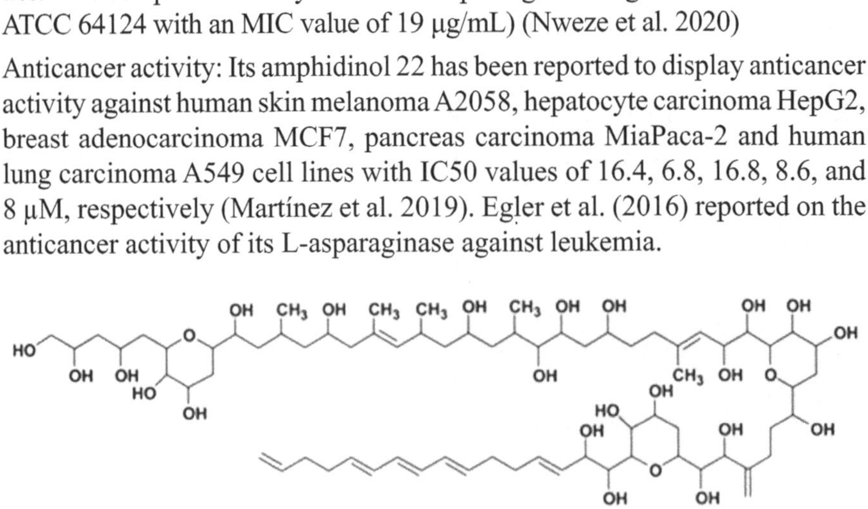

Amphidinol 2

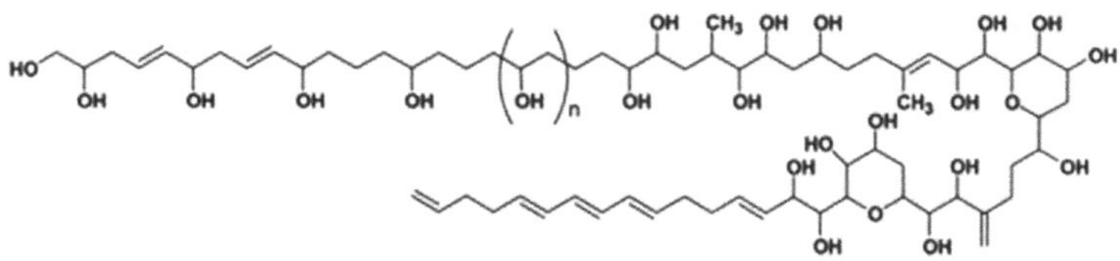

N=0

Amphidinol 4

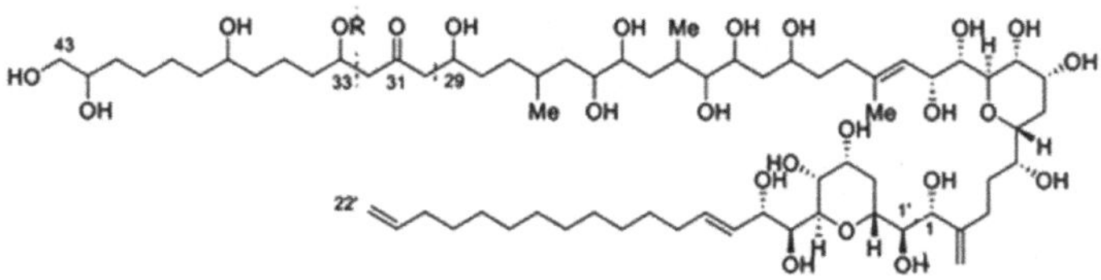

Amphidiniol A: R= H; Amphidiniol B: R= SO3H

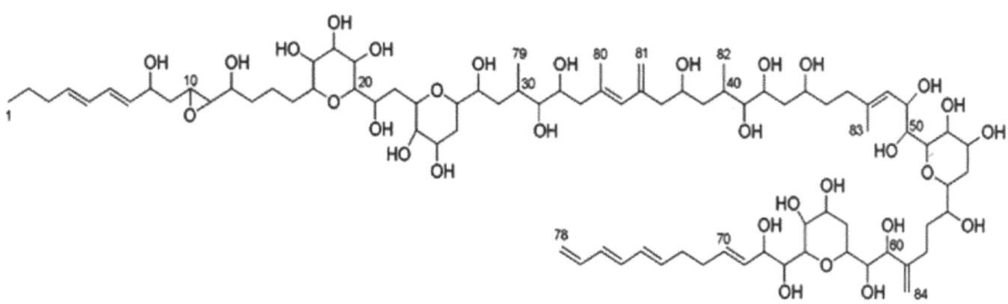

Amphidinol 22

Amphidinium gibbosum

Global distribution: South China Sea

Ecology: This marine, benthic organism is commonly found in the intertidal sandy ecosystems.

Medicinal compounds and their bioactivities: The polyol-polyene compounds of this species viz. gibbosol A-C have been reported to possess anti-osteoclastic, anti- tumor, antifungal, and analgesic properties. However, Li (2020) reported that gibbosol B displayed significant inhibitory effects on the expression of vascular cell adhesion molecule 1 (VCAM-1) which is known to function as a cell adhesion molecule involved in inflammation and associated immunological disorders.

Amphidinium klebsii

Global distribution: Atlantic and Arctic oceans and Mediterranean Sea

Ecology: It is found in neritic waters and salt water marshes

Medicinal compounds and their bioactivities:

Antifungal activity: its polyene-polyhydroxy compound, amphidinol 2 has shown significant antifungal activity (Paul et al., 1995). Pradhan and Ki (2022) reported on the antifungal activity of its compound amphidinol 1. Paul et al. (1997) stated the antifungal activity of its Amphidinol 1-6 on *Aspergillus niger.* The values of minimal effective concentration (MEC) values for these compounds were found varying between 4 and 6 μg/disk.

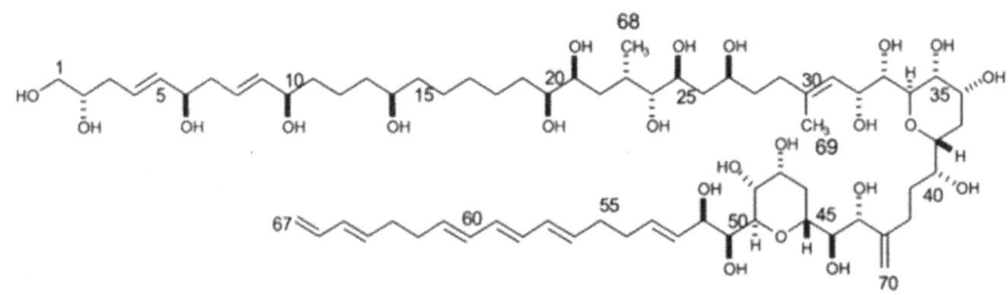

Amphidinol 3

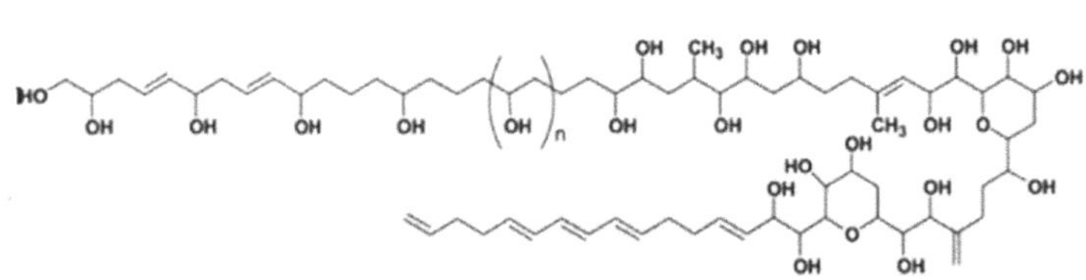

Amphidinol 6 (n=1)

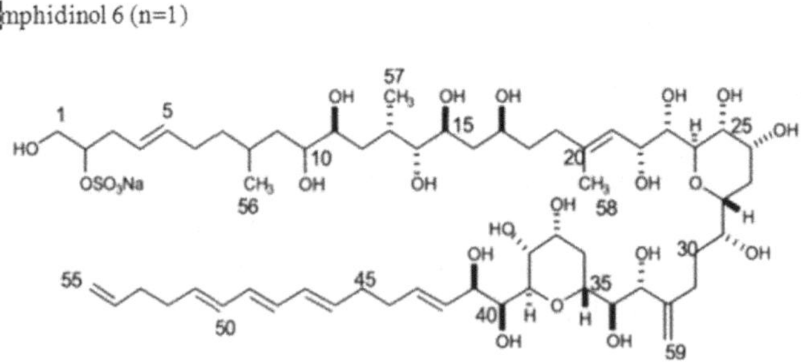

Amphidinol 7

Amphidinium operculatum AA60

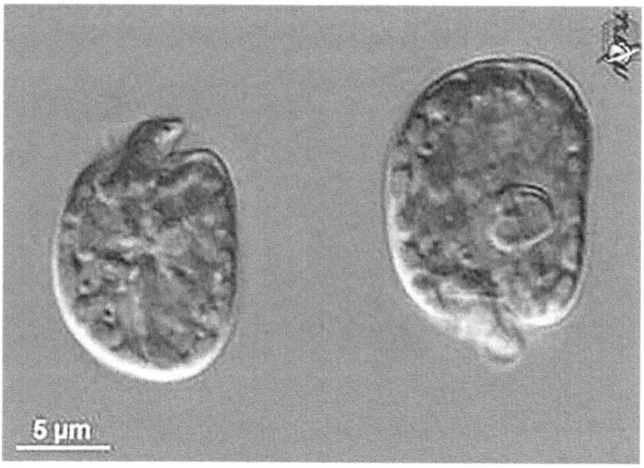

Image credit: Mrponrut. https://creativecommons.org/licenses/by-sa/3.0/

Global distribution: It has been recorded from Japan, Australia, Norway, and USA,

Ecology: It is a neritic species.

Medicinal compounds and their bioactivities:

Anticancer activity/ cytotoxicity: The ethanolic extract of this species displayed the highest inhibitory activity against MCF-7 (human mammary adenocarcinoma) and SKLU-1 (human lung adenocarcinoma) with 60% cell inhibition; 37.1% inhibition with U251 (human glioblastoma); 20% with HCT-15 (human colorectal adenocarcinoma); 14.5% in K562 (human erythroleukemia); and 13.2% with PC-3 (human prostatic adenocarcinoma). On the other hand, the methanolic crude extract of this species displayed cytotoxicity against HL-60 human leukemia cell line with an inhibition between 40 and 60% at a concentration of 50 $\mu g\ mL^{-1}$ and 30% inhibition at 25 $\mu g\ mL^{-1}$. Further, its methanolic extract inhibited (40%) the growth of A549 human lung adenocarcinoma cell at a dosage of 50 µg mL−1 (Mejía-Camacho et al. 2021).

***Amphidinium* sp.** S1-36-5

Medicinal compounds and their bioactivities:

Cytotoxicity: The compound caribenolide-I isolated from the strain S1-36-5 of this unidentified species displayed promising cytotoxic effect against murine tumor P388 and human colon carcinoma cell line (Pradhan and Ki 2022). Baur et al. (1995) reported that this compound had significant cytotoxic effect against HCT 116 human colon tumor cell line and its drug-resistant variant HCT 116/VM 46.

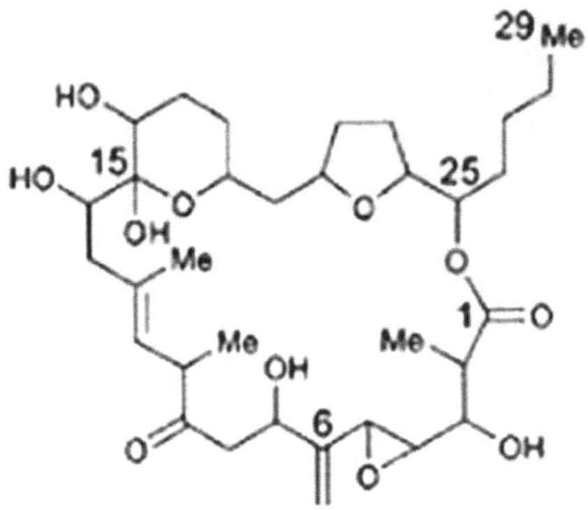

Caribenolide-I

***Amphidinium* sp.** 12012-7-4A

Medicinal compounds and their bioactivities:

Cytotoxicity: The crude cell-free extracts of this unidentified potent growth-inhibitory activity against breast and lung cancer cell lines (MCF-7 and SKLU-1) (Mejía-Camachoet al. 2021)

Antimicrobial activity: The amphidinin C–F and amphidinolide Q extracted from the 2012-7-4A strain of this species sp. (displayed antimicrobial activity against *Staphylococcus aureus*, *Bacillus subtilis*, *Escherichia coli*, *Aspergillus niger*, *Candida albicans* and *Trichophyton mentagrophytes* (Kubota et al. 2014).

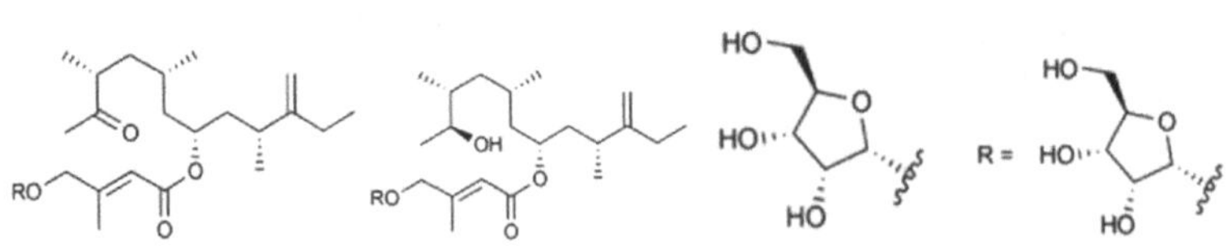

1 2 3 4

1: AmphidiinC; R=H; 2: AmphidiinE: R= H; 3: Amphidiin D: R= HO; 4: AmphidiinF: R= HO

***Amphidinium* sp.1**

Medicinal compounds and their bioactivities:

Anticancer and antifungal activities: The polyketide derivatives karatungiols or amphirionin-2 isolated from this unidentified species have shown bioactivity against human cancer cells and pathogenic fungi and (Washida et al. 2006; Kumagai et al. 2015)

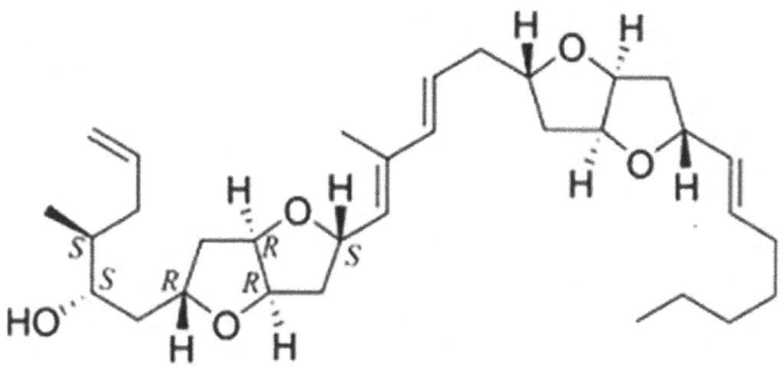

Amphirionin-2

***Amphidinium* sp. 2**

Medicinal compounds and their bioactivities:

Antifungal and cell proliferating activities: The compounds amphidinol 3,4 and 5 derived from this unidentified species have shown significant antifungal activity. Further, its amphidinol 5 has been reported to promote the proliferation of murine stromal ST-2 cells and osteoblastic MC3T3-E1 cells in the bone marrow. Furthermore, its amphidinol 4 helped in the highest proliferation of murine bone marrow stromal ST-2 cells and improved the immune system's ability by inducing TNF-α (Pradhan and Ki 2022).

***Amphidinium* sp. 3**

Medicinal compounds and their bioactivities:

Cytotoxicity and antitumor activity: The 40 amphidinolides (AMPs) produced by this unidentified species have been reported to display in vitro cytotoxicity against human epidermoid carcinoma KB cells and murine lymphoma L1210 cells. Further, the amphidinolide- N (AMP-N) of this species displayed highest antitumor activity (Pradhan and Ki 2022).

***Amphidinium* sp.** 4

Medicinal compounds and their bioactivities:

This unidentified species of *Amphidinium* has been reported to yield gonyautoxins (GTXs), analogues of STXs. These compounds are believed to be useful as a pain reliever and as a safe and potent therapeutic agent for chronic anal fissures (Pradhan and Ki 2022)

***Amphidinium* sp.** 5

Medicinal compounds and their bioactivities:

Cytotoxicity: The amphidinol-related compound, iriomoteolide- 13, derived from this unidentified species displayed cytotoxicity against HeLa human cervical cancer cells with an IC50 value of 0.5 µg/mL (Akakabe et al. 2014; Pradhan and Ki 2022).

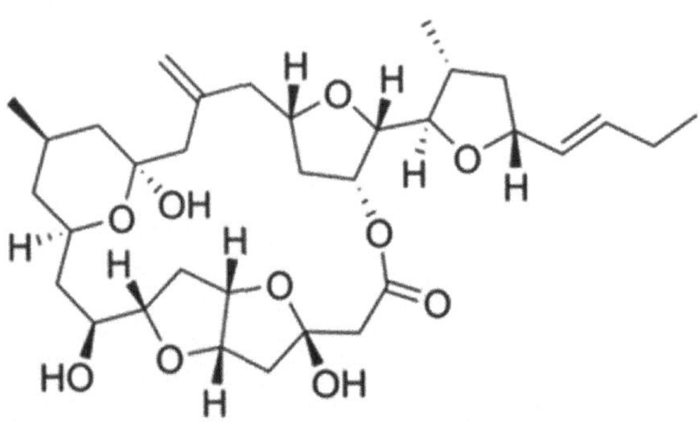

Iriomoteolide- 13 a

***Amphidinium* sp.**6

Medicinal compounds and their bioactivities:

Antifungal and antiprotozoal activities: The polyol compounds, karatungiol A and B of this unidentified species have shown antiprotozoal activity against *Trichomonas foetus* at 1 µg/mL and antifungal activity against NBRC4407 *Aspergillus niger* at 12 µg/disc (Washida et al. 2006)

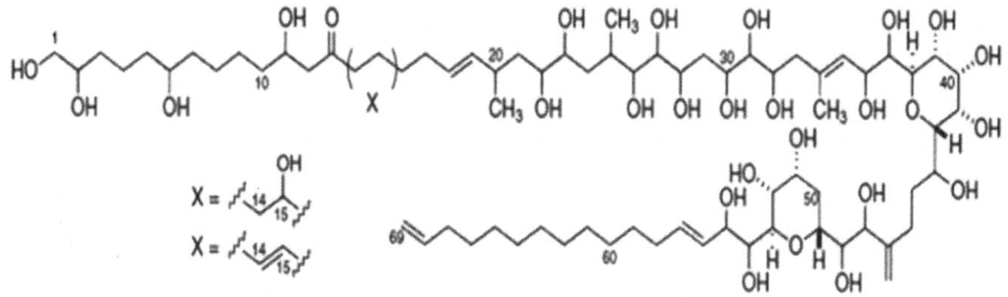

Karatungiols A,B

***Amphidinium* sp.** 7

Medicinal compounds and their bioactivities:

Cytotoxicity: The linear polyketide, amphidinin A which is related to amphidinolides has shown cytotoxicity (Iwai and Kubota 2014)

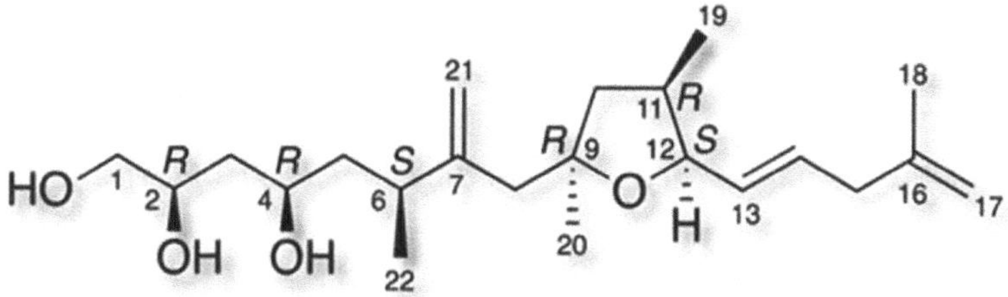

Amphidinin A

***Amphidinium* sp.** 8

Medicinal compounds and their bioactivities:

This unidentified species of *Amphidinium* has been reported to produce gonyautoxins (GTXs) which are important analogues of STXs (Pradhan and Ki 2022). The therapeutic activities of the latter have already been dealt with.

Order: Gymnodiniales Family: Gymnodiniaceae

Gymnodinium catenatum

Global distribution: It has wide distribution form tropical to cool-temperate waters.

Ecology: it lives largely in sheltered embayments and coastal estuaries

Medicinal compounds and their bioactivities:

The cells of this species possess saxitoxin (STX) the therapeutical applications of which have already been dealt with.

Gymnodinium impudicum *(= Gyrodinium impudicum)*

Global distribution: it has worldwide distribution and is commonly distributed in Atlantic, Pacific and Indian Oceans

Ecology: It is found mainly in brackish water

Medicinal compounds and their bioactivities:

Immunostimulation activity: The sulfated exopolysaccharides of the strain KG03 of this species have shown immunostimulation activity and prevented the growth of cancer cells by enhancing the tumor-killing effects of macrophages and natural killer cells (Wu et al. 2021). Riccio and Lauritano (2020) reported that the sulfated exopolysaccharides of this species possessed immunomodulatory properties and activated macrophages at concentrations of 0.1–10 μg/mL

Margalefidinium polykrikoides *(= Cochlodinium polykrikoides)*

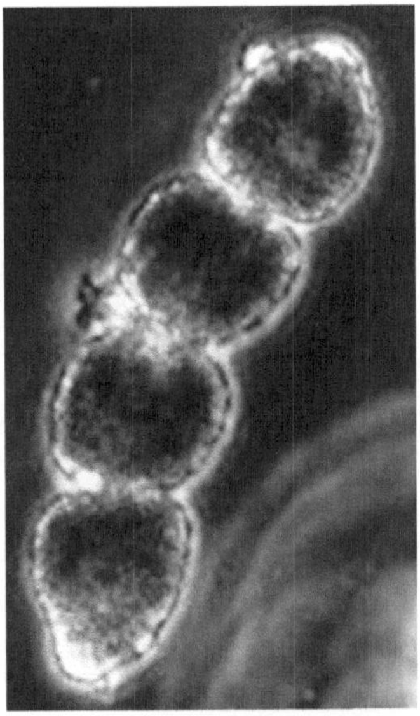

Image credit: Open University, https://creativecommons.org/licenses/by-sa/4.0/

Global distribution: It is widely distributed and its geographic range includes Central America, North America, Southwestern Europe, and Eastern Asia and Western India

Ecology: It is a coastal water species. It is a massive bloom-forming organism and has been reported to cause massive fish kills and damage to coral reefs.

Medicinal compounds and their bioactivities:

Antiviral activity: The sulfated polysaccharides produced by the cells of this species have shown antiviral activity by suppressing the cytopathic impacts of A and B types of influenza virus on canine kidney cells; A and B types of respiratory syncytial virus on HEp-2 cells; and type 1 human immune deficiency virus on MT-4 cells (Wu et al. 2021).

Order: Gymnodiniales Family: Kareniaceae

Karlodinium veneficum.

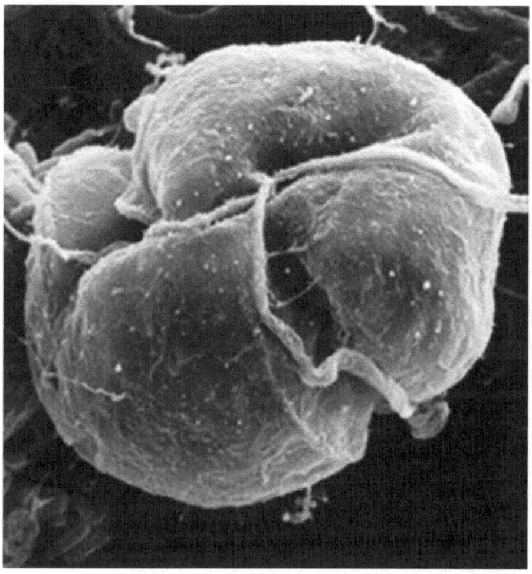

Image credit: National Centers for Coastal Ocean Science, https://creativecommons.org/licenses/by-sa/4.0/

Global distribution: it has world-wide distribution and is very common in Xiangshan Bay of East China Sea

Ecology: It dwells in the surface and bottom waters of coastal ecosystems.

Medicinal compounds and their bioactivities:

Cytotoxicity: Two linear polyketide karlotoxins viz. 4,5-dihydro-KmTx 2 and 4,5-dihydro-dechloro-KmTx 2 derived from this species have shown cytotoxicity against 1 human monocytic leukemia cell line (THP-) with IC50 vlaues of 15 and 36 μM, respectively (Cai et al. 2016)

Others: As the karlotoxin (KmTx) of this species develops pores in cell membranes containing cholesterol, this toxin may be of great use in treating several human diseases including coronary heart disease. Further, KmTx has been reported to induce cell death by the depletion of cholesterol. This toxin may have the potential to be used as a chemotherapeutic drug to control prostate and breast cancer cells (Pradhan and Ki 2022).

Karlodinium sp.

Medicinal compounds and their bioactivities:

Li et al. (2020) reported that the polyol-polyene compounds of *Karlodinium* spp. displayed antitumor, antiosteoclastic, antifungal, and analgesic activities.

Order: Gymnodiniales Family: Kareniaceae

Karenia brevis (= *Ptychodiscus brevis; Gymnodinium breve*)

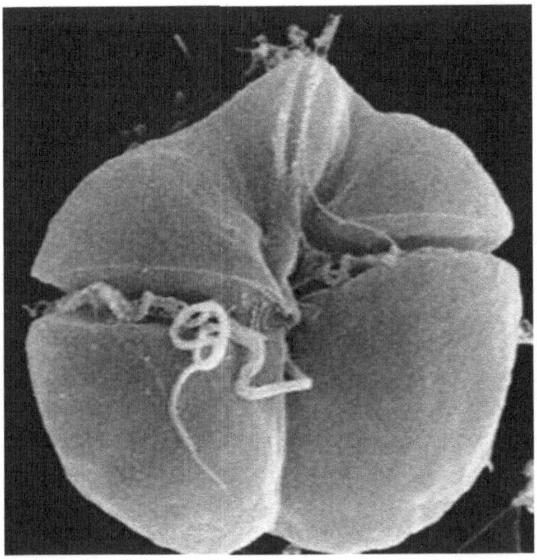

Image credit: Florida Fish and Wildlife Conservation Commission. Public domain.

Global distribution: it is found distributed in the warm temperate to tropical seas with countries such as Greece, Spain, New Zealand, and Japan

Ecology: It is a oceanic species, but blooms of this species have been reported in coastal and estuarine waters

Medicinal compounds and their bioactivities:

Anti-inflammatory activity: The compound, brevenal isolated from this species, has been reported to possess anti-inflammatory activity by reducing the proinflammatory cytokine secretion by 20%–35% from lung epithelial cells. Further, this compound was known to increase mucociliary clearance in animal models associated with chronic respiratory disease (Keeler et al. 2019). The brevenal is therefore believed to be a potential dual-action therapeutic for treating pulmonary diseases, such as asthma, cystic fibrosis (CF) and chronic obstructive pulmonary disease (COPD)

Others: Brevetoxin has also been reported to possess genotoxic properties in leukemic T-cell lines and Jurkat E6-1 cells and cause cell death via apoptosis. Its BTX-2 displayed neuro-activation qualities and improved neuronal plasticity and BTX-2 may therefore be a potential drug for treatments in restoring brain function following traumatic brain damage. Further, this toxin has also been reported to reduce respiratory irritations associated with bronchoconstriction, cough, congestion, and asthma attacks in humans (Pradhan and Ki 2022).

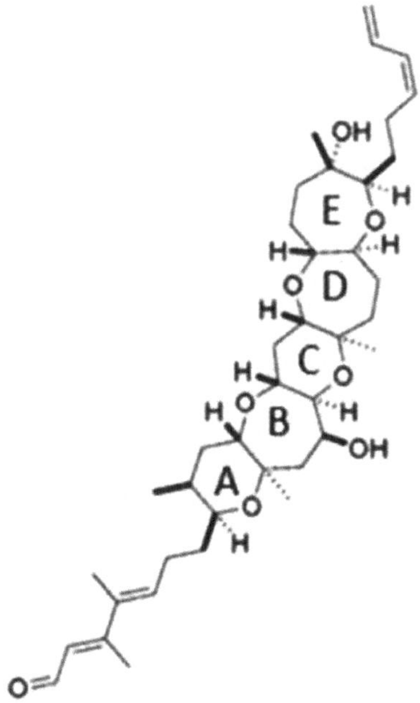

Brevenol

Karenia mikimotoi *(= Gymnodinium mikimotoi)*

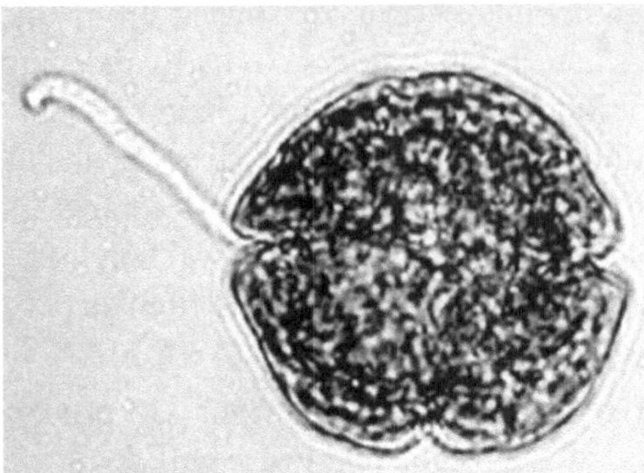

Image credit: Mark Vandersea (Reproduced with his kind permission)

Global distribution: It is a common species in Korea and Japan

Ecology: It is a coastal water species

Medicinal compounds and their bioactivities:

Cytotoxicity: The compound gymnocin-A derived from this species exhibited cytotoxicity on P388 murine leukemia cells. Further, its gymnocin-B displayed similar activity in many cancer cell lines (Pradhan and Ki, 2022).

Others: This species has also yielded a polyether macrolide viz. goniodomin-A, which is known to induce the modulation of actomyosin ATPase activity. However, the therapeutic properties of this compound are yet to be known (Pradhan and Ki 2022)

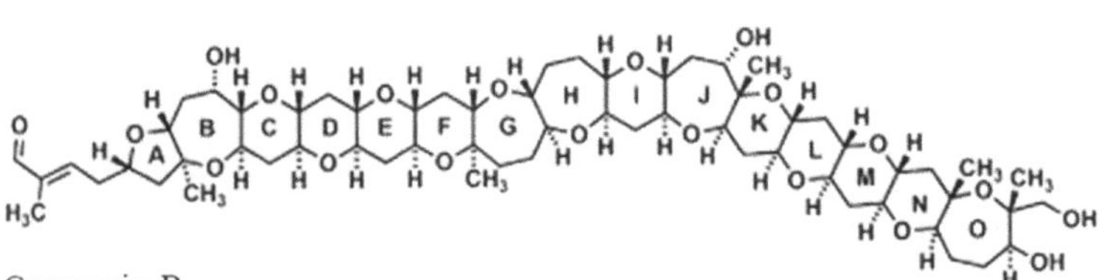

Gymnocin-B

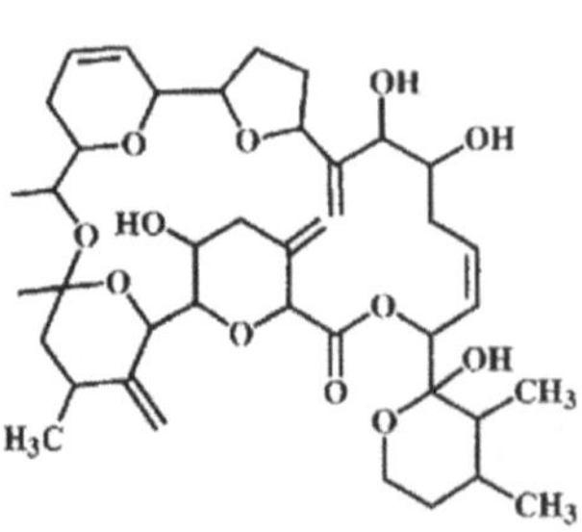

Goniodomin-A

Karenia selliformis *(= Gymnodinium selliforme)*

Global distribution: Pacific Ocean and the Mediterranean Sea.

Ecology: It is a coastal water species

Medicinal compounds and their bioactivities:

Cytotoxicity and neuroprotective activity: The spirolides (SPX) of this species have shown cytotoxic effects. Further, these compounds have also been reported to serve as a neuroprotector in Alzheimer's disease (Pradhan and Ki, 2022)

Others; Its compounds viz. gymnodimine A,B and C have been reported to play an important role in lowering tau phosphorylation and amyloid levels which may help in treating degenerative illnesses (Pradhan and Ki 2022).

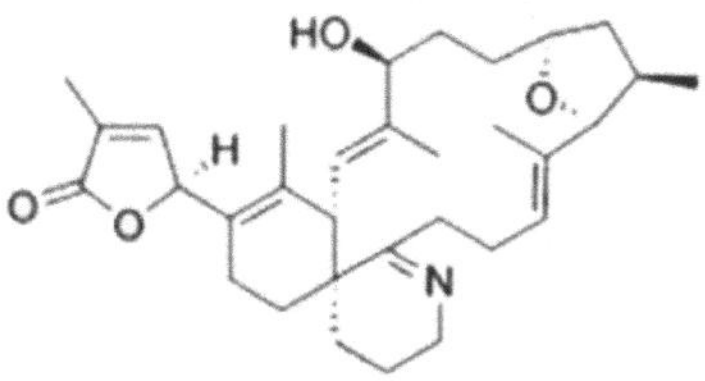

Gymnodimine

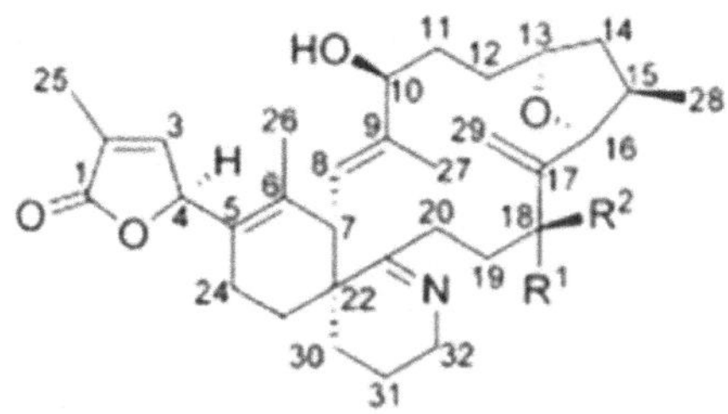

A R^1= H, R^2= OH
R^1= OH, R^2 = H
Gymnodimine B,C

Order: Peridiniales Family: Peridiniales incertae sedis

Vulcanodinium rugosum

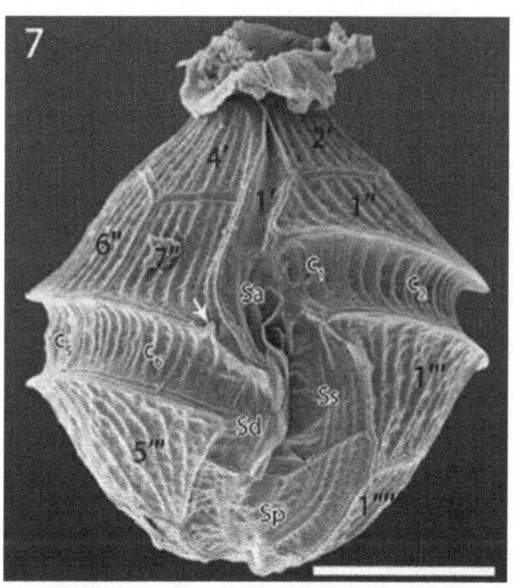

Image credit: Lesley Rhodes (Reproduced with his kind permission)

Global distribution: It has a global distribution and is reported to be common in Mediterranean Sea and warm Pacific regions

Ecology: This benthic species has been reported to be common in warm saline environments and in surface sediments

Medicinal compounds and their bioactivities:

Cytotoxicity and apoptotic activity: The polycyclic ether toxin viz. protamine derived from this species has shown significant toxicity to P388 cells with an LC50 value of 2.7 nM. Further, this toxin compound has also been reported to activate caspases, indicating apoptotic activity (Selwood et al. 2013; Nezan and Chomérat 2011). Geiger et al. (2013) reported that its macrocyclic imine phycotoxins viz pinnatoxins (PnTX) E, F, and G have not shown cytotoxicity. Further, according to them, the crude extracts of this species displayed potent cytotoxicity against KB and Neuro2A cells with IC50 values of 0.19 μg mL^{-1} and 0.38 μg mL^{-1}, respectively.

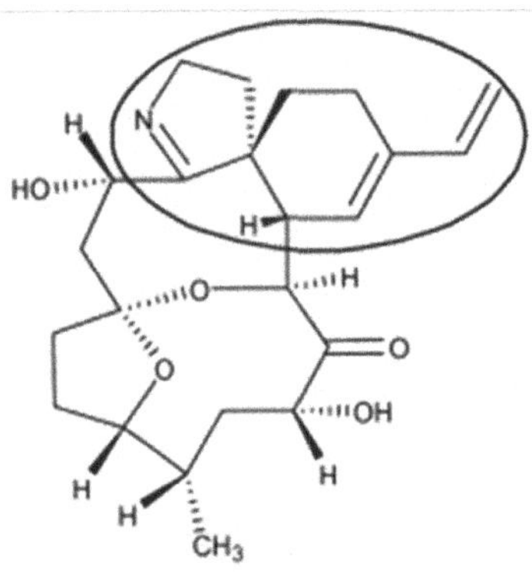

Portimine

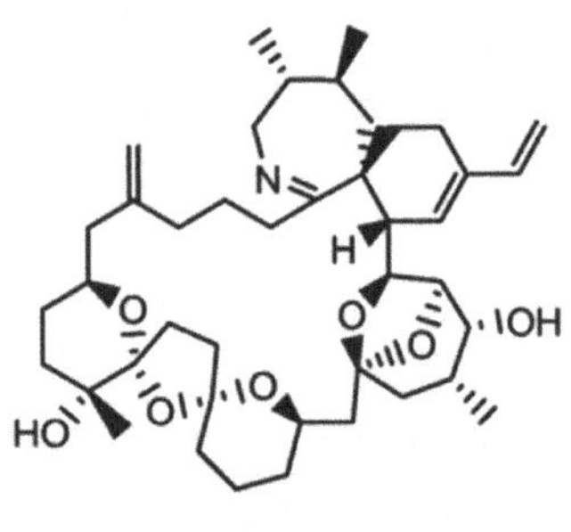

Pinnatoxin G

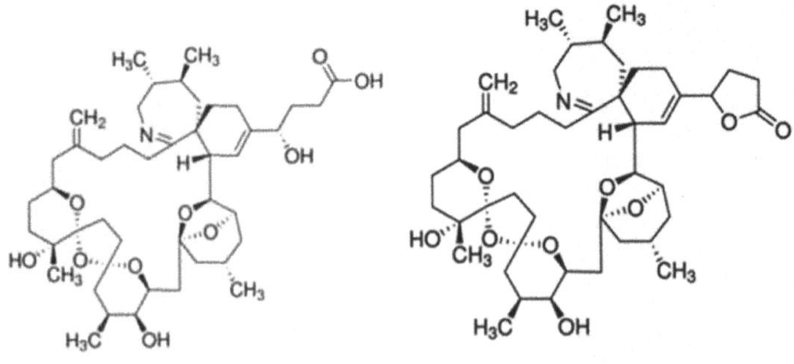

Pinnatoxins E,F

Order: Peridinales Family: Crypthecodiniaceae

Cripthecodinium chonii

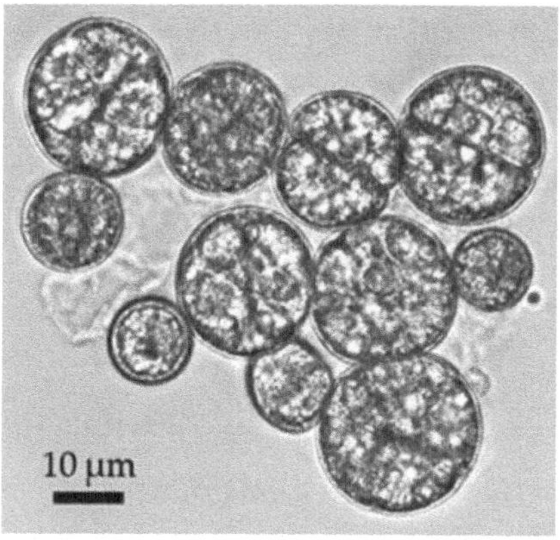

Image credit: Abt et al., 2020. https://creativecommons.org/licenses/by/

Global distribution: This species has been reported mainly from USA

Ecology: This species is often found associated with macrophytes like Fucus spp

Medicinal compounds and their bioactivities:

It is a high-volume producer of DHA amongst all the other marine species of dinoflagellates and this PUFA is well known for its pharmaceutical and nutraceutical applications. Further, DHA has been reported to be an essential compound for the functional development and growth of infant and adult brain. Ferrara (2020) reported that the DHA is a potential therapeutic agent with anti-inflammatory, antioxidant, immunomodulating, anticoagulant, anti-adhesive, anticancer and antiviral and properties. Yu et al. (2017) reported that the n-butanol and water fractions of this species showed scavenging activity.

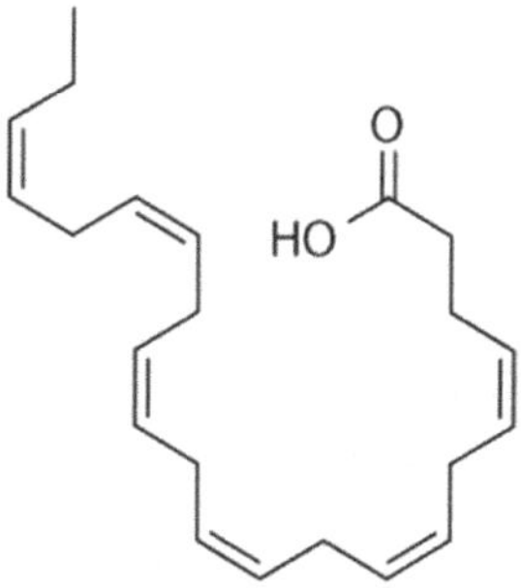

Order: Peridinales Family: Heterocapsaceae

Heterocapsa circularisquama

Global distribution: It is mainly reported in Japan

Ecology: it is a coastal water species

Medicinal compounds and their bioactivities: The compound porphyrin isolated from this species has shown light-based cytotoxicity against cancer cells (Kim et al. 2008).

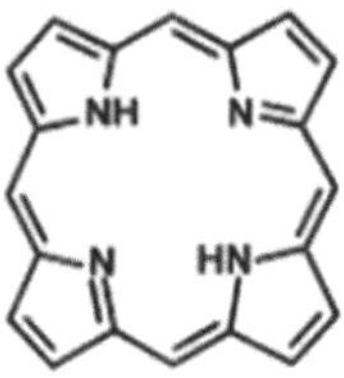

Porphyrin

Heterocapsa triquetra

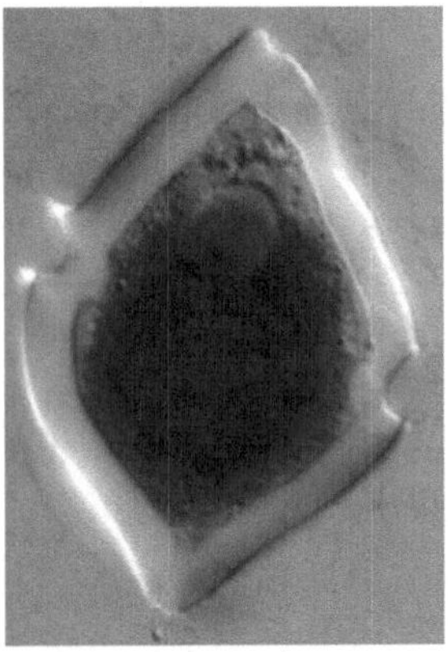

Image credit: Encyclopedia of Life, https://creativecommons.org/licenses/by-sa/4.0/

Global distribution: It has world- wide distribution and is very common in the Atlantic Ocean

Ecology: It is a shallow water, neritic species

Antiproliferative activity: This species has been reported to possess several pigments such as P457 (carotenoid lactoside), peridininol, peridinin, pyrrhoxanthin, diadinochrome, dinoxanthin diadinoxanthin, and chlorophyll c2 which inhibited the growth and proliferation of human melanoma cells (A2058) on this cell line. Further, the high pigment (Ht) ethanol extract of this species has also shown similar activity at 100 μg·mL^{-1} after a 3-day treatment (Quentin et al. 2017).

Apoptotic effect: The carotenoid pigment, peridinin, of this species exhibited a significant reduction of the viability of the DLD-1 human colorectal cancer cells at 0–20 microM and subsequently induced apoptosis by activating caspase-9 and caspase-8 (Sugawara et al. 2007).

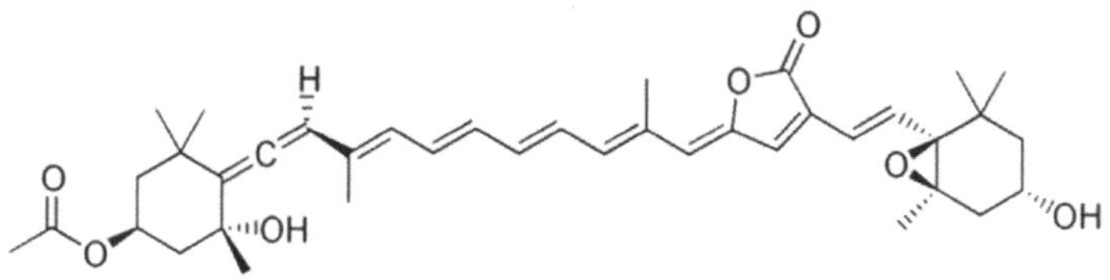

Peridinin

Order: Peridiniales Family: Peridiniaceae

***Durinskia* sp.**

Global distribution: It has been reported from North America, South America, Europe, Pacific Islands and Oceania

Ecology: This marine, symbiotic dinoflagellate is often found associated with the molluscan nudibranch, Chelidonura fulvipunctata A pennate diatom is found in side the cell of this species as endosymbiont.

Medicinal compounds and their bioactivities:

The long carbon-chain polyol compounds, durinskiol A, B, have been derived from this unidentified species. The bioactivities of these compounds are however yet to be known (Kita et al. 2010)

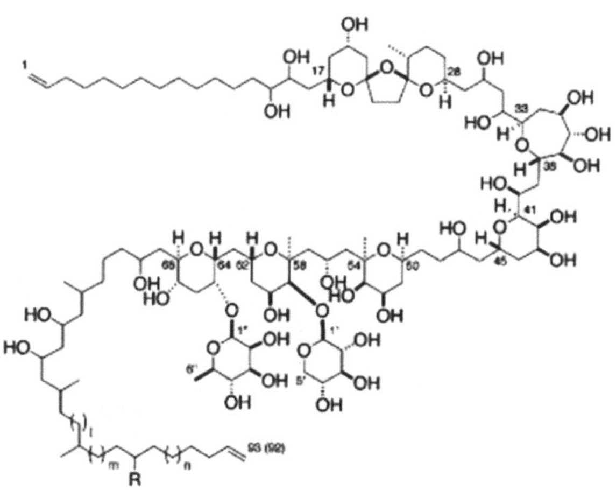

l + m+ n = 4, R= H
l+ m + n = 3, R= Me
Durinskiols A,B

Order: Prorocentrales Family: Prorocentraceae

Prorocentrum arenarium

Global distribution: SW Indian Ocean and tropical e Caribbean Sea

Ecology: It is a benthic and epiphytic species by attaching to coral rubble or sand.

Medicinal compounds and their bioactivities:

Antiproliferative activity: The extracts of this species have shown antiproliferative activity either at concentrations of 5 or 50 μg/mL (de Vera et al. 2018)

Prorocentrum hoffmannianum *(= Prorocentrum belezeanum; Prorocentrum maculosum)*

Global distribution: It is mostly found in the tropical coastal waters of the Caribbean Sea

Ecology: It is a benthic species, and its cells are either motile or attached to floating detritus by mucilage

Medicinal compounds and their bioactivities:

Cytotoxicity and antiproliferative activity: The extracts of this species exhibited potent cytotoxicity and antiproliferative activity against many mammalian cell lines with the inhibition values of 87–92 % (de Vera et al., 2018)

Apoptotic Activity: Extracts of this species have been reported to possess the highest apoptotic potential on the human hepatocellular carcinoma cell line HepG2 (ATCC® HB-8065™) at a concentration of 100 μg/mL (de vera et al. 2018)

Antimicrobial activity: The extracts of this species displayed antimicrobial activity against *Escherichia coli* and *Candida albicans* with inhibition values of 100% and 98%, respectively.

Antiviral activity: The cellular extracts of this species displayed prominent inhibition activity against Chikungunya virus with an inhibition value of about 90% at 100 μg/mL concentration (de Vera et al. 2018).

Others: The Okadaic acid (OA) and its derivatives viz. dinophysistoxins (DTX)-1, 2, and 3 obtained from this species have shown several bioactivities such as cytotoxic effects on the immune system, neurological system and on embryonic development; it is a promising neurotoxin for analysing therapeutic effects of antipsychotic medications in the treatment of schizophrenia and other neurodegenerative diseases; OA has served as a potent tool in the study of Alzheimer's disease and as a model biotoxin for studying AIDS, diabetes and cancer; OA showed a potential immunomodulatory effect; and OA had antifungal activity against *Candida albicans* (Pradhan and Ki 2022).

Prorocentrum concavum

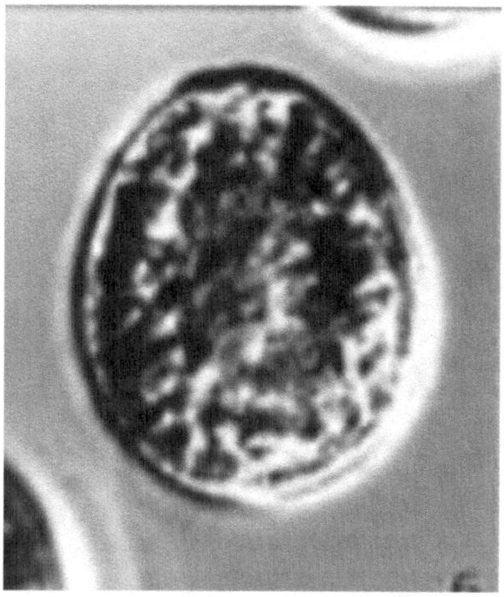

Image credit: Encyclopedia of Life. Public domain.

Global distribution: It is found distributed in all tropical seas of the world

Ecology: It is a benthic and tycoplanktonic. Species occurring in neritic waters. Cells of this species are either motile or often associated with sediments or floating mangrove detritus

Medicinal compounds and their bioactivities:

Like other species, this species also produces Okadaic acid (OA) and its derivatives, like dinophysistoxins (DTX)-1, 2, and 3, the bioacitvities of which have already been dealt with (Pradhan and Ki 2022).

Prorocentrum lima

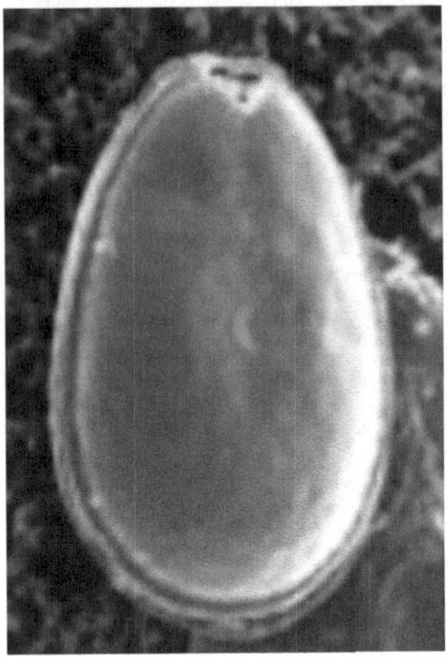

Image credit: Encyclopedia of Life. Public domain..

Global distribution: It has world-wide distribution. *Prorocentrum lima* is a neritic, estuarine species with world-wide distribution

Ecology: This benthic species is seen in neritic, estuarine waters. This is often associated with floating detritus in mangrove habitats and coral reefs and is epiphytic on the surface of red and brown algae

Medicinal compounds and their bioactivities:

Cytotoxicity: It has been reported to produce Okadaic acid (OA) derivattive, dinophysistoxin 1 derivative and limaol. Among these compounds, limaol displayed moderate cytotoxicity against 3 cancer cell lines (Yang et al. 2017)

Antifungal activity: Okadaic acid and its two congeners and desulfated yessotoxin produced by this species has promising antifungal activities against *Candida rugosa, Candida albicans, Aspergillus niger* and *Penicillium funiculosum* (Nagai et al. 1990; Pradhan and Ki 2022).

Others: This species has also been reported to produce Okadaic acid (OA) and its derivatives, like dinophysistoxins (DTX)-1, 2, and 3, the bioactivities of which have already been dealt with (Pradhan and Ki 2022).

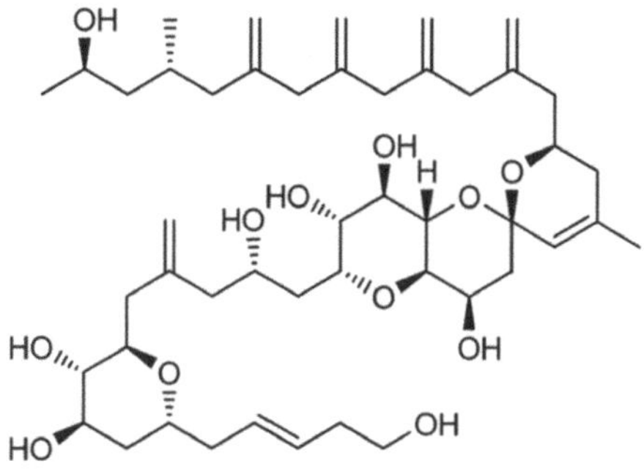

Limaol

Prorocentrum reticulatum

Global distribution: It is often recorded from tropical coastal waters of the Caribbean Sea

Ecology: This benthic species is often found attached to coral rubble

Medicinal compounds and their bioactivities:

Apoptotic Activity: Extracts of this species have been reported to possess the highest apoptotic potential on the human hepatocellular carcinoma cell line HepG2 (ATCC® HB-8065™) at a concentration of 100 μg/mL (de vera et al. 2018)

Others: As other species of *Prorocentrum,* this species has also been reported to produce DSP toxins and OA the bioactivities of which have already been dealt with (Camacho-Muño et al. 2021)

Order: Suessiales Family: Symbiodiniaceae

Symbiodinium sp.

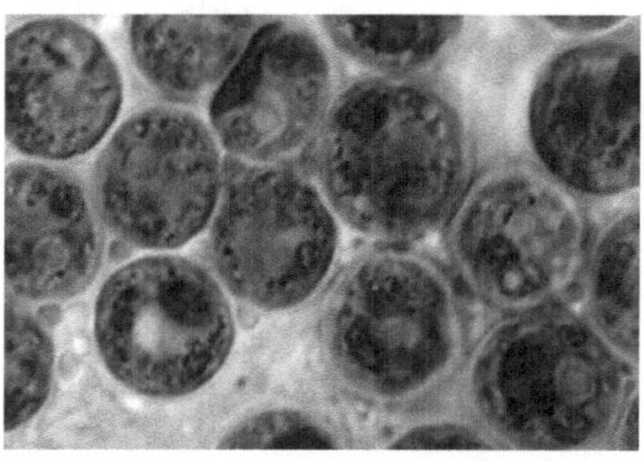

Image courtesy: Allisonmlewis, https://creativecommons.org/licenses/by-sa/4.0/

Global distribution: The species of this genus are distributed in all tropical seas

Ecology: The species of this genus, members of zooxanthellae, are often dominating in the benthic communities. These are endosymbionts and are found associated with several groups of marine organisms such as cnidarians such as jellyfish, corals and sea anemones; demosponges, flatworms and mollusks like giant clams; and ciliates and foraminiferans (soritids)

Medicinal compounds and their bioactivities:

Vasoconstrictive activity: The cpompounds, Zooxanthellatoxins (ZTs) A, B, and C, of this unidentified species showed vasoconstrictive activity (Camacho et al. 2019).

Anti-inflammatory activity: Among its amphoteric iminium compounds viz. symbioimine and neosymbioimine, symbioimine has been reported to be a potential compound in developing a new anti-inflammatory drug towards the treatment of cyclooxygenase2-associated diseases. Further, the compound, symbiodinolide of this species showed anti-inflammatory activity by inhibiting cyclooxygenase-1,2. (COX-1,2) (Sánchez-Suárez et al. 2021). The compound, symbiospirol A, of this species behaved as an antagonist-like compound to suppress inflammation-related diseases. Furthermore, this compound could also serve as a promising antiresorptive drug for the treatment of osteoporosis especially in postmenopausal women. The phenolic diterpenes compounds viz. pseudopterosin A–D of this species displayed promising anti-inflammatory, antioxidant, analgesic and neuroprotective activities. The compound fuscol also showed shown anti-inflammatory properties (Sánchez-Suárez et al. 2021).

Antileukemic activity: Pradhan and Ki (2022) reported that its ceramide compound viz. symbioramide showed antileukemic activity against L-1210 murine leukemia cells.

Antineoplastic activity: The compounds, zooxanthellamide D and zooxanthellactone, have been reported to show antineoplastic potential against A432 and Nakata cell lines with inhibitory activity through voltage-sensitive $Ca2+$ channels at concentrations of more than 10 μM (Sánchez-Suárez et al. 2021)

Others: The compounds zooxanthellatoxin A,B of this species have been reported to display vasoconstriction at concentrations of more than 0.70 μM. Further, zooxanthellatoxin A also induced platelet (blood cell) aggregation (Sánchez-Suárez et al. 2021)

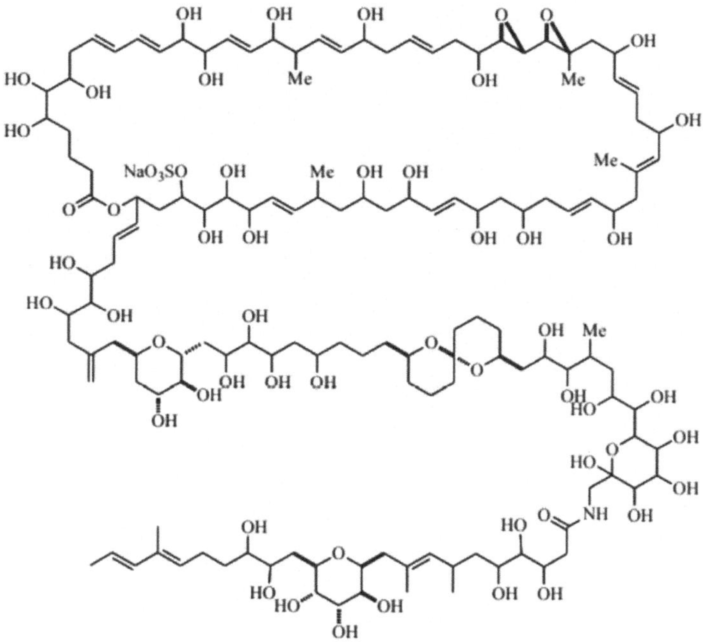

Zooxanthellatoxin A

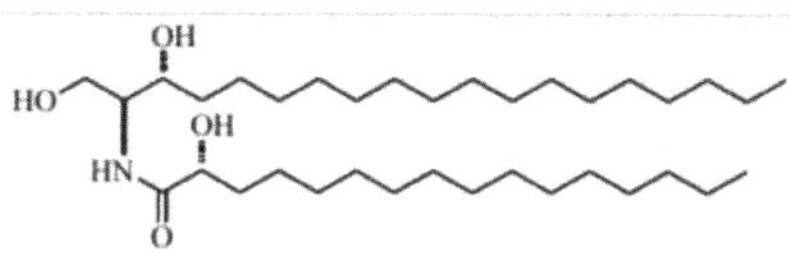

Symbioramide-C16

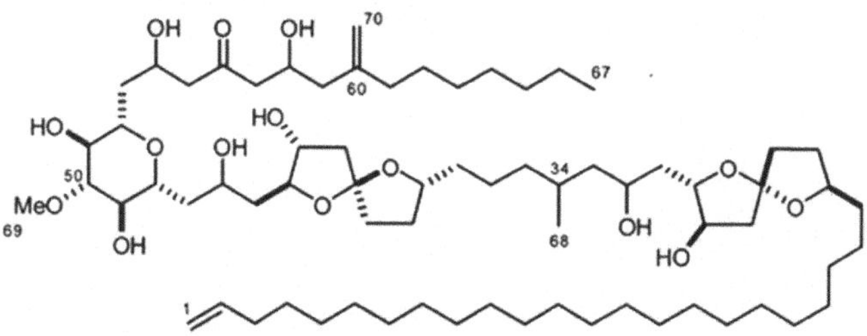

Symbiospirol A

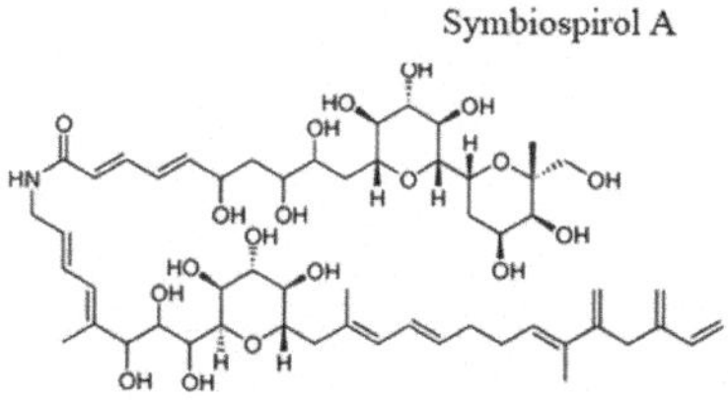

Zooxanthellamide D

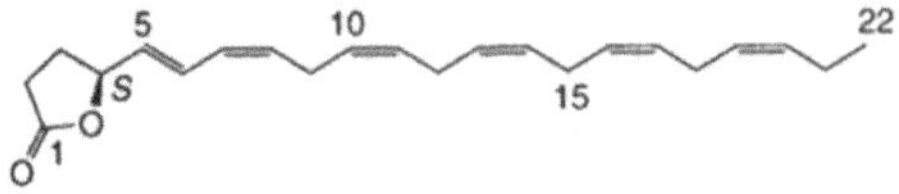

Zooxanthellactone

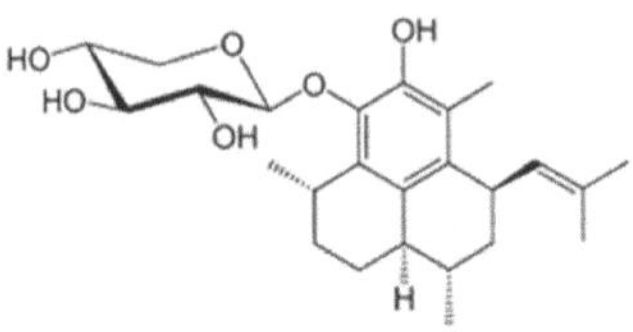

Pseudopterosin A

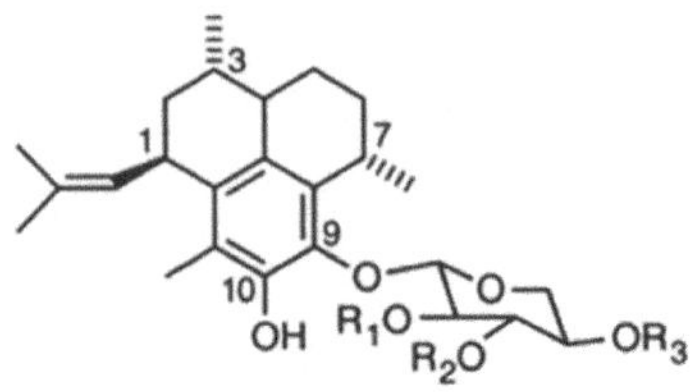

R3 = Ac, R1 = R2 = H

Pseudopterosin D

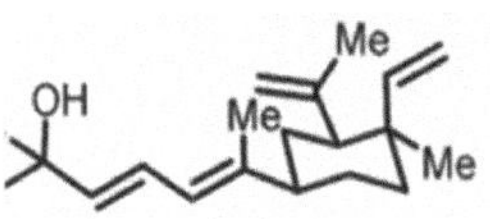

Fuscol

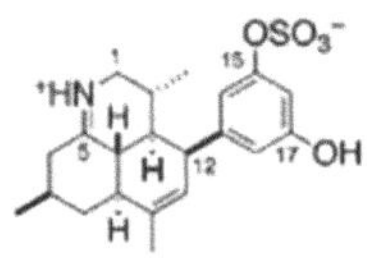

Neosymbiomine

RED ALGAE

Phylum: Rhodophyta Class: Porphyridiophyceae
Order: Porphyridiales Family: Porphyridiacea

Porphyridium cruentum *(= Porphyridium purpureum)*

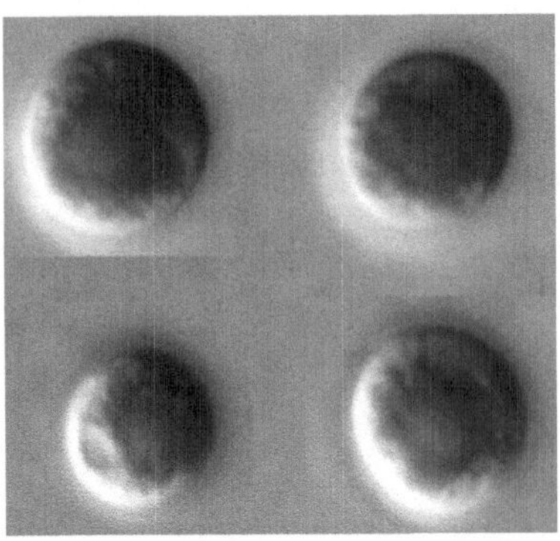

Image credit: Plankton Net https://creativecommons.org/licenses/by-sa/4.0/

Global distribution: It is commonly found in the coastal waters of Northwest Atlantic

Ecology: It resides in the upper intertidal zone and in soils of sea cliffs

Medicinal compounds and their bioactivities: This species has been reported to possess a broad range of pharmaceutical and industrial and applications

Anticancer activity/ cytotoxicity: The sulfated polysaccharides of this species have shown cytotoxicity on several cancer cell lines such as breast cancer (MCF-7), lung cancer (NCI-H460), colorectal carcinoma (HTC-116), human leukemia (U-937 and HL-60), and human gingival fibroblasts (HGF-1) cell lines with IC50 value of 1089.6 μg mL^{-1}, 5498.1 μg mL^{-1}, 2311.2 μg mL^{-1}, 1676.7 μg mL^{-1}, and 2861.5 μg mL^{-1}, respectively (Casas-Arrojo et al. 2021). Wu et al. (2021) reported that the sulfated polysaccharides from the cell wall of this species has shown anticancer activity against Graffi myeloid tumor in hamsters. Based on its unique activity, it is suggested that these compounds may act as a promising candidate antitumor drug

Antiviral activity: The polysaccharides of this species have been reported to exhibit antiviral activity on the Ectromelia virus and Vaccinia virus (Wu et al. 2021).

Antimicrobial activity: The extracts of this species containing polysaccharides showed antimicrobial activity against pathogenic bacteria and fungi (Wu et al. 2021).

Immunomodulatory effects: The sulfated polysaccharides of this species have been reported to show immunomodulatory effects as they are the promising inducers of IL-6 cytokines and TNF-α (Casas-Arrojo et al. 2021)

Class: Rhodellophyceae Order: Rhodellales Family: Rhodellaceae

Rhodella reticulata

Global distribution: It has been reported mostly from the Baltic Sea

Ecology: This marine and brackish water species is commonly seen in the intertidal sand and coastal waters

Medicinal compounds and their bioactivities:

This species has been reported to possess the ω-3 fatty acid eicosapentaenoic acid (EPA) as a major constituent apart from sulfated polysaccharides. The EPA has served as a dietary fiber in experimental rats. When these rats were fed with 10% biomass of this species, there was an increase in EPA content from 0.1 % (control) to 0.7%. Further, there was a considerable reduction in serum cholesterol and triglyceride levels between the control (7.5%) and alga-fed experimental rats (1.8%) (Yaron et al. 1995).

Antioxidant activity: The sulfated exopolysaccharide of this species displayed an antioxidation effect on linoleic acid auto-oxidation. It is also reported that the extracellular crude polysaccharides of this species removed superoxide anion radicals and suppressed linoleic acid auto-oxidation. Further, it is also learnt that even after the degradation of polysaccharides, the components of this species could effectively scavenge free radicals (Wu et al. 2021).

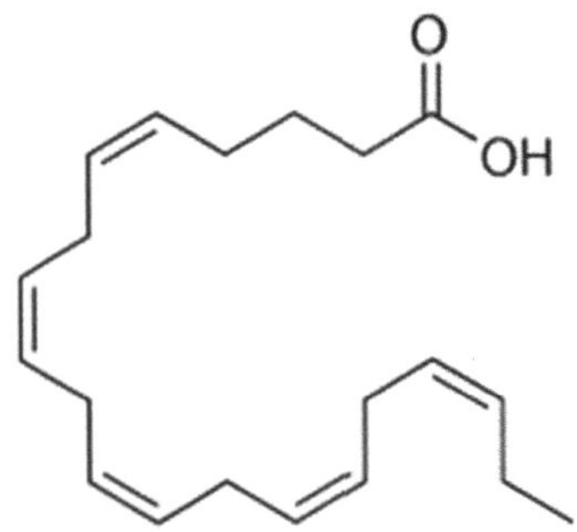

Eicosapentaenoic acid

CHAPTER 4

Medicinal Bacteria Associated with Marine Plants and Animals: Their Compounds and Bioactivities

BACTERIA ASSOCIATED WITH MARINE PLANTS

SEAWEEDS (MARINE MACROALGAE)

Phylum (Division): Chlorophyta Class: Ulvophyceae
Order: Bryopsidales Family: Codiaceae

Codium fragile

Image credit: Flyingdream, Public domain.

Common name (s): Green sea finger, [green sponge, green fleece, oyster thief, dead man's finger, felty finger [forked felt-alga, stag seaweed, sponge seaweed

Global distribution: Pacific ocean and Northern Atlantic Ocean.

Habitat: It occurs in subtidal and intertidal zones

Associated bacteria and their pharmaceutical compounds and activities

Medicinal Marine Microbiota. Santhanam Ramesh, Ramasamy Santhanam, Arumugam Uma & Raju Mohanraju (Authors)

Antibacterial activity: The organic extracts of the bacterial strains Cf1 and Cf2 associated with this green alga showed antibacterial activity against *Proteus mirabilis* with MIC values of 140 and 196 μg ml^{-1}, respectively (Singh et al. 2014).

Order: Ulvales Family: Ulvaceae

Ulva pertusa

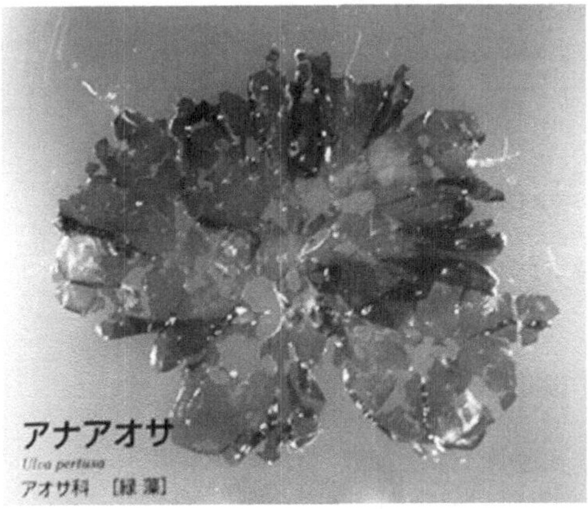

Image credit: Daderot, https://creativecommons.org/publicdomain/zero/1.0/

Common name (s): Sea lettuce, sea green alga

Global distribution: Indo-Pacific Ocean.

Habitat: It grows in the upper subtidal and lower littoral zones of a wide variety of habitats

Associated bacteria and their pharmaceutical compounds and activities

Antibacterial activity: The bacterium, *Streptomyces* sp. HZP2216E associated with this macroalga produced indolizinium alkaloid compounds viz. streptopertusacin A, D, and bafilomycin D which displayed antibacterial activity against several pathogenic bacteria with MIC 40, 13, and 13 g/mL, respectively (Nweze et al. 2020)

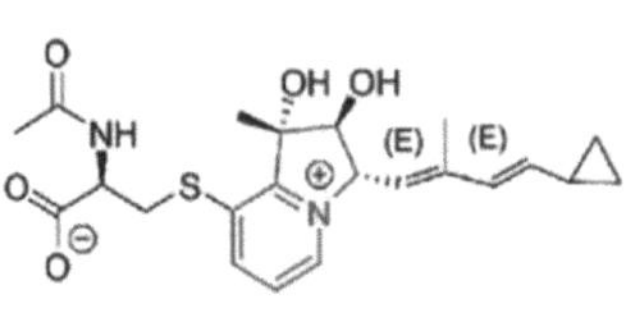

Streptopertusacin A

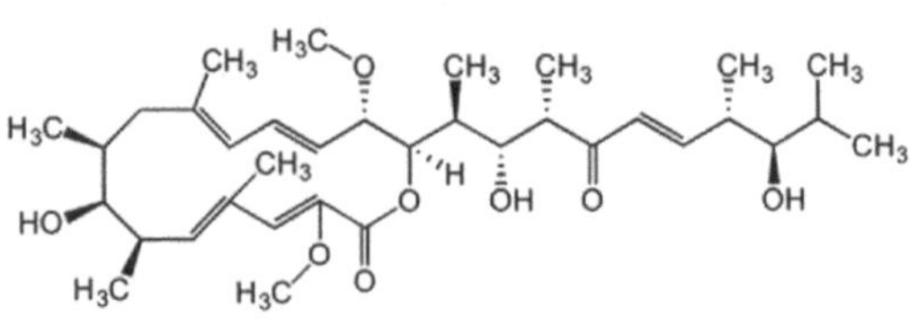

Bafilomycin D

Ulva prolifera *(= Enteromorpha prolifera)*

Common name (s): Grass kelp

Global distribution: Worldwide distribution; common in Africa, Europe, and Central America

Habitat: It is more common in brackish or salty waters

Associated bacteria and their pharmaceutical compounds and activities

Antidiabetic activity: A polyketide compound, Wailupemycin I, derived from this marine green alga-associated *Streptomyces* sp. OUCMDZ-3434 displayed moderate antidiabetic activity by inhibiting α-glucosidase with an IC50 value of 8.3 μM (Mayer et al. 2021).

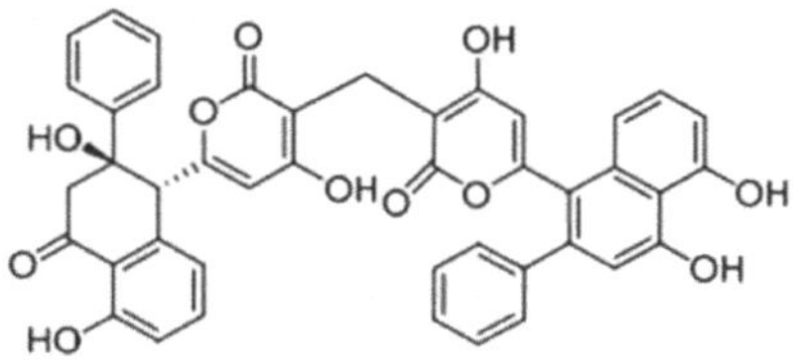

Wailupemycin I

***Ulva* sp.**

Associated bacteria and their pharmaceutical compounds and activities

Antimicrobial activity: The actinomycete, *Streptomyces althioticus*, MSM3 associated with this species produced the compound desertomycin G which possessed antimicrobial activity by inhibiting the growth of Multidrug-resistant (MDR), pathogenic microorganisms including *Mycobacterium tuberculosis*. (Nweze et al. 2020).

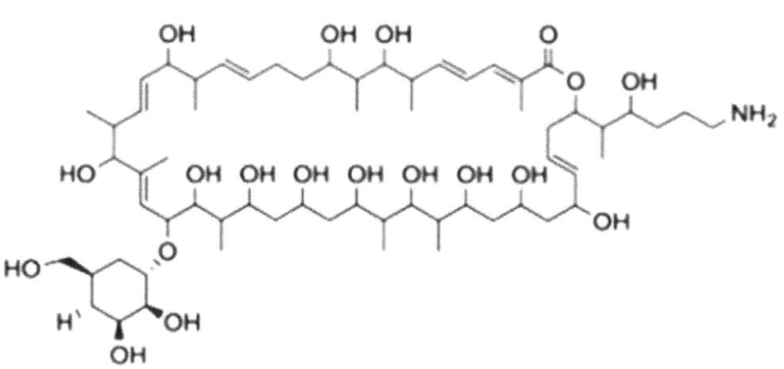

Desertomycin G

Phylum (Division): Phaeophyta Class: Phaeophyceae
Order: Ectocarpales Family: Scytosiphonaceae

Endarachne binghamiae

Common name (s): No common names

Global distribution: Warm temperate seas

Habitat: It occurs in the hard substrates in the upper mid-tidal region

Associated bacteria and their therapeutic compounds and bioactivities

Antibacterial activity: The organic extracts of the bacterial strains Eb1, Eb4, Eb5, and Eb6 associated with this green alga showed antibacterial activity against Proteus mirabilis with MIC values of 179, 188, 210, and 179 μg ml^{-1}, respectively (Singh et al. 2014.

Anticancer activity: The organic extracts of the bacterial strains Eb6 associated with this green alga showed anticancer activity against HCT-116 colorectal cancer cell line, and the IC50 value was 2.8 μg ml^{-1} (Singh et al. 2014).

Order: Fucales Family: Fucaceae

Pelvetia canaliculata

Image credit: Stemonitis https://creativecommons.org/licenses/by-sa/2.5/

Common name (s): Channelled wrack

Global distribution: Europe

Habitat: It is commonly seen on the rocks of the upper shores

Associated bacteria and their pharmaceutical compounds and activities

Antimicrobial activity: The actinobacterial strain *Kocuria marina* CMG S2 harbored by this brown alga has been reported to produce a promising antibiotic viz. kocumarin, which displayed antimicrobial activity by inhibiting the growth of pathogenic fungi and bacteria like *Staphylococcus typhi*, *Staphylococcus pyogenes*, *Pseudomonas aeruginosa* and MDR bacteria strains of MRSA ATCC 33591, with an MIC value of 10 g/mL (Nweze et al. 2020)

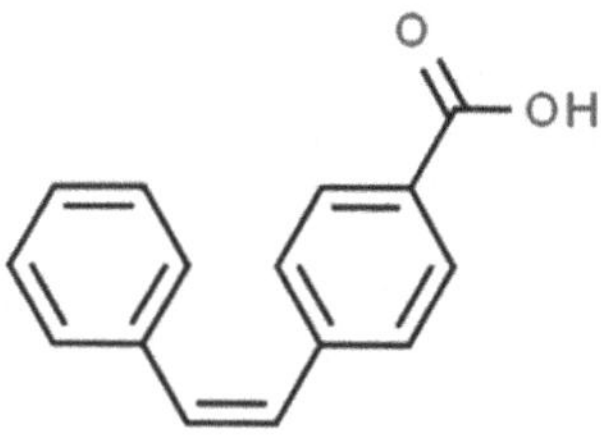

Kocumarin

Order: Fucales Family: Sargassaceae

Sargassum muticum

Image credit: Lamiot, https://creativecommons.org/licenses/by-sa/4.0/

Common name (s): Japanese wireweed

Global distribution:

Habitat: It occurs largely in infralittoral areas to a depth of about 10 m. It is also found fixed on solid substrates such as stones, shells and rocks.

Associated bacteria and their pharmaceutical compounds and activities

Antibacterial activity: The organic extracts of the bacterial strains Sm2, Sm4, and Sm6 associated with this green alga showed antibacterial activity against *Proteus mirabilis* with MIC values of 140, 166, and 174 μg ml^{-1}, respectively (Singh et al. 2014)

Anticancer activity: The organic extracts of the bacterial strains Sm6 associated with this green alga showed anticancer activity against HCT-116 colorectal cancer cells with an IC50 value of 5.5 μg ml^{-1} (Singh et al. 2014).

Order: Laminariales Family: Lessoniaceae

Egregia menziesii

Image credit: Leslie Seaton, https://creativecommons.org/licenses/by/2.0/

Common name (s): Feather boa kelp

Global distribution: Western North America

Habitat: It is found from the lower intertidal to subtidal zone of the coastal waters

Associated bacteria and their pharmaceutical compounds and activities

Antibacterial activity: The organic extracts of the bacterial strain Em4 associated with this green alga showed antibacterial activity against *Proteus mirabilis* with an MIC values of 212 μg ml^{-1}; and strain Em1 and Em4 acted against *Klebsiella pneumoniae* with MIC values of 212 and 213 μg ml^{-1}, respectively (Singh et al., 2014)

Anticancer activity: The organic extracts of the bacterial strains Em6 associated with this green alga showed anticancer activity against HCT-116 colorectal cancer cells with an IC50 value of 58 μg ml^{-1} (Singh et al. 2014)

UNIDENTIFIED BROWN ALGA

Associated bacteria and their pharmaceutical compounds and activities

Antibacterial activity: A terpenoid, polyketide viz. furanoterpenoid derived from this unidentified brown seaweed-associated bacterium *Bacillus subtilis* MTCC 10403 showed antibacterial activity against *Vbrio vulnificus* and *V. parahaemolyticus* with IC50 values of 3 and 12 μg/mL, respectively (Mayer et al. 2021).

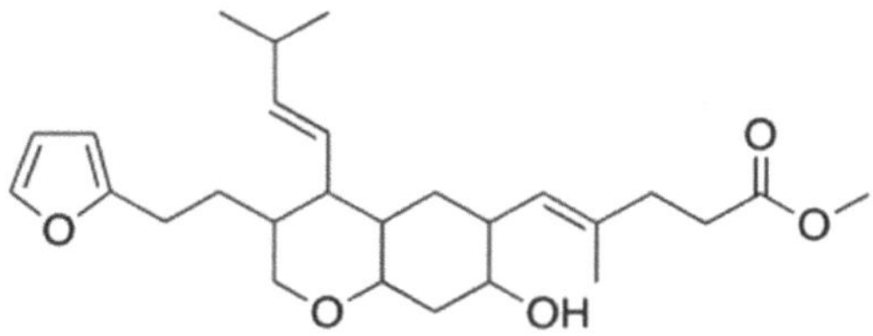

Furanoterpenoid

Phylum (Division): Rhodophyta Class: Florideophyceae

Order: Ceramiales Family: Ceramiaceae

Centroceras clavulatum

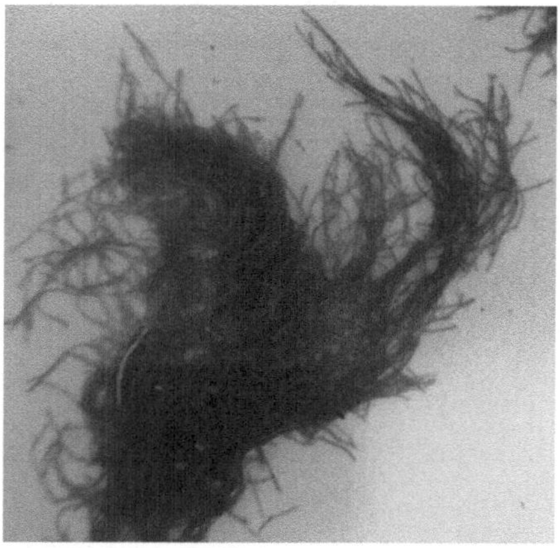

Image credit: Auckland Museum, https://creativecommons.org/licenses/by/4.0/

Common name (s): No common names

Global distribution: It has pantropical to warm-temperate distribution

Habitat: It is found mainly in strongly wave-exposed shorelines and on rocks in middle to lower intertidal zones where the depth is about 2 m. It is epiphyte upon other species

Associated bacteria and their pharmaceutical compounds and activities

Antibacterial activity: The organic extracts of the bacterial strains Cc1, Cc2, Cc3 and Cc7 harbored by this green alga displayed antibacterial activity with Proteus mirabilis with MIC values of 181, 217,153, and 159 μg ml^{-1}, respectively (Singh et al. 2014).

Anticancer activity: The organic extracts of the bacterial strains Cc1, Cc3, and Cc5 associated with this green alga showed anticancer activity against HCT-116 colorectal cancer cells with an IC50 value of 6, 102, and and 57 μg ml^{-1}, respectively (Singh et al. 2014).

Order: Ceramiales Family: Rhodomelaceae

Laurencia pacifica

Laurencia sp.

Image credit: Channer, https://creativecommons.org/licenses/by-sa/3.0/

Common name (s): Not known

Global distribution: It is commonly distributed from California, to Baja California.

Habitat: It is seen on rocks in cool quiet water in the lower intertidal zone

Associated bacteria and their pharmaceutical compounds and activities

Antibacterial activity: The organic extracts of the bacterial strains Lp2 and Lp3 associated with this green alga showed antibacterial activity against *Proteus mirabilis* with MIC values of 185 and 288 $\mu g\ ml^{-1}$, respectively (Singh et al. 2014).

Order: Gigartinales Family: Cystocloniaceae

Hypnea valentiae

Acanthostrongylophora ingens.

Image credit: Leiodermatium pfeifferae-FAO (https://creativecommons.org/licenses/by/4.0/)

Common name (s): Not known

Global distribution: It is widely distributed in subtropical and tropical waters

Habitat: This benthic species is known to occur in sheltered localities with adequate current flow

Associated bacteria and their pharmaceutical compounds and activities

Antibacterial activity: The extracts of the bacterial strains viz. *Bacillus amyloliquefaciens* MTCC 12716 and *Shewanella algae* MTCC 12715 (Gamma-proteobacteria), which were associated with this red alga showed significant antibacterial activity against VRE (vancomycin-resistant *Enterococcus faecalis*) and MRSA (methicillin-resistant *Staphylococcus aureus*) with MIC values of 12.5 and 6.3 μg/mL. respectively. The values of zone of inhibition for these bacterial strains were found to be 30 mm (Kizhakkekalam and Chakraborty 2020).

UNIDENTIFIED RED ALGA

Associated bacteria and their pharmaceutical compounds and activities

Antibacterial activity: The organic extract of a red alga-associated bacterium, *Streptomyces* sp. MC025 has been reported to contain the compound, collismycin C which had antibacterial and anti-biofilm activity of several strains of *Staphylococcus aureus* including MRSA ATCC 33591 with an IC50 value of 10 μg/mL (Nweze et al. 2020; Mayer et al. 2021).

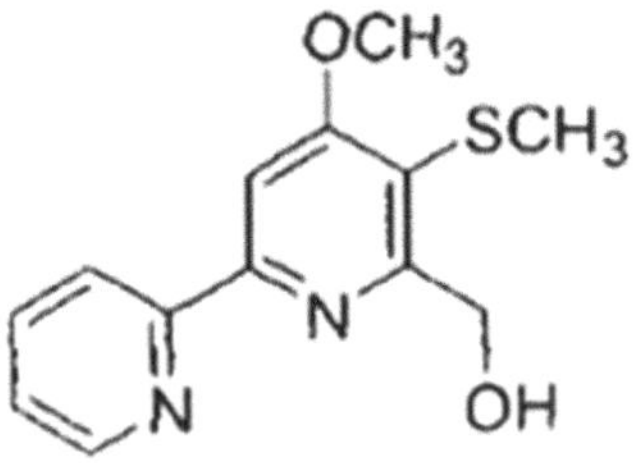

Collismycin C

UNIDENTIFIED MACROALGA

Associated bacteria and their pharmaceutical compounds and activities

Antibacterial and anti-ESBL activity: An actinomycete, *Nocardiopsis* sp. GRG 2 (KT 235641) harboured by this macroalga has been reported to produce the compound, DDIBN (4diaza2, 5dioxo3isobutyl bicyclo[4.3.0] nonane) which displayed potent antiMDR and antiESBL (extended spectrum beta lactamase) activity bearing pathogens like strains of MDR ESBL *Pseudomonas aeruginosa* and *Klebsiella pneumonia* and the MIC value recorded was 75 g/mL. (Nweze et al. 2020)

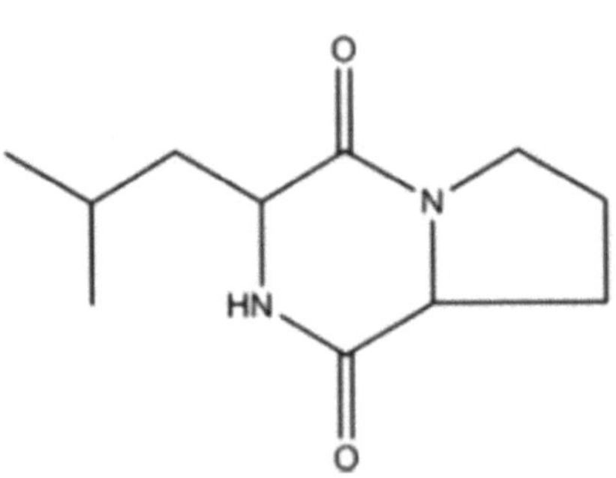

1, 4-diaza-2, 5-dioxo-3-isobutyl bicyclo[4.3.0]nonane

SEAGRASSES

Phylum (Division): Tracheophyta Class: Magnoliopsida
Order: Alismatales Family: Cymodoceaceae

Cymodocea serrulata

***Cymodocea* sp.**
Image credit: Sylvain Le Bris, https://creativecommons.org/licenses/by-sa/4.0/

Common name (s): Serrated ribbon seagrass

Global distribution: Warm oceans.

Habitat: It occurs in high intertidal areas

Associated bacteria and their pharmaceutical compounds and activities

Antibacterial activity: It was noted that the CS 102 103 and 105 bacterial strains of this unidentified species had a significant growth reduction on *Bacillus subtilis* to 3.5 mm and the strains CS 321 and 323 inhibited the growth of *Staphylococcus aureus* to 2.5 mm. The strain CS321 inhibited at least five of the human pathogens with zone size of 2.5 mm. However, CS278 and 279 inhibited *Micrococcus* sp. to 5 and 4 mm, respectively. Also, the strain CS105 inhibited *Proteus mirabilis* to 4 mm. However, the strains CS6,

10 and 286 did not show much inhibition (HYPERLINK "http://shodhganga.inflibnet.ac.in/bitstream/10603/61084/14/14_chapter%209.pdf" http://shodhganga.inflibnet.ac.in/bitstream/10603/61084/14/14_chapter%209.pdf).

Cymodocea rotundata

Common name (s): Smooth ribbon seagrass

Global distribution: Throughout tropical Indo-West Pacific

Habitat: It occurs in pollution-free coral reef areas

Associated bacteria and their pharmaceutical compounds and activities

Antibacterial activity: The bacterial species *Pseudomonas aeruginosa* strain DM1 (C1M7) and *Shewanella algae* strain MARS14 (C2M3) isolated from this halophytic species showed antibacterial activity against MRSA and *Bacillus subtilis* with the clear zone diameter (mm) values of 167 and 124 mm (MRSA); and 1827 and 1653 mm (*Bacillus subtilis*), respectively (Fitri et al. 2017)

Order: Alismatales Family: Hydrocharitaceae

Enhalus acoroides

Enhalus sp.

Image credit: Ria Tan, https://creativecommons.org/licenses/by-sa/2.0/

Common name (s): Tape seagras

Global distribution: Coasts of E Africa through Arabia, India, China, southeast Asia to Australia and the western Pacific.

Habitat: Along the shore; usually near coral rubble and growing in muddy soils; submerged in shallow sea water along the coast.

Associated bacteria and their pharmaceutical compounds and activities

Antibacterial activity: *Bacillus flexus* (EED15) which is an endophytic bacterial strain of this alga showed strong antibacterial activity against MRSA and *Bacillus subtilis*. Similarly, another endophytic bacterial strain, *Streptomyces lienomycini* (EED 16) of this species showed activity against MRSA and MDR *E. coli*; MRSA and vancomycin-resistant *S. aureus*; and MDR *Pseudomonas aeruginosa* (Cristianawati et al. 2019).

Fitri et al. (2017) reported that the bacterial isolates viz. E2M 1,3,4,7, and 8 from the leaves of this species showed antibacterial activity against MRSA and *Bacillus subtilis*, with the values of clear zone diameter ranging from 235 to 915 mm; and from 65 to 278 mm, respectively (Fitri et al. 2017)

Order: Alismatales Family: Cymodoceaceae

Halodule uninervis

Image credit: Paul Asman and Jill Lenoble, https://creativecommons.org/licenses/by/2.0/

Common name (s): Needle seagrass

Global distribution: Indo-West Pacific

Habitat: Sessile; depth range 2 - 38 m

Associated bacteria and their therapeutic compounds and bioactivities

Antibacterial activity: Producer bacterial strains of *Halodule uninervis* inhibited the human pathogen, *Escherichia coli* with zone sizes of 4.5 (HU381), 3.5 (HU33) and 2.5 mm (HU 158). A maximum inhibitory zone size of 4.5 mm was conferred by the strain HU280 against *P. mirabiis* and *Serratia marcescens*. HU10 and HU 79 inhibited *Aeromonas hydrophila* and HU 103 inhibited *P. mirabilis* with zone sizes of 2.5 mm. It was also observed that the strain HU 280 inhibited three of the pathogens (*A. formicans, P. mirabilis* and *S. marcescens*) with inhibition of 4 mm and above (http://shodhganga.inflibnet.ac.in/bitstream/10603/61084/14/14_chapter%209.pdf).

Order: Alismatales Family: Cymodoceaceae

Syringodium isoetifolium

Image credit: Philippe Bourjon, https://creativecommons.org/licenses/by-sa/3.0/

Common name (s): Noodle seagrass

Global distribution: It is found distributed in the warm waters of Indo-West Pacific

Habitat: Just below low tide level; rarely found in shallow intertidal areas.

Associated bacteria and their pharmaceutical compounds and activities

Antibacterial activity: The crude extracts of the associated microbes of this species showed antibacterial activity against *S. aureus, K. pneumoniae, S.epidermis, P. aeruginosa* and *Streptococcus pneumoniae* with MIC values of 256, 32,128, 32 and 128 μg/ml, respectively (Ravikumar et al. 2012).

Order: Alismatales Family: Hydrocharitaceae

Halophila ovata

Halophila sp.

Image credit: Hemprich F.G. & Ehrenberbg C.G., Public domain.

Common name (s): Spoon grass, dugong grass or paddle weed

Global distribution: Indo-Pacific.

Habitat: It is a benthic species occurring in sea beds and other related areas

Associated bacteria and their therapeutic compounds and bioactivities

Antibacterial activity: The epibiotic bacterial strains of this plant species viz. Strains H0 519, 520,16,18,426 and 428 showed antibacterial activity against human pathogens. Strains H0 519 and 520 inhibited *V. cholerae* with the zone size of 6.5 and 5 mm, respectively. H016 and 18 inhibited *S. aureus* (4.5 mm) and *Enterobacter aerogens* (3.5 mm), whereas H0 426 and 428 inhibited *Kiebsiella pneumoniae* with 3.5 mm. (http://shodhganga.inflibnet.ac.in/bitstream/10603/61084/14/14_chapter%209.pdf).

Order: Alismatales Family: Hydrocharitaceae

Thalassia hemprichii

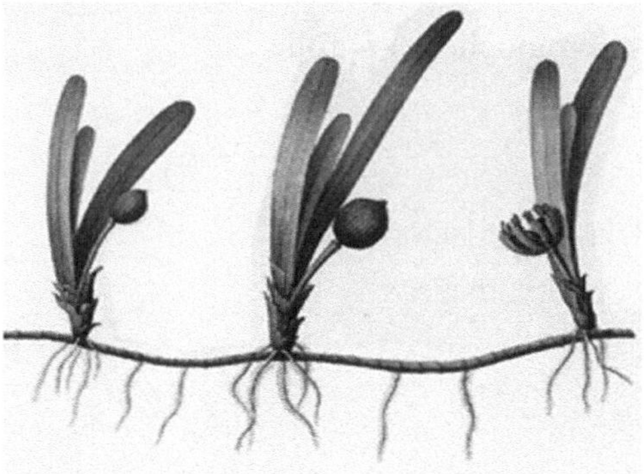

Image credit: Hemprich F.G. & Ehrenberbg C.G., Public domain.

Common name (s): Sickle seagrass, turtle grass

Global distribution: Throughout the tropical Indo-West Pacific region.

Habitat: This sessile species occurring mainly in the middle part of the lagoon with seagrass

Associated bacteria and their pharmaceutical compounds and activities

Antibacterial activity: The bacterial species *Paracoccus zeaxanthinifaciens* strain F71025 isolated from this halophytic species showed antibacterial activity against MRSA with the clear zone diameter (mm) value of 563 mm (Fitri et al. 2017).

Thalassia sp.

Associated bacteria and their pharmaceutical compounds and activities

Antibacterial activity: Two isolates of *Pseudomonas bromoutilis* associated with this marine plant species produced a brominated pyrolle viz. 2,3,4-tribromo-5(1'-hydroxy-2',4'-dibromopheny)-pyrolle which exhibited inhibitory activity against gram-positive bacterial pathogens viz. *Mycobacterium tuberculosis*, *S.aureus*, *Streptococcus pneumoniae*, and *S.pyrogenes*. The inhibitory concentrations of the above compound for these pathogens were found to be 0.2, 0.0063, 0.0063, and 0.0063 μg/ml respectively (Fingerman 2002).

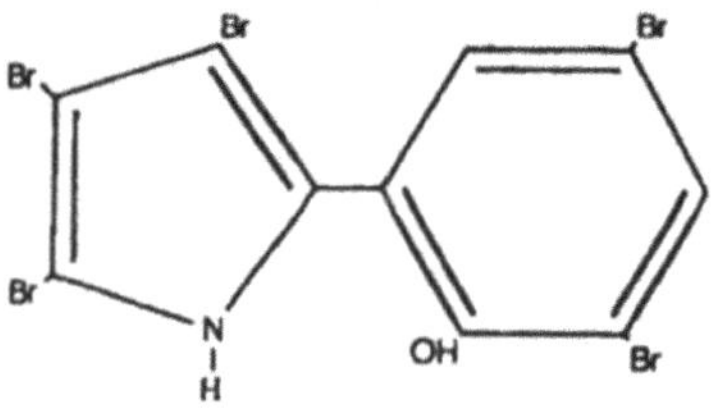

2,3,4-tribromo-5(1'-hydroxy-2',4'-dibromopheny)-pyrolle

HALOPHYTES

Phylum (Division): Tracheophyta Class: Equisetopsida

Order: Caryophyllales Family: Amaranthaceae

Halopeplis perfoliata

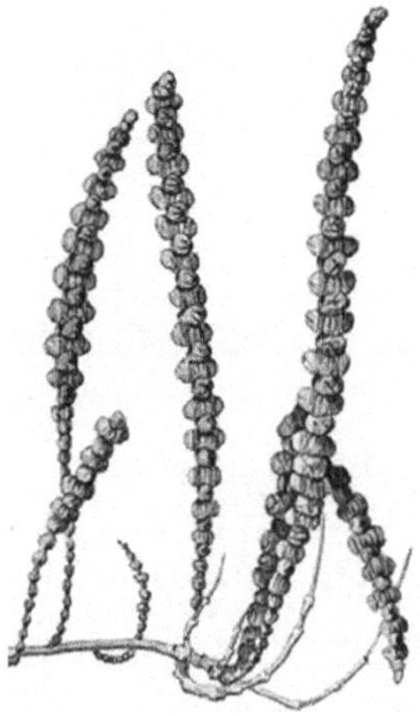

Image credit: Peter Forsskål, Public domain.

Common name (s): String of beads or, Al Kheriza beach plant

Global distribution: Native range is Egypt to Eritrea and SW. Pakistan; N.E. Africa, Nubia, Arabia to Pakistan, Baluchistan extreme S.W. coast

Habitat: Coastal marsh halophyte

Associated bacteria and their pharmaceutical compounds and activities

Antifungal activity: The bacterial strain *Inquilinus alexandrii* EA 220 isolated from this species showed activity against pathogenic fungi, i.e., *Pythium ultimum, Phytophthora capsici*, and *Magnaporthe grisea* with an inhibition zone value of 3 mm (Bibi et al. 2018b).

Order: Caryophyllales Family: Amaranthaceae

Salsola imbricata Forssk.

Common name (s): Not designated

Global distribution: N Africa; Arabia, Syria, Iraq, Iran, Pakistan, India.

Habitat: On many soil types, preferably on ruderal sites, fallows, and overgrazed areas surrounding the settlements, but also in naturally disturbed habitats like eroded slopes, runnels, washes, dry river beds, and sea-shore cliffs; depth range 0–2100 m.

Associated bacteria and their pharmaceutical compounds and activities

Antifungal activity: This species has been reported to harbor both endophytic and rhizospheric bacterial communities belonging to *Streptomyces* sp. (EA64), *Arthrobacter* sp. (EA59), *Nocardiopsis dassonvillei* (EA52), and *Pseudonocardia* sp. (EA58) which showed strong antifungal activity (Bibi et al. 2018a).

Bibi et al. (2018b) also reported that the secondary metabolites viz. sulfamonomethoxine, metronidazole-oh, Ibuprofen, and dimetridazole from the bacterial species. *Streptomyces enissocaesilis*. (EA61) associated with this plant species showed activity against four fungal pathogens viz. *Pythium ultimum, Phytophthora capsici, Magnaporthe grisea*, and *Altenaria mali.* The values of inhibition zone (mm) for *Pythium ultimum* and *Altenaria mali* were 7–9 mm; and for *Phytophthora capsici* and *Magnaporthe grisea*, 3 mm.

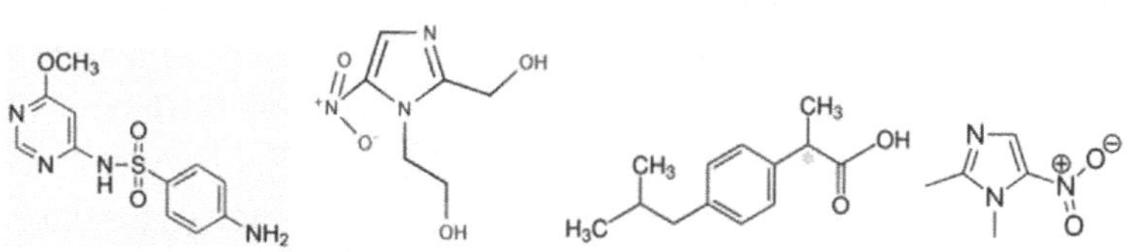

Sulfamonomethoxine Metronidazole-oh Ibuprofen Dimetridazole

Order: Caryophyllales Family: Amaranthaceae

Suaeda nudiflora

Suaeda sp.
Image credit: Roman, Public domain.

Common name (s): Seepweeds or sea-blites

Global distribution: India, Sri Lanka, Malaysia, Australia, N. Africa and N. America.

Habitat: Back-mangroves zone; common on salt marshes and tidal blanks.

Associated bacteria and their pharmaceutical compounds and activities

Antimicrobial activity: The bacterial species *Bacillus anthracis* associated with this plant species showed antibacterial activity against *Burkholderia gladioli* and *Erwinia amylovora* and antifungal activity against *Fusarium oxysporium* and *Sclerotium rolfsii* (Makadia and Panchal 2016)

Suaeda salsa

Common name (s): Not designated

Global distribution: Europe to E. Asia.; China, Korea, Mongolia; Asia, and Europe

Habitat: It is found on beaches and lake shore with saline and alkaline soils.

Associated bacteria and their pharmaceutical compounds and activities

Antimicrobial activity: The bacterium *Bacillus marinus* associated with this plant produced macrolides such as macrolactins A, B, D, O, S, T, and U. Among these compounds, macrolactins T and B showed inhibitory activity against the pathogenic bacteria *Staphylococcus aureus* and fungi, *Pyricularia oryzae* and *Alternaria solani* with MIC values of 5.5, 0.8, and 2.8 μg/mL; and 4.5, 7.5, and 20.1 μg/mL, respectively (Xue et al. 2008).

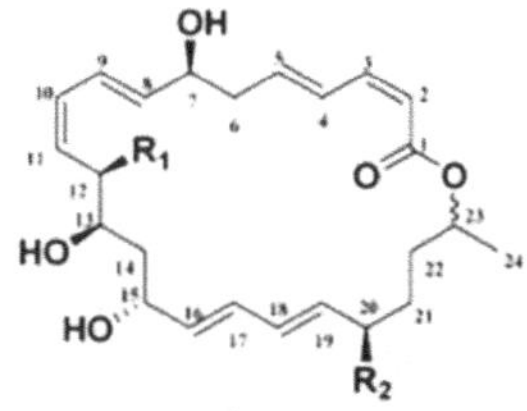

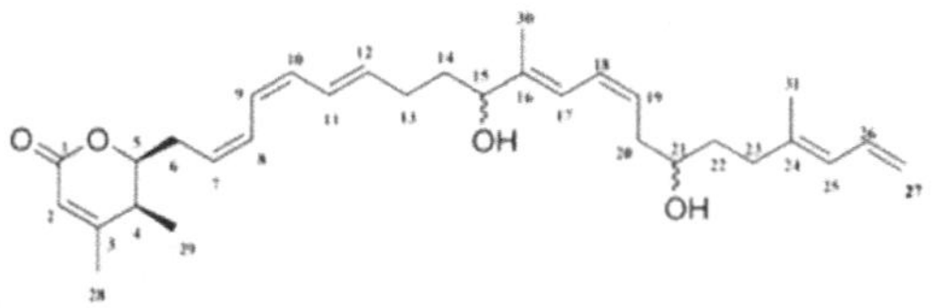

Macrolactin T: R1+H,R2=OH

Macrolactin U

Macrolactin S: R1=OH,R2=H

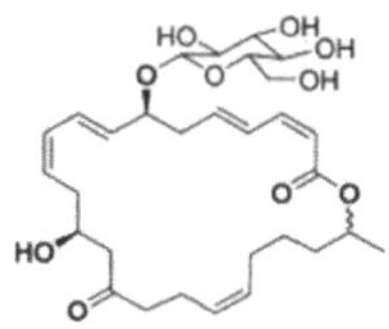

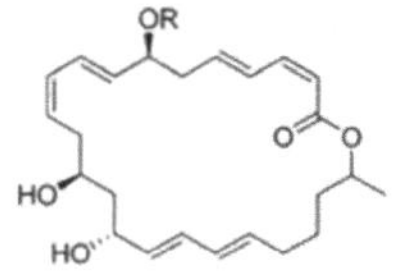

Macrolactin D Macrolactin A (R=H); B (R= β gluc); D (R= 6-O-succinyl- β gluc)

BACTERIA ASSOCIATED WITH MARINE INVERTEBRATE ANIMALS

SPONGES

Phylum: Porifera Class: Demospongiae
Order: Agelasida Family: Agelasidae

AGELAS OROIDES

Image credit: Esculapio, https://creativecommons.org/licenses/by/3.0/

Common name: Maltese sponge

Global distribution: Mediterranean sea

Habitat: Shallow coastal waters of 50 m depth with little light and on rocky bottoms

Associated bacteria: Pharmaceutical compounds and activities

Antibacterial activity: The alkaloid, ageloline A isolated from an unidentified species of *Streptomyces* which was harboured by this sponge has been reported to possess antibacterial activity by inhibiting the growth of *Chlamydia trachomatis* inclusions with an IC50 value of 9.5 μM (Mayer et al. 2021).

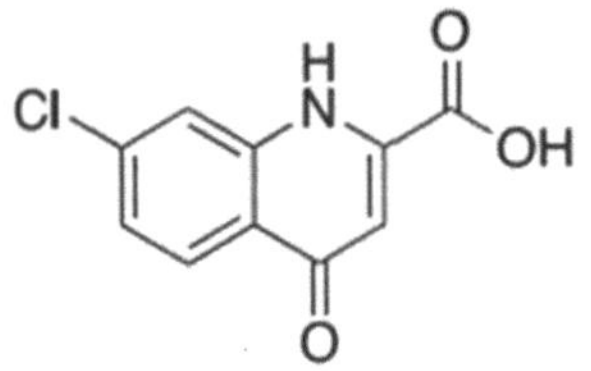

Ageloline A

Order: Axinellida Family: Axinellidae

Axinella polypoides

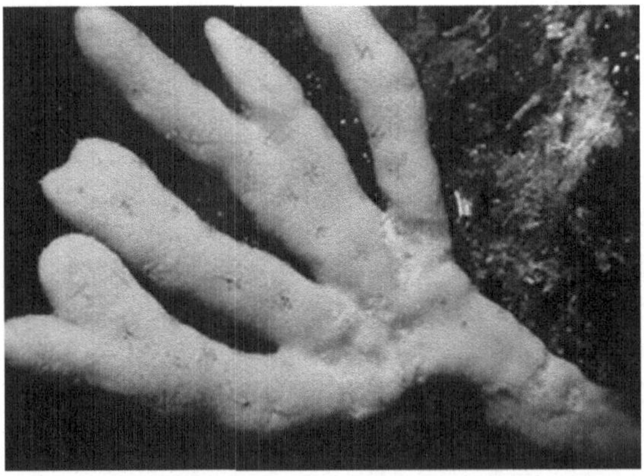

Image credit: Géry PARENT, Public domain.

Common name: Common antlers sponge

Global distribution: Western Mediterranean Sea, and on the Atlantic coast of Europe and Africa from the Bay of Biscay to Cape Verde and west as far as the Azores

Habitat: Rocky bottoms at depths of 20 to 100 m

Associated bacteria: Pharmaceutical compounds and activities

Anti-parasitic activity: The bacterial strain *Streptomyces* sp. strain 34 isolated from this species produced valinomycin, which exhibited significant inhibitory activities against the parasites *Leishmania major* (IC50 < 0.11 μM) and *Trypanosoma brucei brucei* (IC50 0.0032 μM) (Pimentel-Elardo et al. 2010).

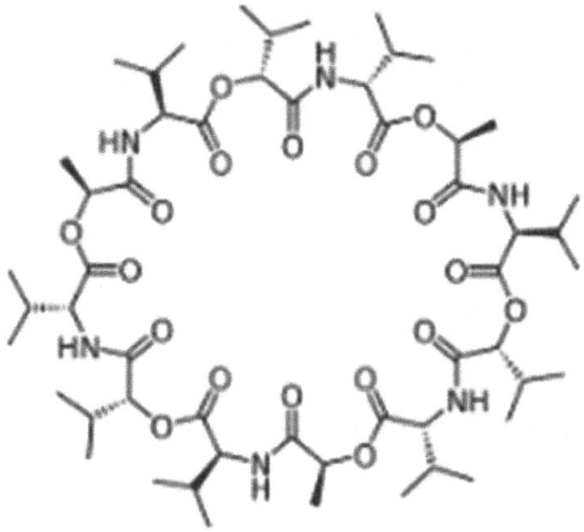

Valinomycin

Cymbastela concentrica (Lendenfeld, 1887)

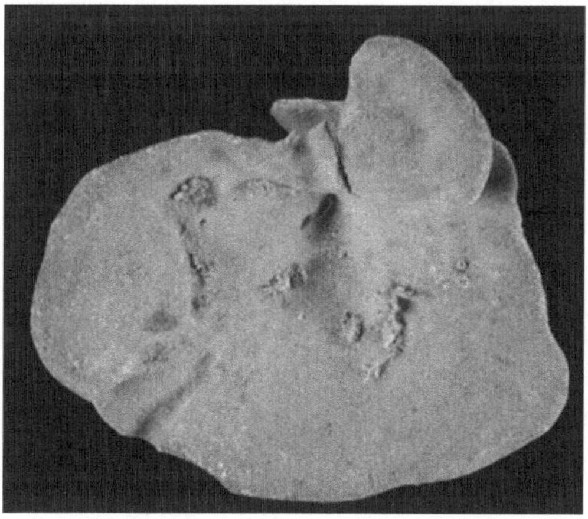

Cymbastela sp.

Image credit: Auckland Museum. https://creativecommons.org/licenses/by/4.0/

Commmon names: Australian sponge, temperate marine sponge

Global distribution: Western Central Pacific: Australia and New Caledonia

Habitat: Sessile; depth range 0 - 100 m

Associated bacteria: Pharmaceutical compounds and activities

Antibacterial activity: Two clones (CcAb1 and AcAb2) belonging to Proteobacteria, deltaproteobacteria which are associated with this sponge species showed antibacterial activity against *Bacillus* strain Cc6. Further, the clone CcAb1 showed activity against *Staphylococcus aureus* and *Alteromonas* sp. strain CCSH174 (inhibition zones of 0.5 and 0.6 cm, respectively) (Yung et al. 2011).

Dragmacidon agariciforme

Global distribution: No information

Habitat: No information

Associated bacteria: Pharmaceutical compounds and activities

AChE inhibition activity: The bacterial isolates of this species showed inhibition activity against the enzyme acetylcholinesterase which is believed to be responsible for promoting the neurodegenerative Alzheimer's and Parkinson's disease (Pandey et al. 2014).

Order: Axinellida Family: Stelligeridae

Higginsia bidentifera

Common name: Not available

Global distribution: endemic to South Africa

Habitat: Coastal waters

Associated bacteria: Pharmaceutical compounds and activities

Antibacterial activity: The bacterial strain *Bacillus marisflavi* TF-11 (PE07-7) associated with this sponge species displayed activity against Gram-positive strains, *M. smegmatis*, *B. cereus*, and *S. epidermidis*. Further, its strains *Bacillus vietnamesis* PE07-207 and PE07-144 displayed activity against *M. smegmatis*, *B. cereus* and *S. epidermidis*; and *E. coli* 1699 and *P. putida* (Matobole et al. 2017)

Order: Bubarida Family: Dictyonellidae

Acanthella acuta

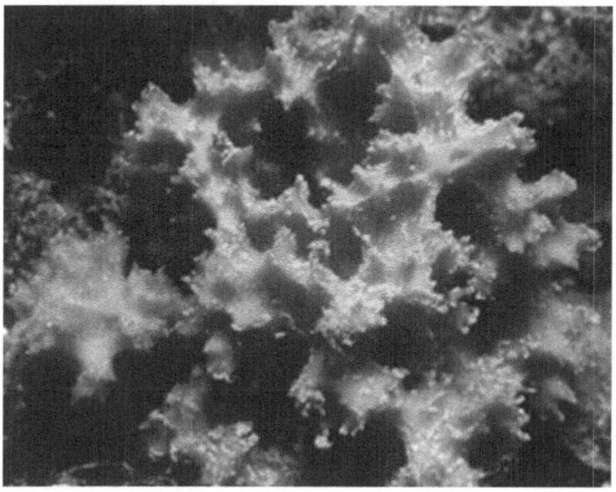

Image credit: Parent Géry, Public domain.

Common name: Orange lumpy sponge

Global distribution: Mediterranean and the Atlantic, between the Cantabrian and the coasts of Morocco.

Habitat: Between 20 and 30 m deep; on almost any type of hard substrate: detritus, rocky, or coral.

Associated bacteria: Pharmaceutical compounds and activities

Anti-tumor activity: The marine strain *Bacillus pumilus* strain AAS3, isolated from this species, produced a diglucosyl-glycerolipid, 1,2-O-diacyl-3-[beta-glucopyranosyl-(1-6)-beta-glucopyranosyl)]glycerol, with 14-methylhexadecanoic acid and 12-methyltetradecanoic acid as the main fatty acid moieties (GGL11). Anti-tumor activity on this class of diglucosyl glycerol products showed that the carbohydrate/glycerol backbone (GG11) has a significant inhibitory activity than the acylated compounds. The diglucosyl-glycerol GG11 strongly inhibited the growth of HM02 and Hep G2 (tumor cell lines) (50% inhibition at approximately 1 μg/ml), while the glycerolipids GGL11 and GGL12 were less active or had no effect (Ramm et al. 2004).

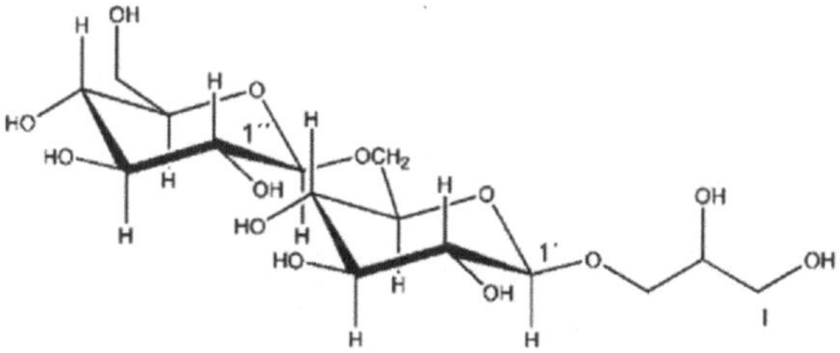

Diglucosyl-glycerol

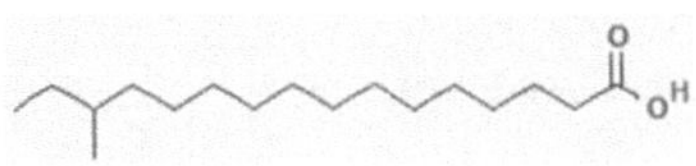

14-methylhexadecanoic acid

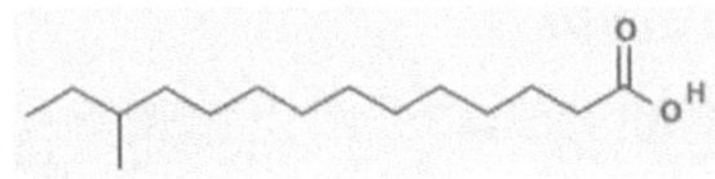

12-methyltetradecanoic acid

Acanthella cavernosa

Common name: Yellow or red lattice sponge

Global distribution: Indo-West Pacific

Habitat: Sessile; depth range 10–20 m

Associated bacteria: Pharmaceutical compounds and activities

Antibacterial activity: The extracts of the associated bacterial strains (49) of this species showed antibacterial activity against *Staphylococcus aureus* ATCC 25923 and *Bacillus subtilis* ATCC 6633, and wild strain *Escherichia coli* with the values of diameter inhibition (mm) from 7.7 to 14.9; 7.4 to 6.6; and from 0 to 10.5, respectively. Further, its another bacterial strain has been reported to possess 7 potent fractions viz. 7, 8, 9, 10, 11, 14, and 15; and the fraction 7 had phenolic compounds such as tri(2-ethylhexyl) trimellitate, oleamide, 4-nonylphenol, 2,4-bis(1-phenylethyl)phenol, 1,2-benzenedicarboxylic acid, mono(2- ethylhexyl) ester, methyl 3-(3,5-di-tert-butyl-4-hydroxyphenyl) propionate, and acetosyringone (Murniasih et al. 2016).

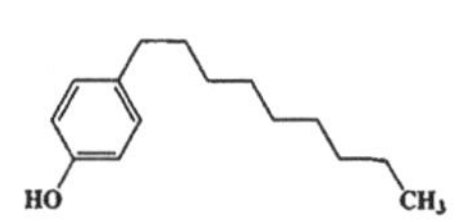

1

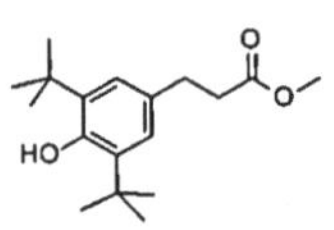

2

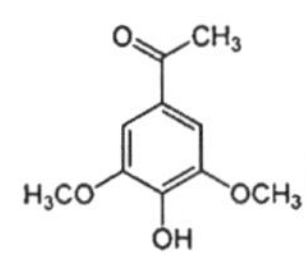

3

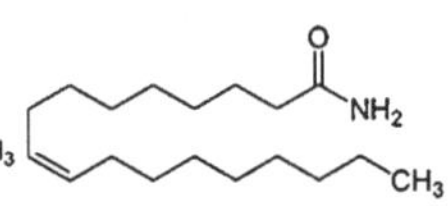

8

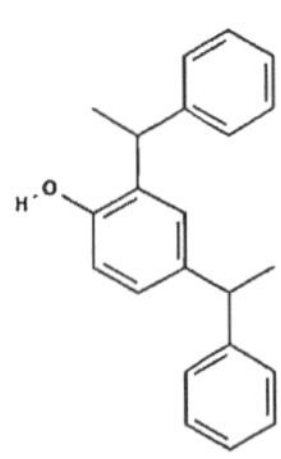

4

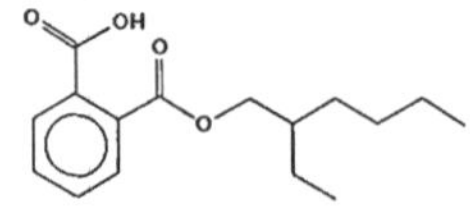

6

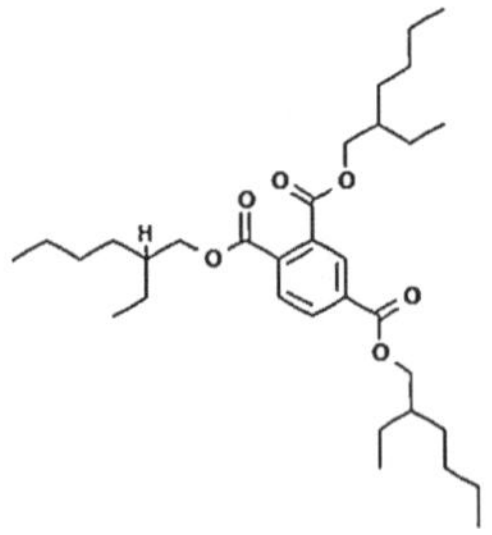

7

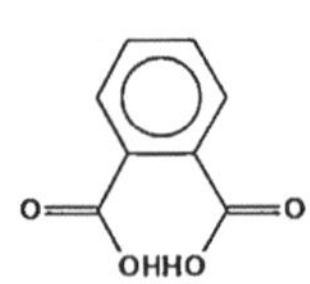

5

1: 4-nonylphenol, 2: methyl 3-(3,5-di-tert-butyl-4-hydroxyphenyl)propionate, 3: acetosyringone, 4: 2,4-bis(1-phenylethyl)phenol, 5: 1,2-benzenedicarboxylic acid, 6: mono(2- ethylhexyl) ester, 7: tri(2-ethylhexyl) trimellitate, 8: oleamide

AChE inhibition activity: The bacterial isolates of this sponge species exhibited >10% AChE inhibition activity (Pandey et al. 2014)

Acanthella ramosa

Common name: No Data.

Global distribution: Bay of Bengal

Habitat: 29–31 m depth

Associated bacteria: Pharmaceutical compounds and activities

The bacterial strain *Streptomyces* sp. DA18 isolated from this species yielded the dikeropiperazine cyclo-(D-Pro-D-Phe). The pharmaceutical activity of this compound is to be known (Atta-ur-Rahman 2016).

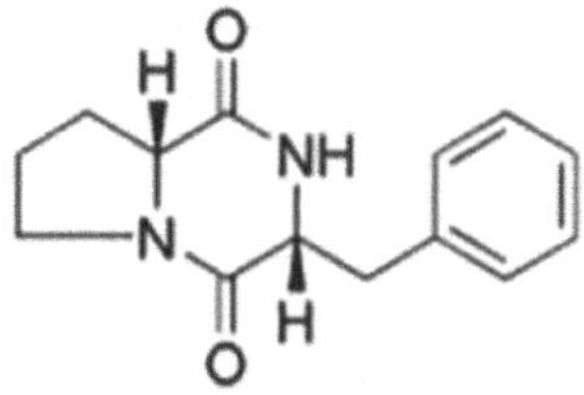

Cyclo-(D-Pro-D-Phe)

Order: Chondrillida Family: Chondrillidae

Chondrilla nucula (Schmidt 1862)

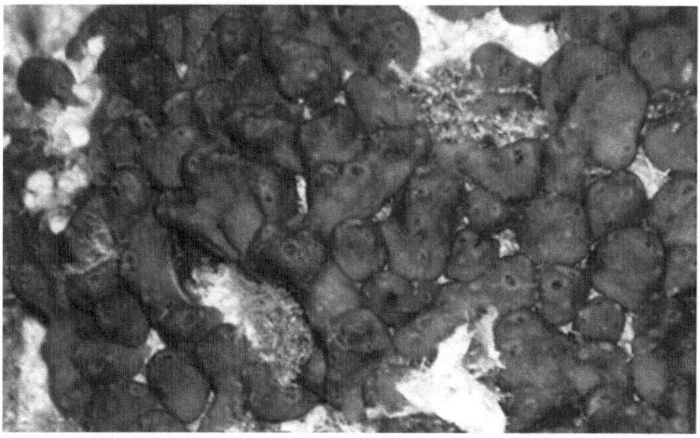

Image credit: Esculapio, https://creativecommons.org/licenses/by/3.0/

Common name: Chicken-liver sponge

Global distribution: Tropical and temperate; Mediterranean Sea

Habitat: It occurs in a variety of habitats such as coastal waters, coral reefs, and mangrove roots

Associated bacteria and their therapeutic compounds and bioactivities

Antibacterial activity: A total of 416 different bacterial strains were isolated, 60 (14.4%) of which displayed variable degrees of antimicrobial activity. Of the bioactive strains, 58.3% were able to inhibit *Staphylococcus aureus*, 6.7% were active against *Bacillus subtilis*, 11.7% against both *Enterococcus faecalis* and *Escherichia coli*, 38.3% against *Pseudoalteromonas atlantica* and 33.3% against *Pseudomonas elongata* A polyketide antibiotic viz. anthracycline keyicin isolated from the marine bacterium *Rhodococcus* sp. (Family: HYPERLINK "http://en.wikipedia.org/wiki/Nocardiaceae" Nocardiaceae) and *Micromonospora* sp.(Family: Micromonosporaceae) which were associated with this sponge showed antibacterial activity on MR *Staphylococcus aureus* and *Bacillus subtilis* with IC50 values ranging from 2.5 to 9.9 μM (Chelossi et al. 2007).

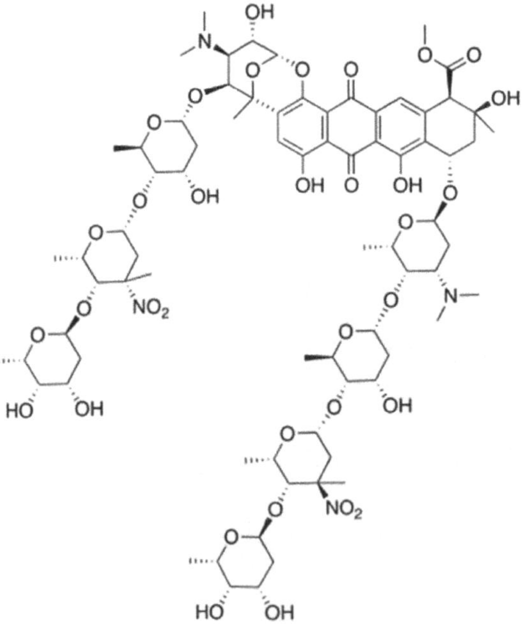

Keyicin

Order: Choristida Family: Craniellidae

Craniella australiensis

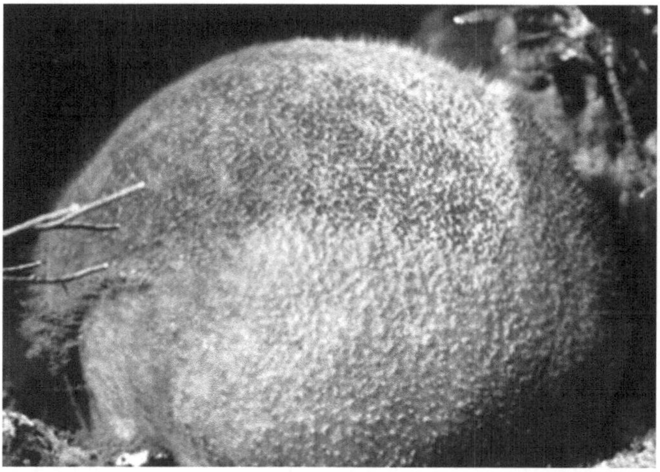

Craniella **sp.**

Image credit: SIMoN/MBNMS, Public domain.

Common name: Not available

Global distribution: Australia, Western Indo-Pacific, South China Sea,

Habitat: No information

Associated bacteria: Pharmaceutical compounds and activities

Antifungal activity: The predominant phylum associated with this sponge species was Proteobacteria followed by Bacteroidetes. The enzyme chitinase isolated from the bacterial extracts of *Streptomyces* sp. DA11 (Actinobacteria) associated with this sponge exhibited antifungal activity (Anbuchezhian and Li 2014).

Order: Clionaida Family: Clionaidae

Cliona sp.

Associated bacteria: Pharmaceutical compounds and activities

Antiviralactivity: The bacterial strains *Comamonas testosteroni* and *Citrobacter freundii* isolated from this sponge species yielded alkaloids with O–H, C–H, C=O, and N–H groups which showed antiviral activity against VP40 target of Ebola virus (Skariyachan et al. 2014).

Pione vastifica

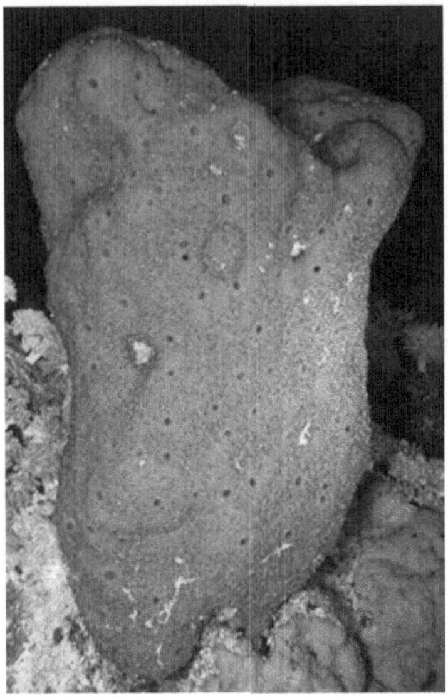

Image credit: Alexander Vasenin, https://creativecommons.org/licenses/by-sa/3.0/

Common name: Red/yellow boring sponge

Global distribution: Worldwide; common in Red Sea to western Pacific Ocean.

Habitat: Sessile; depth range 0–15 m; It bores 5 cm (2.0 in) into coralline substrate

Associated bacteria: Pharmaceutical compounds and activities

Antimicrobial activity: The isolated bacterial genera from this sponge species were *Vibrio, Halomonas, Alteromonas, Staphylococcus, Pseudoalteromonas, Tenachibaculum, Spongiobacter, Bacillus,* and *Pseudovibrio* which belong to three different classes (γProteobacteria, Firmicutes, and Flavobacteria) where dominant class was Actinobacteria (42%). Of 57 bacterial isolates identified from this sponge species, 24 isolates showed antimicrobial activity (agonistic activity) against bacterial strains (*Escherichia coli* ATCC 8739, *Enterococcus faecalis* ATCC 29212, *Enterococcus faecium* ATCC 27270 and *Pseudomonas aeruginosa* ATCC 27853) and pathogenic oomycetes (*Phytophthora capsici* and *Pythium ultimum*) (Bibi et al. 2018).

Spheciospongia vagabunda (= *Spirastrella vagabunda*)

Common name: Chocolate sponge

Global distribution: Australia

Habitat: Muddy bottoms of lagoons near the mid-water mark

Associated bacteria: Pharmaceutical compounds and activities

Antibacterial activity: The compound, microluside A [4 (19-para-hydroxy benzoyloxy-O–D-cellobiosyl), 5 (30-para-hydroxy benzoyloxy-O–D-glucopyranosyl) xanthone isolated from its associated bacterial strain viz. *Micrococcus* sp. EG45 displayed antibacterial activity against *Staphylococcus aureus* NCTC 8325 and *Enterococcus faecalis* JH212 with MIC values of 13 and 10 M, respectively (Eltamany et al. 2014).

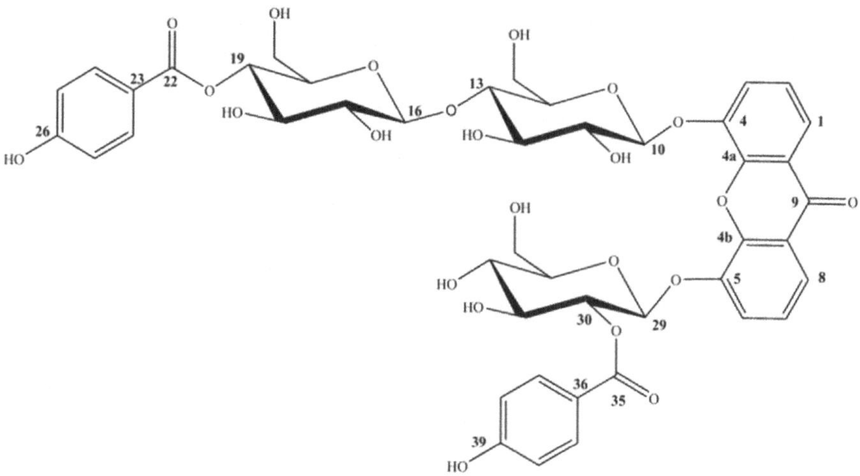

Microluside A

Abdelmohsen et al. (2014) reported that the bacterial strain *Actinokinespora* sp. EG49 (Actinobacteria) isolated from this sponge produced the compound actinosporin A, which displayed antiparasitic activity against *Trypanosoma brucei.*

Order: Dendroceratida Family: Dictyodendrillidae

Spongionella nigra *(= Dendrilla nigra)*

Common name: Black sponge

Global distribution: Arabian coast and from southern India.

Habitat: Not reported

Associated bacteria: Pharmaceutical compounds and activities

Antibacterial activity: The bacterial species *Streptomyces dendra* sp. nov. MSI051 associated with this sponge species yielded an unidentified compound which showed antibacterial activity. Similarly its *Streptomyces* sp. BLT7 also showed similar activity with an unidentified compound (Selvin et al. 2004).

Selvin et al. (2009) reported that the symbiotic Actinobacteria, *Nocardiopsis dassonvillei* MAD08 of this species produced acetic acid, butylester, ethanol, 2-(octyloxy)-, oxalic acid, allylnonyl ester, 2-isopropyl-5-methyl-1-heptanol, butylatedhydroxytoluene, cyclohexanecarboxylic acid, hexyl ester, Diethylphthalate, pentadecanal, 1-tridecanol, and 9-octadecenal. All these compounds have shown antimicrobial activity.

Antioxidant activity: *Nocardiopsis dassonvillei* MAD08 of this species also produced compounds viz. hexadecanoic acid, methyl-ester, hexadecanoic-acid, ethyl ester, and n-hexadecanoic-acid and all these compounds displayed antioxidant, hypocholesterolemic, nematicide, antiandrogenic, and hemolytic activities (Selvin et al. 2009)

Anti-inflammatory activity: *Nocardiopsis dassonvillei* MAD08 of this species also produced 9-octa-decenamide-(Z)-, (E)-9-octadecenoicacid ethyl ester, oleic acid, and 9-ctadecenoicacid- (Z)-, methyl-ester and all these compounds displayed anti-inflammatory, antiandrogenic, cancerpreventive, dermatitigenic, hypo-cholesterolemic, and anemiagenic activities (Selvin et al. 2009).

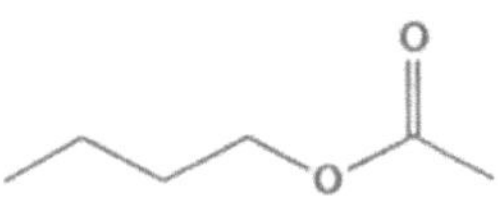

Acetic acid, butyl ester

Ethanol, 2-(octyloxy)-

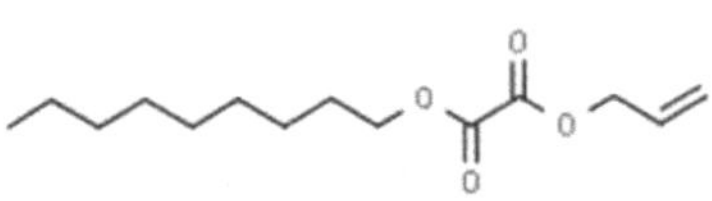

Oxalic acid, allyl nonyl ester

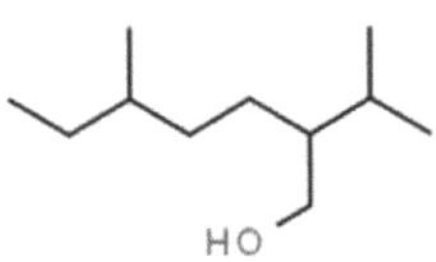

2-Isopropyl-5-methyl-1-heptanol

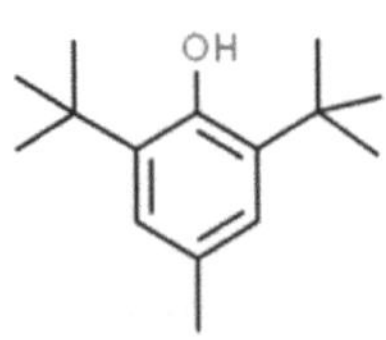

Butylated hydroxytoluene

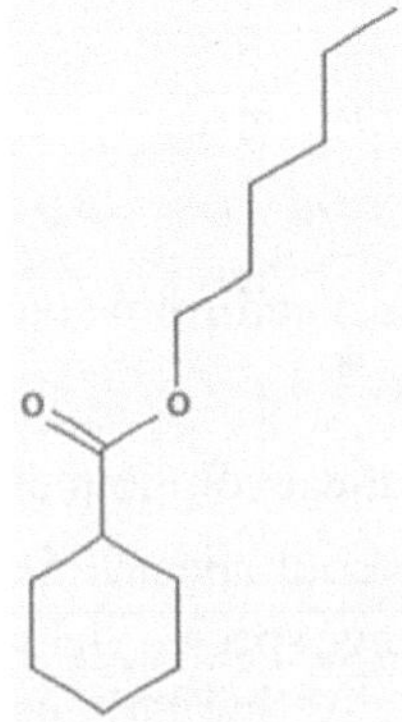

Cyclohexanecarboxylic acid, hexyl ester

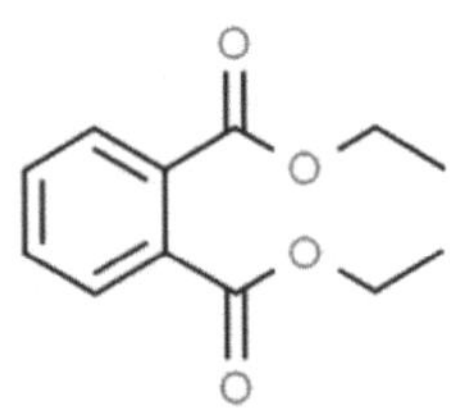

Diethyl phthalate

Pentadecanal-

1-Tridecanol

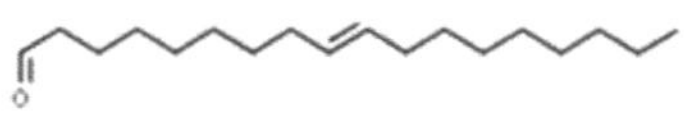

9-Octadecenal

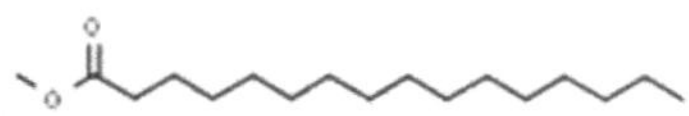

Hexadecanoic acid, methyl ester

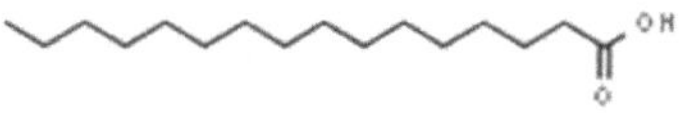

n-Hexadecanoic acid

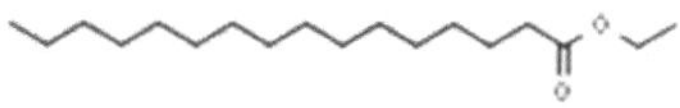

Hexadecanoic acid, ethyl ester

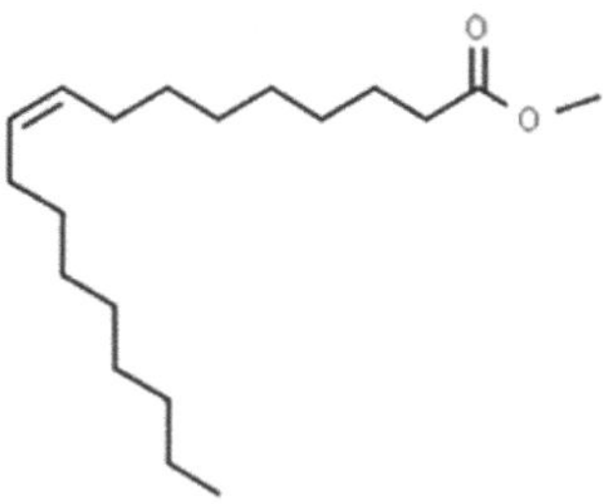

9-Octadecenoic acid (Z)-, methyl ester

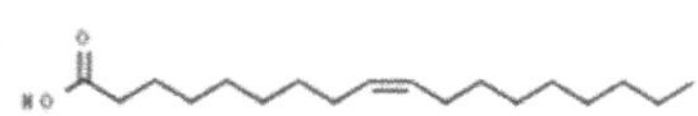

Oleic acid

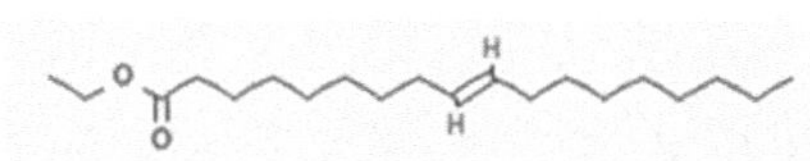

(E)-9-Octadecenoic acid ethyl ester

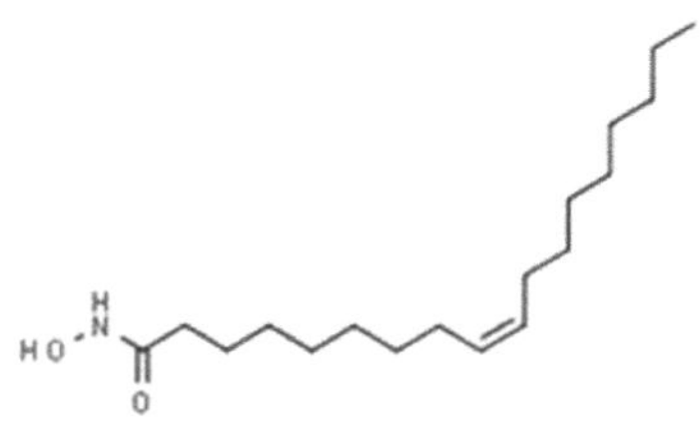

9-Octadecenamide, (Z)-

Order: Dictyoceratida Family: Dysideidae

Dysidea avara

Common name: Mediterranean sponge

Global distribution: Western Europe, Galicia; Mediterranean.

Habitat: In caves and on vertical cliffs, from 1 m downwards (to at least 100 m)

Associated bacteria: Pharmaceutical compounds and activities

Antimicrobial and antiangiogenic activities: The compound 2-methylthio-1,4-naphthoquinone isolated from its associated bacterial strain viz. Proteobacteria MBIC3368 showed both antiangiogenic and antimicrobial activities (Thakur et al. 2005).

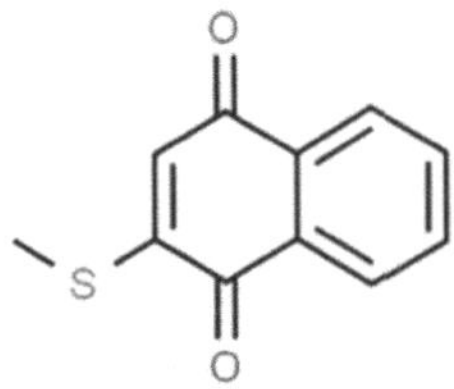

2-Methylthio-1,4-naphthoquinone

Li (2019) reported that the bacterial strain *Bacillus vallismortis* C89 associated with this sponge species produced compounds viz. neobacillamide A and bacillamide C. The pharmaceutical activities of these compounds are to be known.

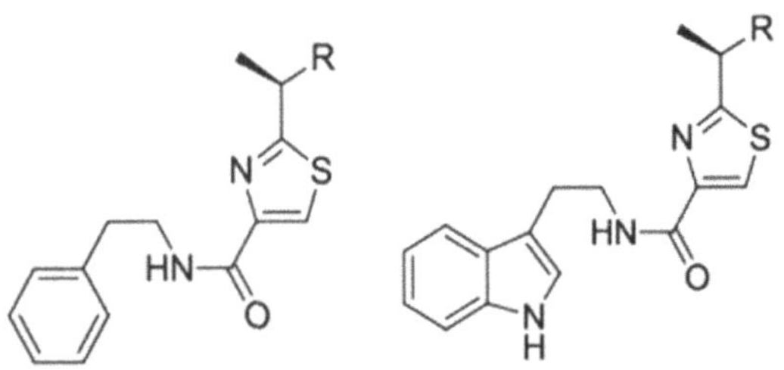

R = NHAc

Neobacillamide A Bacillamide C

Dysidea fragilis (Schulze, 1879)

Image credit: Auguste Le Roux, https://creativecommons.org/licenses/by-sa/4.0/

Common name: Goosebump sponge

Global distribution: Worldwide.

Habitat: Sessile; brackish; depth range 13–2300 m

Associated bacteria: Pharmaceutical compounds and activities

Antibacterial activity: The compound Pyrrolo[1,2-a]pyrazine-1,4-dione, hexahydro- isolated from its associated bacterial strain viz. *Bacillus* sp SEB32 displayed antibacterial activity against several pathogenic bacteria such as *Vibrio* spp. viz. *Vibrio vulnificus*, *Vibrio alginolyticus*, and *Vibrio parahaemolyticus* with inhibition zone diameter values of 16 mm, 18 mm and 33 mm, respectively. Other pathogens inhibited by the aforesaid compound were *Aeromonas salmonicida* (21 mm), *Flavobacterium* sp. (19 mm), *Edwardsiella* sp. (19mm), *Proteus mirabilis* (18 mm) and *Citrobacter brackii* (16 mm) (Mohan et al. 2016)

***Dysidea* sp.**

Associated bacteria: Pharmaceutical compounds and activities

Cytotoxic and antibacterial activities: The *Vibrio* sp. (γProteobacteria) associated with this sponge produced tetrabromodiphenyl ethers, which possess cytotoxic and antibacterial activities (Thomas et al. 2010).

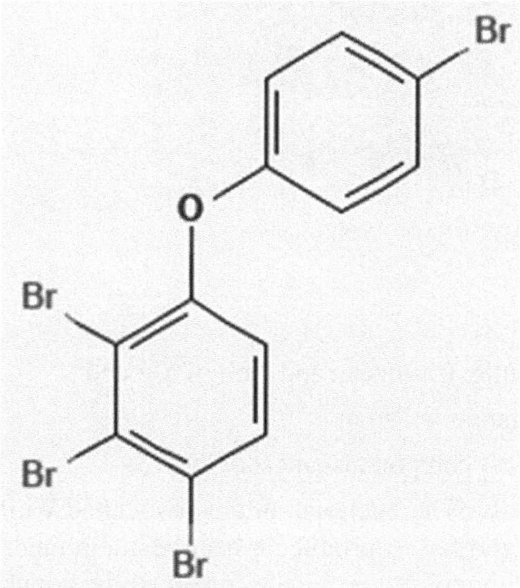

Tetrabromodiphenyl ether

Lamellodysidea herbacea *(= Dysidea herbacea)*

Common name: Silvery blue sponge

Global distribution: Indo-West Pacific

Habitat: Mostly in the lagoon, on the inner reef barrier slope or on the pinnacles; sessile; depth range 0 - 100 m

Associated bacteria: Pharmaceutical compounds and activities

Antibacterial activity: Ethyl acetate extracted metabolites of this sponge-associated strain *Bacillus amyloliquefaciens* (MK135790) (CAB1) showed significant antibacterial activity against four pathogens *B. cereus, B. subtilis, S. aureus* and *S. entrica typhimurium* and another strain *Alcaligenes faecalis* (MK135791) (CAB38) showed activity against three pathogens *B. subtilis*, *E. coli*, and *S. entrica typhimurium* (Sawhney and Mishra 2019).

Order: Dictyoceratida Family: Irciniidae

Ircinia felix

Image credit: NURC/UNCW and NOAA/FGBNMS., Public domain.

Common name: Stinker sponge

Global distribution: Western Atlantic: Caribbean and Gulf of Mexico.

Habitat: Sessile; brackish; depth range 1–100 m

Associated bacteria: Pharmaceutical compounds and activities

Quorum sensing inhibiting activity: The bacterial strains associated with this sponge species have been reported to produce bioactive compounds such as manoalide, manoalide monoacetate and secomanoalide which possess quorum sensing (bacterial communication) inhibiting activity) (Kalia 2018).

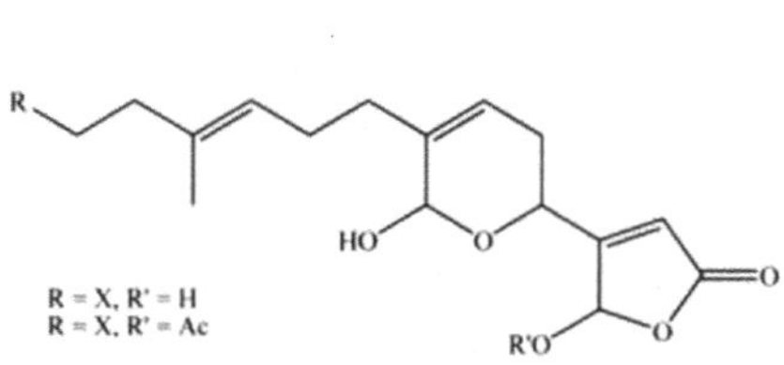

Manoalide, Secomanoalide

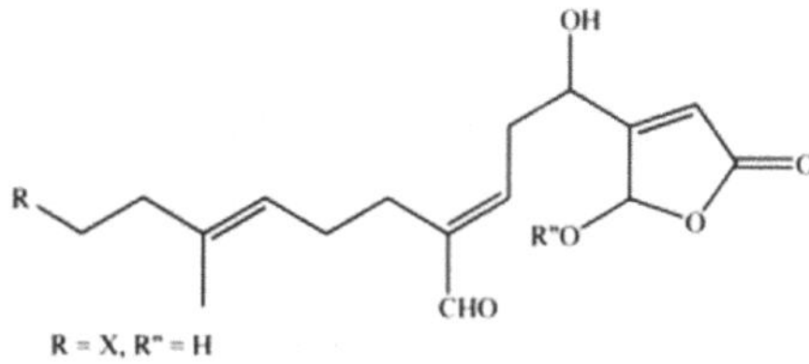

Manoalide monoacetate

Ircinia fusca

Common name: Not designated

Global distribution: Tropical, Indo-West Pacific: India and Indonesia.

Habitat: Sessile

Associated bacteria: Pharmaceutical compounds and activities

Anticancer activity: The bacterial species *Micrococcus* sp. and *Bacillus* sp associated with this sponge species showed antibacterial activities (Hardoim and Costa 2014)

Ircinia variabilis *(Schmidt 1862)*

Common name: Not designated

Global distribution: Portugal, Galicia; Mediterranean.

Habitat: From the low water mark downwards to at least 150 m; common on vertical cliffs between 7 and 25 m

Associated bacteria: Pharmaceutical compounds and activities

Antimicrobial activity: The isolates *Microascus* sp. 44 and 27 of this sponge species showed antibacterial activity towards *E. coli* and *S. aureus*, respectively. Further, its strain K14 produced the compound fungerin, which showed antifungal activity(Hardoim and Costa 2014).

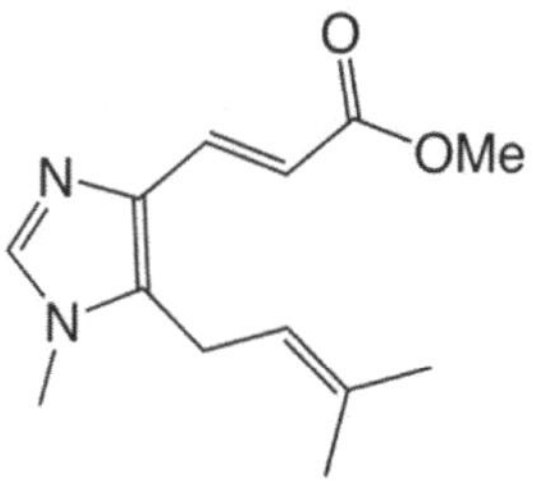

Fungerin

Ircinia sp.

Associated bacteria: Pharmaceutical compounds and activities

Anti-tumor and anti-HIV activities: *Bacillus pumilus* strain which is isolated form this sponge species produced cyclic acyldepsipeptides or bacircines, bacircines (BI) 2, 3, 4, 5, and 5A which exhibited anti-tumour activity against Ehrlich ascites carcinoma and anti-HIV activity (Prokof'eva et al. 1999).

Sarcotragus fasciculatus (= *Ircinia fasciculata*)

Common name: Stinking sponge

Global distribution: Subtropical: Western Central Pacific, Atlantic, and the Mediterranean.

Habitat: Sessile; depth range 7–9 m

Associated bacteria: Pharmaceutical compounds and activities

Antimicrobial activity: The Proteobacteria isolated from this sponge species displayed antimicrobial activity against *Staphylococcus lentus*, *Escherichia coli*, *Bacillus subtilis*, *Candida* sp., and *Mycobacterium* sp. (Thomas et al. 2010).

β-glucosidase inhibitory activity: Five Firmicutes isolates and one Proteobacteria isolate inhibited β-glucosidase activity [69]. Glucosidase inhibitors may be of use in the treatment of viral infections, cancer, and genetic disorders (Hardoim and Costa 2014).

Sarcotragus spinosulus

Common name: Not reported

Global distribution: Portugal, Galicia; Mediterranean.

Habitat: It is commonly occurring in rocky substrates at depths of 1–300 m. Depths of abundance have been reported as 8–25 m.

Associated bacteria: Pharmaceutical compounds and activities

Antimicrobial activity: Antagonistic activity of *Escherichia coli* NCTC 9001 (Gram-negative) and *Staphylococcus aureus* NCTC 6571 (Gram-positive). A few isolates ($n = 18$) were active against both strains, whereas 44 and 27 isolates showed antimicrobial activity towards *E. coli* and *S. aureus*, respectively (Hardoim and Costa 2014).

Order: Dictyoceratida Family: Spongiidae

Hyattella cribriformis

Global distribution: Australia and New Zealand

Habitat: Not reported

Associated bacteria: Pharmaceutical compounds and activities

Antibacterial activity: The bacterial strain *Bacillus cereus* SBS02 associated with this sponge species displayed antibacterial activity against pathogens such as *Klebsiella pneumoniae, Pseudomonas aeroginoasa, E.coli, Staphylococcus aureus, Salmonella typhi,Salmonella paratyphi, Klebsiella oxytoca* and *Vibrio cholerae* (Brammavidhya and Usharani 2013).

Antimicrobial activity: The associated bacteria of this sponge species viz. *Bacillus cereus, Bacillus megaterium, Bacillus subtilis, Citrobacter freundii, Neisseria mucosa, Pseudomonas putida* and *Lactobacillus plantarum* displayed antibacterial activity against *Escherichia coli, Proteus mirabilus, Salmonella typhi, Salmonella paratyphi, Vibrio cholerae, Klebsiella oxytoca, Klebsiella pneumonia, Staphylococcus aureus, Lactobacillus vulgaris,* and *Agrobacterium tumefaciens*; and its aforesaid bacteria along with *Vibrio cholerae* showed antifungal activity against *Alternaria alternata, Botrytis cinerea, Cercospora theae, Fusarium udum, Fusarium oxysporum, Macrophomina phaseolina, Poria hypolateritia, Phomopsis thae*, and *Rhizoctonia solani* (Vasanthabharathi and Jayalakshmi 2012).

Hyatella sp.

Associated bacteria: Pharmaceutical compounds and activities

Antibiotic activity: The bacterial strain *Vibrio* sp. M22-1 (γ- Proteobacteria) isolated from this sponge species produced the compound andrimid which displayed antibiotic activity (Oclarit et al. 1994).

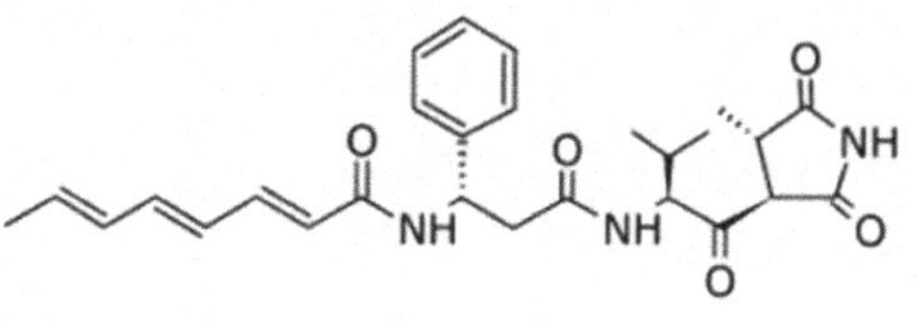

Andrimid

Rhopaloeides odorabile

Common name: Great Barrier Reef sponge

Global distribution: French Polynesia: Australes archipelago.

Habitat: On the costal slope or on the outer reef slope.

Associated bacteria: Pharmaceutical compounds and activities

Antimicrobial activity: The isolates of this sponge speices belonging to the genera *Pseudomonas, Pseudovibrio, Bacillus, Kocuria*, and *Kytococcus* showed antimicrobial activity against human pathogenic microbial reference strains: *Escherichia coli* (ATCC® BAA-196™), *E. coli* (ATCC® 13706™), *E. coli* (ATCC® 25922™), *Klebsiella pneumoniae* (ATCC® BAA-1705™), *Enterococcus faecalis* (ATCC® 51575™), *Bacillus subtilis* (ATCC® 19659™), *Staphylococcus aureus* (ATCC® 29247™), *Candida albicans* (ATCC® 10231™), and *Aspergillus niger* (ATCC® 16888™) (Brinkmann et al. 2017)

Spongia (Spongia) ceylonensis (= *Spongia officinalis* var. *ceylonensis)*

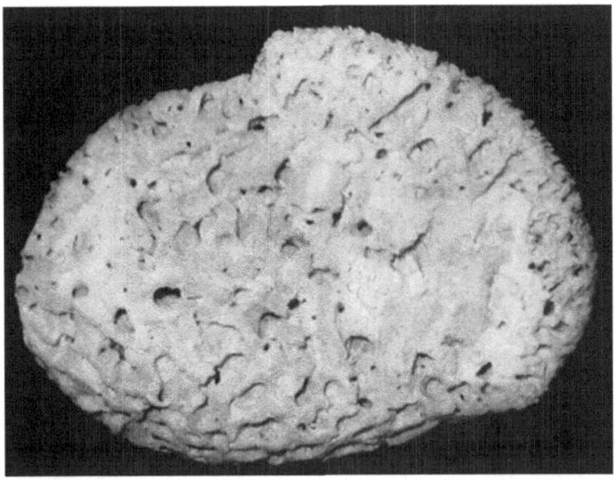

Image credit: H. Zell, https://creativecommons.org/licenses/by-sa/3.0/

Common name: Bath sponge

Global distribution: Mediterranean Sea, Caribbean, and Micronesian areas

Habitat: Shallow water, on littoral rocky or sandy bottoms

Associated bacteria: Pharmaceutical compounds and activities

Antimicrobial activity: The associated bacteria of this sponge species viz. *Bacillus megaterium, Bacillus brivis, Bacillus subtilis, Vibrio parahaemolyticus,* and *Escherichia coli* displayed antibacterial activity against *Escherichia coli, Proteus mirabilus, Salmonella typhi, Salmonella paratyphi, Vibrio cholera, Klebsiella oxytoca, Klebsiella pneumonia, Staphylococcus aureus, Lactobacillus vulgaris,* and *Agrobacterium tumefaciens*; and the aforesaid associated bacteria along with *Lactobacillus fermentum* showed antifungal activity against *Alternaria alternata, Botrytis cinerea, Cercospora theae, Fusarium udum, Fusarium oxysporum, Macrophomina phaseolina, Poria hypolateritia, Phomopsis thae*, and *Rhizoctonia solani* (Vasanthabharathi and Jayalakshmi 2012).

Spongia sp.

Associated bacteria: Pharmaceutical compounds and activities

Antimicrobial activity: The extracts of the isolates of *Dietzia* sp., *Bacillus marisflavi,* and *Salinicoccus roseus* of this sponge species showed antibacterial activity against *Bacillus subtilis* with inhibition zone values of 7.5 and 7.5 mm, respectively. Its another isolate *Staphylococcus equorum* showed activity toward both *Bacillus subtilis* and *Candida albicans* with values of trace and 10 mm, respectively (Anand et al. 2006).

Order: Dictyoceratida Family: Thorectidae

Candidaspongia flabellata

Common name: No data

Global distribution: Great Barrier Reef, Australia, Philippines

Habitat: No informaiton

Associated bacteria: Pharmaceutical compounds and activities

Antimicrobial activity: The isolates of this sponge speices belonging to the genera *Pseudomonas, Pseudovibrio, Bacillus, Kocuria* and *Kytococcus* showed antimicrobial activity against human pathogenic microbial reference strains: *Escherichia coli* (ATCC® BAA-196™), *E. coli* (ATCC® 13706™), *E. coli (*ATCC® 25922™*), Klebsiella pneumoniae* (ATCC® BAA-1705™), *Enterococcus faecalis* (ATCC® 51575™), *Bacillus subtilis* (ATCC® 19659™), *Staphylococcus aureus* (ATCC® 29247™), *Candida albicans* (ATCC® 10231™), *and Aspergillus niger* (*ATCC® 16888*™) (Brinkmann et al. 2017)

Fascaplysinopsis reticulata

Common name: Not reported

Global distribution: Western Central Pacific: Indonesia and New Caledonia

Habitat: Sessile; depth range 0–100 m

Associated bacteria: Pharmaceutical compounds and activities

Cytotoxicity: Two brominated yellow main pigments, bromoalterochromide A and A′, isolated from *Pseudoalteromonas maricaloris* KMM 636T (γ-Proteobacteria) associated with this sponge species have shown cytotoxicity (Speitling et al. 2007).

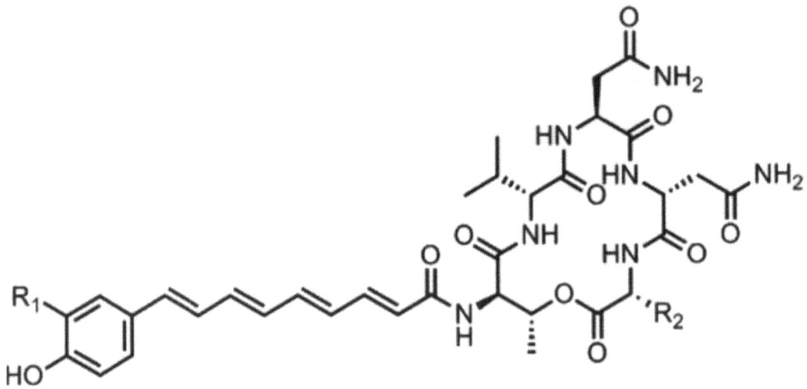

Bromoalterochromide A R1= Br,R2 = Isoleucine

Bromoalterochromide A’: R1 = Br,R2 = Leucine

Fasciospongia cavernosa

Common name: Not reported

Global distribution: Northern Adriatic Sea, Southern Adriatic Sea, Aegean Sea

Habitat: Cave, coralligenous community, rocky/detritic/muddy bottom, *Posidonia oceanica* meadow. Sometimes it presents a burrowing behavior. Bathymetric range 1–367 m.

Associated bacteria: Pharmaceutical compounds and activities

AChE inhibiting activity: The bacterial isolate *Bacillus subtilis* M18SP4 of this species showed significant (54%) acetylcholinesterase (AChE) inhibiting activity. This strain may have important therapeutic use in the treatment of dementia associated with Alzheimer’s disease (Pandey et al. 2014).

Antibacterial activity: The bacterial strain *Halobacteria* sp. GUVFFM-1 isolated from this sponge species displayed antibacterial activity against *S. marcesans, A. aerogens, E. coli*, and *P. vulgairs* with zone of inhibition values of 10, 1, 1, and 17 mm, respectively (Velho-pereira and Furtado 2012).

Hyrtios altus

Global distribution: Indo-Pacific

Habitat: Not reported

Associated bacteria: Pharmaceutical compounds and activities

Antibacterial activity: The compound trisindole derivative, viz. trisindoline produced by the associated *Vibrio* sp., of this sponge species showed antibacterial activity against *Escherichia coli* and *Bacillus* sp. (Thomas et al. 2010).

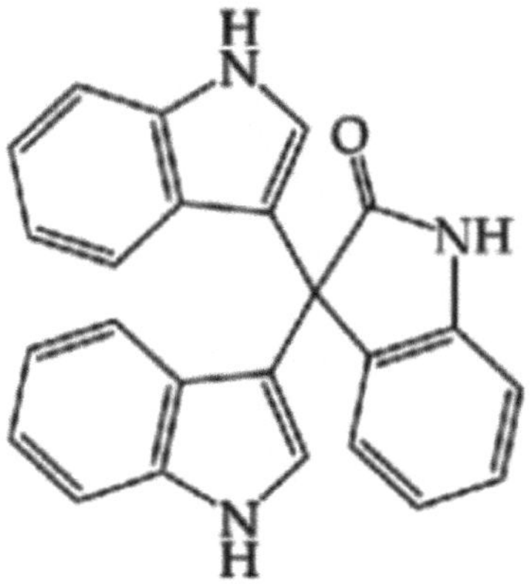

Trisindoline

Hyrtios erectus (= *Heteronema erectus)*

Global distribution: Indo-Pacific

Habitat: Not reported

Associated bacteria: Pharmaceutical compounds and activities

Antibacterial activity: The bacterial strains *Micromonospora* sp. S6 (HF674982), *Salinispora arenicola* strain SCSIOZ-SH11 (KC747479) and *Saccharomonospora* sp. have been found associated with this sponge species. Among these strains, the Actinomycete *Saccharomonospora* sp. showed antibacterial activity against *Escherichia coli* K12 and *Bacillus* sp. P25 (Abdelmohsen et al. 2010).

Velho-pereira and Furtado (2012) reported that the bacterial strain *Pseudomonas* sp. GUVFHEM-4 isolated from this sponge species displayed antibacterial activity against *E. coli*, *S. citreus*, and *P. vulgaris* with zones of inhibition 10, 20, and 18 mm, respectively.

Hyrtios sp.

Associated bacteria: Pharmaceutical compounds and activities

Antibacterial activity: An Actinomyces isolate HRHa.5 of this sponge displayed antibacterial activity against several pathogenic bacteria (Dita 2017).

Order: Hadromerida Family: Raspailiidae

Echinodictyum gorgonoides

Common name: No information

Global distribution: No information

Habitat: No information

Associated bacteria: Pharmaceutical compounds and activities

Antimicrobial properties: The bacterial strain *Staphyloccous* sp. MB2 associated with this sponge species showed antibacterial activity against bacterial pathogens *E. coli, Pseudomonas, S. aureus*, and *B. subtilis* (Dhinakaran et al. 2012).

Anticancer activity: The strain *Staphyloccous* sp. MB2 has also been reported to possess anticancer activity against against glioma C6 cell line (Dhinakaran et al. 2012).

Echinodictyum sp.

Associated bacteria: Pharmaceutical compounds and activities

Antimicrobial activities: The bacterial species *Bacillus megaterium* and *Vibrio* sp. isolated from this species showed antibacterial activity toward pathogenic *Bacillus subtilis* with values of inhibition zone diameter 8 and 3 mm, respectively. Further, its isolate *Idiomarina baltica* had activity against both *Bacillus subtilis* and *Candida albicans* with 7 and 20 mm, respectively (Anand et al. 2006).

Order: Haplosclerida Family: Callyspongiidae

Callyspongia diffusa

Common name: Champagne sponge.

Global distribution: Indo-West Pacific.

Habitat: Sessile; depth range 0–100 m

Associated bacteria: Pharmaceutical compounds and activities

Antibacterial activity: The compound, pyrrolo[1,2-a] pyrazine-1,4-dione, hexahydro isolated from the bacterial strain *Bacillus tequilensis* MS145 associated with this sponge showed antibacterial activity against MDR *Staphylococcus aureus* with an MIC value of 15 mg/L (Nweze et al. 2020).

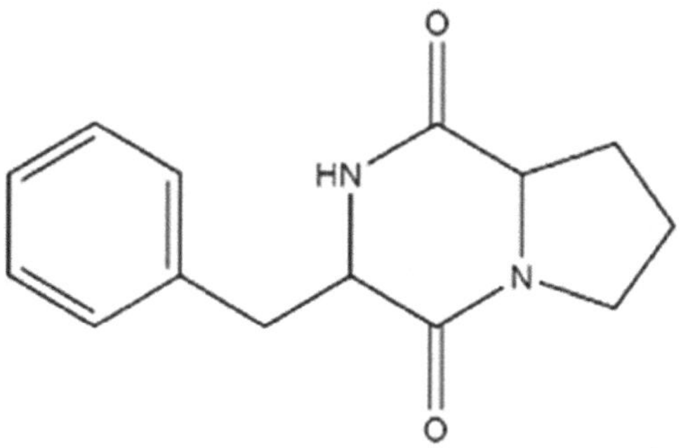

Pyrrolo[1,2-a] pyrazine-1,4-dione, hexahydro

The bacterial strain *Virgibacillus* sp. associated with this sponge displayed activity against pathogenic bacteria viz. *E.coli, Staphylococcus aureus, Vibrio mimicus, Vibrio alginolyticus*, and *Aeromonas hydrophila*. Further, this isolation also acted as potential immunomodulators (Kalirajan et al. 2013).

Antimicrobial activity: The associated bacteria of this sponge species viz. *Alcaligenes faecalis, Aeromonas hydrophila, Bacillus licheniformis, Bacillus subtilis, Pseudomonas aeruginosa, Pseudomonas fluorescens* and *Pseudomonas putida* displayed antibacterial activity against *Escherichia coli, Proteus mirabilus, Salmonella typhi, Salmonella paratyphi, Vibrio cholera, Klebsiella oxytoca, Klebsiella pneumonia, Staphylococcus aureus, Lactobacillus vulgaris,* and *Agrobacterium tumefaciens*; and antifungal activity against *Alternaria alternata, Botrytis cinerea, Cercospora theae, Fusarium udum, Fusarium oxysporum, Macrophomina phaseolina, Poria hypolateritia, Phomopsis thae*, and *Rhizoctonia solani* (Vasanthabharathi and Jayalakshmi 2012).

Callyspongia fibrosa

Common name: No data

Global distribution: Indo-West Pacific: India and Indonesia.

Habitat: Sessile; depth, 20 m.

Associated bacteria: Pharmaceutical compounds and activities

Antibacterial Activity: The bacterial strain *Corynebacterium* sp. GUVFCFM-3 isolated from this sponge species displayed antibacterial activity against *S. typhi* with a 1mm zone of inhibition (Velho-pereira and Furtado 2012).

Callyspongia reticutis

Common name: No data

Global distribution: south-west coast of India

Habitat: Coastal waters

Associated bacteria: Pharmaceutical compounds and activities

Antibacterial activity: The bacterial strain *Halobacteria* sp. GUVFCM isolated from this sponge species displayed antibacterial activity against *S. typhi*, *S. marcesans, S. citreus*, and *P. vulgaris* with zone of inhibition values of 2, 4, 3 and 6 mm, respectively (Velho-pereira and Furtado 2012)

Callyspongia (Callyspongia) siphonella (= *Siphonochalina siphonella)*

Image credit: Alexander Vasenin, https://creativecommons.org/licenses/by-sa/3.0/

Common name: Colonial tube-sponge

Global distribution: Indian Ocean: Red Sea.

Habitat: Inhabits reef slopes

Associated bacteria: Pharmaceutical compounds and activities

Antimicrobial activity: The isolated bacterial genera from this sponge species viz. *Pseudoalteromonas, Vibrio, Photobacterium, Psychrobacter, Halomonas, Kocuria, Bacillus,* and *Carnobacterium* belonged to three different classes (γ-Proteobacteria, Actinobacteria, and Firmicutes) and the dominant class was γ-Proteobacteria (34%). Of 38 bacterial isolates identified from this sponge species, 13 isolates showed antimicrobial activity (agonistic activity) against bacterial strains (*Escherichia coli* ATCC 8739, *Enterococcus faecalis* ATCC 29212, *Enterococcus faecium* ATCC 27270 and *Pseudomonas aeruginosa* ATCC 27853); and pathogenic oomycetes (*Phytophthora capsici* and *Pythium ultimum*) (Bibi et al., 2018).

***Callyspongia* sp.**

Antibacterial activity: Two phenazines viz. 6-hydroxymethyl-1-phenazine-carboxamide (1) and 1,6-phenazinedimethanol (2) isolated from the bacterial strain *Brevibacterium* sp. KMD 003, which is harbored by this sponge species, showed antibacterial activity against *Enterococcus hirae* and *Micrococcus luteus* with 5M MIC values (Choi et al. 2009).

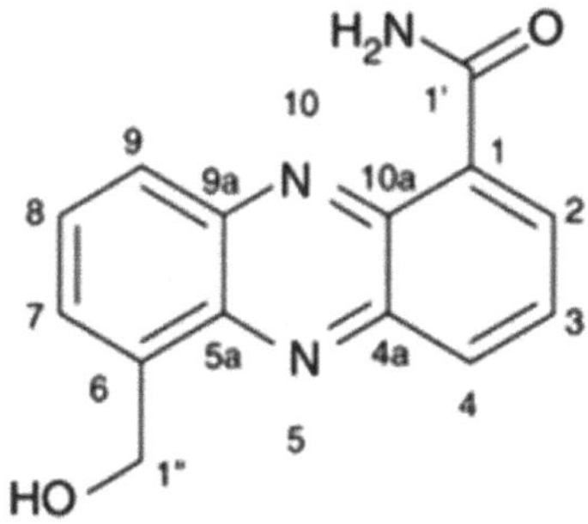

6-hydroxymethyl-1-phenazine-carboxamide

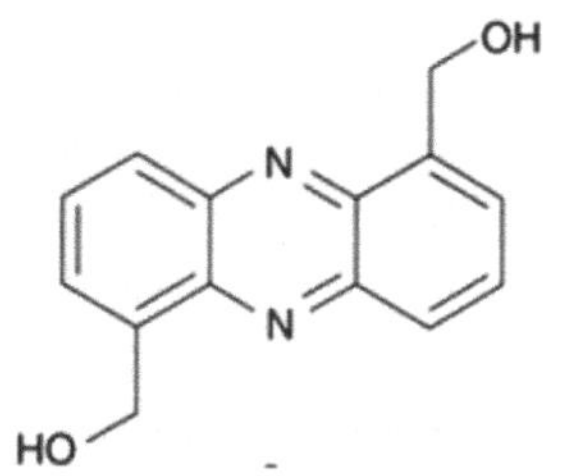

1,6-phenazinedimethanol

Cytotoxic effects: The compounds, nocardiotide A, tryptophan, kynurenic acid, and 4-amino-3-methoxy benzoic acid have been derived from the bacterial strain *Nocardiopsis* sp. UR67 strain associated with this sponge species. Among these compounds, nocardiotide A exhibited significant cytotoxicity against the human multiple myeloma, human HeLa cervix carcinoma, and murine CT26 colon carcinoma with the IC50 values of 8, 11, and 12 μM/mL, respectively (Ibrahim et al. 2018).

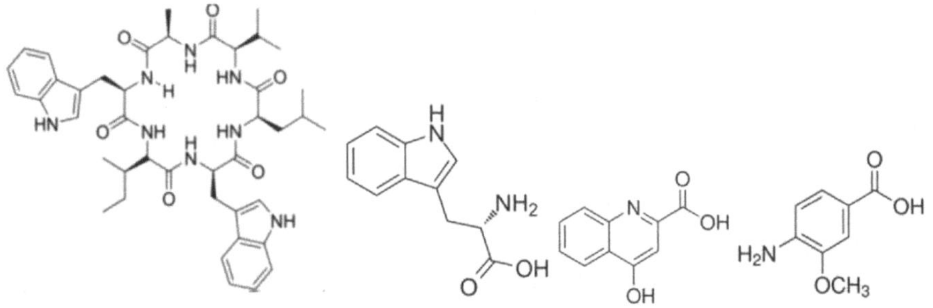

Nocardiotide A Tryptophan Kynurenic acid 4-amino-3-methoxy benzoic acid

Order: Haplosclerida Family: Family Chalinidae

Chalinula **sp.**

Associated bacteria: Pharmaceutical compounds and activities

Antibacterial activity: The actinomycete, *Kocuria* sp. isolated form this sponge displayed antibacterial activity against *Bacillus sp.* P25 and *Escherichia coli* K12 (Abdelmohsen et al. 2010).

Gelliodes carnosa *(= Sigmadocia carnosa)*

Gelliodes sp.

Image credit: Bernard Dupont from France, https://creativecommons.org/licenses/by-sa/2.0/

Common name: Not designated

Global distribution: Indo-West Pacific: India and New Caledonia.

Habitat: Sessile; depth range 0–100 m

Associated bacteria: Pharmaceutical compounds and activities

Antimicrobial activity: The associated bacteria of this sponge species viz. *Bacillus cereus, Bacillus macerans, Bacillus megaterium, Bacillus brivis, Lactobacillus casei, Serratia marcescens, Pseudomonas aeruginosa,* and *Pseudomonas putida* displayed antibacterial activity against *Escherichia coli, Proteus mirabilus, Salmonella typhi, Salmonella paratyphi, Vibrio cholera, Klebsiella oxytoca, Klebsiella pneumonia, Staphylococcus aureus, Lactobacillus vulgaris,* and *Agrobacterium tumefaciens*; and the associated bacteria of this sponge viz. *Bacillus cereus, Bacillus macerans, Bacillus megaterium, Bacillus brivis, Lactobacillus acidophilus, Serratia marcescens, Pseudomonas aeruginosa,* and *Pseudomonas putida* showed antifungal activity against *Alternaria alternata, Botrytis cinerea, Cercospora theae, Fusarium udum, Fusarium oxysporum, Macrophomina phaseolina, Poria hypolateritia, Phomopsis thae*, and *Rhizoctonia solani* (Vasanthabharathi and Jayalakshmi 2012).

Haliclona (Gellius) fibulata *(= Sigmadocia fibulata)*

Common name: Not designated

Global distribution: British Isles, Atlantic coasts of France, Spain, Mediterranean, Spitsbergen, Greenland.

Habitat: In the infralittoral, to ca. 60 m. On stones, conglomerates of shells, calcareous algae, sponges etc.

Associated bacteria: Pharmaceutical compounds and activities

Antimicrobial activity: The extracts of the isolate, *Bacillus* sp. of this sponge species showed antimicrobial activity against *Bacillus subtilis* (inhibition zone, 26 mm), *Escherichia coli* (26 mm), and *Candida albicans* (15 mm); isolate *Marinobacter* sp. had activity against *Bacillus subtilis* (5 mm) and *Escherichia coli* (9 mm); isolate *Bacterium* K54 had activity against *Bacillus subtilis* (6 mm); isolate *Pseudomonas* sp. showed activity against *Bacillus subtilis* (10 mm) and *Candida albicans* (7 mm); and its isolate *Ruegeria* sp. showed activity against *Bacillus subtilis* (5 mm) (Anand et al. 2006).

Haliclona oculata

Image credit: Nudibranch Marine Discovery Productions, https://creativecommons.org/licenses/by-sa/3.0/nl/

Common name: Mermaid's glove

Global distribution: Atlantic coast of Canada and North America

Habitat: This benthic species occurs mainly in sandy and rocky bottom at depths up to 100m where it is often found attached to stones or mollusc, *Mytilus sp*.

Associated bacteria: Pharmaceutical compounds and activities

Antibacterial activity: The bacterial strain *Bacillus licheniformis* T6-1 (Firmicutes) isolated from this sponge species produced Fluorophore compound which showed antibacterial activity against *Salmonella typhi* (Brinkmann et al. 2017).

Haliclona simulans

Common name: Encrusting tubular sponge

Global distribution: Indo-West Pacific and Northeast Atlantic

Habitat: Under intertidal stones and in the infralittoral, to 30 m; sheltered microhabitats.

Associated bacteria: Pharmaceutical compounds and activities

Antimicrobial activity: Certain unidentified compounds of the Actinomycetes; and *Bacillus* and *Pseudovibrio* isolates associated with this sponge displayed activity against important pathogens *Pseudomonas aeruginosa*, *Clostridium difficile*, multi-drug-resistant *Staphylococcus aureus*, and pathogenic yeast strains (Kennedy et al. 2009).

Haliclona verrucosa

Common name: Not known

Global distribution: Antarctic

Habitat: Not reported

Associated bacteria: Pharmaceutical compounds and activities

Antibacterial activity: The bacterial species *Pseudoalteromonas haloplanktis* isolated from this sponge species exhibited almost 100% inhibition against 21 Gram negative Bcc species by producing the bioactive volatile organic compounds (VOCs) identified as benzenamine-N-ethyl and 3-buten-1-ol-3-methyl. Other bacterial strains of the genera associated with this sponge are *Arthrobacter, Pseudoalteromonas, Psychrobacter*, and *Shewanella* (Buchholz and Duncan 2017).

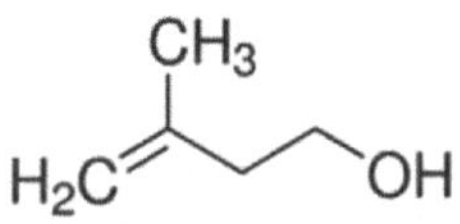

3-Methyl-3-buten-1-ol

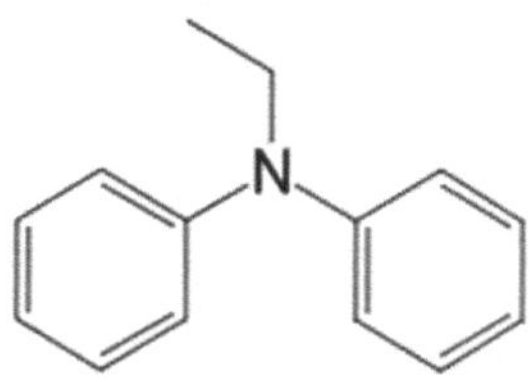

Benzenamine-N-ethyl

Haliclona sp.

Associated bacteria: Pharmaceutical compounds and activities

Antibacterial Activity: The bacterial strain *Chromohalobacter* sp. GUVFHM-2 isolated form this sponge species displayed antibacterial activity against *S. typhi, A. aerogens*, and *S. citreus* with zone of inhibition values of 7, 8, and 7 mm, respectively (Velho-pereira and Furtado 2012).

Asagabaldan et al. (2017) reported that the bacterial strain *Chromohalobacter salixigens* strain DSM3043 associated with this sponge showed positive antibacterial activities against *P. aeruginosa* (zone of inhibition diameter, 21.20 mm), *E. cloaceae* (17.13 mm), *A. baumannii* (15.90 mm), and *S. aureus* (13.10 mm) and shown negative against *K. pneumonia* and *E. coli.*

Santos et al. (2015) and Brinkmann et al. (2017) reported that the bacterial strains *Pseudomonas fluorescens* H40, H41 and *Pseudomonas aeruginosa* H51 (Proteobacteria) associated with this sponge produced the compound, diketopipe-razine, cyclo-(L-Leu-L-Pro) which displayed antibacterial activity against *S. aureus*.

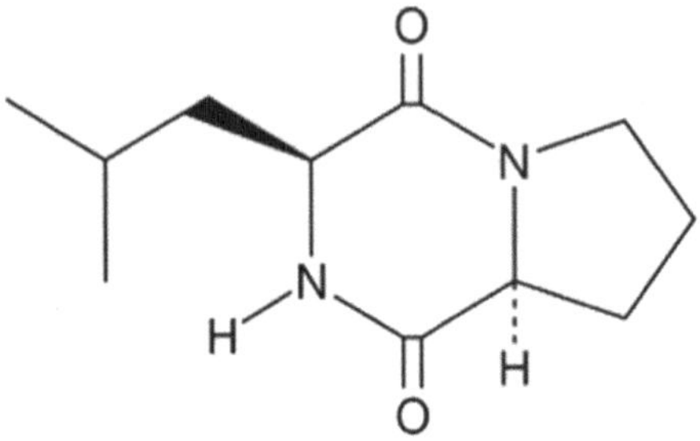

Cyclo-(L-Leu-L-Pro)

Murniasih et al. (2018) reported that the potential extract of bacterial strain, *Agromyces tropicus* UJ.17.10-4 (actinobacteria) associated with this sponge showed the highest anti-Gram-positive bacteria. Further this strain plays an important role in synthesizing Halistanol sulfate and cyclic bis1,3-dialkyl-pyridinium metabolite derived from the host.

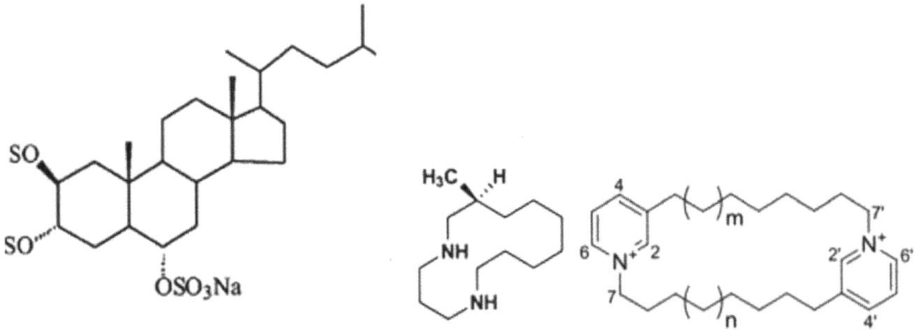

Halistanol sulfate C Haliclorensin cyclic bis1,3-dialkylpyridinium

Cytotoxicity: The anthracycline compound viz. Tetracenoquinocin isolated from its associated strain *Streptomyces* sp. Sp080513GE-26 exhibited weak activity against myelogenous leukemia LH-60 and human cervical carcinoma HeLa cell lines with IC50 values of 210 and 120 μM, respectively (Zhang et al. 2017).

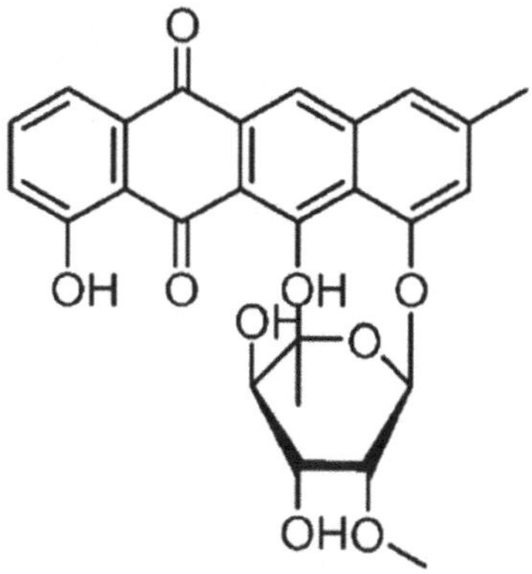

Tetracenoquinocin

Order: Haplosclerida Family: Niphatidae

Haliclonissa verrucosa

Common name: Not designated

Global distribution: Antarctica

Habitat: Depth range 5–40 m

Associated bacteria: Pharmaceutical compounds and activities

Antibacterial activity: *Arthrobacter* strains (TB23, TB26, and CAL618), which were isolated from this sponge species specifically inhibited the growth of a panel of 40 proteobacterial *Burkholderia cepacia* strains which are known to cause cystic fibrosis in humans (Orlandini et al. 2014).

Cryoprotective effect: Caruso et al.(2018) reported that the bacterial strain *Shewanella* sp. CAL606 which is associated with this sponge species produced compounds such as galactose, mannose, galactosamine, glucuronic acid, and galacturonic acid. This strain showed a cryoprotective effect.

Hemigellius pilosus

Commmon name: Antarctic sponge

Global distribution: Antarctica

Habitat: Not reported

Associated bacteria: Pharmaceutical compounds and activities

Cryoprotective effect: Caruso et al. (2018) reported that the bacterial strain *Colwellia* sp. GW185 which is associated with this sponge species, produced compounds such as galactose, mannose, galactosamine, glucuronic acid, and galacturonic acid. This strain showed a cryoprotective effect.

Order Haplosclerida Family Petrosiidae

Acanthostrongylophora ingens

Common name: Indo-Pacific sponge

Global distribution: Western Central Pacific: Indonesia.

Habitat: Sessile; depth range 8–30 m

Associated bacteria: Pharmaceutical compounds and activities

Antitumor activity: The alkaloid compound manzamine A isolated from the associated bacterium *Micromonospora* sp. of this sponge speices has shown antitumor activity. Treatment with manzamine A impedes metastasis in cancer cell lines and sensitizes them to TNF-related apoptosis-inducing ligand (TRAIL)-induced apoptosis (Waters et al. 2014).

Antimalarial activity: Its compound, manzamine A has shown significant activity against the malaria parasite *Plasmodium falciparum* including a 2–7 fold greater activity compared to artemisinin in vitro and a single dose clearance of the parasite from the host (Waters et al., 2014).

Others: The above compound has also promise as a suppressor of hyperlipidemia and atherosclerosis in vivo (Waters et al. 2014).

Acanthostrongylophora sp.

Associated bacteria: Pharmaceutical compounds and activities

Antimalarial and antileukemic properties: An anti-infective alkaloid manzamine A isolated from its associated actinobacterium *Micromonospora* sp. has shown antitumor properties against mouse leukemia cells and antimalarial activity on the rodent malarial parasite viz. *Plasmodium berghei* (Thomas et al. 2010).

Others: The compounds, manzamine A and 8-hydroxy-manzamine derived from its associated *Micromonospora* sp. strain M42 displayed antimicrobial, antiparasitic, and anti-tumor activities. Owing to the promising antimalarial activity, manzamine could become a drug class in the near future (Peraud 2006).

Petrosia (Petrosia) ficiformis

Image credit: Guido Picchetti, https://creativecommons.org/licenses/by-sa/3.0/

Common name: Stony sponge

Global distribution: North Atlantic, Adriatic Sea, Aegean Sea, Ionian Sea, Levantine Sea, and Mediterranean Sea

Habitat: Underside of rocks, on overhangs and in caves between 5 and 70 m deep

Associated bacteria: Pharmaceutical compounds and activities

Antimicrobial activity: The bacterial strains of *Rhodococcus* sp. and *Pseudomonas* sp. isolated from this sponge species showed antimicrobial activity (Thomas et al. 2010).

Chelossi et al. (2007) reported on the isolation of a total of 416 different bacterial strains from this sponge species. Of the bioactive strains, 58.3% were able to inhibit *Staphylococcus aureus*, 6.7% were active against *Bacillus subtilis*, 11.7% against both *Enterococcus faecalis* and *Escherichia coli,* 38.3% against *Pseudoalteromonas atlantica* and 33.3% against *Pseudomonas elongata*.

Petrosia sp.

Associated bacteria: Pharmaceutical compounds and activities

Antibacterial activity: The bacterial genera isolated from this sponge species include *Acidimicrobium, Ferrimicrobium, Ferrithrix, Actinomyces, Iamia, Propionibacterium, Solirubrobacter, Streptomyces, Thermomonospora,* and *Microthrix*. The acetone extracts of this sponge showed antibacterial activity against *Micrococcus luteus* ATCC 9341, *Streptomyces niveus* NBRC 12804 and *Streptomyces paucisporeus* NBRC 102072 with zone of inhibition values of 1.0 cm, 1.0 cm and 0.9 cm respectively (Khan et al., 2012).

Xestospongia testudinaria

Image credit: Bernard Dupont from France, https://creativecommons.org/licenses/by-sa/2.0/

Common name: Barrel sponge

Global distribution: Widely distributed around the world, from Indo-Pacific to Caribbea.

Habitat: Thrive under various substrates, such as sand, rock, and even dead coral rubble; found in depth of more than 10 m from the sea surface

Associated bacteria: Pharmaceutical compounds and activities

AChE inhibition activity: The bacterial isolates of this sponge species exhibited >10% AChE inhibition activity (Pandey et al. 2014).

Antibacterial activity: The gammaproteobacterial strains Xp 4.1 to 4.6 associated with this sponge species showed antibacterial activity against *E. coli*, *B. subtilis*, and *K. pneumoniae* with the inhibition zone diameter values of 18–24 mm, 6–17 mm, and 8–22 mm, respectively (Cita et al. 2017).

The bacterial strain *Chromohalobacter* sp. GUVFPM-1 isolated from this sponge species displayed antibacterial activity against *S. typhi, A. aerogens*, and *S. citreus* with zone of inhibition values of 6, 6, and 5 mm, respectively; *Corynebacterium* sp. GUVFPM-2 showed activity against *A. aerogens* with 2 mm; and its another isolate *Pontibacillus sp.* GUVFPM-6 showed activity against *P. vulgaris* with 2 mm (Velho-pereira and Furtado 2012).

***Xestospongia* sp.**

Associated bacteria: Pharmaceutical compounds and activities

Antibacterial activity: The crude extracts derived from its associated bacterial strain *Micrococcus luteus* R-1588-10 yielded two compounds such as 2,4,4-trichloro-2-hydroxydiphenylether (triclosan) and acyl-1-(acyl-6-mannobiosyl)-3-glycerol (lutoside) showed antibacterial activity (Thomas et al. 2010).

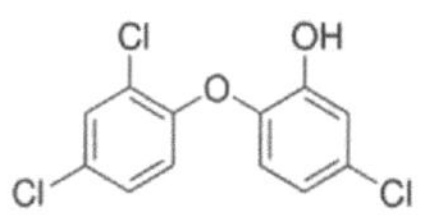

2,4,4′-trichloro-2′-hydroxydiphenylether

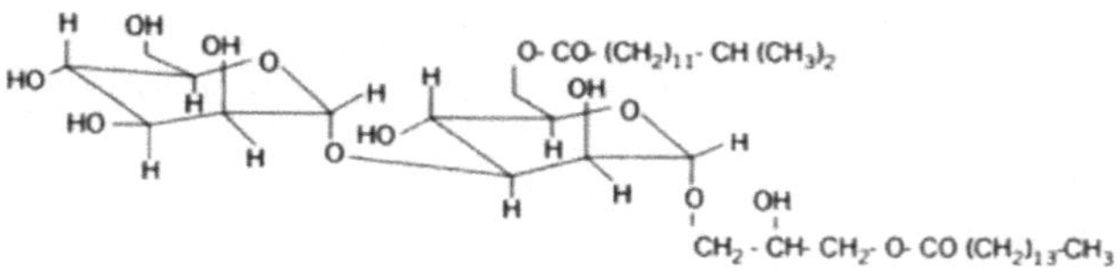

Acyl-1-(acyl-6′-mannobiosyl)-3-glycerol

Order: Haplosclerida Family: Phloeodictyidae

Siphonodictyon coralliphagum

Image credit: iNaturalist user: thibaudaronson, https://creativecommons.org/licenses/by-sa/4.0/

Common name: Variable boring sponge

Global distribution: Western Central Atlantic: Caribbean.

Habitat: Reef-associated; brackish; depth range 3–5 m

Associated bacteria: Pharmaceutical compounds and activities

AChE inhibition activity: The bacterial isolates of this sponge species exhibited >10% AChE inhibition activity (Pandey et al. 2014).

Order: Poecilosclerida Family: Coelosphaeridae

Lissodendoryx (Ectyodoryx) nobilis

Commmon name: Not available

Global distribution: Polar; Antarctic Indian Ocean

Habitat: It is a sessile species

Associated bacteria: Pharmaceutical compounds and activities

Antibacterial activity: The bacterial strain *Arthrobacter* sp. TB23 isolated from this species completely inhibited the pathogenic bacteria, *Burkholderia cepacia* complex which are considered to be the most important pathogens in cystic fibrosis which is the genetic disorder that affects mostly the lung (Giudice and Rizzo 2018).

Buchholz and Duncan (2017) reported that the bacterial species *Pseudoalteromonas haloplanktis* isolated from this sponge species produced benzenamine-N-ethyl and 3-buten-1-ol-3-methy which exhibited almost 100% inhibition against 21 Gram207 negative Bcc species.

Order: Poecilosclerida Family: Crambeidae

Crambe crambe

Image credit: Hectonichus, https://creativecommons.org/licenses/by-sa/4.0/

Common name: Orange-red encrusting sponge or oyster sponge

Global distribution: Mediterranean, North Atlantic Ocean, English Channel and North Sea.

Habitat: It occurs in on hard substrates at depths of 5-30 m

Associated bacteria: Therapeutic compounds and their bioactivities

The bacterial community harbored by this species are believed to be Betaproteobacteria which yielded pentacyclic guanidine alkaloids viz. crambescidines and crambescines, the bioactivities of which are however, yet to be known (Croué et al. 2013).

Monanchora sp.

Image credit: Nhobgood Nick Hobgood. https://creativecommons.org/licenses/by-sa/3.0/

Associated bacteria: Pharmaceutical compounds and activities

Antibacterial activity: Bacterial strains such as *Rhodococcus* sp. L-15 (JN006270), *Mycobacterium* sp. CNJ879 PL04 (DQ448780), *Mycobacterium* sp. CNJ879 PL04 (DQ448780) and *Nocardia* sp. have been isolated from this sponge species. Among these microbes, *Nocardia* sp. has been reported to possess antibacterial activity against *Bacillus* sp. P25 and *Escherichia coli* K12 (Abdelmohsen et al. 2014).

Order: Poecilosclerida Family: Crellidae

Crella cyathophora

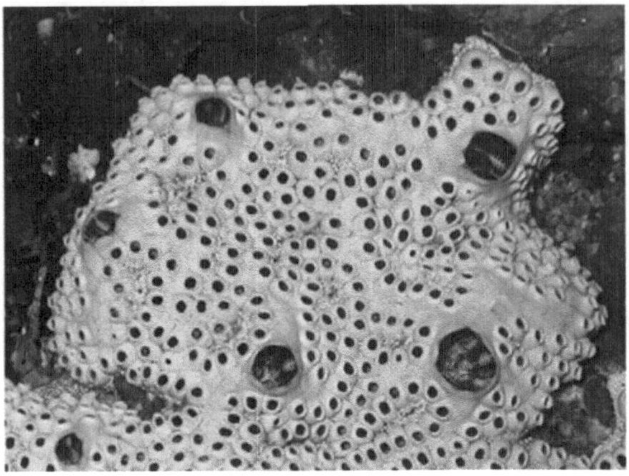

Image credit: Matthieu Sontag, https://creativecommons.org/licenses/by-sa/3.0/

Common name: Honeycomb Sponge

Global distribution: Tropical; Western Central Pacific: Indonesia

Habitat: Sessile.

Associated bacteria: Pharmaceutical compounds and activities

Antibacterial activity: The actinomycete, *Micromonospora* sp, isolated from this sponge species, showed antibacterial activity against *Bacillus* sp. P25 and *Escherichia coli* K12. The other bacterial species associated with this sponge is *Micrococcus* sp. X-48(JX997909) (Abdelmohsen et al. 2014).

Order: Poecilosclerida Family: Desmacellidae

Neofibularia sp.

Global distribution: Off North Carolina, United States, North Atlantic Ocean

Habitat: Depth, 79 m

Associated bacteria: Pharmaceutical compounds and activities

Antibacterial activity: *Streptomyces* spp. (isolate NOHa.2) from this sponge had the highest inhibitory activity (inhibition zone diameter, 14 mm) against *Shigella dysenteriae* (Dita et al., 2017).

Order: Poecilosclerida Family: Esperiopsidae

***Ulosa* sp.**

Image credit: Bernard Picton, https://creativecommons.org/licenses/by/4.0/

Associated bacteria: Pharmaceutical compounds and activities

Antimicrobial activity: The extracts of the associated bacterial species of this sponge showed antimicrobial activity against *Escherichia coli*, *Micrococcus luteus*, *Candida albicans* and *Streptomyces* spp., with the values of zone of inhibition 2.0, 2.5, 1.0, and 0–2.3 mm, respectively (Khan et al. 2012).

Order: Poecilosclerida Family: Hymedesmiidae

Phorbas tenacior

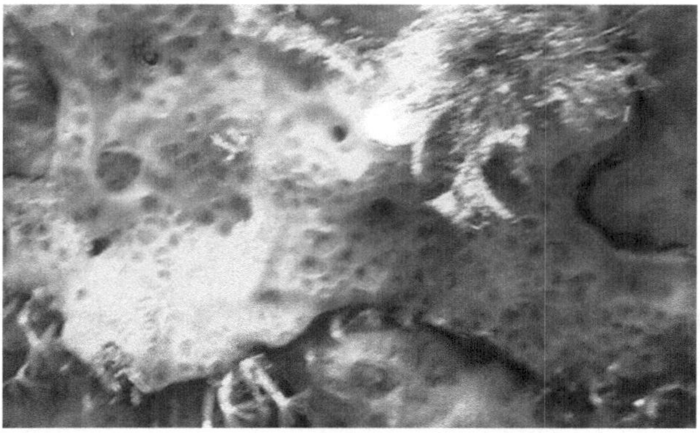

Image credit: Esculapio, https://creativecommons.org/licenses/by/3.0/

Common name: Blue Encrusting Sponge

Global distribution: Northern Atlantic Ocean: Mediterranean.

Habitat: Inhabits shaded reef ledges and crevices in areas where current is strong.

Associated bacteria: Pharmaceutical compounds and activities

Bioactivities: Among eleven bacterial strains which were isolated from this sponge species, three isolates *Pseudovibrio* P1Ma4, Vibrio P1MaNal1 and *Citricoccus* P1S7 revealed antimicrobial activity; *Citricoccus* P1S7 and *Vibrio* P1MaNal1 isolates also exhibited antiplasmodial activity, while two *Vibrio* isolates P1Ma8 and P1Ma5 displayed antioxidant activity (Dupont et al. 2014).

Order: Poecilosclerida Family: Isodictyidae

Isodictya compressa

Common name: Not designated

Global distribution: Worldwide

Habitat: Wide range of habitats

Associated bacteria: Pharmaceutical compounds and activities

Antibacterial activity: The bacterial strain *Arthrobacter citreus* DSM 20133 (PE09-119) associated with this sponge species displayed activity against Gram-positive strains,*M.smegmatis, B.cereus* and *S.epidermidis*. Further, its *Bacillus zhangzhouensis* DW5-4 (PE09-72) and PE09-105 showed activity against both Gram-positive and Gram-negative test strains, viz. *M.smegmatis, B.cereus* and *S.epidermidis;* and *E.coli* 1699 and *P.putida* (Matobole et al. 2017).

Isodictya setifera

Common name: Not available

Global distribution: Antarctica

Habitat: Depth 50 m.

Associated bacteria: Pharmaceutical compounds and activities

Antibacterial activity: The compound cyclo-(l-proline-l-methionine) isolated from the symbiotic bacterium *Pseudomonas aeruginosa* (γProteobacteria) obtained from this sponge showed antimicrobial activity against *Bacillus subtilis, Staphylococcus aureus,* and *Micrococcus luteus* (Thomas et al. 2010).

Brinkmann et al. (2017) reported that the compound phenazine-1-carboxylic acid isolated from the associated bacterial species, *Pseudomonas aeruginosa* showed antibacterial activity against *Staphylococcus aureus.*

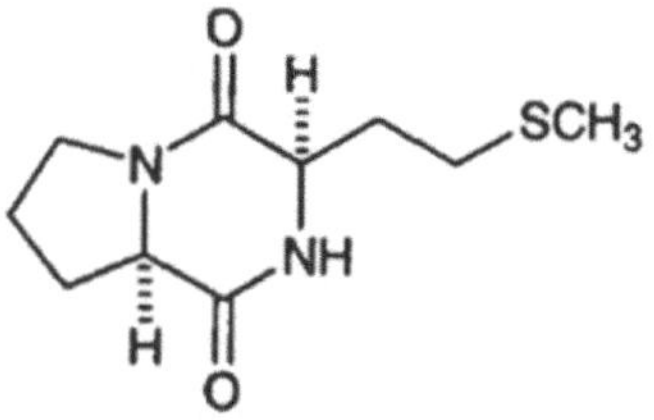

Cyclo-(l-proline-l-methionine)

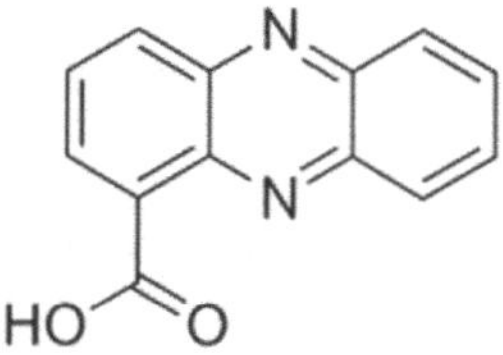

Phenazine-1-carboxylic acid

Order: Poecilosclerida Family: Mycalidae

Mycale (Aegogropila) adhaerens

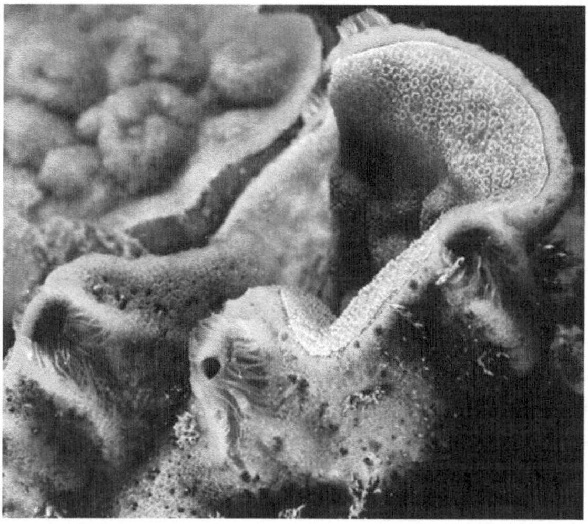

Mycale sp.

Image credit: Nhobgood Nick Hobgood, https://creativecommons.org/licenses/by/3.0/

Common name: Smooth scallop sponge

Global distribution: northern to southern California

Habitat: It is moderately common on rocky substrate where moderate water motion and the bivalve mollusc, *Chlamys* ae present.

Associated bacteria: Pharmaceutical compounds and activities

Antimicrobial activity: The Gram-positive bacterial epibionts of this sponge displayed antibacterial activities against *Pseudoalteromonas* (N2 and N43) and Vibrio (N11, N31, and N40) and the values of growth inhibition zones were found to range from 0.2 to 6.3 mm (Lee and Qian 2004).

Mycale mannarensis

Global distribution: South East coast of India.

Habitat: Not reported

Associated bacteria: Pharmaceutical compounds and activities

Antimicrobial activity: The bacterial isolates, *Vibrio parahaemolyticus, Bacillus subtilis, Vibrio fortis,* and *Vibrio* sp. of this sponge species showed antibacterial activity against *Escherichia coli* with inhibition zone values of 10, 16, 6, and 5 mm, respectively (Anand et al. 2006).

Mycale (Carmia) microsigmatosa

Common name: Not reported

Global distribution: Western Central Atlantic; Bahamas, Belize, Bonaire, Cuba, Curaçao, Margarita, Florida, Jamaica

Habitat: Sessile; encrusting mangrove roots, also found on pipes or piles.; brackish; depth range 4–25 m

Associated bacteria: Pharmaceutical compounds and activities

Antibacterial activity: A strain of *Pseudomonas putida* (Mm3) isolated from this sponge species showed antibacterial activity toward multidrug-resistant bacteria (MDR) such as *Staphylococcus aureus* 42AE, *Staphylococcus epidermidis* 57AE (inhibition zone diameter, 11 mm), *Enterococcus faecalis* V583, *Escherichia coli* 54AE (10 mm), *Escherichia coli* 2AE (12 mm) and *Enterobacter cloacae* 43AE (12 mm) (Marinho et al. 2009).

Mycale (Aegogropila) plumosa

Common name: Not reported

Global distribution: Tropical; Indo-West Pacific

Habitat: Sessile

Associated bacteria: Pharmaceutical compounds and activities

Cytotoxic activities: An actinomycete strain *Saccharopolyspora* sp. nov. associated with this sponge species produced two prodigiosins analogs viz. metacycloprodigiosin (1) and undecylprodigiosin (2) which showed cytotoxic activities against cancer cell lines such as P388 (mouse lymphoma cell), HL60 (human peripheral blood promyeloblast cell), A-549 (human lung carcinoma cell), BEL-7402 (human hepatic carcinoma cell), and SPCA4 (human lung carcinoma cell) as given below (Liu et al. 2005).

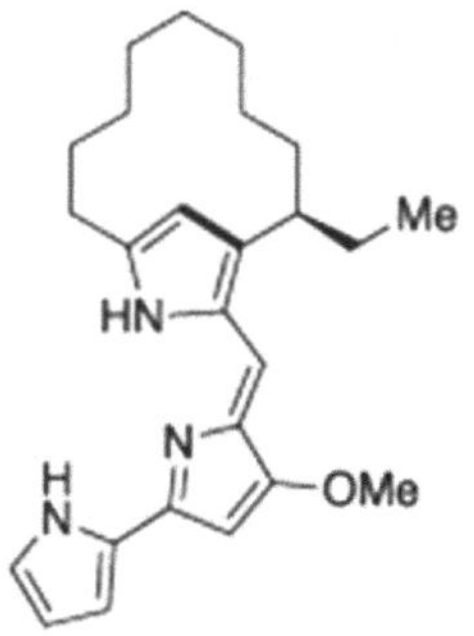

Metacycloprodigiosin

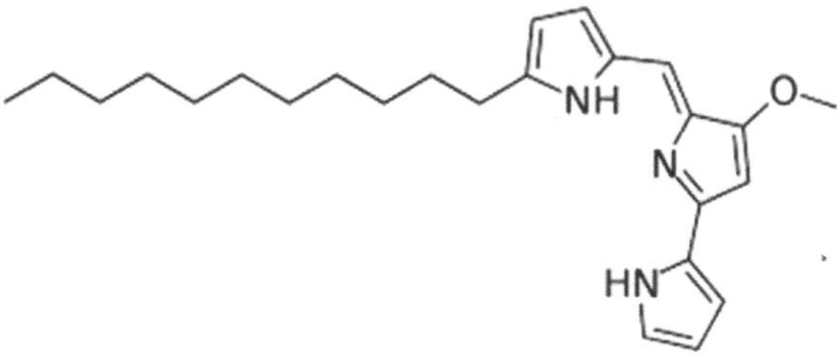

Undecylprodigiosin

Order: Poecilosclerida Family: Tedaniidae

Tedania (Tedania) anhelans

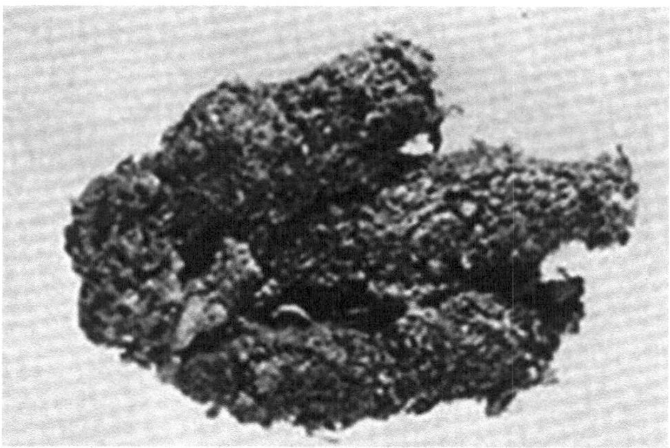

Tedania sp.

Image credit: Dickinson, Malcolm Gibson, https://creativecommons.org/licenses/by/4.0/

Common name: Not reported

Global distribution: Northeast Atlantic and Indo-West Pacific; Bretagne, Galicia, Portugal, Mediterranean, West Africa, Canary Islands, and Azores

Habitat: Subtidal localities down to 40 m

Associated bacteria: Pharmaceutical compounds and activities

Antimicrobial activities: A total of 25 Gram-positive bacterial species have been found to be associated with this sponge. Among these isolates 10 isolates showed inhibitory activity against Gram-negative bacteria, Gram-positive bacteria, antitumor indicator species viz. *Escherichia coli* 343/591 and pathogenic regular species, *E. coli* 343/636; and disease-causing fungi (Zeng et al. 2013).

Tedania (Tedaniopsis) charcoti

Common name: Not reported

Global distribution: East Antarctic; Chilean Coast

Habitat: Abundant on the shallow rock wall, between 5 and 9 m.

Associated bacteria: Pharmaceutical compounds and activities

Cryoprotective effects: The bacterial strains *Winogradskyella* sp. CAL384 (Flavobacteriaceae) and CAL396 isolated from this sponge species produced compounds such as mannose, arabinose, galacturonic acid, glucuronic acid, galactose, glucose, glucosamine, and displayed cryoprotective effects (Giudice and Rizzo 2018).

Tedania (Tedania) ignis

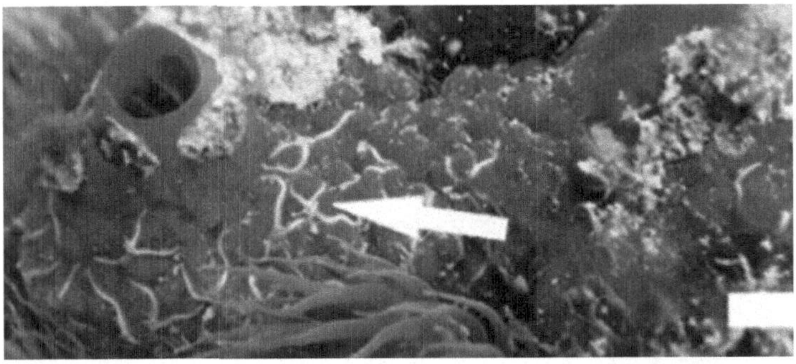

Image credit: Marcelo Checoli Mantelatto, et al., https://creativecommons.org/licenses/by/4.0/

Common name: Fire sponge

Global distribution: Caribbean to Brazil, and Florida.

Habitat: As a member of the fouling community, this species is commonly seen in lagoons, bays, and mangroves. It has also been recorded from algae or eelgrasses.

Associated bacteria: Therapeutic compounds and their bioactivities

Antibacterial activity: Stierle et al. (1988) reported that the bacterial isolate, *Micrococcus* sp of this sponge produced diketopiperazines which possess antibacterial activity.

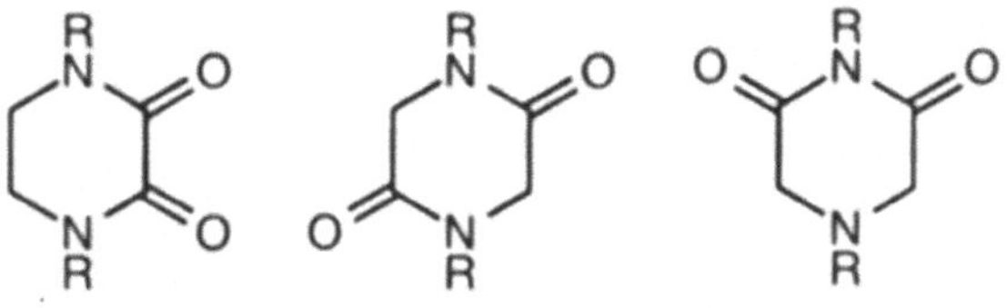

Diketopiperazines

***Tedania* sp.**

Associated bacteria: Pharmaceutical compounds and activities

Antiprotozoal and cytotoxic activities: The *Streptomyces* sp isolated form this species displayed antiprotozoal activity against *Leishmania major* and *Trypanosoma brucei brucei*; and cytotoxicity against kidney cancer cells 293T (Dita 2017; Pimentel-Elardo 2010).

Order: Scopalinida Family: Scopalinidae

Stylissa carteri

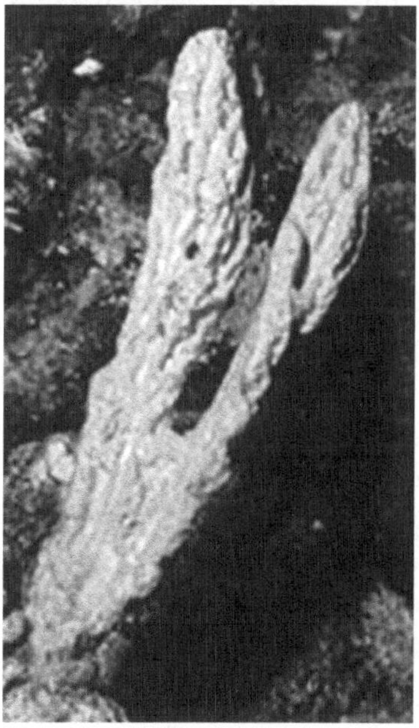

Image credit: Alexander Vasenin. https://creativecommons.org/licenses/by-sa/3.0/

Common name: Elephant ear sponge

Global distribution: Indo-West Pacific; Red Sea to Australia

Habitat: Sessile; inhabits coastal rocky shelf areas where currents are strong; depth range 0–100 m;

Associated bacteria: Pharmaceutical compounds and activities

Antiplasmodial activity: The ethyl acetate extracts of the bacterial isolates of this this sponge species displayed antiplasmodial activity against *Plasmodium falciparum*. The activity of isolate THB17 (IC50 20.56 μg/ mL) extract was found to be highly comparable with the positive control chloroquine (IC50 19.59 μg/mL) (Inbaneson and Ravikumar 2012).

Stylissa massa *(= Stylotella aurantium)*

Common name: Mango sponge

Global distribution: Indo-Pacific

Habitat: Sessile; depth range 0–100 m

Associated bacteria: Pharmaceutical compounds and activities

Anticancer activity: The bacterial strains *Streptomyces setonensis* SpC080624 SC-11 and SpA080624GE-02 associated with this sponge yielded alkaloids viz. JBIR 46 –48 which displayed weak cytotoxicity against human acute myelogenous leukemia HL-60 cell line with IC50 values of 189, 226, and 96 M, respectively (Zhang et al. 2017).

***Stylissa* sp.**

Associated bacteria: Pharmaceutical compounds and activities

Antimicrobial activities: The bacterial isolates of the genera *Halomonas, Cytophagas, Marinomonas, Alteromonas, Marinobacter,* and *Chromohalobacter* derived from this sponge displayed antimicrobial properties against the pathogenic *Bacillus subtilis*, *Enterococcus durans, Aeromonas hydrophila*, *Streptococcus lentus*, *Rolstonia solanacearum*, and *Klebsiella pneumoniae*. It was also reported that 67% of the bacterial isolates possessed broad spectrum to species-specific properties. Further, the isolate S8 had 95% potency with the pathogenic *K. pneumoniae* (Krishnan et al. 2014).

Order: Spirophorida Family: Tetillidae

Cinachyra cavernosa

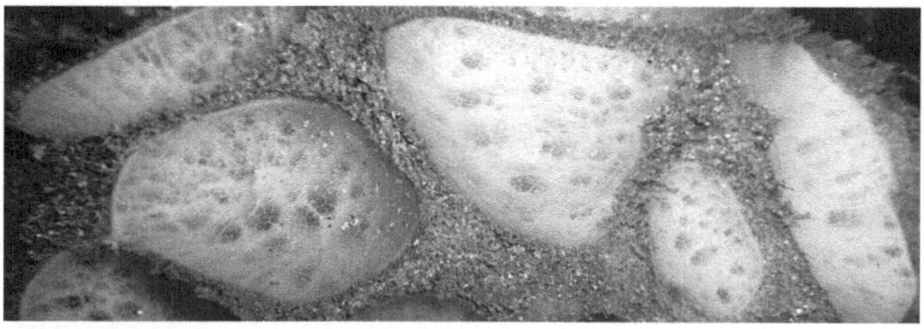

Cinachyra sp.

Image credit: Nhobgood Nick Hobgood, https://creativecommons.org/licenses/by-sa/3.0/

Global distribution: Red sea, Indian Ocean, Australian region, Pacific Ocean and Atlantic Ocean.

Habitat: No data

Associated bacteria: Pharmaceutical compounds and activities

Antibacterial activity: The bacterial strain *Chromohalobacter* sp. GUVFCCM-2 isolated form this sponge species displayed antibacterial activity against *S. typhi, A. aerogens, E. coli, S. citreus*, and *P. vulgaris* with zone of inhibition values of 8, 20, 30, 12, and 10 mm, respectively (Velho-pereira and Furtado 2012).

Cinachyra sp.

Associated bacteria: Pharmaceutical compounds and activities

Cytotoxic activity: Three alkaloids (salicylamide derivatives) JBIR 46-48 were detected in the cultures of bacterium *Streptomyces setonensis* SpC080624SC-11 and SpA080624GE-02, which were isolated from this unidentified species. Cytotoxic tests indicated that these three compounds possessed weakly cytotoxic activities against human acute myelogenous leukemia HL-60 cells with IC50 values of 189, 226, and 96 μM, respectively (Zhang et al. 2017).

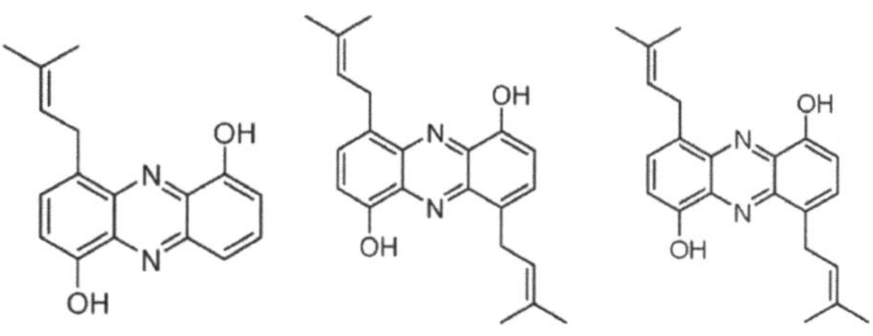

JBIR 46-48

Order: Suberitida Family: Halichondriidae

Halichondria japonica

Common name: Not reported

Global distribution: Japan

Habitat: Rocky shore

Associated bacteria: Pharmaceutical compounds and activities

Antibacterial activity: The bacterial strain *Bacillus cereus* QNO3323 (Firmicules) isolated from this sponge species produced thiopeptide YM-266183 which showed antibacterial acitivity against *Staphylococcus aureus* (Brinkmann et al. 2017).

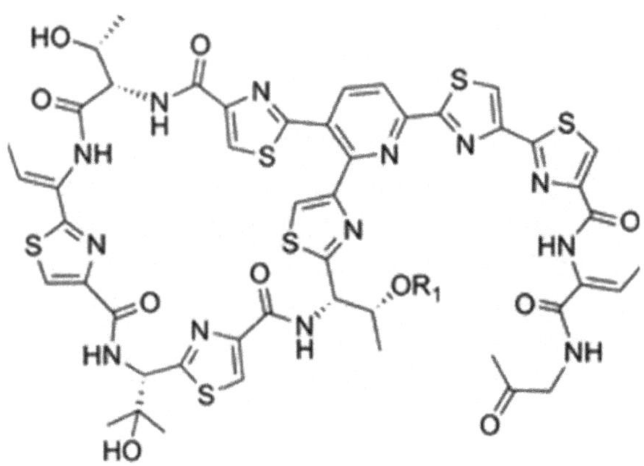

R1=H

QNO3323

Halichondria okadai

Common name: Encrusting sponge

Global distribution: Japan

Habitat: Intertidal rocky shore areas; depth, 27 m

Associated bacteria: Pharmaceutical compounds and activities

Antioxidant activity: The diapolycopenedioic acid xylosyl esters A, B and C have been extracted from the Gram-negative bacterium *Rubritalea squalenifasciens* which is a symbiont of this sponge. Among these compounds, compound A showed O_2 suppression activity with an IC50 of 5.1 μM (Shindo and Misawa 2014).

Cytotoxic and antimicrobial properties: The bacterial strains belonging to *Pseudomonas* sp. KK10206C and *Alteromonas* sp. associated with this sponge species have yielded a 50-carotenoid, okadaxanthine and lactam alteramide A, respectively. While the former compound is a well-known source of okadaic acid and a potent singlet oxygen quencher, the latter (alteramide A) displayed cytotoxicity against epidermal carcinoma KB, leukemia P-388 and lymphoma L-1210 cell lines (Thomas et al. 2010).

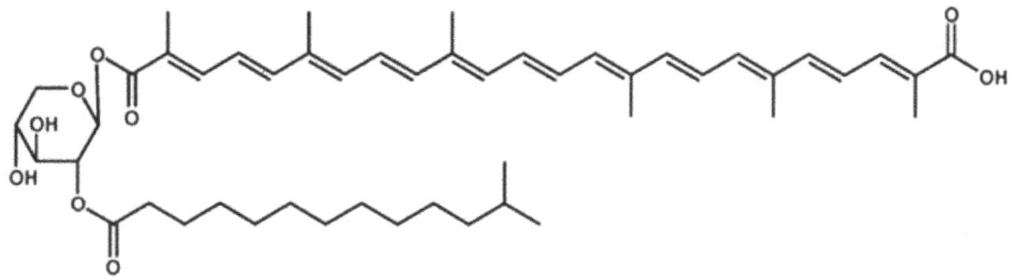

Diapolycopenedioic acid xylosyl ester

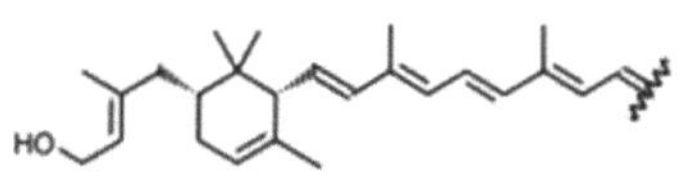

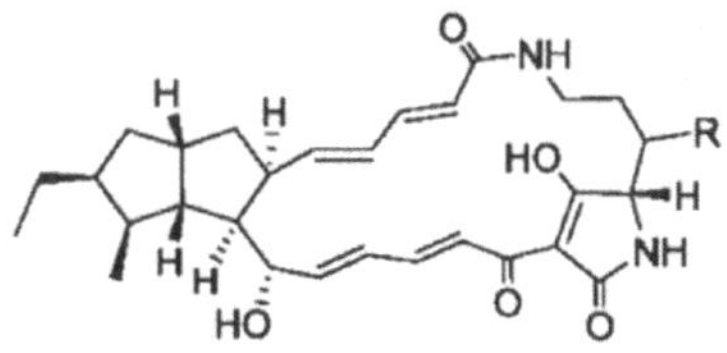

Okadaxanthine R=H

Alteramide A

Halichondria panicea

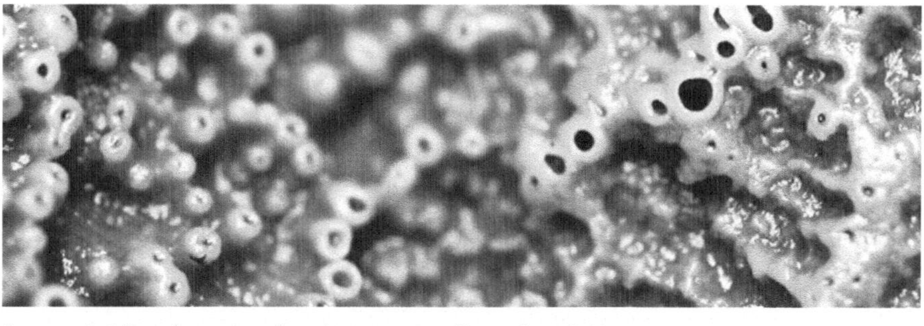

Image credit: Minette Layne, https://creativecommons.org/licenses/by-sa/2.0/

Common name: Breadcrumb sponge

Global distribution: North Atlantic and the Mediterranean Sea

Habitat: It occurs from the intertidal zone to depths of about 500 m

Associated bacteria: Therapeutic compounds and their bioactivities

Antitumor activity: Organic extracts of the associated bacteria *Rhodobacter* of this sponge displayed cytotoxicity against leukemia cells. Further, the glycoglycerolipid, 1-O-acyl3-[R-glucopyranosyl-(1-3)- (6-O-acyl-R-mannopyranosyl)]-glycerol derived from its actinobacterium, Microbacterium sp. showed positive results for antitumor activities (Thomas et al. 2010).

Cytotoxic and antibacterial activity: The bacterial strain, *Streptomyces* sp. strain HB202 associated with this sponge produced mayamycin, a benz[a] anthracene derivative which exhibited potent cytotoxic activity against eight human cancer cell lines and showed activity against several bacteria including antibiotic-resistant strains (Schneemann et al. 2010a). R-mannopyranosyl)]-glycerol showed positive results for antitumor activities (Thomas et al. 2010).

Neuroactivity: The bacterial species associated with this sponge viz. *Antarcticum vesiculatum* and *Psychroserpens burtonensis* displayed neuroactivity by agonizing the NMDA receptor (Perovic et al. 1998).

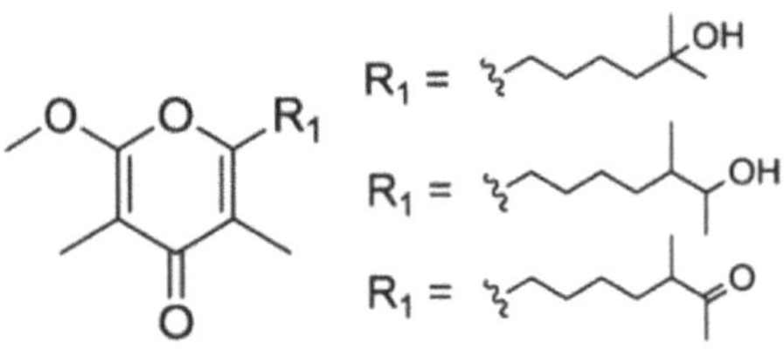

Nocapyrone A, C, D

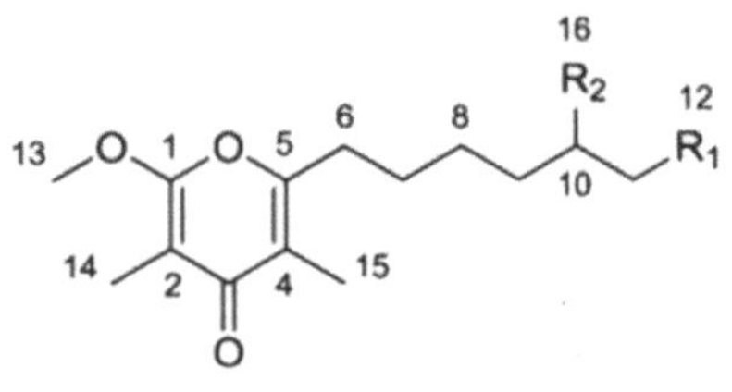

R_1=H, R_2=CH_3

Nocapyrone B

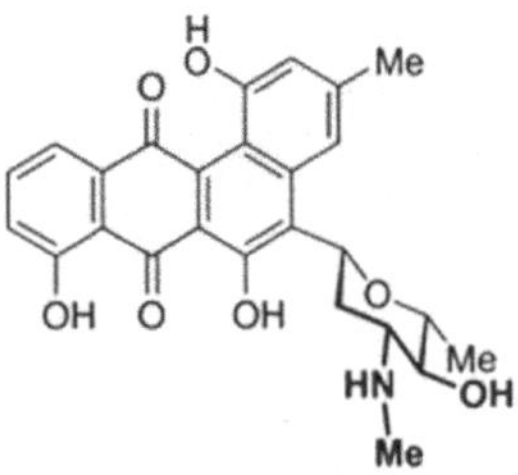

Mayamycin

Antifungal activity: The bacterial strain, *Streptomyces* sp. G248 associated with this sponge species has been reported to yield lavandulylated flavonoids, which displayed antifungal activity by inhibiting the growth of MDR *Candida albicans* 10231 (Nweze et al. 2020).

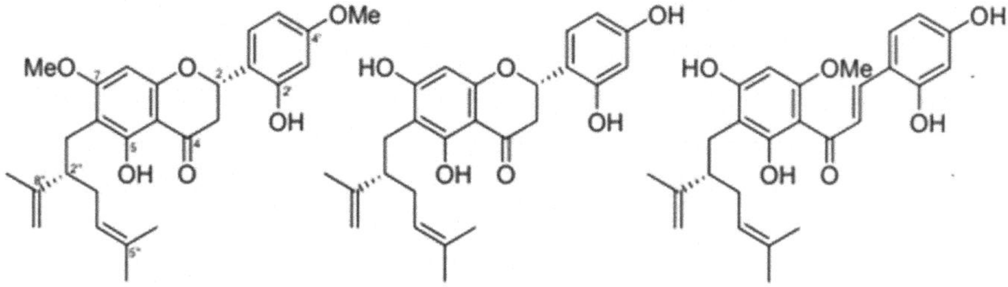

Lavandulylated flavonoids

Halichondria sp.

Associated bacteria: Pharmaceutical compounds and activities

Antimicrobial activity: The bacterial strain *Bacillus licheniformis* SAB1 (Firmicute) isolated from this sponge produced compounds viz. indole (1) 3-phenylpropionic acid (2) and 4,4-oxybis[3-phenylpropionic acid] (3) which showed antimicrobial activity against several species of pathogenic bacteria. While the compounds 1 and 2 acted against MRSA, *Rhodotorula* sp., *Pseudomonas aeruginosa, Staphylococcus aureus* and *Salmonella typhi,* compound 3 was against *Salmonella typhi, Vibrio cholerae,* and *Aspergillus fumigatus* only (Devi et al. 2010).

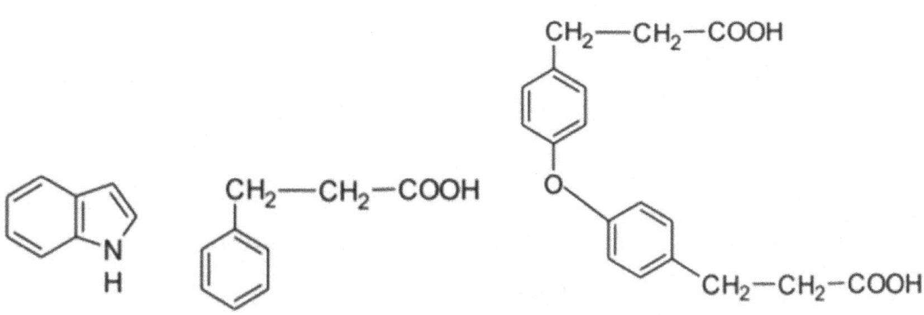

Indole 3-phenylpropionic acid 4,4′-oxybis[3-phenylpropionic acid]

Hymeniacidon perlevis

Global distribution: Arctic; British Isles; Belgium; France; Spain; Mediterranean etc

Habitat: Intertidal and shallow sub-littoral; on the shore it is found on the sides of boulders, under weed

Associated bacteria: Pharmaceutical compounds and activities

Antimicrobial activity: The bacterial strains *Bacillus subtilis* NJ6-3-2 and *Pseudoalteromonas piscicida* NJ6-3-1 associated with this species showed antimicrobial activity against *Bacillus subtilis* CMCC 63003 (inhibition zone, 5 mm and 3–5 mm, respectively) *Staphylococcus aureus* CMCC 26001 (3-5 mm), *Escherichia coli* CMCC 44102 (1-3 mm and nil), *Agrobacterium tumefaciens* AS1.1416 (3-5 mm and 5 mm) and fungus *Saccharomyces cerevisiae* ACCC 2.1882(3–5 mm). Fermentation of the marine bacterial strain *Pseudoalteromonas piscicida* NJ6-3-1 yielded the antimicrobial metabolite, norharman (9H-Pyrido(3,4-b)indole) (Zheng et al. 2005).

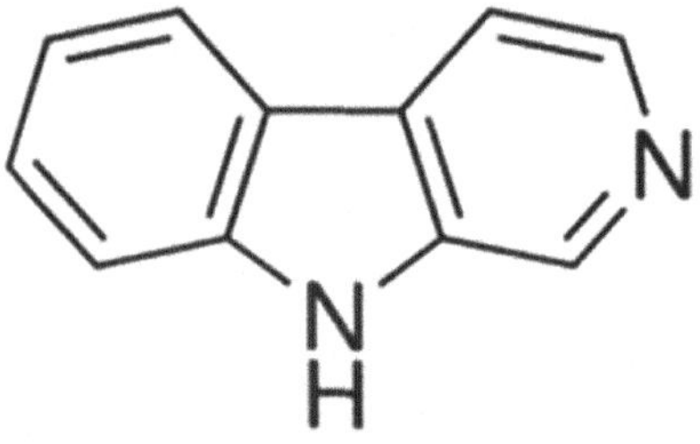

Norharman (9H-Pyrido(3,4-b)indole)

Hymeniacidon sp.

Compounds: The bacterial strain *Pseudogymnoascus* sp. (class, Leotiomycetes) associated with this Antarctic sponge yielded nitroasteric acid derivatives such as Pseudogymnoascins A-C, 3-nitroasterric acid, nitrodiketopiperazine pyriculamide, and anthraquinone question which however did not show any significant antibacterial and antifungal activities (MIC > 64 μg·ml−1) (Figueroa et al. 2015).

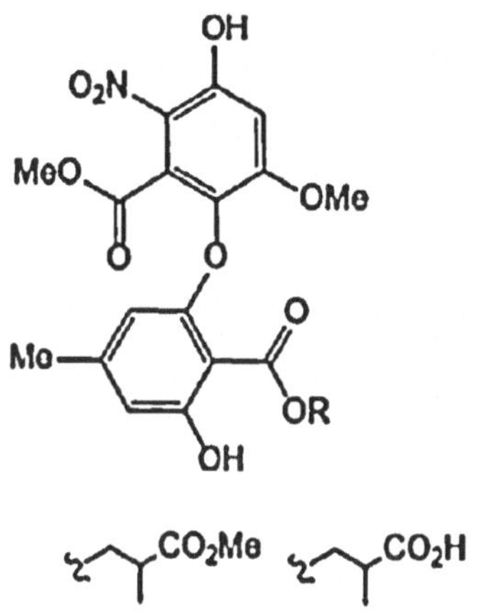

Pseudogymnoascins A,B

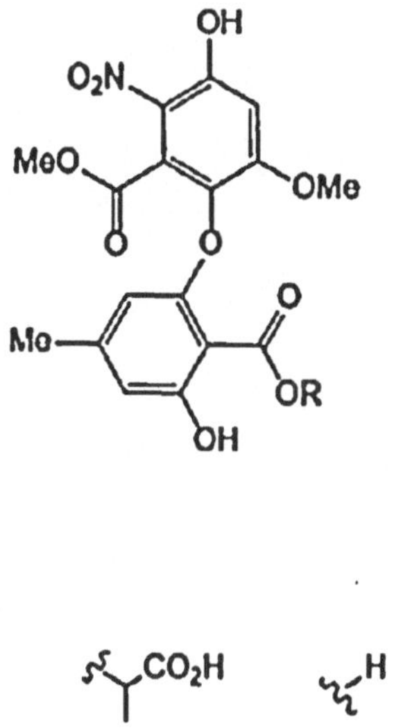

Pseudogymnoascin C, 3-nitroasterric acid

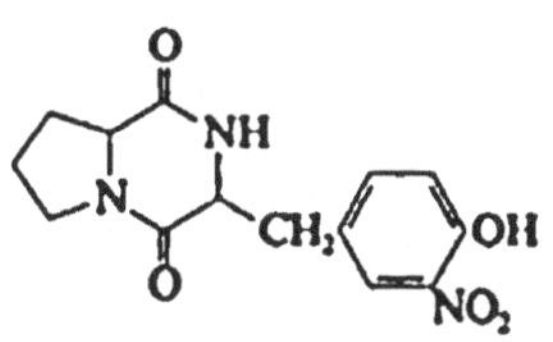

Pyriculamide

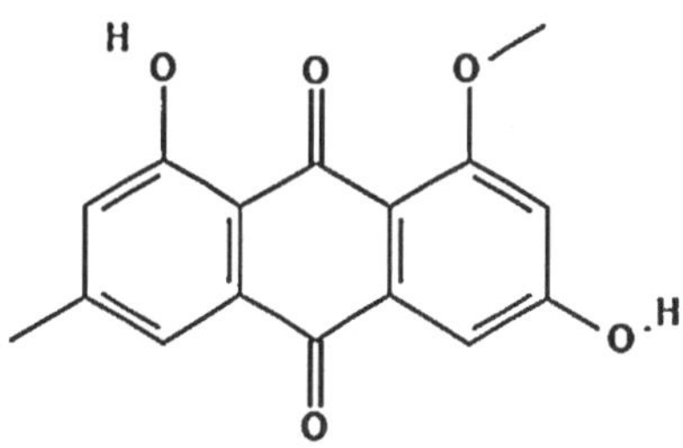

Questin

Associated bacteria: Pharmaceutical compounds and activities

Anticancer activity: The actinomycete, *Streptomyces carnosus* strain AZS17 isolated from tis sponge species produced lobophorin C and D. Among these compounds, lobophorin C displayed a significant cytotoxicity against 7402 hepatoma cells, with an IC50 value of 0.6 g/mL. On the other hand, lobophorin D exhibited pronounced inhibitory activity on thuman breast cancer cell line, MDA-MB 435 with an IC50 value of 7.5 M. Further lobophorin D was found to induce apoptosis in breast adenocarcinoma cells (Wei et al. 2011).

Order: Suberitida Family: Suberitidae

Suberites domuncula

Image credit: Guido Picchetti, https://creativecommons.org/licenses/by-sa/3.0/

Common name: Sea-orange

Global distribution: Arctic and the Mediterranean; Subtropical to polar.

Habitat: Colonizing gastropod shells occupied by hermit crabs; also attached to the shells of live gastropods and the carapace of other crustaceans.; sessile; depth range 0 – 1100 m

Associated bacteria: Pharmaceutical compounds and activities

Antiangiogenic activity: The crude extract of its associated Proteobacterium MBIC3368- SB2 strain displayed inhibition of angiogenesis at 10–20 μg /ml concentrations (Thakur et al. 2005).

Antibacterial activity: The bacterial isolate SB2 of this species displayed activity against *Staphylococcus epidermidis* strains and MRSA, which are believed to be responsible for hospital-acquired infections associated with surgical interventions (Thakur et al. 2005).

A strain of *Ruegeria* associated with this sponge species exhibited its antibacterial properties toward *Bacillus subtilis* (Mitova et al. 2004).

Cytotoxicity: The extract obtained from the sponge surface bacterium SB2 was found to be toxic against PC12 cells (Thakur et al. 2005).

Others: Thomas et al. (2010) reported that an identified compound of the bacterial isolate *Pseudomonas* sp. (isolate 1) (γ-Proteobacteria) of this sponge species showed hemolytic and cytotoxic activities. Further its isolation with its unidentified compound has shown antiangiogenic, antimicrobial, hemolytic, and cytotoxic activities.

Suberites japonicus

Associated bacteria: Pharmaceutical compounds and activities

Cytotoxic effect: The bacterial strain, *Actinomadura* sp. SBMs009 associated with this marine sponge yielded a sterol viz. bendigole which exhibited moderate cytotoxicity on L929 cell line with an IC50 value of 30 μM (Zhang et al. 2017).

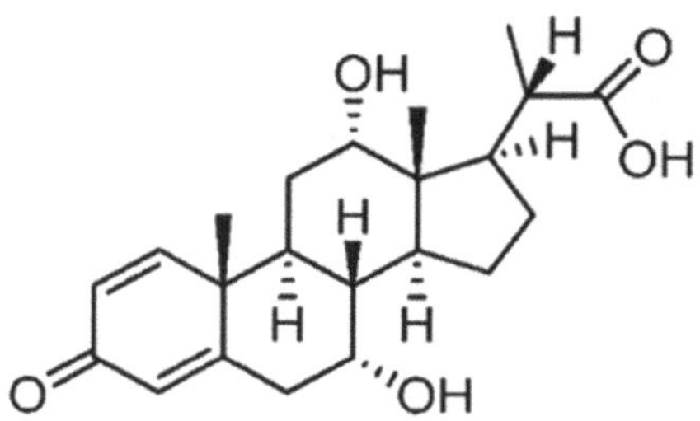

Bendigole

Order: Tetractinellida Family: Ancorinidae

Jaspis sp.

Global distribution: Off Florida, United States, North Atlantic Ocean

Associated bacteria: Pharmaceutical compounds and activities

The associated deltaproteobacterium *Chondromyces crocatus* Cm c5 of this sponge species produced the compound chondramide D, which showed cytotoxic activity (Rachid et al. 2006).

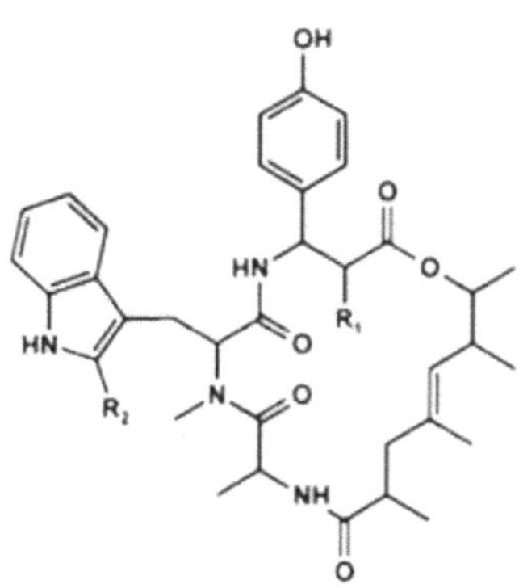

D = R 1 = H, R 2 4 Cl

Chondramide D

Rhabdastrella globostellata

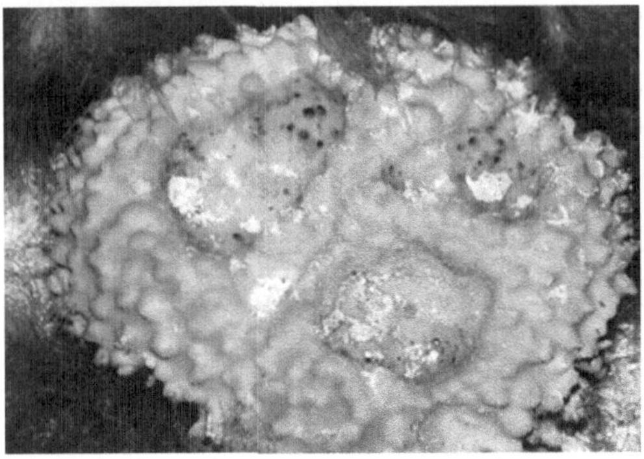

Image credit: Ria Tan, https://creativecommons.org/licenses/by-sa/2.0/

Common name: Yellow pot sponge

Global distribution: Native to many regions of the Indian Ocean including the shores of Madagascar, the Seychelles, and Australia as well as the Malayan Peninsula and Singapore.

Habitat: Intertidal to 53m in depth

Associated bacteria: Pharmaceutical compounds and activities

AChE inhibition activity: The bacterial isolates of this sponge species exhibited >10% AChE inhibition activity (Pandey et al. 2014).

Stelletta tenuis

Stelletta sp.

Image credit: Peter Southwood, https://creativecommons.org/licenses/by-sa/4.0/

Common name: Not designated

Global distribution: South China Sea, Southeast Asia, Java

Habitat: Seamounts and knolls

Associated bacteria: Pharmaceutical compounds and activities

Antibacterial activity: The bacterium *Alcaligenes faecalis* A72 associated with this sponge speices yielded L,L-Diketopiperazine (cyclo-(l-Pro-l-Phe) which displayed moderate antimicrobial activity against *Staphylococcus aureus* (Li et al. 2008).

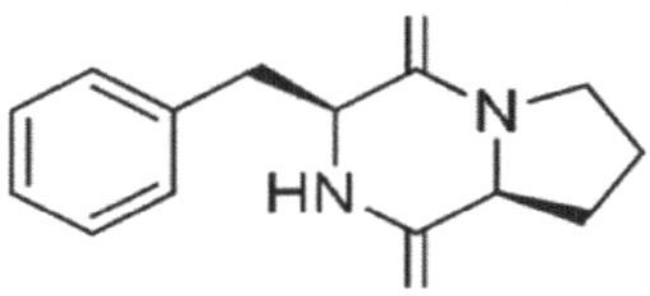

Cyclo-(l-Pro-l-Phe

Stelletta sp.

Associated bacteria: Pharmaceutical compounds and activities

Antitumor activity: The indole oligomer compound derived from its Gram negative Proteobacterium viz. *Psychrobacter* strain exhibited antitumor activity against human tumor cell lines viz. XF-498, HCT-15, A-549, SK-OV-3, and SK-MEL-2 with EC50 values of 2.39, 3.13, 2.34, 1.57, and 3.44 mg/mL, respectively (Zhang et al. 2017).

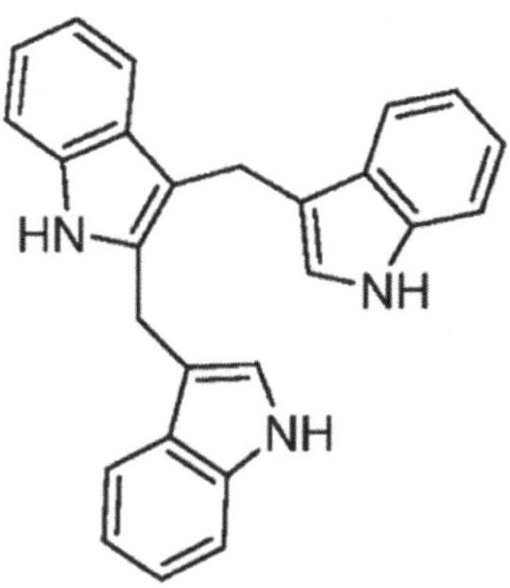

Indole oligomer

Order: Tetractinellida Family: Azoricidae

Leiodermatium pfeifferae

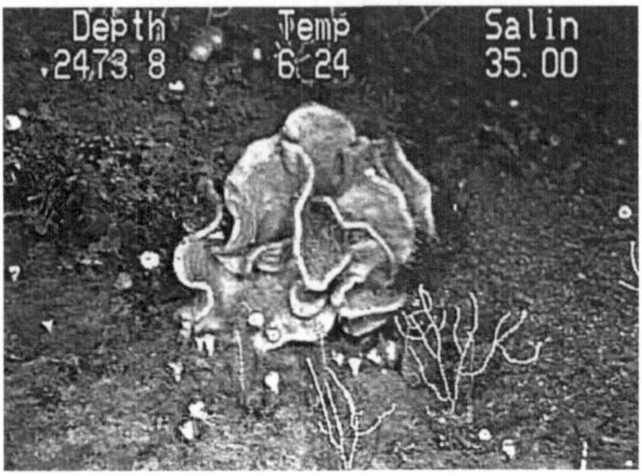

Image credit: NOAA (Public domain.).

Common name: It has no common name

Global distribution: Tropical; Indo-West Pacific.

Habitat: Sessile

Associated bacteria: Pharmaceutical compounds and activities

AChE inhibition activity: The bacterial isolates of this sponge species exhibited >10% AChE inhibition activity (Pandey et al. 2014).

Order: Tetractinellida Family: Geodiidae

Erylus deficiens

No other information

Associated bacteria: Pharmaceutical compounds and activities

Antimicrobial activity: The bacterial genera isolated from this sponge species include *Pseudoalteromonas* (42.1%), *Microbacterium* (14.0%) and *Proteus* (12.3%) which showed antimicrobial activity against *C. albicans, V. anguillarum, E. coli* and *B. subtilis*. *Microbacterium* was the only genus able to produce bioactive compounds effective against *B. subtilis* (Graca et al. 2015).

Order: Tetractinellida Family: Neopeltidae

Homophymia sp.

Associated bacteria: Pharmaceutical compounds and activities

Antimalarial and anti-HIV activities: The bacterial strain *Pseudomonas* species strain, 1537–E7 isolated from from the surface of this sponge, produced 2-undecyl-4-quinolone which was active against the malarial parasite *Plasmodium falciparum* and HIV-1 (Thomas et al. 2010).

Antimicrobial and cytotoxic activity: The above bacterial strain associated with this species also produced 2-undecen-1′-yl-4-quinolone and 2-nonyl-4-hydroxyquinoline N-oxide. Among these compounds, the latter showed antimicrobial activity against *Staphylococcus aureus* as well as cytotoxicity (Thomas et al. 2010).

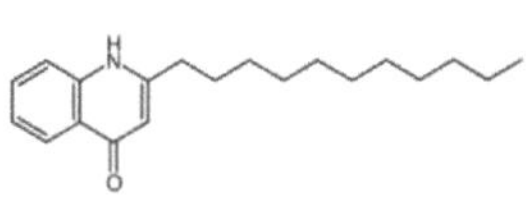
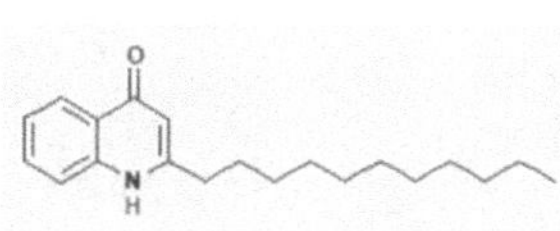
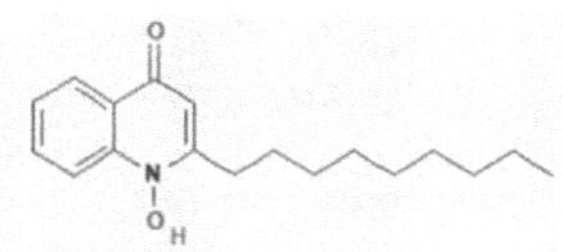

2-Undecyl-4-quinolone 2-undecen-1′-yl-4-quinolone 2-nonyl-4-hydroxyquinoline N-oxide

Order: Tetractinellida Family: Theonellidae

Theonella swinhoei

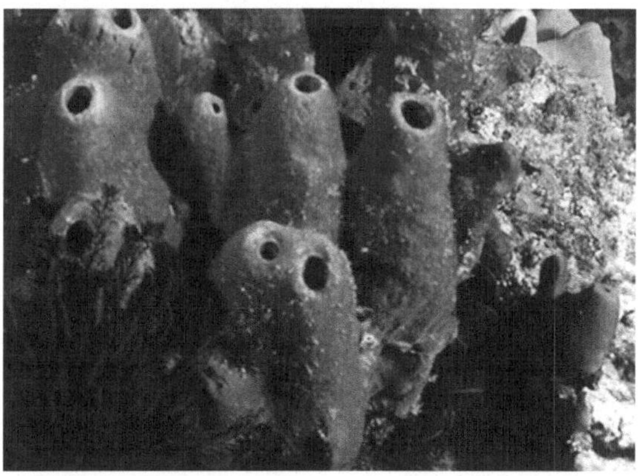

Theonella sp.

Image credit: Nhobgood Nick Hobgood, https://creativecommons.org/licenses/by-sa/3.0/

Common name: Palauan sponge

Global distribution: Tropical; Indo-West Pacific:, Andaman Island and Palau Philippines and Indonesia.

Habitat: Sessile; depth range 10–188 m

Associated bacteria: Pharmaceutical compounds and activities

Antifungal activity: This sponge species has been reported to harbor the filamentous bacterial symbiont, *Entotheonella palauensis* which contains the bicyclic glycopetide viz. theopalaumide. This compound showed antifungal activity (Schmidt et al. 2000).

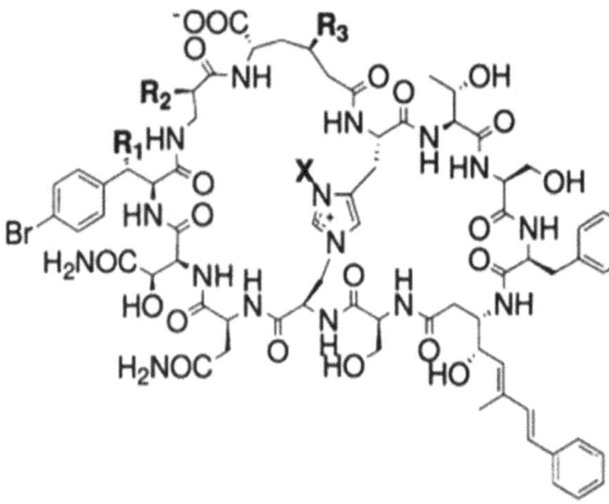

R1=CH3; R2 OH; R3 H; X β-D-galactose

Theopalaumide

Antimicrobial activity: The bacteria associated with *T. swinhoei* were found to be highly diverse—as many as 12 different phyla of bacteria were identified. However, in terms of population evenness, the community was dominated by two phyla—Acidobacteria (71.54%) and Chloroflexota (19.60%). A total of 700 bacterial strains were isolated and cultured from the samples of this species. Within these culturable strains, only 12% were Actinomycetes. Despite the low percentage of Actinobacteria from the samples, among the 51 strains of culturable bacteria that showed high antimicrobial activity, a great majority (62%) were Actinomycetes (30 strains of *Streptomyces* and 1 strain each of *Micromonospora* and *Brevibacterium*). The remaining isolates that produced antimicrobial compounds were Gammaproteobacteria (10 strains of P*seudoalteromonas*) and Firmicutes (8 and 1 strains of *Bacillus* and *Paenibacillus*, respectively) (Kuo et al. 2019).

Antitumor activity: The mixed polyketide-non ribosomal peptides, theopederin A, and onnamide A of the epibiont, *Pseudomonas* of this sponge species showed antitumor activity (Piel et al. 2004).

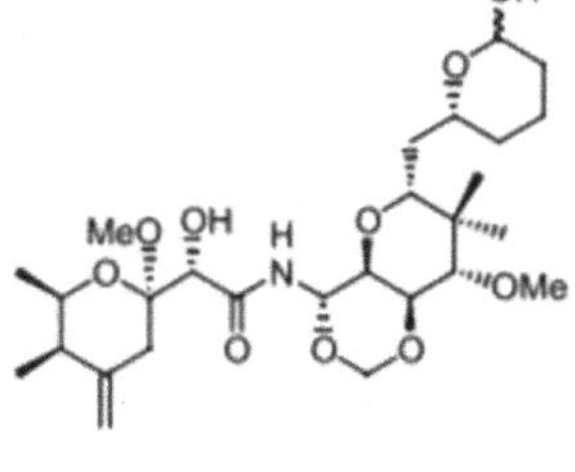

Theopederin A

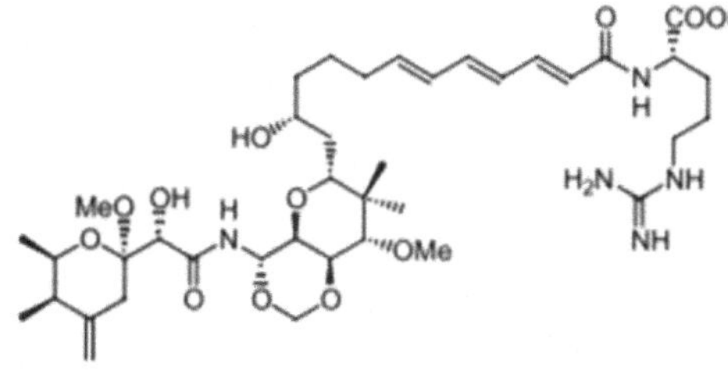

Onnamide A

Antifungal and cytotoxic activities: The polyketide swinholide A has been reported to be produced by heterotrophic unicellular bacteria associated with this sponge species.This compound has shown several bioactivities including antifungal activity and cytotoxic activity against a number tumor cell lines (Nicolaou et al. 1996; Thomas et al. 2010).

Antifungal activity: The filamentous δ-Proteobacterium, *Entotheonella palauensis* associated with this sponge species has produced compounds theonegramide and theonellamide F, which displayed antifungal activity (Matsunaga et al. 1989; Bewley and Faulkner1994; Thomas et al. 2010).

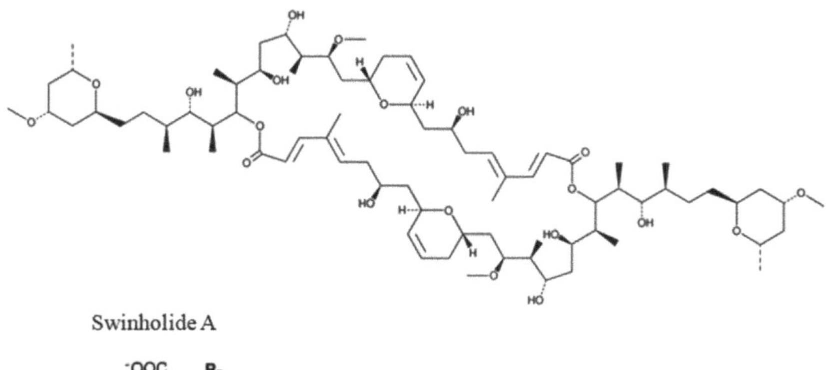

Swinholide A

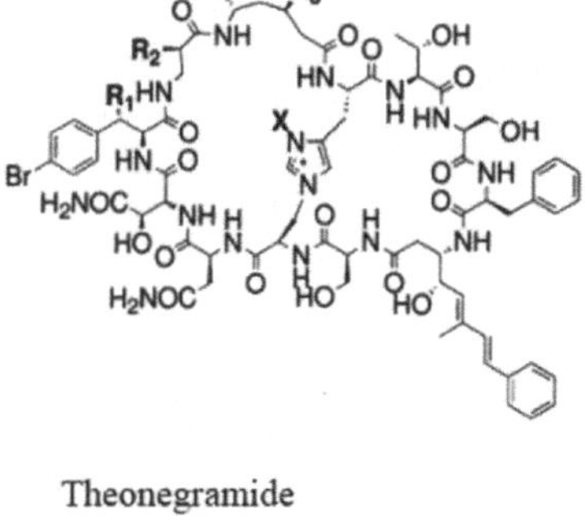

R_1 CH_3; R_2 OH; R_3 H; X β D- arabinose;

Theonegramide

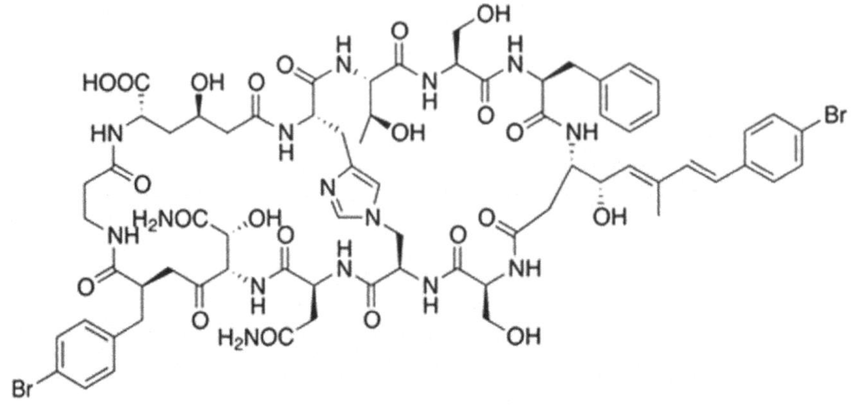

Theonellamide F

***Theonella* sp.**

Associated bacteria: Pharmaceutical compounds and activities

Antibacterial activity: The polysaccharide-producing bacteria *Shewanella putrefaciens*, *Capnocytophaga* sp., *Burkholderia cepacia* and *Brevundimonas diminuta* derived from this sponge displayed antibacterial activity against *Klebsiella pneumoniae* (13.67 mm), *Bacillus subtilis* (7.33 mm), and *Salmonella typhimurium* (15.67 mm) (Din et al. 2018).

Nweze et al.(2020) reported that the compounds nocardiopsistins A-C identified from the associated bacterium, *Nocardiopsis* sp. HB-J378 of this sponge showed antibacterial activity by inhibiting the growth of MRSA ATCC 700787 with MIC values ranging from 3.12 to 12.5 µg/mL.

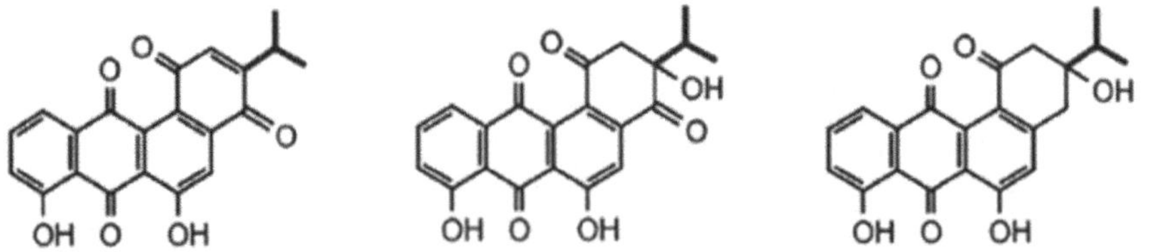

Nocardiopsistins A-C

Order Verongida Family Aplysinidae

Aplysina aerophoba

Image credit: Esculapio, https://creativecommons.org/licenses/by-sa/3.0/

Common name: Yellow Tube Sponge; golden sponge; Mediterranean sponge

Global distribution: Mediterranean; Atlantic along the coasts of Portugal and NW Spain.

Habitat: Depths down to 20m; more likely to be found in areas exposed to direct sunlight

Associated bacteria: Pharmaceutical compounds and activities

Antimicrobial activity: This sponge has been reported to harbor *Bacillus subtilis* strains A184, A190,A 202, and A586. Among these strains, A184, A190, and A202 displayed antifungal activity against *Candida albicans*; A184 also exhibited antibacterial activity against pathogenic MRSA and *Staphylococcus epidermidis*; and A586 which produced plumilacidin containing hydroxy fatty acid showed significant activity against *Staphylococcus aureus* (Thomas et al. 2010).

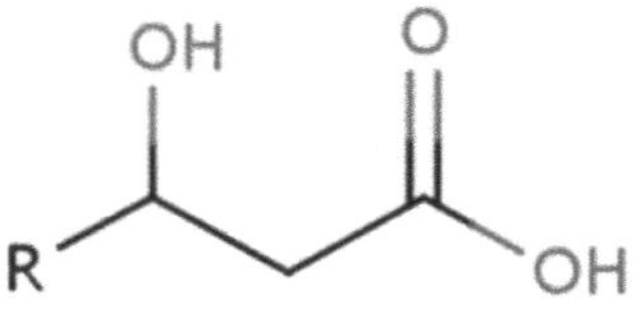

β-hydroxy fatty acid

Anti-parasitic activity: The bacterial strain *Streptomyces* sp. strain 22 isolated from this species produced valinomycin which exhibited significant inhibitory activities against the parasites *Leishmania major* (IC50 < 0.11 μM) and *Trypanosoma brucei brucei* (IC50 0.0032 μM) (Pimentel-Elardo et al. 2010).

Aplysina cavernicola

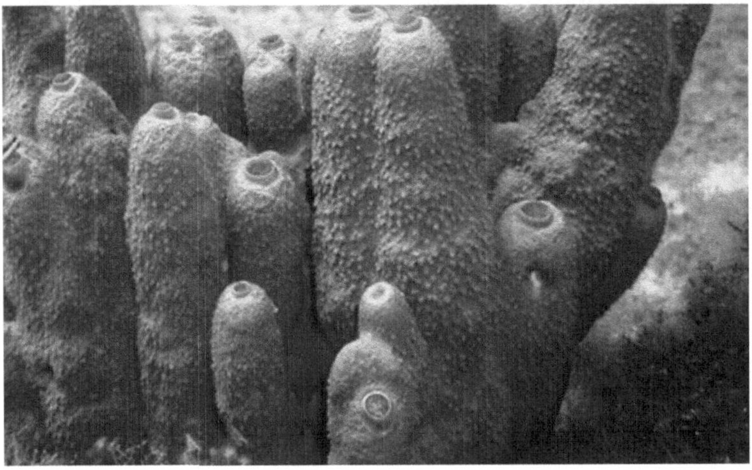

Image credit: Christian-wittmann, https://creativecommons.org/licenses/by-sa/4.0/

Common name: Mediterranean sponge

Global distribution: Temperate; Northeast Atlantic: Spain.

Habitat: Sessile.

Associated bacteria: Pharmaceutical compounds and activities

Antimicrobial activity: The bacterial isolates of this species were affiliated with the low (*Bacillus*) and high G+C Gram-positive bacteria (*Arthobacter* and *Micrococcus*), as well as the alpha-Proteobacteria (unknown isolate) and gamma-Proteobacteria (*Vibrio* and *Pseudoalteromonas*). Isolates with antimicrobial activity were numerically most abundant in the genera *Pseudoalteromonas* and the alpha-Proteobacteria. These sponge isolates showed antimicrobial activities against Gram-positive and Gram-negative reference strains. Clinical isolates viz multiresistant *Staphylococcus aureus* and *Staphylococcus epidermidis* strains isolated from hospital patients were also found inhibited by the bacterial isolates of this sponge (Hentschel et al. 2001).

Aplysina fistularis

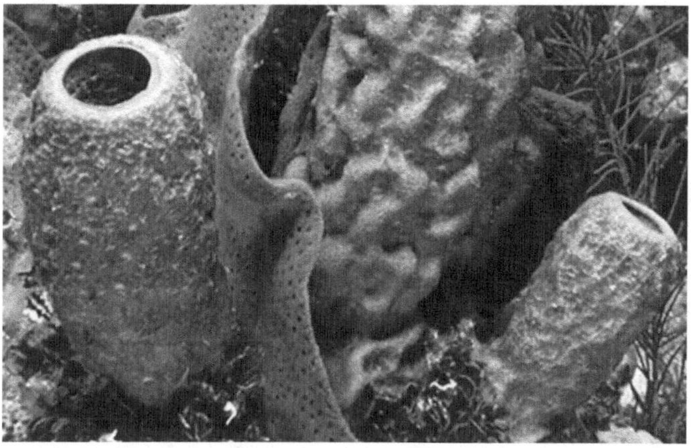

Image credit: Nhobgood Nick Hobgood, https://creativecommons.org/licenses/by-sa/3.0/

Common name (s): Yellow Tube Sponge

Global distribution: Caribbean: from Florida to Venezuela.

Habitat: Open water areas; often growing on vertical walls.

Associated bacteria: Pharmaceutical compounds and activities

Antimycotic activity: The endophytic Streptomyces sp. Hedaya48 isolated from this species yielded compounds viz. vanillin, 5,7-dimethoxy-4-p-methoxylphenylcoumarin and saadamycin. Among these compounds, 5,7-dimethoxy-4-p-methoxylphenylcoumarin and saadamycin. Have been reported to show potent antimycotic activity against clinical fungal isolates such as *Aspergillus fumigatus* and *Candida albicans*; and other dermatophytes (El-Gendy et al. 2010).

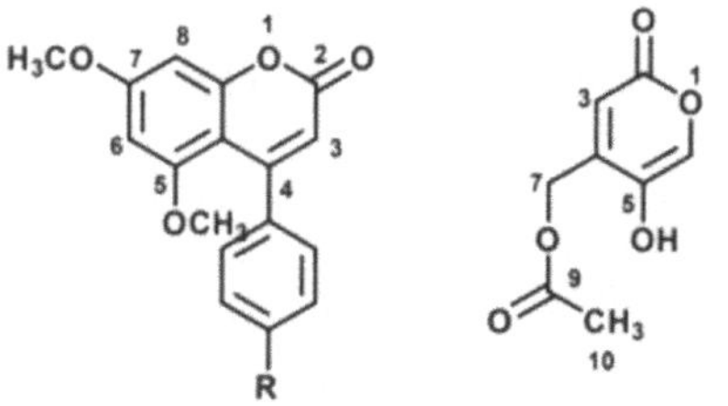

1,2 3

1. Vanillin,R= CHO; 2. 5,7-dimethoxy- 4-p-methoxylphenylcoumarin, R= OCH3; 3 Saadamycin

Order: Verongiida Family: Aplysinellidae

Suberea clavata *(= Pseudoceratina clavata)*

Common name: Not reported

Global distribution: Great Barrier Reef of Australia

Habitat: Reef-associated

Associated bacteria: Pharmaceutical compounds and activities

Antibiotic activity: The compounds rifamycin B and rifamycin SV are derived from its actinobacteria viz. *Salinospora* displayed antibiotic activities. Further, the bioactivities of the Proteobacteria, Bacteriodetes and Firmicutes derived from this sponge are yet to be known (Thomas et al. 2010).

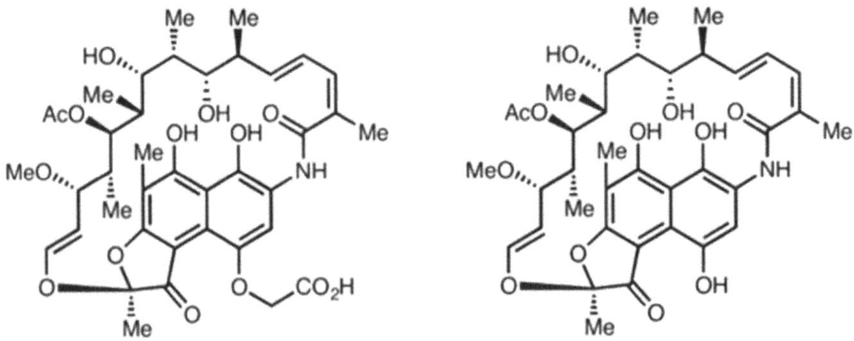

Rifamycin B, SV

Suberea creba

Common name: Not reported

Global distribution: New Caledonia; Coral Sea

Habitat: Occurs spreading over stable coral substrates; depth range, 30-55 m

Associated bacteria: Pharmaceutical compounds and activities

Antibacterial activity: A marine bacterium (pseudomonad) isolated from this sponge species has produced compounds like 2-n-heptylquinol-4-one, 2-n-nonylquinol-4-one, 2-n-(1'E-nonenyl)quinol-4-one, phenazine-alpha-carboxamide and 3-n-heptyl-3-hydroxyquinolin-2,4-dione. Among these compounds, 2-n-Heptylquinol-4-one has shown significant antibacterial activity (Debitus et al. 1998).

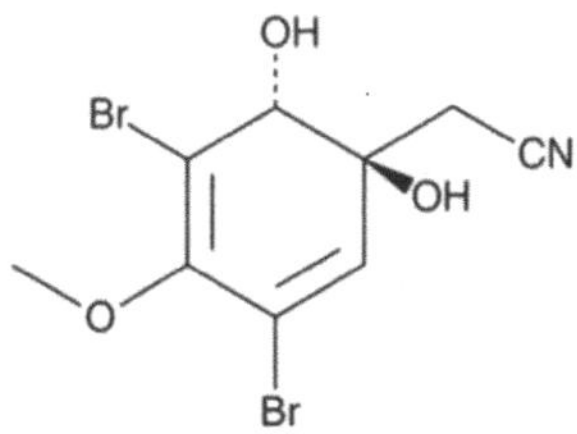

Aeroplysinin-1

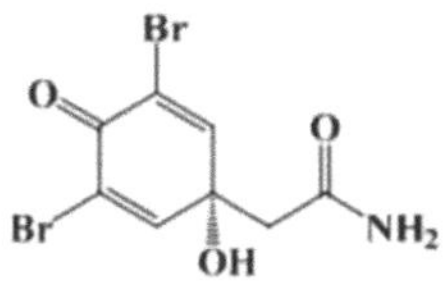

Dibromoverongiaquinol

***Suberea* sp.**

Associated bacteria: Pharmaceutical compounds and activities

Antibacterial activity: The actinomycete, *Nocardia* sp. isolated form this sponge species showed antibacterial activity against *Bacillus* sp. P25 and *Escherichia coli* K12. Other strains isolated from this sponge include *Salinispora pacifica* strain S34, *Salinispora arenicola* strain SCSIOZ-SH19 and *Saccharomonospora azurea* strain RR1 (Abdelmohsen et al. 2010).

Order: Verongiida Family: Pseudoceratinidae

Pseudoceratina purpurea

Pseudoceratina sp.

Image credit: Nhobgood (talk) Nick Hobgood, https://creativecommons.org/licenses/by-sa/3.0/

Common name: Yellow prickly branching sponge

Global distribution: Great Barrier Reef.

Habitat: Commonly seen on coral rubble; sometimes, this sponge is inhabited by Sponge finger oysters (*Vulsella* sp.).

Associated bacteria: Pharmaceutical compounds and activities

Antibacterial activity: The bacterial isolates of the genera *Bacillus* and *Virgibacillus* showed antibacterial activity against *Vibrio alginolyticus and Vibrio fischeri* with values of inhibition zone diameter 12.0 and 2.5 mm, respectively (Kanagasabapthy et al. 2005).

Class: Hexactinellida Order: Lyssacinosida Family: Rossellida

Anoxycalyx joubini (Topsent, 1916)

Image credit: Paul K. Dayton, et. al., https://creativecommons.org/licenses/by/4.0/

Common name: Giant volcano sponge

Global distribution: Throughout Antarctica and the South Shetland Islands

Habitat: Depths from about 15 m to more than 400 m.

Associated bacteria: Pharmaceutical compounds and activities

Antibacterial activity: The bacterial speices *Pseudoalteromonas haloplanktis* isolated from this sponge species produced benzenamine-N-ethyl and 3-buten-1-ol-3-methy which exhibited almost 100% inhibition against 21 Gram207 negative Bcc species (Buchholz and Duncan 2017).

Class: Homoscleromorpha Order: Homosclerophorida
Family: Plakinidae

Plakortis simplex

Plakortis sp.

Common name: Not reported

Global distribution: Western Central Pacific and Northeast Atlantic; Mediterranean, West Africa, Roscoff, SW Ireland, Norway.

Habitat: Sesile; on and under rocks and boulders from the littoral to at least 15 m.

Associated bacteria: Pharmaceutical compounds and activities

Antimalarial activity: As there are 89 times more bacterial cells than sponges cells, this species is classified as a bacteriosponge' or high microbial abundance sponge'. The microbial symbionts of this species yielded compounds like hopanoids, glycolipids and polyketide plakortin. Its highest amounts of plakortin were found in the bacterial cell fraction, which points to a symbiotic origin of this polyketide. Although its biological role is still unknown, plakortin is the most abundant secondary metabolite from this species, and its antimalarial activity makes it a candidate lead compound for new antimalarial drugs (Buchholz and Duncan 2017).

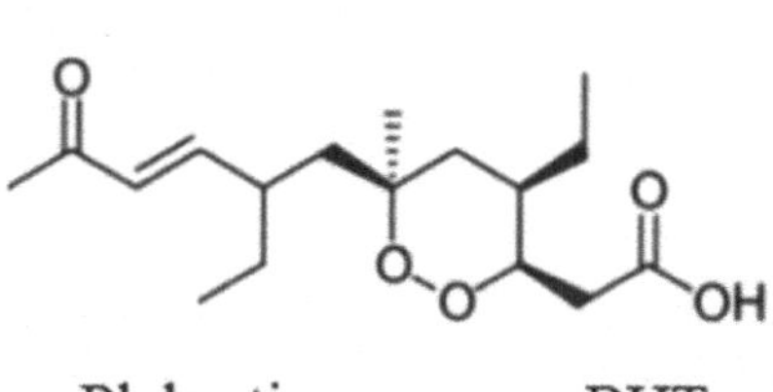

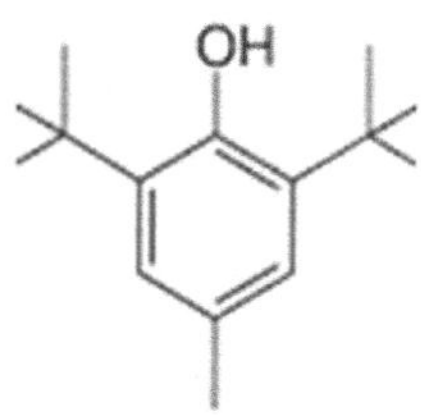

Plakortin BHT

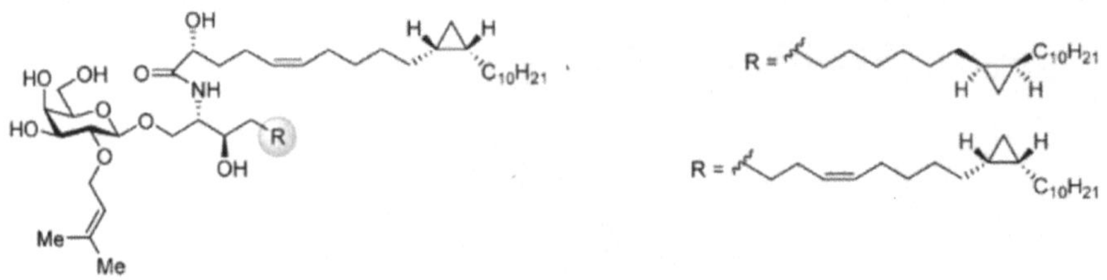

Plakoside A,B

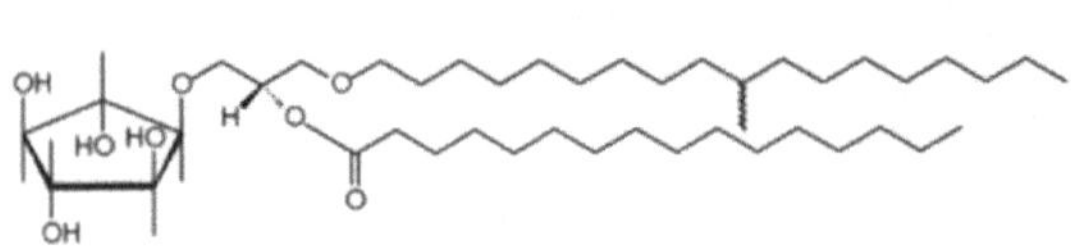

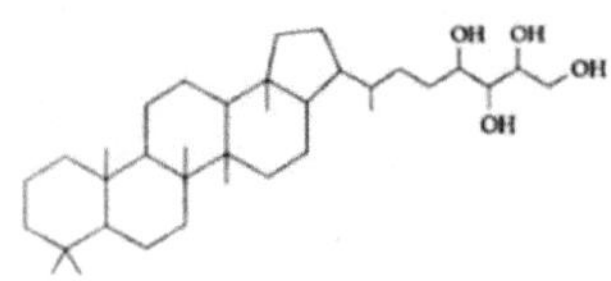

Simplexide Crasseride

Antioxidant activity: Two exocellular diketopiperazines, cyclo-(L-prolyl-L-tyrosine) and L-tyrosyl-L-valyl-L-prolyl-L-leucin were extracted from *P. haloplanktis* and showed antioxidant activity with free radical scavenging properties (Buchholz and Duncan 2017).

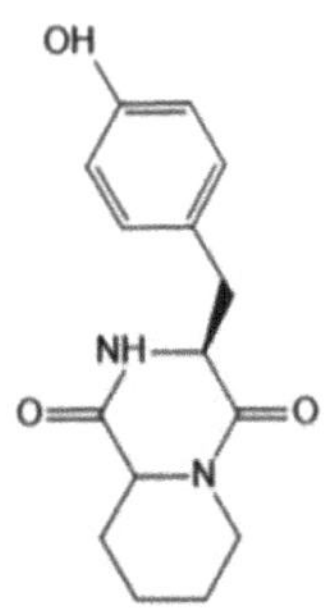

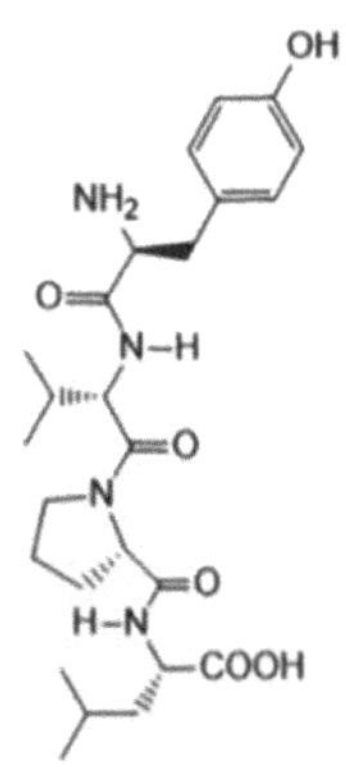

Cyclo-(L-prolyl-L-tyrosine L-tyrosyl-L-valyl-L-prolyl-L-leucin

Others: *Sphingomonas aquatilis* PS193 (Alphaproteobacteria; Sphingomonadales; Sphingomonadaceaea) which is associated with this sponge species produces glycosphingolipid GSL-1. This compound is involved in a variety of important biological processes including cell recognition, proliferation, and differentiation, signal transduction, interaction with bacterial toxins, and modulation of immune responses (Memon et al. 2001; Laroche et al. 2007).

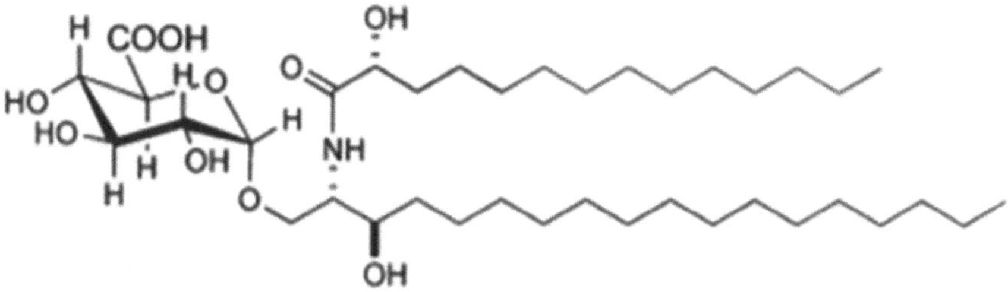

Glycosphingolipid GSL-1

Class: Calcarea Order: Clathrinida Family: Clathrinidae

Clathrina clathrus

Image credit: Elapied. https://creativecommons.org/licenses/by-sa/2.0/fr/

Common name: White clathrina

Global distribution: British Isles, west coasts of France and Spain; Mediterranean. New Zealand and Ternate (Indonesia).

Habitat: Sessile; depth range 0–20 m

Associated bacteria: Pharmaceutical compounds and activities

Antimicrobial activity: The bacterial strains (more than 99.5% sequence similarity with *Pseudoalteromonas citrea* NCIMB 18T and *Pseudoalteromonas aurantia* ATCC 33046T (Gamma-Proteobacteria, Alteromonadales) isolated from this species produced compounds viz. glycerolipids monopalmitin and monostearin which exhibited antimicrobial activity against pathogenic *Staphylococcus aureus* ATCC 6538 and the yeast *Candida albicans* ATCC 66029 (Quévrain et al. 2014).

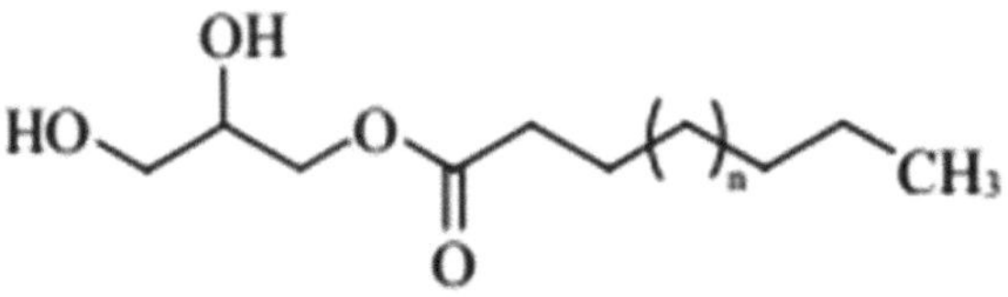

Monopalmitin, n = 10; Monostearin, n = 12

Clathrina coriacea

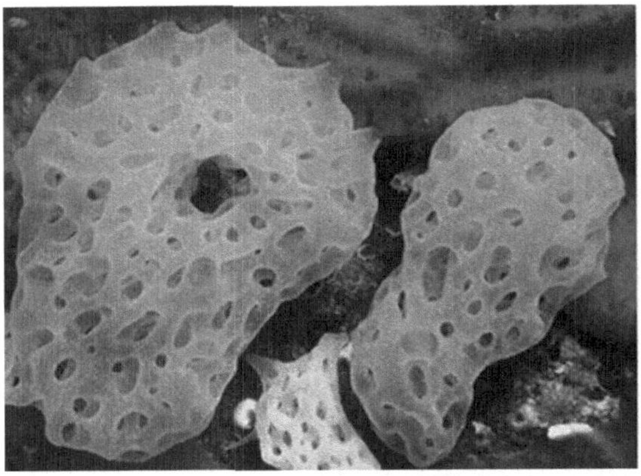

Image credit: Parent Géry, Public domain.

Common name: calcareous sponge

Global distribution: NE Atlantic

Habitat: This ascidian-associated sponge is seen in the sublittoral areas with clean rock surface

Associated bacteria: Pharmaceutical compounds and activities

Cytotoxic activity: 4-N-methyl-5-hydroxystaurosporine and 5-hydroxystaurosporine obtained from its associated bacterial strain of *Micromonospora* sp. L-31-CLCO-00 displayed significant cytotoxicity against tumor cell lines viz. SK-MEL-28 (ATCC HTB-72), A549 (ATCC CCL-185), P388D1 (ATCC CCL-46), and HT-29 (ATCC HTB-38) (Zhang et al. 2017).

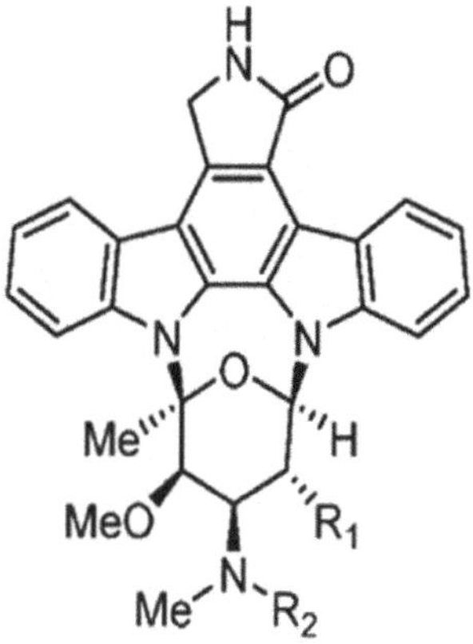

4′-N-methyl-5′-hydroxystaurosporine, R1=H, R2=Me; 5′-hydroxystaurosporine, R1=H, R2=H

Order: Baerida Family: Baeriidae

Leuconia johnstoni

Common name: Tube Sponge Clusters

Global distribution: British Isles (extending to N. of Scotland, E coast of UK). Channel coasts of France; Spain.

Habitat: Inhabits rocky overhangs and shipwrecks in wave-exposed areas

Associated bacteria: Pharmaceutical compounds and activities

Antimicrobial activity: The bacterial strain *Microbulbifer arenaceous* L4n2 isolated from this sponge species produced nine compounds belonging to the esters of para-hydroxybenzoate which exhibited antimicrobial activity. Among these compounds, compound 4 displayed very significant activity against *Staphylococcus aureus* with an MIC values of 2.8–5.6 μM. Further, all these nine compounds have also been reported to act against the *Bacillus* sp. strain L6-3e (Quévrain et al. 2014).

p- hydroxybenzoate esters

Unidentified sponge sp.1

Associated bacteria: Pharmaceutical compounds and activities

Cytotoxicity: Three secondary metabolites, amycofuran, amycocyclopiazonic acid, and amycolactam were isolated from this sponge-associated rare actinomycete *Amycolatopsis* sp. Among these compounds, amycolactam exhibited potent cytotoxicity against colon cancer cell line HCT116 and gastric cancer cell line SNU638 and (Kwon et al. 2014).

Unidentified sponge sp.2

Associated bacteria: Pharmaceutical compounds and activities

Antibacterial activity: The bacterial strain *Kocuria* sp. F-276,345 associated with this unidentified species of sponge produced the thiazolyl peptide compound, kocurin which exerted extremely potent activity against Gram-positive bacteria with MIC values of 0.25–0.5 μg/mL against MRSA and no activity against the Gram-negative bacterial pathogens *A. baumannii, P. aeruginosa and E. coli* (Palomo et al., 2013).

Anti-parasitic activity: The extracts of the isolate *Kocuria* sp. F-276,345 of this sponge showed anti-parasitic activity against *Trypanosoma brucei* (Palomo et al. 2013).

Unidentified sponge sp.3

Associated bacteria: Pharmaceutical compounds and activities

Anti-tumor activity: *Tenacibaculum mesophilum* (phylum Bacteroidetes) associated with this unidentified species of sponge produced a compound bisucaberin B which showed direct cytostasis (inhibition of cell growth and multiplication) for tumor cells (Kameyam et al. 1987) is attributable to the specific inhibition of DNA synthesis in tumor cells (Kameyam et al. 1987).

Unidentified sponge sp.4

Associated bacteria: Pharmaceutical compounds and activities

Antifungal activity: The bacterial strain *Streptomyces* sp. Ni-80 associated with this unidentified sponge yielded urauchimycins A and B. These antimycin antibiotics exhibited inhibitory activity against *Candida albicans* (Thomas et al. 2010).

Unidentified sponge sp.5

Associated bacteria: Pharmaceutical compounds and activities

Anticancer activity: Compounds ulbactins F, G isolated from its associated bacterium, *Brevibacillus* sp. Displayed significant inhibitory activity against epidermoid carcinoma cell line A431 with IC50 values of 6.4 and 6.1 μM, respectively (Zhang et al. 2017).

Unidentified sponge sp.6

Associated bacteria: Pharmaceutical compounds and activities

Antibacterial activity: The bacterial strain *Corynebacterium* sp. GUVFUM-1 isolated form this sponge species displayed antibacterial activity against *S. typhi, A. aerogens, S. citreus*, and *P. vulgaris* with zone of inhibition values of 10, 8, 5, and 11 mm, respectively (Velho-pereira and Furtado 2012).

Unidentified sponge sp.7

Associated bacteria: Pharmaceutical compounds and activities

Antibacterial activity: The bacterial strain *Halobacteria* sp GUVFSM-1 isolated from this unidentified sponge species displayed antibacterial activity against *S. typhi, S. marcesans, A. aerogens, E. coli* and *P. vulgaris* with zone of inhibition values of 7, 6, 5, 3, and 14 mm, respectively; its another isolate, *Chromohalobacter* sp. GUVFSM-2 acted agsint *S. typhi, A. aerogens, E. coli,* and *S. citreus* with 12, 9, 11, and 10 mm, respectively (Velho-pereira and Furtado 2012).

Unidentified sponge sp.8

Antitrypanosomal activity: An azepino-diindole alkaloid compound, rhodozepinone isolated from an unidentified marine sponge- associated bacterium, *Rhodococcus* sp. UA13 showed with moderate antitrypanosomal activity on *T. brucei brucei* TC221 with an IC50 value of 16.3 μg/mL (Mayer et al. 2021).

Unidentified sponge sp.9

Antibacterial activity: The medermycin derivatives viz. MDN-0171, medermycin and antibiotic G15-F isolated from the bacterial strain, *Streptomyces albolongus* CA-186053 which was associated with an unidentified sponge species showed antibacterial activity by inhibiting the growth of MRSA MB5393 with an MIC value of 2–4 μg/mL (Nweze et al. 2020).

Unidentified sponge sp.10

Antibacterial activity: The compound, urauchimycin D identified from an deep-sea sponge-associated actinobacterium (WG1-60-61) showed anti-bacterial activity by inhibiting the growth of MRSA ATCC 700787 with an MIC value of 0.125 μg/mL (Nweze et al. 2020).

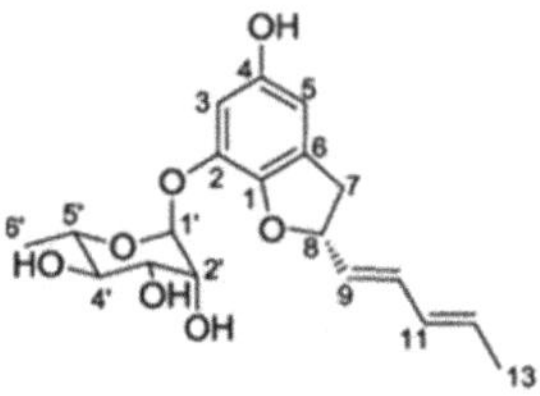
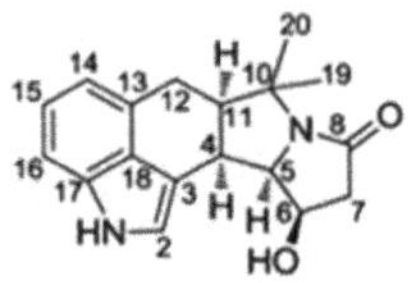
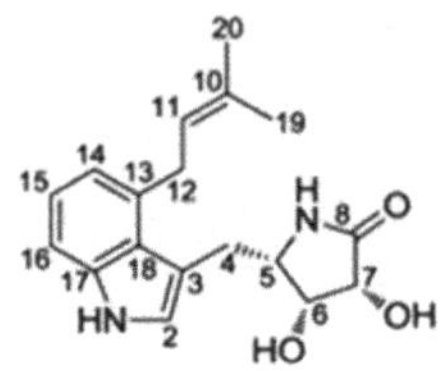

Amycofuran Amycocyclopiazonic acid Amycolactam

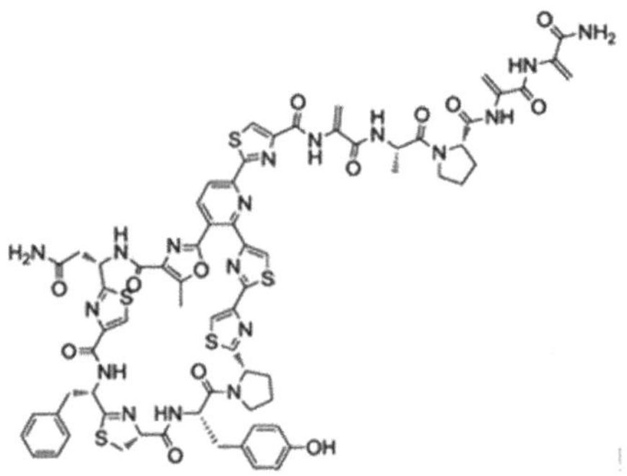
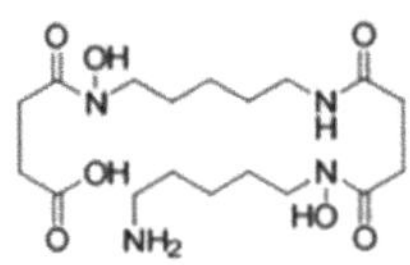

Kocurin Bisucaberin B

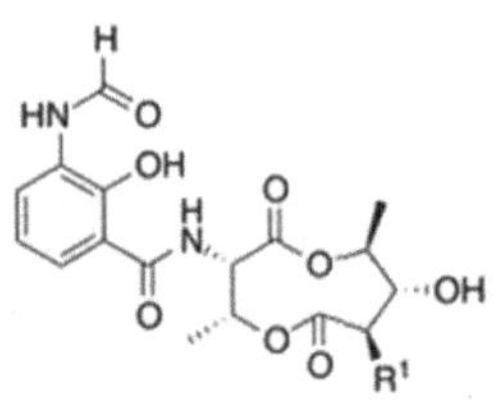
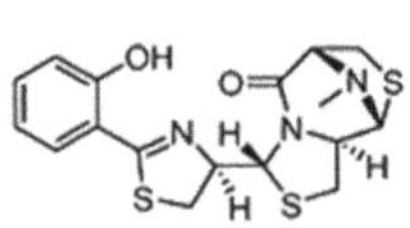
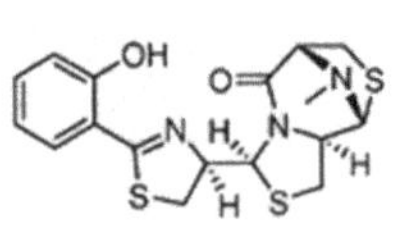

Urauchimycin A: R1= 2-methylbutyl; Ulbactin F,G

Urauchimycin B: R1= Isopentyl

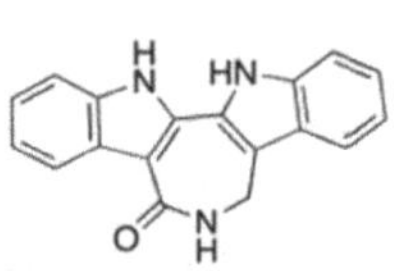
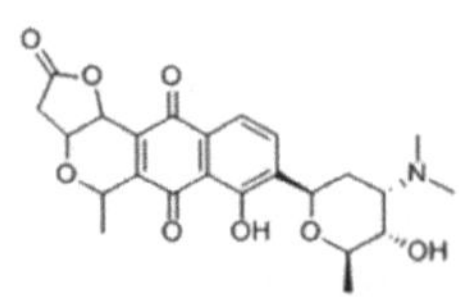
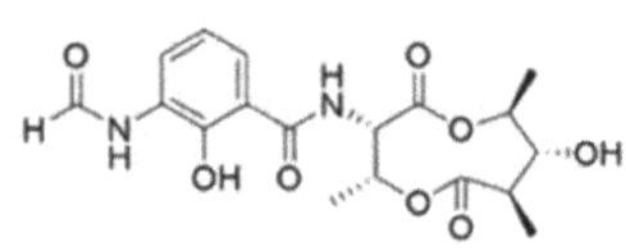

Rhodozepinone Medermycin Urauchimycin D

CNIDARIANS

Class: Scyphozoa Order: Rhizostomeae Family: Cassiopeidae

Cassiopeia xamachana

Common name: Upside-down jellyfish

Global distribution: Warm parts of the western Atlantic Ocean, Gulf of Mexico and Caribbean Sea

Habitat: It is found in large aggregations on the muddy seabed of inshore bays and shallow lagoons; and if disturbed the jellyfish may pulsate their bells and rise in the water, soon flopping back onto the seabed.

Associated bacteria: Pharmaceutical compounds and activities

RNAP-inhibitory, anti-inflammatory and antibacterial activities: The compounds, salinamides A-E have been isolated from from the marine actinomycete, *Streptomyces* sp., strain CNB091 obtained from the surface of this jellyfish. Among these compounds, salinamides A and B showed moderate antibiotic activity against gram-positive bacteria and potent anti-inflammatory activity (Trischman et al. 1994). On the other hand, salinamides C and E showed anti-inflammatory activity (Moore et al. 1999). Further, the salinamide analog, salinamide F isolated from this bacterial strain possessed significant RNAP-inhibitory and antibacterial activity (Hassan et al. 2015).

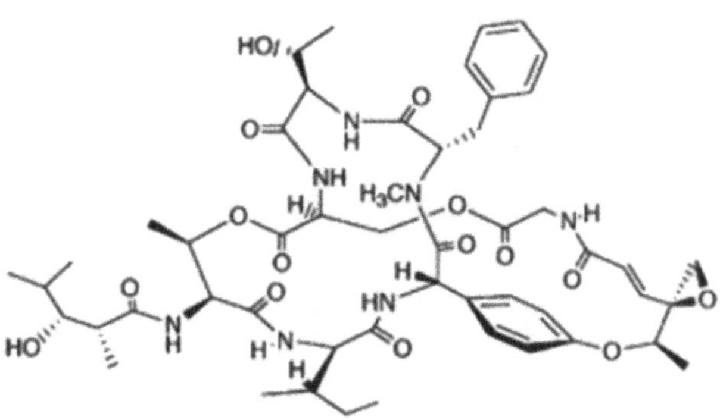

Salinamides

Anti-IL 5 activity: The bacterium *Salinosora pacifica* associated with this species produced a polyketide, salinopyrone A which moderately inhibited

Interleukin-5 (IL-5) with an IC50 vlaue of 10 μg/mL in a mouse splenocyte model which possessed allergic inflammation (Arora et al. 2019).

Class: Anthozoa Order: Alcyonacea Family: Alcyoniidae

Lobophytum crassum

Image credit: Philippe Bourjon, https://creativecommons.org/licenses/by-sa/4.0/

Common name: Dimorphic soft coral; cabbage leather coral

Global distribution: Indo-West Pacific

Habitat: Sessile; inhabits shallow nutrient-rich reef waters experiencing moderate water flow at depths between 6 and 20 m.

Associated bacteria: Pharmaceutical compounds and activities

Cytotoxic and anti-inflammatory activity: The bacterial strain, *Pseudoalteromonas* sp. CGH2XX associated with this species produced a ubiquinone-monoterpenoid derivative viz. pseudoalteromone A which displayed cytotoxicity against human acute lymphoblastic leukemia (MOLT-4) cel line; and anti-inflammatory activity by inhibiting elastase from human neutrophils (Raimundo et al. 2018).

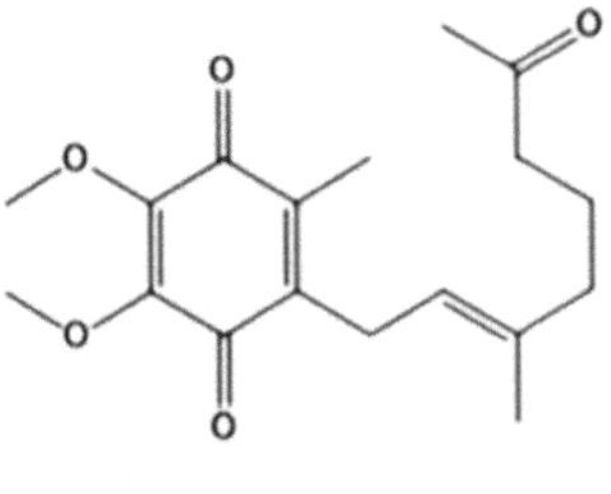

Pseudoalteromone A

Sinularia polydactyla

Image credit: Derek Keats, https://creativecommons.org/licenses/by/2.0/

Common name: Many finger leather coral

Global distribution: Indo-West Pacific Ocean: Red Sea, Madagascar, Seychelles, to Guam.a

Habitat: Inhabits shallow water reef tops, lagoons, and reef slopes from depths of 2 to 10 m

Associated bacteria: Pharmaceutical compounds and activities

Antibacterial activity: The marine bacterium associated with this soft coral viz. *Pseudomonas* sp. displayed antibacterial activity against *Streptococcus equi* subsp. *zooepidemicus* (Radjasa et al. 2007).

Cytotoxic and antibacterial activities: The *Vibrio* species WMBA associated with this cnidarian yielded seven maleimide compounds viz. aqabamycin A-G; 3-nitro-1H-indazole, indazole-3-carbaldehyde, phenyl-2-bis-indolylmethane, turbomycin B, vibrindole A, 1,4-dithiane, 3-(3-nitro-4-hydroxyphenyl)-2-propenoic acid, 3-nitro-4-hydroxybenzaldehyde, phenylacetic acid, benzoic acid, 3-hydroxybenzoic acid, and 4-hydroxycinnamic acid. All these compounds have shown cytotoxicity against cell lines such as L1210, jurkat, MDA-MB321, MCF-7, and Colo-320. Further, these compounds have also shown antimicrobial activity against pathogenic bacteria like *Escherichia coli*, *Micrococcus luteus*, *Bacillus subtilis*, and *Proteus vulgais*; and antifungal activity against *Candida albicans* and *Saccharomyces cerevisiae* (Al-Zereini et al. 2010).

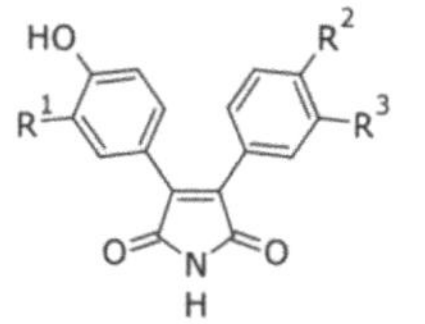

A: R1=H,R2=H,R3=H; B,:R1=H, R2=OH, R3=NC;
C: R1=NO2, R2= H, R3=H; D: R1=NO2,R2=OH,R3= NO2
Aqabamycin A-D

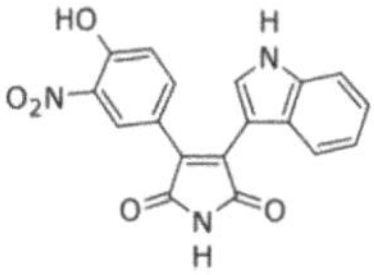

Aqabamycin G

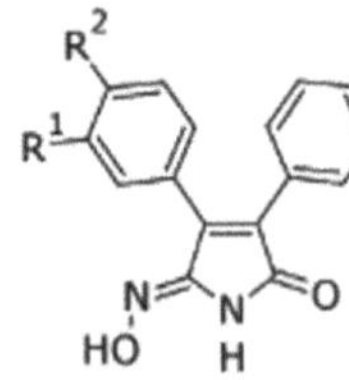

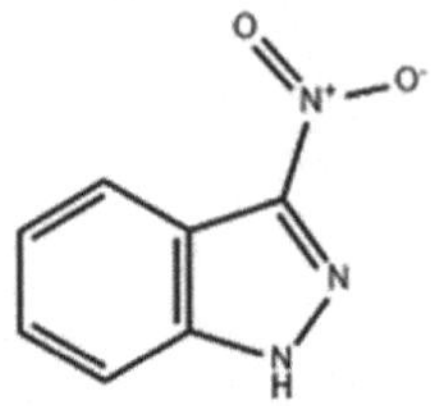

E: R1=NO2, R2=OH,R3=H, R3=h,R4=H

3-nitro-1H-indazole

E': R1=H,R2=H,R3= OH,R4=NO2

F: R1=NO2, R2=OH, R3=OH,R4=NO2

Aqabamycin E,E', F

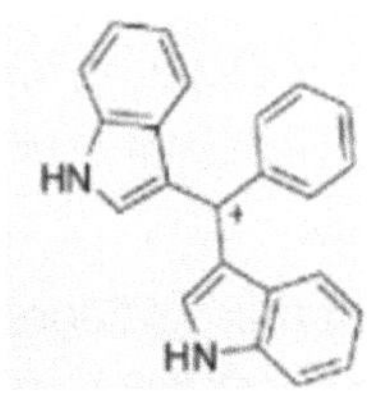

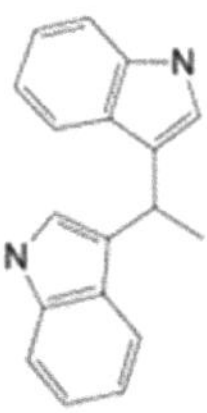

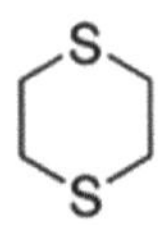

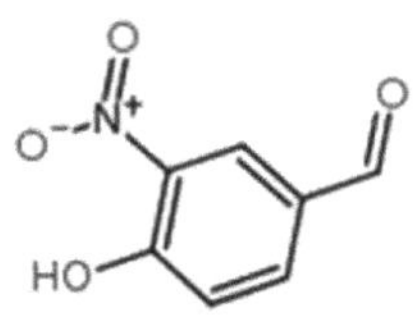

Turbomycin B Vibrindole 1,4-dithiane 3-nitro-4-hydroxybenzaldehyde

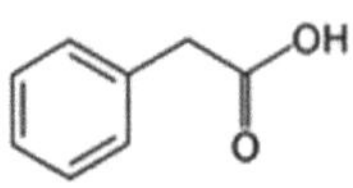

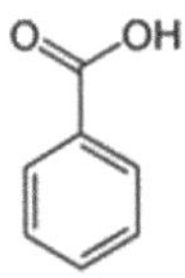

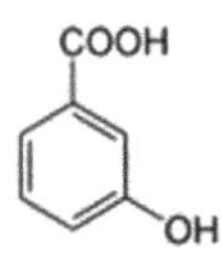

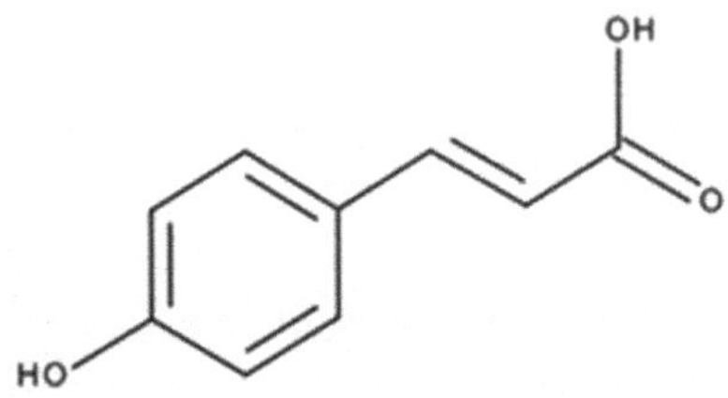

Phenylacetic acid Benzoic acid 3-hydroxybenzoic acid 4-hydroxycinnamic acid

Sinularia sp.

Associated bacteria: Pharmaceutical compounds and activities

Antibacterial antimycobacterial activity: The strain of actinobacteria associated with this cnidarian species viz. *Arthrobacter* sp. SFNB.5 showed antibacterial activity by inhibiting the growth of *Staphylococcus, Vibrio*, and *Tenacibaculum* species. Further, two isolates of Alphaproteobacteria viz. *Pseudovibrio* sp. SC4TGZ3 and strain SC4TGZ4 derived from this cnidarian displayed potent antimycobacterial activity against MDR *Mycobacterium tuberculosis* strains (Raimundo et al. 2018).

Order: Alcyonacea Family: Clavulariidae

***Telesto* sp.**

Associated bacteria: Pharmaceutical compounds and activities

Antibacterial activity: The bacterial species *Bacillus* sp. and *Chromohalobacter* sp. associated with this species showed antibacterial activity. While *Bacillus* sp. acted against *S. typhi* and *A. aerogens* (zone of inhibition, 2 mm), *Chromohalobacter* sp. showed inhibition against *S. typhi, A. aerogens, E. coli, S. citreus* and *P. vulgaris* with the inhibition zone dia. values of 6,15,12,12, and 12 mm, respectively (Velho-Pereira and Furtado 2012).

Order: Alcyonacea Family: Ellisellidae

Junceella juncea (Pallas, 1766)

Image credit: Fernando Herranz Martín, https://www.gnu.org/licenses/gpl-3.0.html

Common name: Not reported

Global distribution: Tropical Indo-Pacific; South China Sea

Habitat: Reef and reef slopes.

Associated bacteria: Pharmaceutical compounds and activities

Antibacterial activity: The marine bacterium *Bacillus amyloliquefaciens* SCSIO 00856 isolated from this species produced compounds, macrolactin V, macrolactin S, 4-butoxyphenol, 4-propoxyphenol, and 4-ethoxyphenol (Gao et al. 2010). Among these compounds, macrolactin V was found to be most active against *Staphylococcus aureus*, *Escherichia coli*, and *Bacillus subtilis* with an MIC value of 0.1 gmL^{-1} (Raimundo et al. 2018).

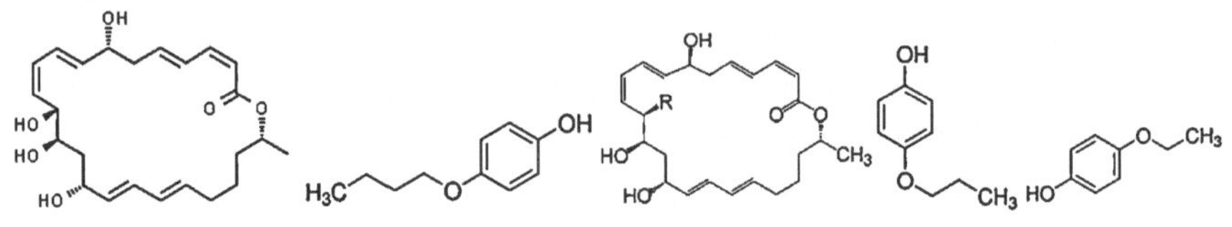

Macrolactin V 4-butoxyphenol Macrolactin S 4-propoxyphenol 4-ethoxyphenol

R=OH

Order: Alcyonacea Family: Gorgoniidae

Leptogorgia alba

Leptogorgia sp.

Image credit: NOAA, Public domain.

Common name: Not designated.

Global distribution: Eastern Pacific

Habitat: Benthic.

Associated bacteria: Pharmaceutical compounds and activities

Antifungal activity: The bacterial strain *Pseudoalteromonas* sp. OT59 isolated from this species produced a polyketide, ateramide A, which displayed light-dependent antifungal activity (Raimundo et al. 2018).

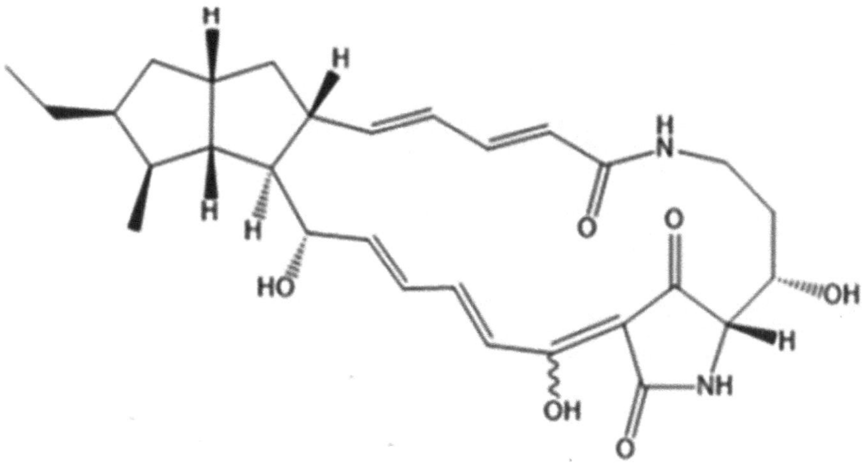

Ateramide A

***Pacifigorgia* sp.**

Associated bacteria: Pharmaceutical compounds and activities

Antitumor activity: Among the compounds octalactins A and B isolated from the associated *Streptomyces* sp. of this gorgonian, octalactin A exhibited potent cytotoxicity against HCT-116 human colon tumor cell line and B-16-FlO murine melanoma cell line with IC50 values of 0.5 and 0.0072 pg/mL, respectively (Tapiolas et al. 1991).

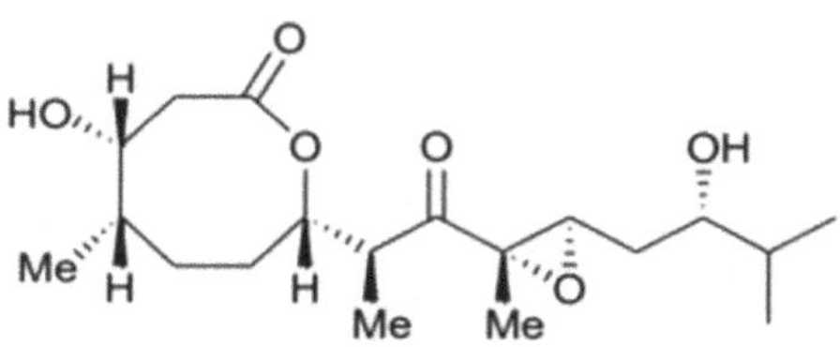

Oclactin A

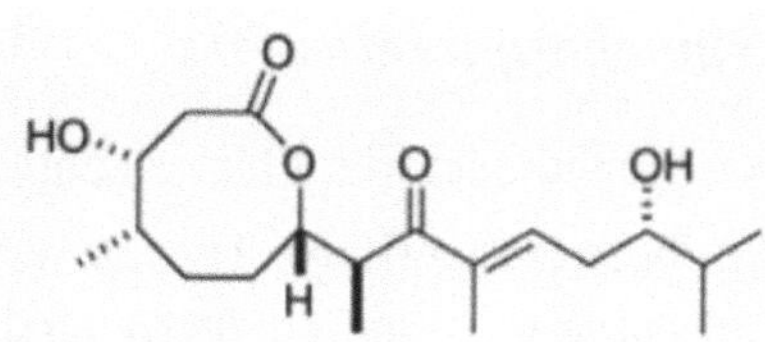

Octalactin B

Order: Alcyonacea Family: Plexauridae

Echinogorgia complexa

Echinogorgias sp.

Image credit: Chaloklum Diving https://creativecommons.org/licenses/by-sa/4.0/

Associated bacteria: Pharmaceutical compounds and activities

Antibacterial activity: The bacterial species viz. *Deinococcus* sp., *Chromohalobacter* sp., *Bacillus* sp, and *Virgibacillus* sp. associated with this species showed Antibacterial activity. *Deinococcus* sp. showed inhibition against *Escherichia coli* and *S. citreus* with the values of zone inhibition 10 and 30 mm respectively; *Chromohalobacter* sp. acted against *S. typhi, S. marcesans, A. aerogens, E. coli*, *S. citreus* and *P. vulgaris* with values of 2, 6, 12, 2, 5, and 15 mm, respectively; and *Bacillus* sp. and *Virgibacillus* sp inhibited only *A. aerogens* with the value of 2 mm (Velho-Pereira and Furtado 2012).

Echinogorgia reticulata

Associated bacteria: Pharmaceutical compounds and activities

Antibacterial activity: The bacterial species viz. *Corynebacterium* sp. and *Chromohalobacter* sp associated with this species showed antibacterial activity against *Aerobactor aerogens* with the values of zone of inhibition, 2.0 and 2.8 mm, respectively (Velho-Pereira and Furtado 2012).

Order: Alcyonacea Family: Plexauridae

Echinomuricea indica

Global distribution: India

Habitat: 24- 65 m depth.

Associated bacteria: Pharmaceutical compounds and activities

Antibacterial activity: The bacterial species viz. *Chromohalobacter* sp. and *Psychrobacter* sp. associated with this species showed antibacterial activity against *Aerobactor aerogens* with the values of zone of inhibition, 2- 30 and 2 mm, respectively (Velho-Pereira and Furtado 2012).

Antipathes dichotoma

Order: Antipatharia Family: Antipathidae

Image credit: Fernándo Herranz Martín, https://www.gnu.org/licenses/gpl-3.0.html

Common name: South China Sea black coral

Global distribution: Temperate waters of western Atlantic Ocean.

Habitat: It is a deep water cnidarian occurring at depths of 200 - 300 m

Associated bacteria: Therapeutic compounds and their bioactivities

Antibacterial activity: The bacteria isolated from this species viz. *Bacillus* spp., s showed antibacterial activity against *Micrococcus luteus* and *Pseudoaltermonas piscida*, *Micromonospora coxensi*, *Staphylococcus equorum* and *Streptomyces* spp. The values of zone of inhibition with respect to

Micrococcus luteus and *Pseudoaltermonas piscida* ranged from 8 to 16.3 mm and from 11.2 to 16.9 mm respectively (Zhang et al. 2012).

Order: Scleractinia Family: Acroporidae

Acropora muricata (= *Acropora formosa*)

Image credit: MDC Seamarc Maldives, https://creativecommons.org/licenses/by-sa/4.0/

Common name: Staghorn coral

Global distribution: Central Indo-Pacific and western Pacific Ocean; and Indian Ocean,

Habitat: It occcurs mainly in shallow reefs and lagoons at depths of 5 – 30 m.

Associated bacteria: Pharmaceutical compounds and activities

Antibacterial activity: The extracts of the bacterial strain, *Pseudoalteromonas luteoviolacea* TAB4.2 derived from this cnidarian species displayed antibacterial activity against *Escherichia coli*, *Vibrio parahaemolyticus*, *V. harvei*, *V. anguillaraum*, and *Staphylococcus aureus*) (Radjasa et al. 2007).

The bacterial species associated with this species viz. *Bacillus* sp., *Pontibacillus* sp., *and Corynebacterium* sp. showed antibacterial activity against *S. typhi*, *S. marcesans*, *A. aerogens*, *E. coli*, *S. citreus*, and *P. vulgaris* (Velho-Pereira and Furtado 2012).

Acropora sp.

Associated bacteria: Pharmaceutical compounds and activities

Antibacterial activity: The bacterial strain *Pseudoalteromonas luteoviolacea* TAB4.2 isolated from this species showed antibacterial activity by inhibiting the growth of *Escherichia coli, Vibrio parahaemolyticus, V. harvei, V. anguillaraum*, and *Staphylococcus aureus*) (Radjasa et al. 2007).

Order: Scleractinia Family: Agariciidae

Pavona sp.

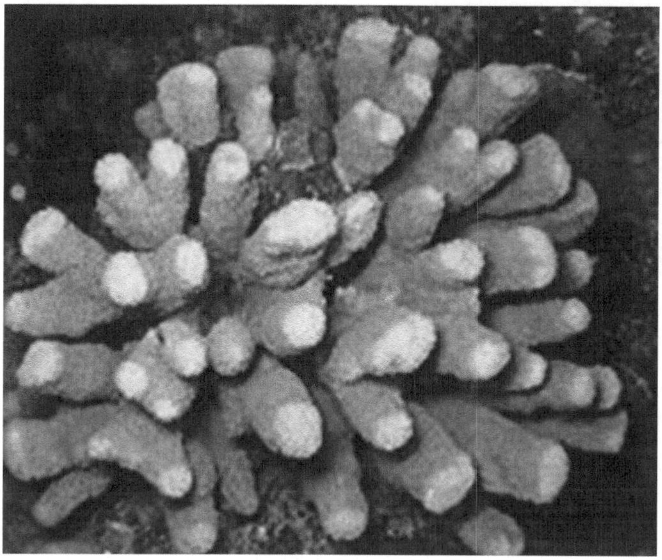

Pavona sp.

Image credit: Benzoni, F., https://creativecommons.org/licenses/by/3.0/

Common name: Not reported

Global distribution: Indo-Pacific region.

Habitat: shallow waters

Associated bacteria: Pharmaceutical compounds and activities

Antimicrobial activity: The genus *Vibrio* and *Pseudoalteromonas* sp. associatd with this species showed antimicrobial activity (Radjas et al. 2008).

Ayuningrum et al. (2016) reported that two bacterial isolates viz. PHC 44/04 and PHC 44/05 associated with this species showed activity against MDR bacteria such as *Enterobacter aerogens, Acinetobacter baumanni*, and *E. coli.*

Order: Scleractinia Family: Euphylliidae

Galaxea fascicularis

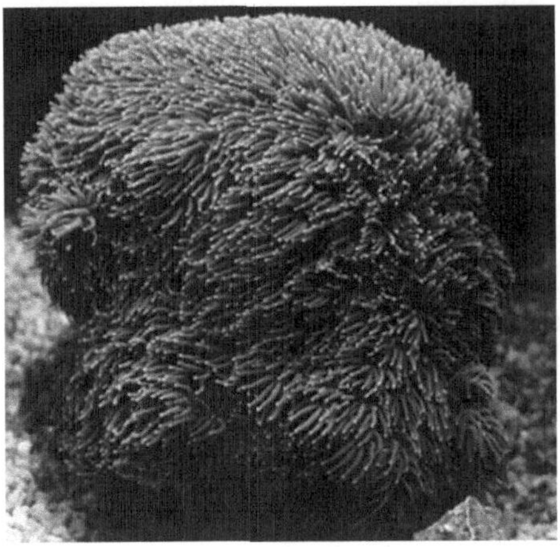

Galaxea sp.

Image credit: Josuevg, https://creativecommons.org/licenses/by-sa/3.0/

Common name: Octopus coral, fluorescence grass coral.

Global distribution: Indo-Pacific region

Habitat: It occurs on the slopes of coral reefs with weak wave action at depths of 2–15 m.

Associated bacteria: Pharmaceutical compounds and activities

Antimicrobial activity: The genus *Vibrio* associated with this species showed antibacterial activity (Radjas et al. 2008).

Order: Scleractinia Family: Poritidae

Porites lutea

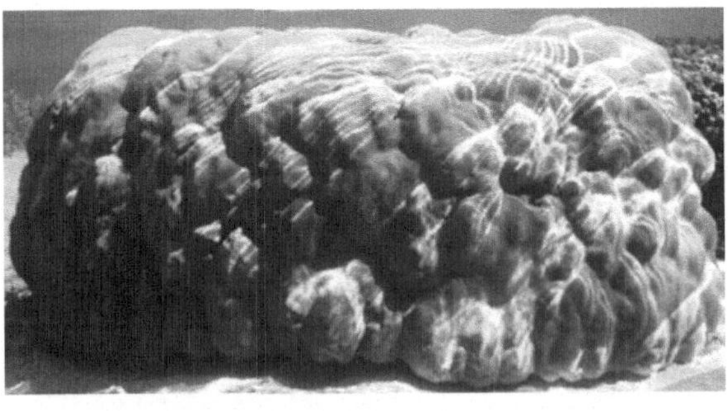

Image credit: Philippe Bourjon. https://creativecommons.org/licenses/by-sa/3.0/

Common name: Mound coral

Global distribution: common coral in Hawaii and throughout the Indo-Pacific Ocean.

Habitat: Occurs with *Porites lobata* and *P. australiensis* on back reef margins, lagoons and fringing reefs; mostly often found in water up to 6 m in depth.

Associated bacteria: Pharmaceutical compounds and activities

The isolates of vibrios, *V. coralliitycus* MJ.5 and *V. parahaemolyticus* MJ.11 obtained from this coral showed antibacterial activity (Radjasa et al. 2008).

Unidentified sp.1

Associated bacteria: Pharmaceutical compounds and activities

Anticancer activity: This soft-coral associated *Streptomyces* sp. OUCMDZ-1703 produced the polyketides viz. streptochloritide A and B that possessed cytotoxicity against breast cancer (MCF-7) cells with (IC50 values of 9.9 and 20.2 M, respectively) (Raimundo et al. 2018).

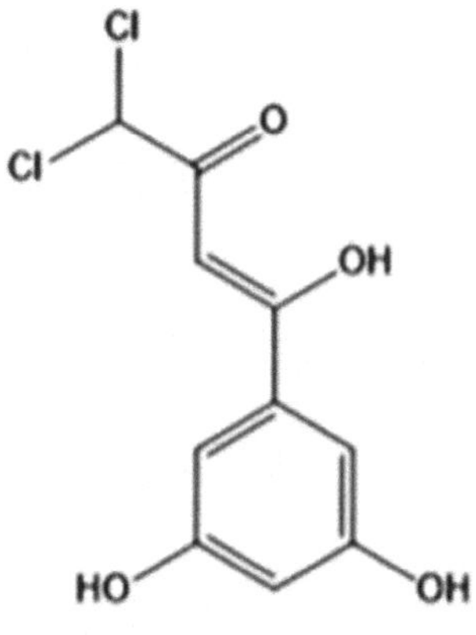

Streptochloritide A

NEMERTEANS

Phylum: Nemertea (ribbon worms)

Class: Palaeonemertea

Cephalothrix rufifrons

Class: Pilidiophora

Lineus longissimus

Lineus ruber

Lineus viridis

Lineus sanguineus (= *Ramphogordius sanguineus*)

Riseriellus occultus

Class: Hoplonemertea

Amphiporus lactifloreus

Global distribution: British nemerteans

Associated bacteria; and their pharmaceutical compounds and activities

The tetrodotoxin (TTX) produced by the associated *Vibrio* sp. has been reported to be a potential therapeutic agent for pain and intramuscular TTX at doses of 30 g twice daily for 4 days was a regimen with an acceptable toxicity and analgesic profile (Nieto et al. 2012). Further, this potent neurotoxin is also believed to serve as a pain reliever especially to treat severe cancer pain (Yu et al. 2017).

Phylum: Nemertea Class: Palaeonemertea

Cephalothrix simula

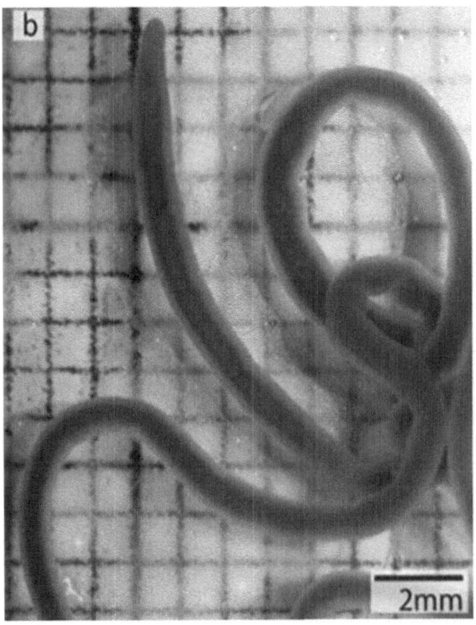

Cephalothrix sp.

Image credit: Patrick Beckers, Rudi Loesel, Thomas Bartolomaeus, https://creativecommons.org/licenses/by/2.5/

Global distribution: Pacific Ocean; and Mediterranean and Atlantic coasts

Associated bacteria; and their pharmaceutical compounds and activities

The bacterial species *Pseudomonas luteola, Bacillus* sp. 1839, and *Vibrio alginolyticus* have been found to be associated with this nemertean, and they are known to produce the neurotoxin viz. TTX the therapeutic uses of which have already been given (Turner et al. 2018).

Class: Pilidiophora

Parborlasia corrugatus

Global distribution: Signy island

Associated bacteria and their pharmaceutical compounds and activities

The bacterium associated with this nemertena species is *Vibrio alginoliticus*, which produced TTX (McEvoy et al. 1998).

Class Hoplonemertea

Malacobdella grossa

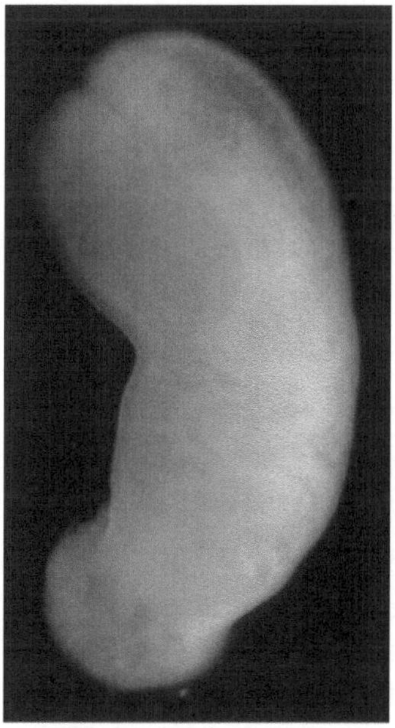

Malacobdella sp..

Image credit: Christopher Laumer. https://creativecommons.org/licenses/by/2.0/

Global distribution: Pacific coasts of North America and coasts of Europe and Atlantic

Associated bacteria: and their pharmaceutical compounds and activities

Bacteria associated with this nemertean species belonged to the genus *Alteromonas (*one strain*), Ruegeria* (two strains), and *Pseudoalteromonas* (three strains). These bacterial strains have been reported to possess only antimicrobial activity (Melnikova et al. 2017).

Class: Hoplonemertea

Quasitetrastemma stimpsoni

Quasitetrastemma sp.
Image credit: Wesley Roswell Coe, Public domain.

Global distribution: Northern Pacific Ocean

Associated bacteria; and their pharmaceutical compounds and activities

The bacterial community of this nemertean species was mainly presented by the genus *Pseudoalteromonas* (71.4% of the total number of strains). Two strains each of the genera *Bacillus, Pseudomonas,* and *Shewanella*, as well as a single *Vibrio* strain, were also found in this ribbon worm. Most bacterial strains of this nemertean species belonging to the genus *Pseudoalteromonas* showed antimicrobial activity (Melnikova et al. 2017).

BRYOZOANS

Phylum: Bryozoa Class: Gymnolaemata

Adeonella calveti

Global distribution: Mediterranean Sea

Habitat: Benthic

Associated bacteria and their pharmaceutical compounds and activities

Antibacterial activity: The bacterial strain *Alteromonas macleodii* 'Deep ecotype' isolated from this species showed antibacterial activity against *Bacillus subtilis*. On the other hand, the strains *Sphingomonas* sp. and *Sphingomonas* sp. SKJH-30 acted only against *Staphylococcus lentus* (Heindl 2011).

Bugula neritina

Image credit: Gall L, https://creativecommons.org/publicdomain/zero/1.0/

Global distribution: Cosmopolitan distribution.

Habitat: Harbors and embayments; intertidal to 5 m; and are attached to any available hard substrate

Associated bacteria; and their pharmaceutical compounds and activities

Anticancer activity: The bacterial symbiont of this species has been characterized as a gamma-proteobacterium, *Candidatus* Endobugula sertula. These

Endobugula symbionts are involved in the production of the polyketides, bryostatins that are also being tested for use in human cancer treatments (Davidson et al. 2001; Lim and Haygood 2004).

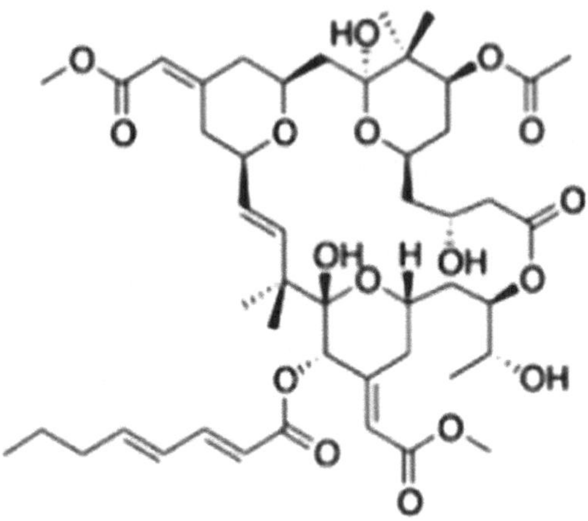

Bryostatin

Bugulina simplex *(= Bugula simplex)*

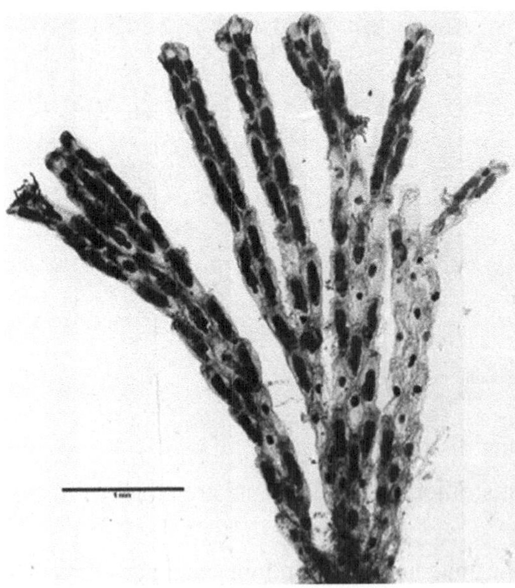

Image credit: Daniel J. Drew - Gall L, https://creativecommons.org/publicdomain/zero/1.0/

Global distribution: Britain, North Irish coast, Norway, Belgium, and Brittany

Habitat: Attached to immersed surfaces of pontoons, supporting poles, fenders, and fouled hulls

Associated bacteria; and their pharmaceutical compounds and activities

Anticancer activity: The extracts of this species and the symbiotic bacterium *Candidatus Endobugula glebosa* have shown the presence of compounds similar to bryostatins which are being rested for human cancer treatment (Lim and Haygood 2004). Further investigations are however needed to characterize bioactive metabolites in *B. simplex/Candidatus* E. glebosa (Lim and Haygood 2004).

Amphiblestrum auritum *(= Callopora aurita)*

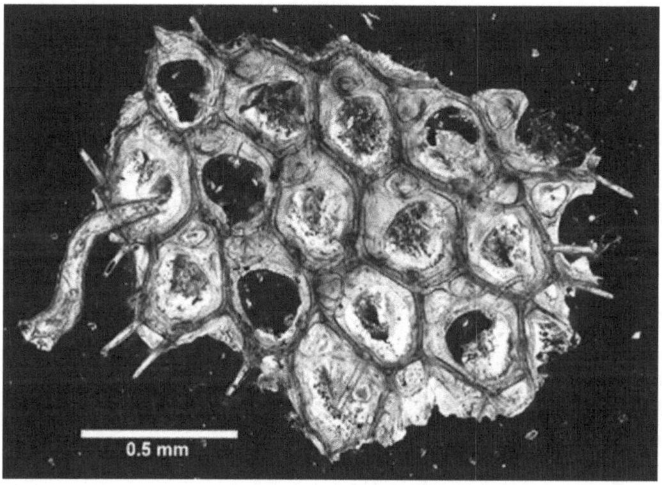

Image credit: Gall L, https://creativecommons.org/publicdomain/zero/1.0/

Global distribution: Northeast Atlantic: Denmark; Baltic and the Mediterranean Sea.

Habitat: Sessile.

Associated bacteria and their pharmaceutical compounds and activities

Antibacterial activity: The bacterial strain *Pseudoalteromonas* sp. BSw20441 isolated from this species showed antibacterial activity against *Staphylococcus lentus, Bacillus subtilis* and *E.coli* (Heindl 2011).

Carbasea papyrea

Global distribution: Baltic and the Mediterranean Sea.

Habitat: Sessile

Associated bacteria: Therapeutic compounds and their bioactivities

Antibacterial activity: The bacterial strain *Sphingomonas paucimobilis* ATCC29837 isolated from this species showed antibacterial activity against *Staphylococcus lentus*. On the other hand, the strain *Pseudoalteromonas* sp. CF14-5 acted against both *Staphylococcus lentus* and *Bacillus subtilis* (Heindl 2011).

Cellaria salicornioides

Global distribution: Baltic and the Mediterranean Sea.

Habitat: Sessile

Associated bacteria: and their pharmaceutical compounds and activities

Antibacterial activity: The bacterial strain *Sphingomonas paucimobilis* ATCC29837 isolated from this species showed antibacterial activity against *Staphylococcus lentus*. On the other hand the strains *Alteromonas* sp. strain KE10 acted against *Staphylococcus lentus* and *Bacillus subtilis* (Heindl 2011).

Cellepora pumicosa

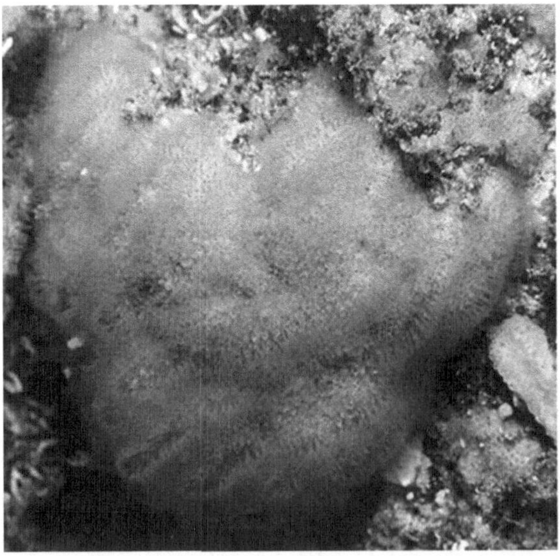

Image credit: Joanne Porter, https://creativecommons.org/publicdomain/zero/1.0/

Global distribution: Northeast Atlantic and the Mediterranean Sea.

Habitat: Sessile; depth range 15 - 32 m

Associated bacteria; and their pharmaceutical compounds and activities

Antibacterial activity: The bacterial isolate *Alteromonas* sp. CF14-2 isolated from this species showed antibacterial activity against *Staphylococcus lentus*. On the other hand, the strain *Pseudovibrio ascidiaceicola* F423 acted on all pathogens viz. *Staphylococcus lentus, Bacillus subtilis* and *E.coli* (Heindl 2011).

Cryptosula pallasiana

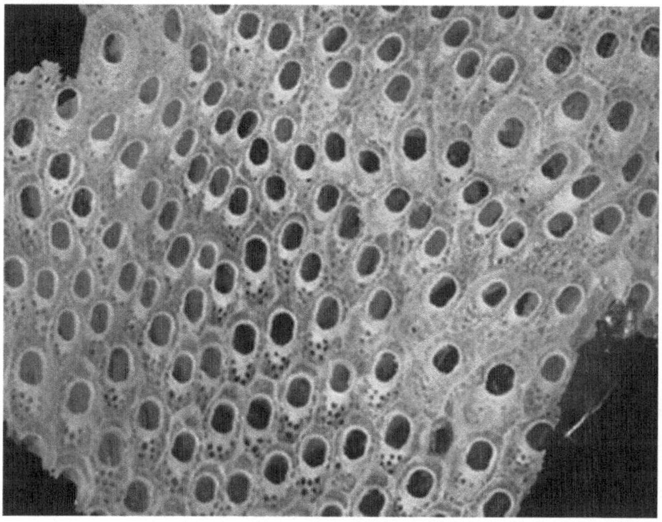

Image credit: Gall L, https://creativecommons.org/publicdomain/zero/1.0/

Global distribution: New Zealand: Norway to the Mediterranean and Black Sea and on the Atlantic coast of Canada and the USA to Florida.

Habitat: Sessile; shallow subtidal waters

Associated bacteria and their pharmaceutical compounds and activities

Antibacterial activity: The bacterial strain *Sphingomonas melonis* strain MPU95 isolated from this species showed antibacterial activity against *Staphylococcus lentus* and *Bacillus subtilis*. On the other hand, the strains *Alteromonas macleodii* DSM 6062 and *Alteromonas* sp. CF14-2 acted against *Bacillus subtilis* (Heindl 2011).

Electra pilosa

Image credit: Michel Langeveld, https://creativecommons.org/licenses/by-sa/4.0/

Common name: Hairy sea-mat

Global distribution: NE and NW Atlantic Ocean

Habitat: This sessile species occurs in the rocky shores from the intertidal zone to about 50 m depth

Habitat: Sessile; found from the intertidal zone of sheltered rocky shores down to depths of about 50 m

Associated bacteria and their pharmaceutical compounds and activities

Antibacterial activity: The bacterial strains *Pseudoalteromonas* sp. M12-3, *Vibrio* sp. S648 and S4734; and *Shewanella* sp. IE9-1 and, CR07-A1 isolated from this species showed antibacterial activity against *Staphylococcus lentus* and *Bacillus subtilis*. On the other hand, the strain *Vibrio splendidus* LGP32 acted only against *Bacillus subtilis* (Heindl 2011).

Membranipora membranacea

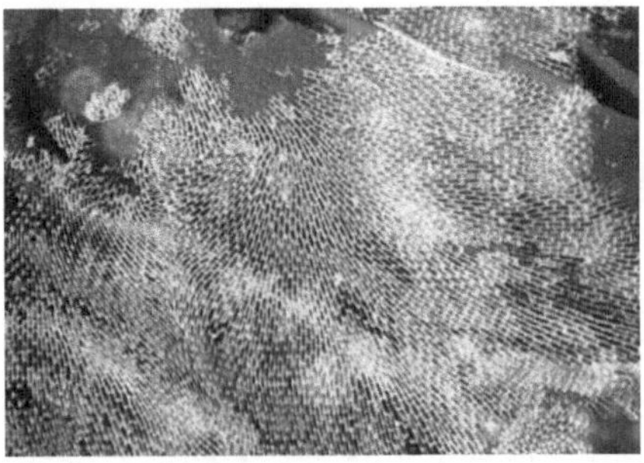

Image credit: USGS, Public domain.

Global distribution: North Pacific coasts of North America; and NW Atlantic

Habitat: It is a benthic species and is found attached to molluscan shells, seaweeds and even artificial substrates between mid intertidal to sublittoral zone. It has also been reported to thrive in brackishwaters.

Associated bacteria and their pharmaceutical compounds and activities

Antibacterial activity: The bacterial strain *Shewanella piezotolerans* WP3 isolated from this species showed antibacterial activity against *Staphylococcus lentus* and *Bacillus subtilis* (Heindl 2011).

Membranipora sp.

Associated bacteria and their pharmaceutical compounds and activities

Antibacterial activity: The bacterial strain *Vibrio* sp. S4738 isolated from this species showed antibacterial activity against *Staphylococcus lentus, Bacillus subtilis*, and *E.coli*. On the other hand, the strains *Vibrio* sp. S3897 and *Vibrio splendidus* LGP32 acted only against *Bacillus subtilis* (Heindl 2011).

Pentapora fascialis

Image credit: Matthieu Sontag, https://creativecommons.org/licenses/by-sa/3.0/

Global distribution: Throughout the Mediterranean, from the Strait of Gibraltar in the west to the Aegean in the east.

Habitat: Colonises boulders or bedrock, with small colonies occasionally found on kelp holdfasts; found in subtidal waters and is most common below 18 m and often abundant between 25-35 m.

Associated bacteria and their pharmaceutical compounds and activities

Antibacterial activity: The bacterial strain isolated from this species showed antibacterial activity against *Staphylococcus lentus, Bacillus subtilis*, and *E.coli* (Heindl 2011).

Reteporella beaniana *(= Sertella beaniana)*

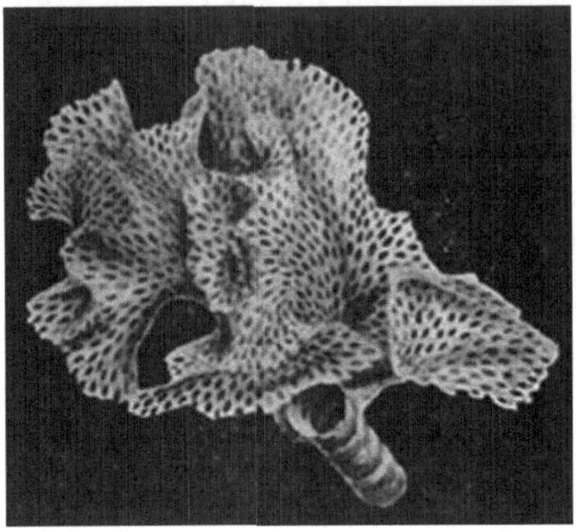

Image credit: British Museum. Public domain.

Global distribution: Atlantic coasts of Europe, from Spain to Norway.

Habitat: Sessile; coastal waters; prefers hard substrate and current exposed locations, usually deeper than 20 m

Associated bacteria; Therapeutic compounds and their bioactivities

Antibacterial activity: The bacterial strain *Sphingomonas paucimobilis* ATCC29837 isolated from this species showed antibacterial activity against *Staphylococcus lentus*. On the other hand, the strain *Pseudoalteromonas* sp. S1650 acted against both *Staphylococcus lentus* and *Bacillus subtilis* (Heindl 2011).

Schizobrachiella sanguinea

Image credit: Massimiliano DeMartino. https://creativecommons.org/licenses/by-sa/3.0/

Global distribution: Mediterranean Sea.

Habitat: Sessile; depth range 27–110 m

Associated bacteria; and their pharmaceutical compounds and activities

Antibacterial activity: The bacterial strain *Sphingomonas paucimobilis* ATCC29837 isolated from this species showed antibacterial activity against *Staphylococcus lentus*. On the other hand, the strains, *Tenacibaculum adriaticum* B390, *Sphingomonas* sp., *Alteromonas* sp. SCSWC25, *Alteromonas* sp. CF14-2 and *Alteromonas macleodii* DSM 6062 acted against *Bacillus subtilis* (Heindl 2011).

Schizoretepora serratimargo *(= Schizotheca serratimargo)*

Schizoretepora sp.

Global distribution: Atlantic Ocean, Mediterranean, Noth Sea and English Channel

Habitat: It is a benthic and sessile species

Associated bacteria and their pharmaceutical compounds and activities

Antibacterial activity: The bacterial strain *Sphingomonas* sp. isolated from this species showed antibacterial activity against *Staphylococcus lentus*. On the other hand, the strain *Alteromonas macleodii* DSM 6062 acted against *Bacillus subtilis* (Heindl 2011).

Phylum: Bryozoa Class: Stenolaemata

***Tubulipora* sp.**

Associated bacteria and their pharmaceutical compounds and activities

Antibacterial activity: The bacterial strains *Pseudoalteromonas* sp. IC013, M12-3 isolated from this species showed antibacterial activity against *Staphylococcus lentus* and *Bacillus subtilis* (Heindl 2011).

CRUSTACEANS

CIRRIPEDES

Phylum: Arthropoda Subphylum: Crustacea Class: Hexanauplia
Order: Sessilia Family: Balanidae

Amphibalanus amphitrite *(= Balanus amphitrite)*

Image credit: James St. John, https://creativecommons.org/licenses/by/2.0/

Common name (s): Acorn barnacle, striped barnacle, the purple acorn barnacle and Amphitrite's rock barnacle

Global distribution: Warm and temperate waters worldwide

Habitat: Common coastal and estuarine organism; found on hard natural surfaces such as bedrock, boulders, mollusc shells and red mangrove roots; also found on artificial surfaces such as the hulls of ships, pilings and seawalls.

Associated bacteria and their pharmaceutical compounds and activities

Antibacterial activity: The ethyl acetate extract with the B3 bacterial strain of this species showed antibacterial activity against human pathogens such as

Salmonella paratyphi, Klebsiella pneumoniae, Klebsiella pneumoniae, Vibrio cholerae, Pseudomonas aeruginosa, Streptococcus pneumoniae, Staphylococcus epidermidis, Escherichia coli, Bacillus cereus, Bacillus subtilis, and *Enterobacter aerogenes* with the values of zone of inhibition varying from 1 to 6 mm (Jebasingh and Murugan 2011).

COPEPODS

Subphylum: Crustacea Class: Hexanauplia Subclass: Copepoda
Family: Caligidae

Caligus lagocephali *(= Caligus fugu; Pseudocaligus fugu)*

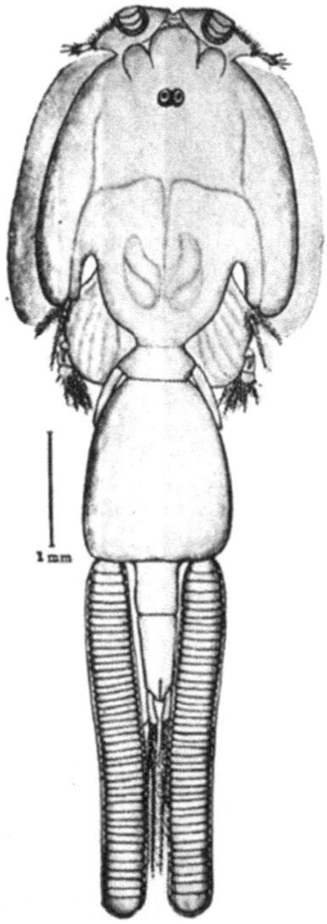

Caligus sp.
Image credit: Charles Branch Wilson. Public domain.

Common name (s): Sea lice
Habitat: Parasite

Associated bacteria; and their pharmaceutical compounds and activities

TTX production: The bacterium *Roseobacter* sp. associated with this copepod has led to the production of TTX and anhydro-TTX (Magarlamov et al., 2017).

SHRIMPS

Subphylum: Crustacea Class: Malacostraca Order: Decapoda
Family: Penaeidae

Penaeus japonicus *(= Marsupenaeus japonicus)*

Image credit: MNHN-Museum National d'Histoire Naturelle, https://creativecommons.org/licenses/by/4.0/

Common name (s): Kuruma shrimp; Japanese or kuruma prawn

Global distribution: Indo-West Pacific: from the Red Sea, E. and S.E. Africa to Korea, Japan and the Malay Archipelago, also reported from Fiji; Eastern Atlantic: eastern Mediterranean and south coast of Turkey.

Habitat: Bottom sandy mud and sand; depth range 0–90 m.

Associated bacteria and their pharmaceutical compounds and activities

Immunomodulatory and bactericidal activity: Fifty-one lactic acid bacteria (LAB) strains were isolated and identified from the intestinal tracts of this kuruma shrimp. Among these strains, the strain *Lactococcus lactis* D1813 exhibited the highest amount of IFN-γ production and also bactericidal activity. Further, this strain seemed to play an immunomodulatory role (Maeda et al. 2014).

Order: Decapoda Family: Palaemonidae

Palaemon macrodactylus

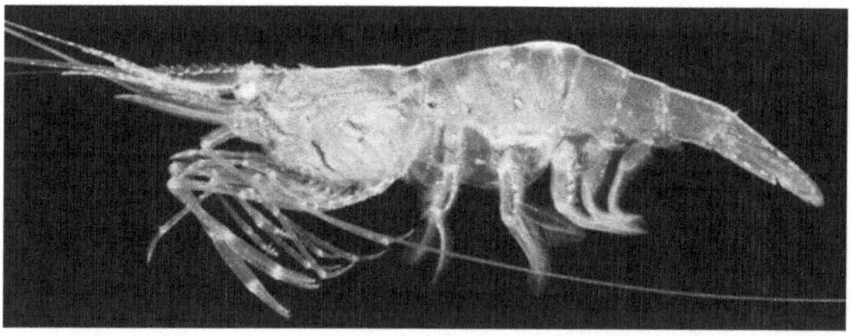

Image credit: Smithsonian Environmental Research Center, https://creativecommons.org/licenses/by/2.0/

Common name (s): Oriental shrimp

Global distribution: Native to Japan; Korea and Northern China

Habitat: Brackish water and estuaries,; favors protected harbors, bays, ponds and tidal creeks.

Associated bacteria; and their pharmaceutical compounds and activities

Antifungal activity: The compound, 2,3-indolinedione produced by the associated bacterium *Alteromonas* sp. of the embryos of the shrimp *Palaemon macrodactylus* sowed antifungal activity against the pathogenic fungus, *Lagenidium callinectes*. It is interesting to note that bacteria-free embryos died quickly and embryos reinoculated with *Alteromonas* sp. or treated with 2,3-indolinedione lived very well (Gil-Turnes et al. 1989).

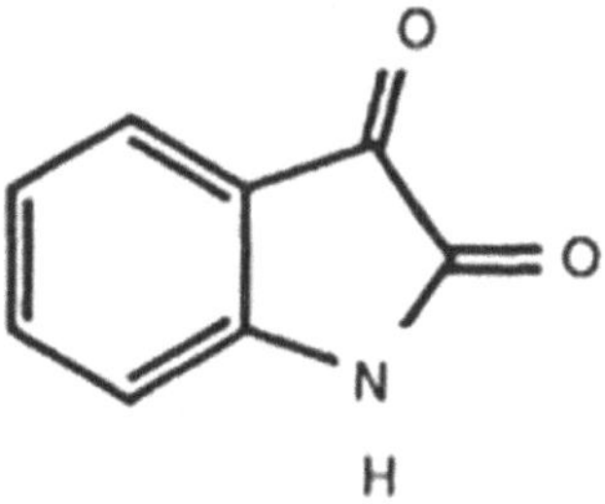

2,3-indolinedione

CRABS

Order: Decapoda Infraorder: Brachyura Family: Portunidae

Scylla olivacea

Image credit: Judgefloro, https://creativecommons.org/publicdomain/zero/1.0/

Common name (s): Orange mud crab.

Global distribution: Tropical; Indo-West Pacific.

Habitat: Demersal.

Associated bacteria and their pharmaceutical compounds and activities

Antimicrobial activities: The body parts of this carab viz. carapace, abdomen, leg and hand have been reported to harbor 43 bacterial isolates and three strains of them namely BPSCRB12, BPSCRB14, and BPSCRB41 displayed potent antimicrobial activity due to their volatile compounds which are believed to be about 20 (Zote et al. 2018).

Infraorder: Brachyura Family: Xanthidae

Atergatis floridus

Image credit: Public domain.

Common name (s): Green egg crab or floral egg crab,

Global distribution: Indo-Pacific waters

Habitat: It is found among coral rubble or reefs; depth range 0 – 6 m

Associated bacteria; and their pharmaceutical compounds and activities

TTX production: The bacterium *Vibrio* species isolated from this crab species is responsible for the production of the neurotoxins, TTX and anhydro-TTX (Magarlamov et al., 2017).

Subphylum: Chelicerata Class: Merostomata Order: Xiphosura Family: Limulidae

HORSESHOE CRAB

Tachypleus gigas

Image credit: Shubham Chatterjee, https://creativecommons.org/licenses/by-sa/3.0/

Common name (s): Indo-Pacific horseshoe crab, Indonesian horseshoe crab,, Indian horseshoe crab, southern horseshoe crab, Pacific horseshoe crab, coastal horseshoe crab.

Global distribution: South and Southeast Asia

Habitat: Coastal water; depths up to 40 m

Associated bacteria; and their pharmaceutical compounds and activities

Antibacterial activities: A total of 62 bacteria were isolated from the gills and mouths of this crab species. Gram-negative bacteria dominated with 60.3%

and 11.1% of bacterial isolates showed antibacterial activities. Two major fractions of bioactive compounds from terpenoid and phenol are said to be responsible for these activities (Ismail et al. 2015).

MOLLUSKS

BIVALVES

Class: Bivalvia Order: Arcida Family: Arcidae

Anadara broughtoni

Image credit: Bioportal Naturalis, Public domain.

Common name: Ark clam

Global distribution: China, Japan, and Korea

Habitat: Not reported

Associated bacteria: Pharmaceutical compounds and activities

Antimicrobial and hemolytic activity: The species has been found associated with more Alphaproteobacteria and Gammaproteobacteria; and less with Cytophaga-Flavobacterium-Bacteroides, Actinobacteria and Firmicutes, Among the different strains of these groups, the genera *Pseudoalteromonas*, *Aeromonas*, and *Bacillus* showed hemolytic activity. The strain *Bacillus pumilus* an 112 which is also responsible for hemolytic activity possessed cyclic depsipeptides. Further, two Gram-negative strains of *Pseudomonas* and *Sphingomonas* and six Gram-positive strains of the genera *Saccharothrix*, *Bacillus*, and *Paenibacillus* displayed antimicrobial activity (Romanenko et al. 2008).

Order: Myida Family: Teredinidae

Teredo navalis

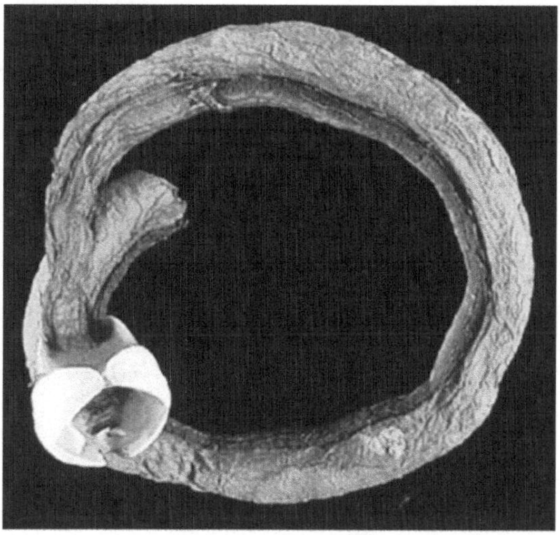

Image credit: United States Geological Survey, Public domain.

Common names: Shipworm; teredo worm; pile worm

Global distribution: It is common in most seas and oceans

Habitat: It is found burrowing in wooden ship hulls

Associated bacteria: Pharmaceutical compounds and activities

Antibacterial activity: The pyoluteorin analogues, mindapyrroles AC produced by this shipworm-associated *Pseudomonas aeruginosa* strain 1682U.R.0a showed antibacterial activity against MRSA ATCC 43300 (Nweze et al. 2020).

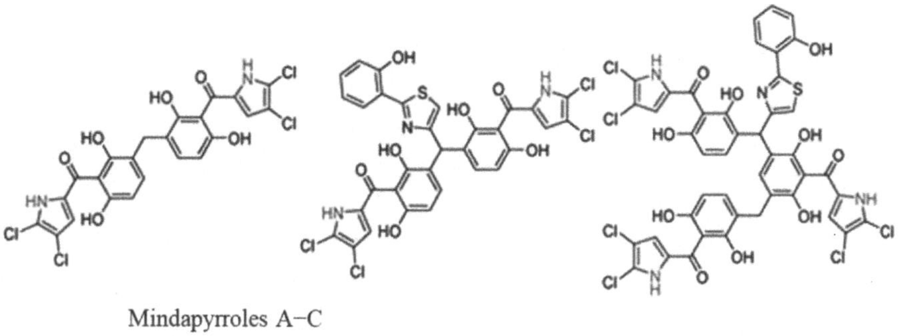

Mindapyrroles A–C

Order: Mytilida Family: Mytilidae

Perna viridis

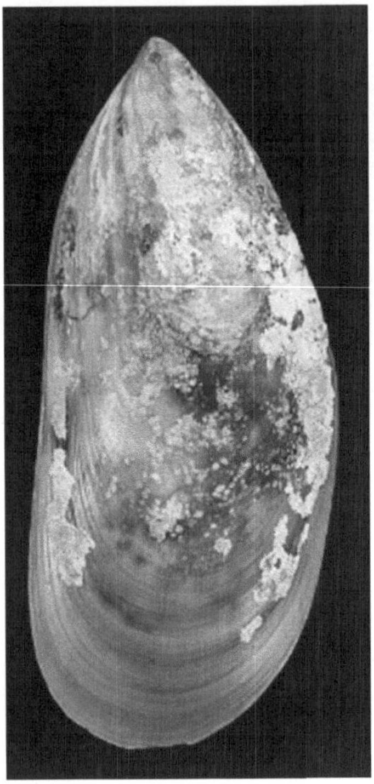

Image credit: H. Zell, https://creativecommons.org/licenses/by-sa/3.0/

Common name: Asian green mussel,

Global distribution: Native in the Asia-Pacific region but has been introduced in the Caribbean, and in the waters around Japan, North America, and South America.

Habitat: Coastal waters; also in estuarine habitats

Associated bacteria: Pharmaceutical compounds and activities

Antibacterial activity: The different strains bacteria *Planococcus* sp., *Bacillus* sp.1, *Bacillus* sp.2, *Bacillus* sp.3 *and Psychrobacter* sp., isolated from this mollusk showed antibacterial activity against *S.typhi, S.marcescens, A.aerogenes, E.coli, S.citreus*, and *P.vulgaris.* The values of zone of inhibition were ranging from 2 to 10 mm in respect of *Planococcus* sp.; from 1 to 20 mm with the species of *Bacillus*; and 3 to 8 mm with *Psychrobacter* sp. (Velho-Pereira and Furtado 2012).

Order: Ostreida Family: Ostreidae

Magallana gigas *(= Crassostrea gigas)*

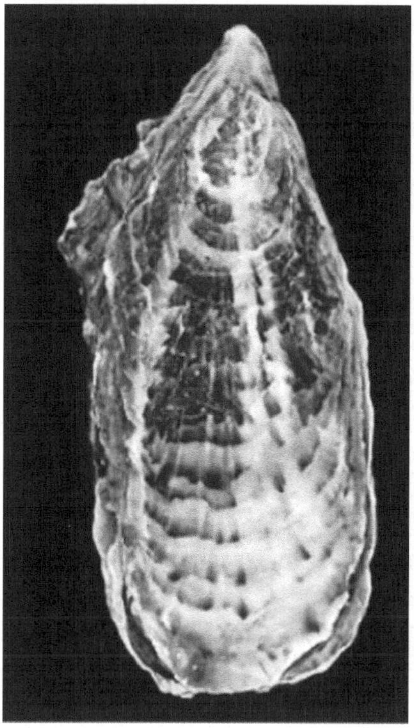

Image credit: Llez, https://creativecommons.org/licenses/by-sa/3.0/

Common name: Miyagi oyster, Pacific oyster, or Japanese oyster

Global distribution: It is native to the Pacific coast of Asia and is found distributed presently in Europe,North America, New Zealand and Australia.

Habitat: This euryhaline species is seen in estuaries and coastal waters of intertidal and subtidal zones; It is also a sessile species found attached to hard or rocky surfaces of muddy or sandy areas

Associated bacteria: Pharmaceutical compounds and activities

TTX production: The bacterial species *Vibrio cholerae* and *Vibrio parahaemolyticus* associated with this mollusk produced the neurotoxin, TTX (Magarlamov et al. 2017).

GASTROPODS

Class: Gastropoda Order: Lepetellida Family: Haliotidae

Haliotis discus hannai

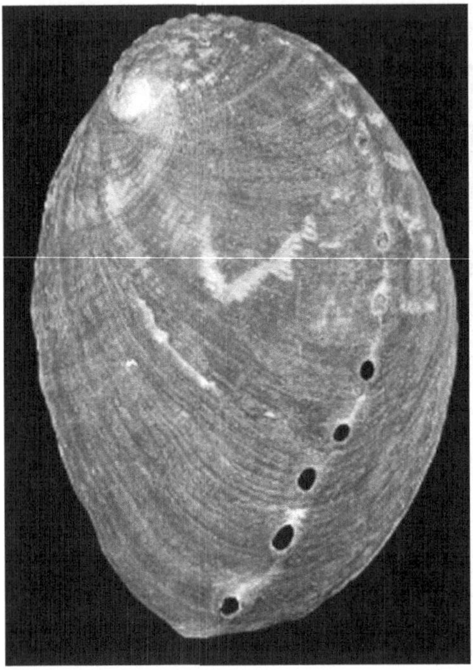

Image credit: H. Zell, https://creativecommons.org/licenses/by-sa/3.0/

Common names: Disk abalone, Japanese abalone, Pacific abalone

Global distribution: Endemic to the waters off Japan and eastern Asia

Habitat: Shallow subtidal habitats, extending from about 1 to 5 m water depth.

Associated bacteria: Pharmaceutical compounds and activities

Antibacterial activity: The strains of the bacterial species isolated from this mollusk viz. An 10 *Pseudomonas fulvus*, An 18 *Sphingomonas asacharolytica*, An 114 *Bacillus licheniformis*, An 118 *Bacillus pumilus* and An 145 *Paenibacillus polymixa*, An 105 *Bacillus licheniformis*, An 112 *Bacillus pumilus*, and An 113 *Saccharothris espanensis*, showed antibacterial activity against *S. aureus*, *B. subtilis*, Xanthomonas sp., *E. faecium*, *E. axysporum*, and *Cirticoccus* sp. (Tanaka et al. 2004).

Order: Littorinimorpha Family: Naticidae

Tanea lineata *(= Natica lineata)*

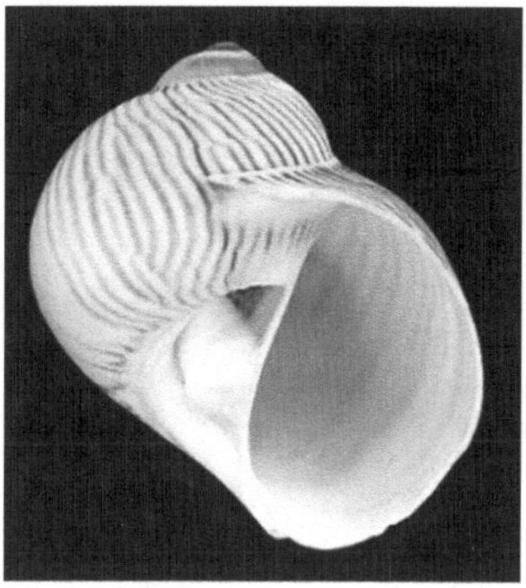

Image credit: H. Zell, https://creativecommons.org/licenses/by-sa/3.0/

Common name: Lined moon snail

Global distribution: Western Central Pacific: Philippines.

Habitat: It is commonly seen in silty sand areas near seagrasses; But it is also found in muddy and sandy bottoms

Associated bacteria: Therapeutic compounds and their bioactivities

TTX production: The bacterial species *Vibrio alginolyticus, Vibrio parahaemolyticus, Aeromonas* sp., and *Pseudomonas* sp. associated with this mollusk produced the neurotoxin, TTX (Magarlamov et al. 2017).

Order: Neogastropoda Family: Conidae

Conus miles

Image credit: Jan Delsing, Public domain.

Common name: Soldier cone

Global distribution: Entire Indo-Pacific

Habitat: It is more commonly found in the intertidal and upper subtidal areas of about 50 m depth

Associated bacteria: Pharmaceutical compounds and activities

Antifungal activity: A polycyclic tetramic macrolactam compound, dihydromaltophilin isolated from the *Conus miles*-associated *Streptomyces* sp. CMB-CS038 possessed antifungal activity on *Candida albicans* with 3 μM as IC50 value (Mayer et al. 2021).

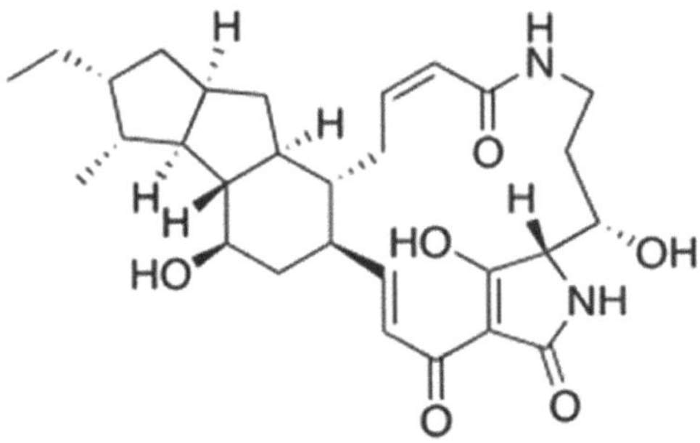

Dihydromaltophilin

Conus pulicarius

Image credit: Almed2 https://creativecommons.org/licenses/by-sa/3.0/

Common name: Flea-bitten cone

Global distribution: Central and Western Pacific waters

Habitat: Inhabits shallow sandy areas; reef areas of lagoons; depth range 1–52 m

Associated bacteria: Pharmaceutical compounds and activities

Antidepressant activity: The bacterial strain, *Streptomyces* sp. CP32 of this mollusc has been reported to produce compounds such as thiazoline and pulicatins A-E which are known to bind selectively to the human serotonin 5-HT(2B) receptor. This 5-HT2B as an agonist induces antidepressant activity (Lin et al. 2010).

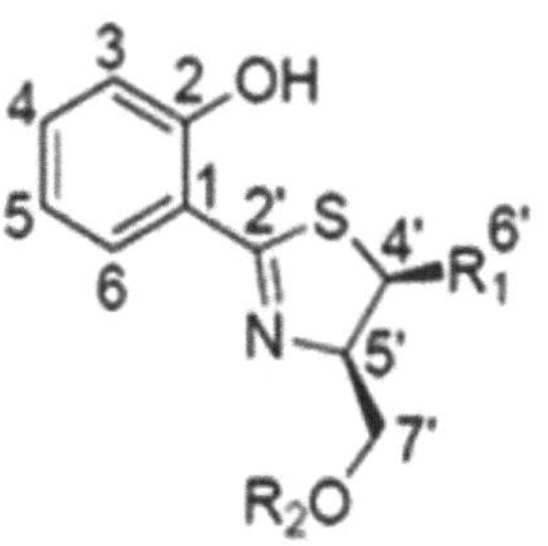

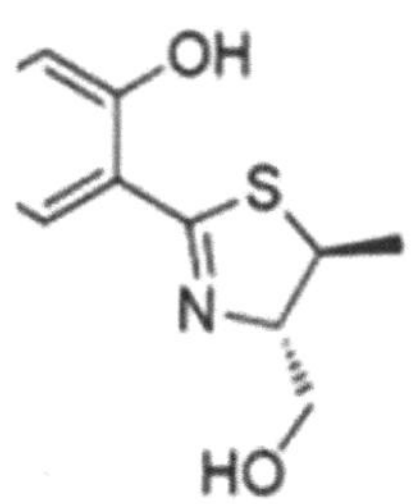

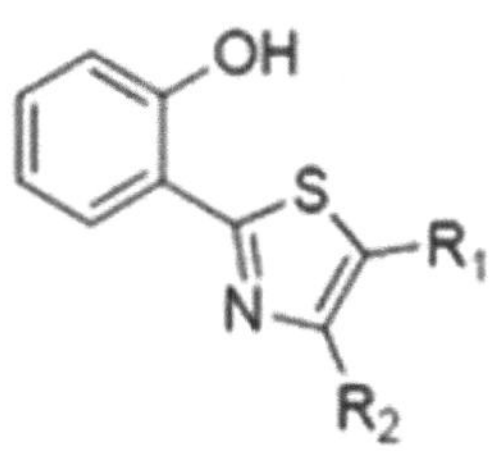

R1=CH3 R2=H

Pulicatin A Pulicatin B Pulicatins C-E

(Pulicatin C: R1=CH3, R2=CH2OH; Pulicatin D: R1=CH3, R2=CHO; Pulicatin E: R1=CH3, R2= CONH2)

Conus rolani

Image credit: Nick Zantop, https://creativecommons.org/licenses/by-sa/3.0/

Common name: Rolan's cone

Global distribution: Western Central Pacific: Philippines; off Taiwan and Papua New Guinea

Habitat: Not reported

Associated bacteria: Pharmaceutical compounds and activities

Neuroactive and anticancer properties: The extract of the bacteria, *Nocardiopsis alba* CR167 associated with this species yielded polyketide pyrones viz. nocapyrones H-Q (1–10), nocapyrones A-C (11–13) which possessed neuroactive properties. Nocapyrones modulated nerve cell depolarization, with the most active compounds achieving an IC50 of 2 μM. Nocapyrones B (12) and H (1) were active against nearly all DRG neuronal cell types at 50 μM. Further, compounds 1, 5, and 12 were able to activate or inhibit Ca2+ flux into those cells with IC50s between 2 and 70 μM. Compounds 1 and 12 also showed cytotoxic activity against MCF-7 with IC50 values of 8.7 and 22.2 μM, respectively, whereas 3 was inactive against MCF-7 (>100 μM), and against AA8 only compound 1 was cytotoxic with an IC50 of 10.2 μM (Lin et al. 2013).

Conus tribblei

Image credit: H. Zell, https://creativecommons.org/licenses/by-sa/3.0/

Common name: Tribble's cone

Global distribution: Western Central Pacific: Fiji and Philippines; off Japan, Taiwan, Vietnam, Solomon Islands and Australia (Western Australia).

Habitat: Not reported

Associated bacteria: Pharmaceutical compounds and activities

Anticancer properties: This mollusk harbored Actinomycete bacteria including actinobacterial genera viz. *Brevibacterium, Streptomyces, Gordonia, Microbacterium*, etc. Its strain *Nocardiopsis alba* CR167 which is capable of synthesizing polyketide pyrones. The latter possessed anticancer properties (Lin et al. 2013; Kalia et al. 2017).

***Conus* sp.**

Associated bacteria: Pharmaceutical compounds and activities

Analgesic properties: Actinomycetes living in close association with the cone snails (*Conus* spp.) contribute to the production of conotoxins, which are peptide-based neurological toxins with analgesic properties (Leal et al. 2014).

Order: Neogastropoda Family: Nassariidae

Nassarius conoidalis *(= Niotha clathrata)*

Image credit: H. Zell, https://creativecommons.org/licenses/by-sa/3.0/

Common names: Cone-shaped Nassa or jewelled dog whelk; Mud Snail, Nassarius Snail

Global distribution: Indo-West Pacific Ocean: Red Sea, East Africa west to South China Sea, Vietnam, Indonesia, and Australia.

Habitat: Inhabits intertidal areas where mud and/or organics accumulate.

Associated bacteria: Pharmaceutical compounds and activities

TTX production: The bacterial species *Vibrio alginolyticus, Vibrio parahaemolyticus, Pseudomonas* sp., *Aeromonas* sp. and *Plesiomonas* sp. associated with this mollusk produced the neurotoxin, TTX (Magarlamov et al. 2017).

Associated bacteria: Pharmaceutical compounds and activities

TTX production: The bacterial species *Vibrio* sp, *Shewanella* sp., *Marinomonas* sp., *Tenacibaculum* sp., and *Aeromonas* sp. associated with this mollusk produced the neurotoxin, TTX (Magarlamov et al. 2017).

Order: Sacoglossa Family: Plakobranchidae

Elysia rufescens

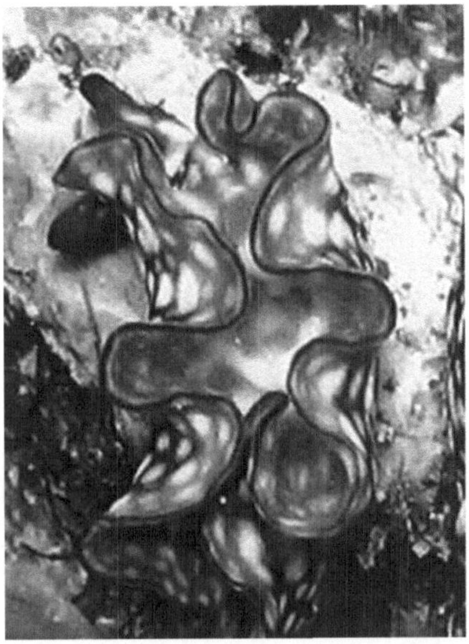

Image credit: Philippe Bourjon, https://creativecommons.org/licenses/by-sa/4.0/

Common name: Not reported

Global distribution: Eastern Central Pacific: Tahiti. Western Indian Ocean: South Africa.

Habitat: Shallow inshore reefs

Associated bacteria: Pharmaceutical compounds and activities

Cytotoxic effects: Bacteria, such as *Mycoplasma* spp. and *Vibrio* spp., that are known to associate with this sacoglossan, produce elisidepsin (Irvalec©), another cyclic peptide belonging to the kahalalide family that is currently under phase II development (Leal et al. 2014). Elisidepsin showed potent and broad cytotoxic effects in cancer cell line panel, being active at concentrations ranging from 0.4 to 2 μM that may be relevant for clinical settings (Serova et al. 2013).

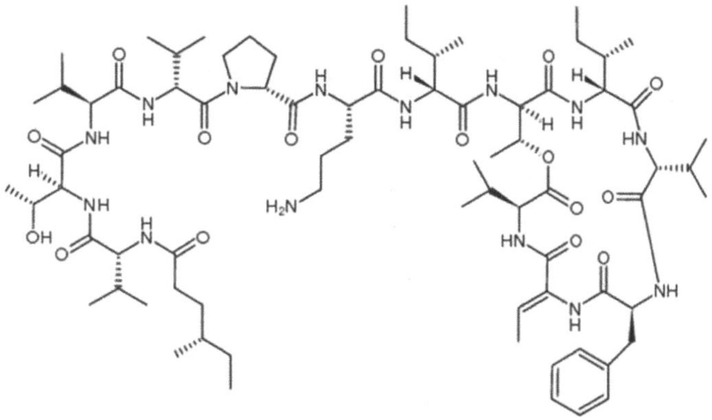

Elisidepsin

CEPHALOPODS

Class: Cephalopoda Order: Octopoda Family: Octopodidae

Hapalochlaena maculosa *(= Octopus maculosus)*

Image credit: Saspotato, Public domain.

Common name: Southern blue-ringed octopus

Global distribution: South coast of Australia.

Habitat: This species occurs in the crevasses and cracks of the rocky reefs and tidal rock pools

Associated bacteria: Therapeutic compounds and their bioactivities

Production of TTX: The bacterial species *Pseudomonas* sp., *Vibrio* sp., *Alteromonas* sp. and *Bacillus* sp, associated with this mollusk produced the neurotoxin, TTX (HYPERLINK "https://www.ncbi.nlm.nih.gov/pubmed/?term=Magarlamov%20TY%5BAuthor%5D&cauthor=true&cauthor_uid=28513564" Magarlamov et al. 2017).

Unidentified sp.1

Antimicrobial and antiproliferative activities: The crude extracts of mollusk-associated bacteria from mollusks collected from various sites in the Philippines showed the presence of *Staphylococcus aureus*, *Klebsiella pneumoniae* and *Pseudomonas aeruginosa*. The compounds yielded by these bacteria were Amicetin (1), Colismycin A (2), Bromophene (3) and Bromoalterochromides A and A' (4,5) which showed antimicrobial and antiproliferative activities. The antimicrobial values (hits) of *K. pneumoniae, P. aeruginosa,* and *S. aueus* were 29,121, and 211, respectively; and the values antiproliferative activity against HCT116 and MCF-7 were 16 and 91 μM, respectively (Robes et al. 2017).

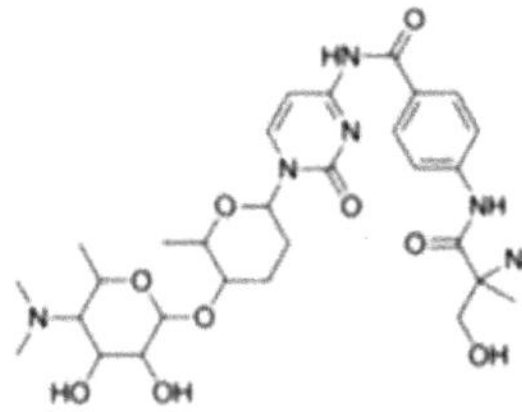

(1)

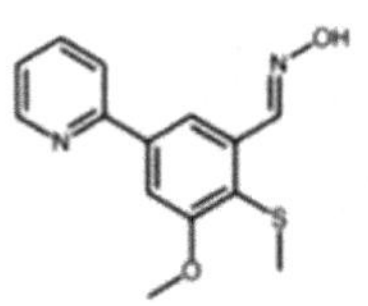

(2)

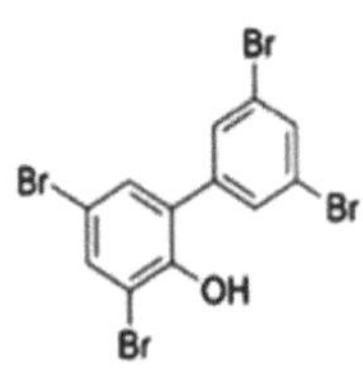

(3)

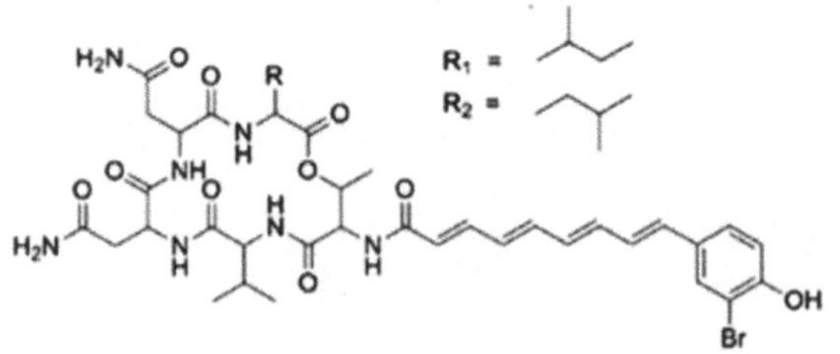

(4, 5)

Amicetin (1), Colismycin A (2), Bromophene (3) and Bromoalterochromides A and A' (4,5)

ECHINODERMS

ASTEROIDS

Class: Asteroidea Order: Paxillosida Family: Astropectinidae

Astropecten polycanthus

Image credit: Chaloklum Diving, https://creativecommons.org/licenses/by/3.0/

Common names: Comb sea star or sand-sifting starfish

Global distribution: Throughout the Indo-Pacific region

Habitat: It is often seen in harbours and estuaries especially on silty sand bottoms at depths down to about 185 m.

Associated bacteria: Pharmaceutical compounds and activities

TTX production: The bacterium *Vibrio alginolyticus* associated with this species has been reported to produce TTX, anhydro-TTX and 4-epi-TTX (Yu et al. 2017).

Order: Valvatida Family: Acanthasteridae

Acanthaster planci

Image credit: JSLUCAS75, https://creativecommons.org/licenses/by-sa/3.0/

Common name: Crown-of-thorns starfish

Global distribution: It is the most common species in Australia. It is also found distributed from the Red Sea to the west coast of Central America. In the Indo-Pacific

Habitat: It is commonly occurring in coral reefs or hard coral communities

Associated bacteria:Therapeutic compounds and their bioactivities

Cytotoxicity: The pyrone compounds viz. Violapyrones B, C, H, and I derived from its symbiotic actinomycete, *Streptomyces* sp, displayed cytotoxic activity on 10 human cancer cell lines with GI50 values of 1.10–26.12 g/mL (Shin et al. 2014).

Violapyrone B,C

Violapyrone H,I

ECHINOIDS

Class: Echinoidea Order: Spatangoida Family: Brissidae

Meoma ventricosa

Image credit: Recolnat & Marie Hennion.

Common name: Cake urchin or red heart urchin,

Global distribution: Caribbean Sea, Bahamas, Florida and Bermuda.

Habitat: It occurs from intertidal zone to a depth of about 200 m. It is known to bury itself in sediment- associated sandy or silty areas.

Associated bacteria: Pharmaceutical compounds and activities

TTX production: The bacterium *Pseudoalteromonas* sp. associated with this sea urchin has been reported to produce TTX (Yu et al. 2017).

HOLOTHUROIDS

Class: Holothuroidea Order: Holothuriida Family: Holothuriidae

Holothuria edulis

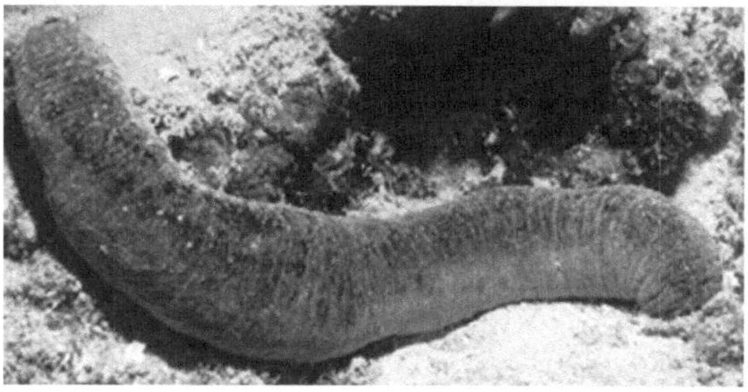

Image credit: Ed Bierman, https://creativecommons.org/licenses/by/2.0/

Common name: Pinkfish

Global distribution: Indo-pacific

Habitat: This reef-associated species lives lives at a depth range 0–45 m

Associated bacteria: Pharmaceutical compounds and activities

Antifungal activity: The associated actinomycete, *Streptomyces* sp. G278 of this echinoderm has been reported to produce compounds viz. 3-hydroxyl-2-methylpyridine and N-phenylnaphthalen-2-amine which displayed antifungal activity against *Candida albicans* ATCC10231 with MIC values of 64 and 128 μg/mL, respectively (Nweze et al. 2020).

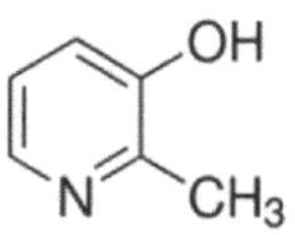

3-hydroxyl-2- methylpyridine

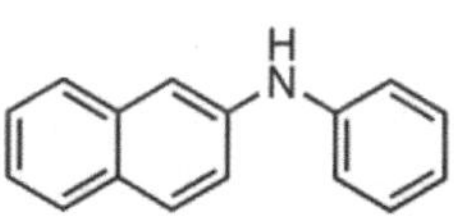

N-phenylnaphthalen-2-amine

BACTERIA ASSOCIATED WITH MARINE CHORDATES

TUNICATES

Order: Aplousobranchia Family: Didemnidae

Didemnum psammatodes

Didemnum sp.

Image credit: Adapted from U.S. Geological Survey/Photo by Dann Blackwood (USGS)

Common name: Colonial ascidian

Global distribution: Indo-Pacific and Atlantic: Subantarctic region, Malaya and West Africa.

Habitat: Intertidal sites; sessile; depth range 0–15 m

Associated bacteria: Pharmaceutical compounds and activities

Antibacterial activity: The associated bacterium *Bacillus firmus* of this species showed antibacterial activity against pathogenic *Escherichia coli* and *Pseudomonas aeruginosa* (Mary et al. 2016).

Didemnum sp.

Associated bacteria: Pharmaceutical compounds and activities

Cytotoxic activities: This tunicate-derived actinomycete, *Streptomyces* sp. Did-27 yielded compounds viz. (S)-6-(sec-butyl)-3-isopropylpyrazin-2 1H)-one, (S)-3-(sec-butyl)-6-isopropylpyrazin-2(1H)-one, (S)-6-(sec-butyl)-3-isobutylpyrazin-2(1H)-one, deoxymutaaspergillic acid, 3,6-diisobutyl-2(1H)-pyrazinone, 3,6-di-sec-butyl-2(1H)-pyrazinone, cyclo(6-OH-d-Pro-l-Phe), bacillusamide B, cyclo(l-Pro-l-Leu) and cyclo(l-Pro-l-Ile). Among these compounds, 3,6-diisobutyl-2(1H)-pyrazinone displayed potent activity against HCT-116 cells with IC50 of 1.5 μg/mL, while compound cyclo(l-Pro-l-Ile exhibited moderate against HCT-116, HepG2, and MCF-7 cell lines with IC50 values of 1.5–35 μg/m, >50 μg/mL and 0–35 μg/mL respectively (Shaala et al. 2016).

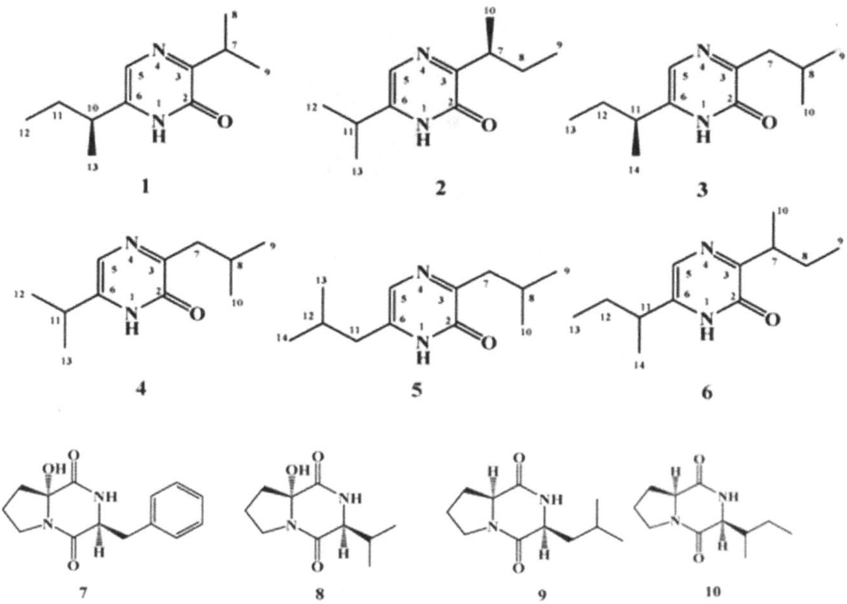

(S)-6-(sec-butyl)-3-isopropylpyrazin-2(1H)-one (1), (S)-3-(sec-butyl)-6-isopropylpyrazin-2(1H)-one (2), (S)-6-(sec-butyl)-3-isobutylpyrazin-2(1H)-one (3), deoxymutaaspergillic acid (4), 3,6-diisobutyl-2(1H)-pyrazinone (5) 3,6-di-sec-butyl-2(1H)-pyrazinone (6), cyclo(6-OH-d-Pro-l-Phe) (7), bacillusamide B (8), cyclo(l-Pro-l-Leu) (9),cyclo(l-Pro-l-Ile) (10)

Order: Aplousobranchia Family: Didemnidae

Diplosoma swamiensis

Common name: Not reported

Global distribution: West Indies, West Pacifc Ocean; Gulf of Mannar, India

Habitat: Intertidal rocky areas, at depth of 2–6 m.

Associated bacteria: Pharmaceutical compounds and activities

Antibacterial activity: The associated *Bacillus* sp. of this species showed antibacterial activity against pathogenic *Pseudomonas aeruginosa*. Further,

Arthrobacter sp. of this species showed similar activity against *Escherichia coli* (Mary et al. 2016).

Order: Aplousobranchia Family: Didemnidae

Lissoclinum patella

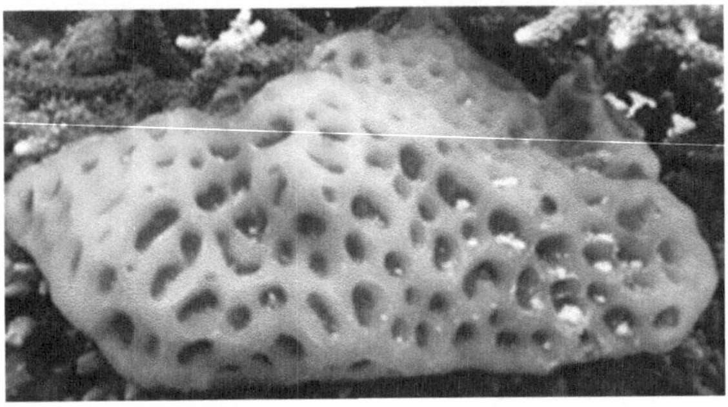

Image credit: Singgih Afifa Putra, https://creativecommons.org/licenses/by-sa/4.0/

Common name: Not reported

Global distribution: Indo-West Pacific; Philippines, Indonesia, and northern Australia

Habitat: coastal areas; depth range is 1–28 m

Associated bacteria: Pharmaceutical compounds and activities

Cytotoxic activity: The bacterium *Candidatus Endolissoclinum faulkneri* isolated from this species yielded patellazoles A–C which cytotoxic activities against HCT-116 tumor cells (Watters 2018; Chen et al. 2018).

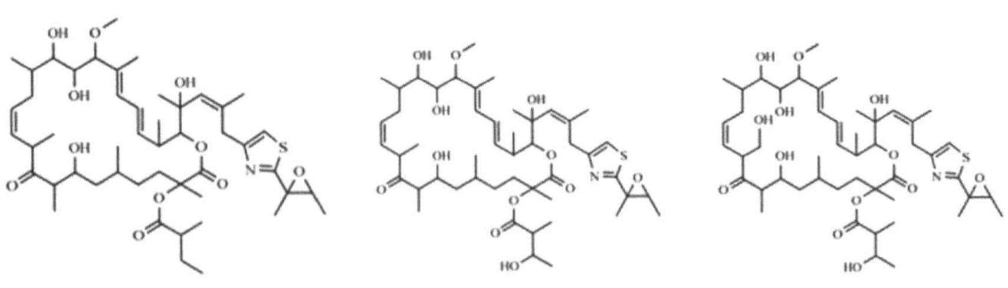

Patellazoles A,B,C

Antibacterial activity: This ascidian-associated bacterium, *Streptomyces* sp. PTY087I yielded naphthoquinone derivatives viz. dihydrogranaticin B, granaticin and granatomycin D which showed significant antibacterial activity against human pathogenic bacteria such as *Pseudomonas aeruginosa*, *Bacillus subtilis*, methicillin-resistant *Staphylococcus aureus*, and methicillin-sensitive *Staphylococcus aureus* (Chen et al. 2018).

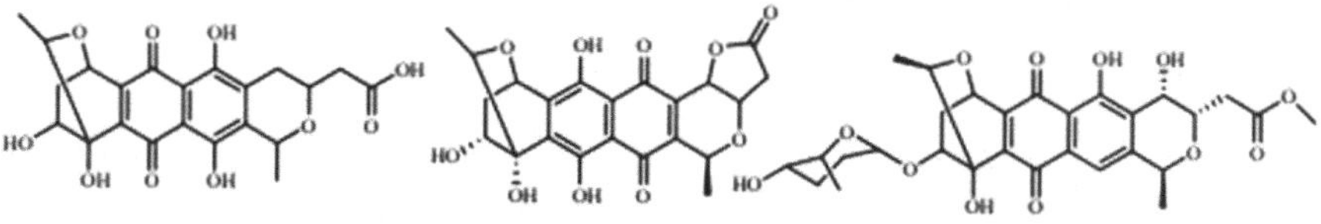

Granaticin Granatomycin D Dihydrogranaticin B

Order: Aplousobranchia Family: Didemnidae

Polysyncraton lithostrotum

Associated bacteria: Pharmaceutical compounds and activities

Antibactrial and cytotoxic activities: The bacterial strains *Salinispora pacifica* strain LL-37I366 and *Micromonospora lomaivitiensi* isolated from this species yielded lomaiviticins A and B which showed antibacterial activities against *Enterococcus faecium*. and *Staphylococcus aureus*. Further, lomaiviticin A also has been reported to display cytotoxic activity against several human cancer cell lines (He 2001; Chen et al. 2018).

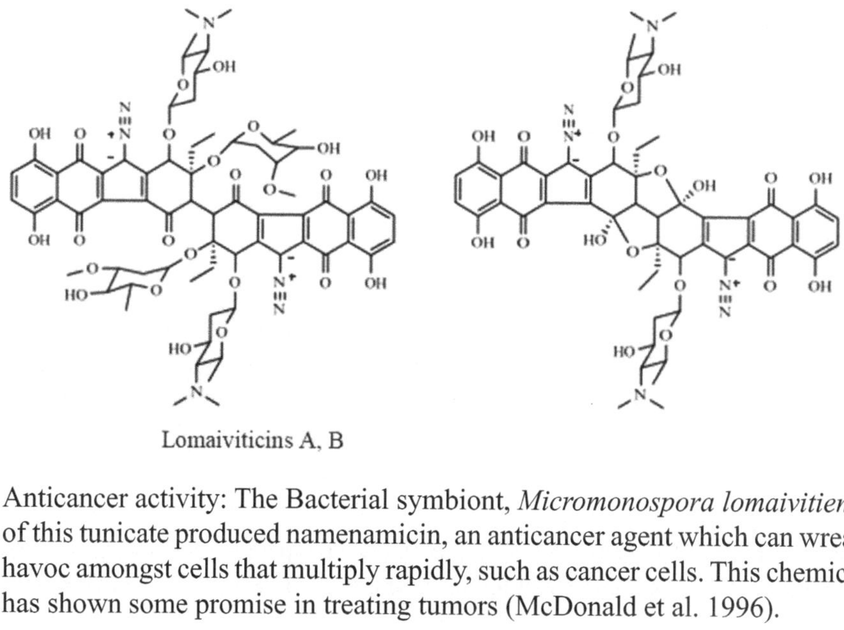

Lomaiviticins A, B

Anticancer activity: The Bacterial symbiont, *Micromonospora lomaivitiensi* of this tunicate produced namenamicin, an anticancer agent which can wreak havoc amongst cells that multiply rapidly, such as cancer cells. This chemical has shown some promise in treating tumors (McDonald et al. 1996).

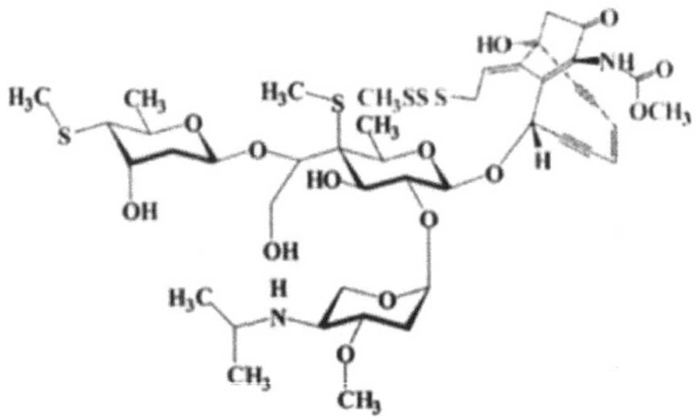

Namenamicin

Order: Aplousobranchia   Family: Didemnidae

Trididemnum orbiculatum

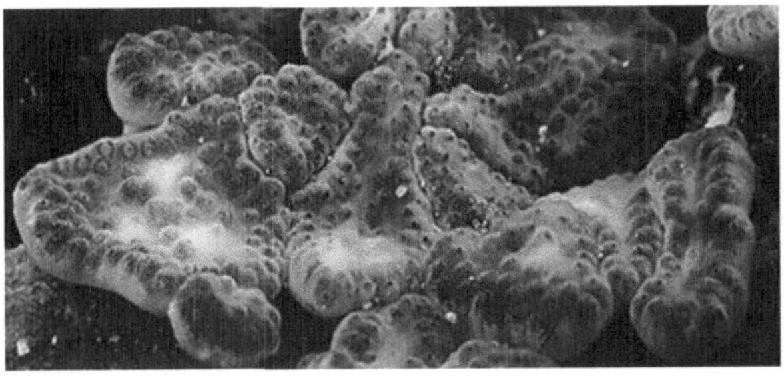

Trididemnum sp.

Image credit: Philippe Bourjon, https://creativecommons.org/licenses/by-sa/4.0/deed.en

Common name: Not reported

Global distribution: Western Atlantic.

Habitat: Sessile; depth range 8–8 m

Associated bacteria: Pharmaceutical compounds and activities

Antibacterial activity: The bacterial strain *Nocardia* sp. isolated from this species produced peptidolipins BF. Among these compounds, peptidolipins B and E showed antibacterial activities against methicillin-sensitive *Staphylococcus aureus* and methicillin-resistant *Staphylococcus aureus* (Chen et al. 2018).

Chen et al. (2018) reported that the bacterial species *Solwaraspora* sp. isolated from this tunicate yielded compounds, solwaric acid A and B; and

- 2,4,6-triphenyl-1-hexene. Among these compounds, solwaric acid A, B exhibited antibacterial activity against MRSA and MSSA (Chen et al. 2018).

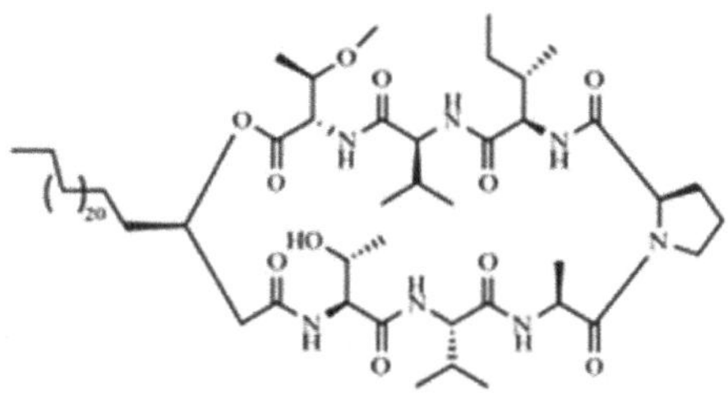

Peptidolipin B

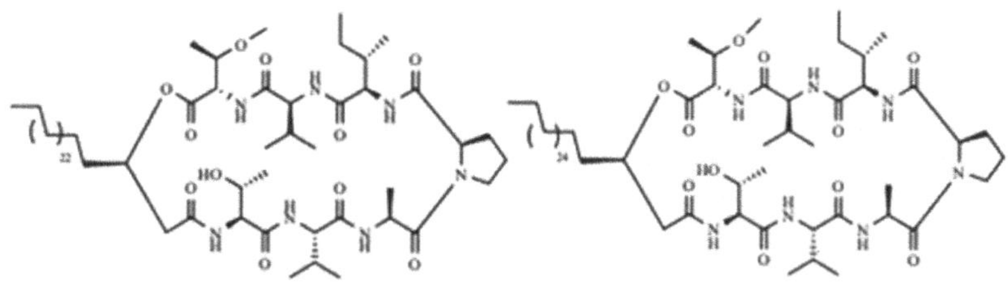

Peptidolipins C,D

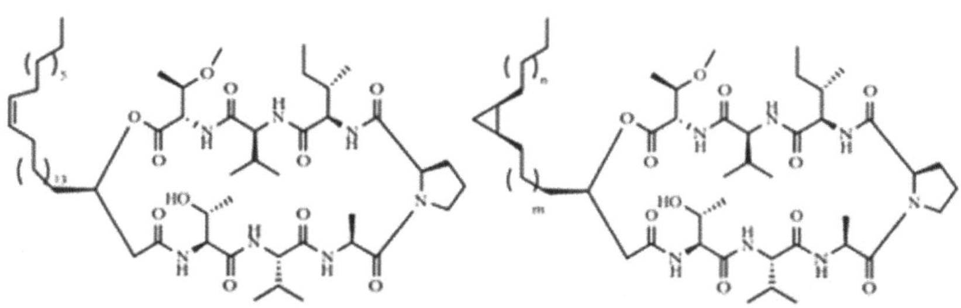

Peptidolipins E,F

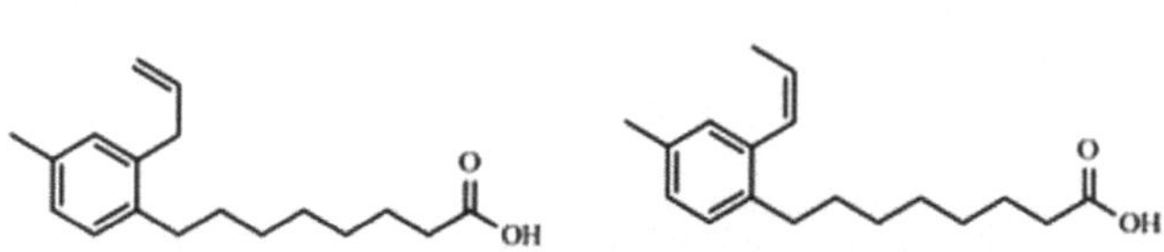

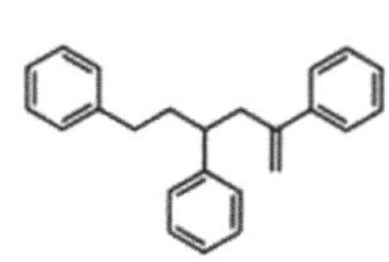

Solwaric acids A,B 2,4,6-triphenyl-1-hexene

Trididemnum solidum

Common name: Caribbean ascidian, overgrowing mat tunicate

Global distribution: Western Atlantic.; throughout the West Indies to Florida, the Bahamas to Venezuela

Habitat: Sessile

Associated bacteria: Pharmaceutical compounds and activities

Anticancer activity: The marine proteobacteria *Tistrella mobilis* associated with this species produced the cyclic depsipeptide, didemnin B which showed significant in vitro cytotoxicity and in vivo anti-tumor activity, and it was also potently active on viruses of DNA and RNA (Xu et al. 2012).

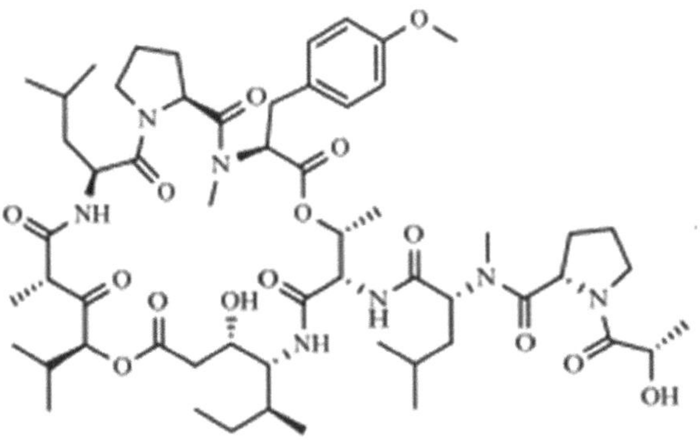

Didemnin B

Order: Aplousobranchia Family: Polyclinidae

Aplidium lenticulum

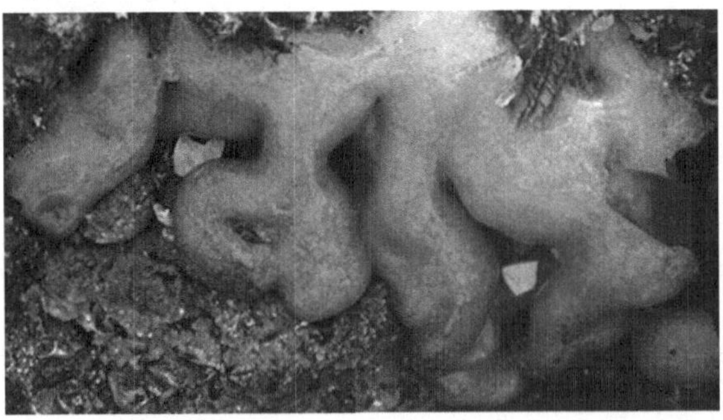

Aplidium sp.

Common name: Not reported

Global distribution: Indo-West Pacific: Australia. Climate: tropical, subtropical.

Habitat: Sessile; depth range 3 - 3 m

Associated bacteria: Pharmaceutical compounds and activities

Anticancer and BCh inhibitory activity: The compound griseorhodin A isolated from the bacterial strain *Streptomyces* sp. JP95 of this tunicate acted against the enzyme, human telomerase. This enzyme is known to be present in human cancer cells at high levels. Further, the strain Streptomyces sp. JP90 of this species produced (S)-cinnamoylphosphoramide which displayed inhibitory activity towards Butyrylcholinesterase (BChE) which is said to be responsible for certain metabolic disorders (Chen et al. 2018).

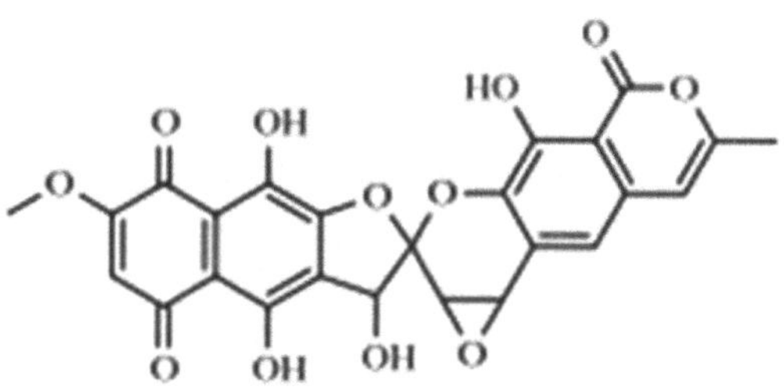

Griseorhodin A

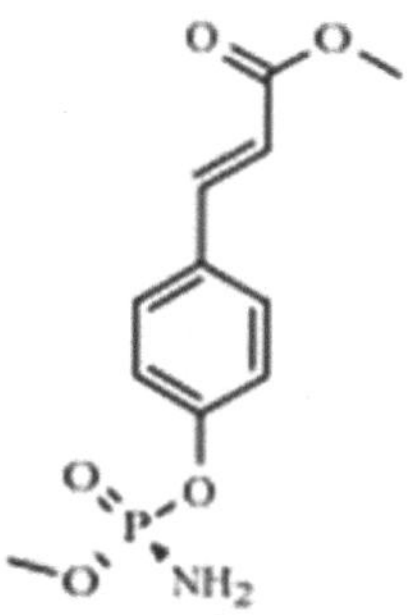

(S)-cinnamoylphosphoramide

Order: Aplousobranchia Family: Polycitoridae

Eudistoma sp.

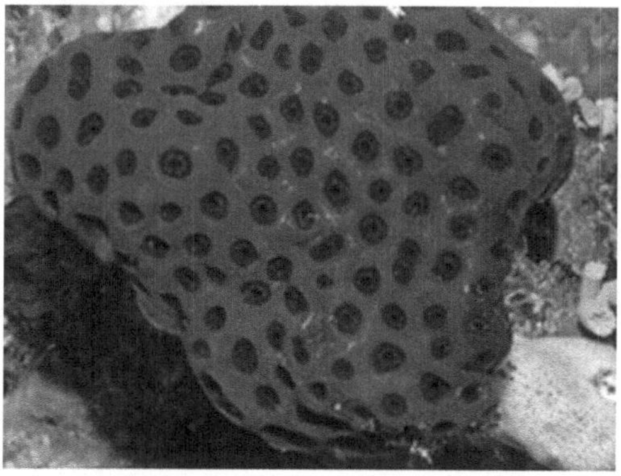

Eudistoma sp.

Image credit: Bernard Dupont, https://creativecommons.org/licenses/by-sa/2.0/

Common name: Strawberry tunicate

Associated bacteria: Pharmaceutical compounds and activities

Antibacterial activity: Two fractions Actinomycete MB0086E and MB0088E isolated from this tunicate displayed antibacterial activity against *Enterococcus faecium*, *Listeria ivanovii* and *Staphylococcus aureus* (Yankova 2014).

Order: Aplousobranchia Family: Polyclinidae

Synoicum adareanum

Common name: Antarctic tunicate

Global distribution: Southwest Pacific, Antarctic and sub-Antarctic.

Habitat: Sessile; depth range 0–1000 m

Associated bacteria: Pharmaceutical compounds and activities

Cytotoxicity: The bacterial genera *Pseudovibrio* and *Microbulbifer* isolated from this tunicate species yielded a macrolide, palmerolide A which displayed potent and selective cytotoxicity toward melanoma (UACC-66 LC50 = 0.018 μM) and appeared to operate via inhibition (IC50 = 2 nM) of V-ATPase. This compound may serve as an effective antimelanoma drug (Diyabalanage et al. 2006; Riesenfeld et al. 2008).

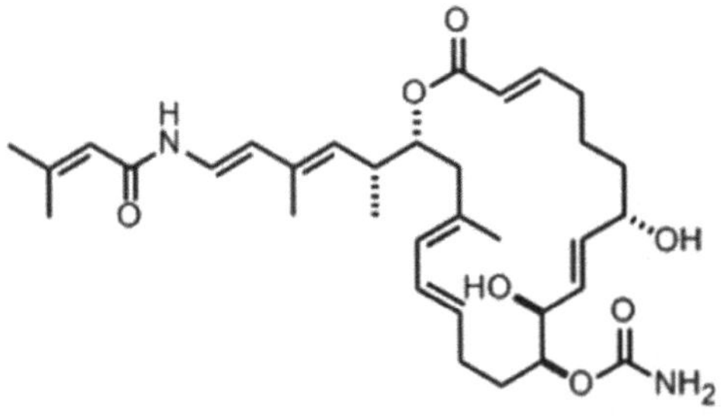

Palmerolide A

Order: Phlebobranchia Family: Perophoridae

Ecteinascidia turbinata

Image credit: Davis, Alison. Public domain.

Common name: Caribbean tunicate, mangrove tunicate

Global distribution: Caribbean Sea, the east coast of Florida, Bermuda and the Gulf of Mexico

Habitat: Shallow waters

Associated bacteria: Pharmaceutical compounds and activities

Common name: Caribbean tunicate, mangrove tunicate

Global distribution: Bermuda Gulf of Mexico, east coast of Florida and Caribbean Sea

Habitat: It inhabits shallow coastal waters

Associated bacteria: Therapeutic compounds and their bioactivities

Anticancer activity: The bacterium *Candidatus* Endoecteinascidia frumentensis associated with this tunicate species has been reported to produce the anticancer agent ecteinascidin 743 (Yondelis (trabectedin, ET-743), a clinically approved chemotherapeutic natural product (PérezMatos et al. 2007). Cytotoxicity: The bacterium, *Agrobacterium* PH-130 isolated from this species yielded an alkaloid compound, sesbanimide A which demonstrated significant cytotoxic activity against KB cell line and P-388 murine leukemia cell line (Chen et al. 2018).

Antibacterial and cytoxic activity: A streptomycete, *Streptomyces* sp. #N1-78-1 associated with this tunicate yielded two antibacterial bisanthraquinone metabolites (1,2) and a derivative (3) which displayed both antibacterial activity against methicillin-resistant *Staphylococcus aureus* and vancomycin-resistant

Enterococcus faecalis (IC(50) = 0.15-130 microM); and cytotoxic activity against HCT-116 cells (Socha et al. 2006; Chen et al. 2018).

Antimicrobial activities: The bacterial strain, *Salinispora arenicola* strain CNR-647- of this tunicate produced arenimycin which demonstrated significant antimicrobial activity against a *Mycobacterium* strain, drug-resistant Staphylococci and some Gram-positive microbes (Chen et al. 2018).

Antibacterial activity: A polyketide antibiotic viz. anthracycline keyicin derived from the marine bacterial species *Rhodococcus* sp. (Family: HYPERLINK "http://en.wikipedia.org/wiki/Nocardiaceae" Nocardiaceae) and *Micromonospora* sp. (Family: Micromonosporaceae) which were associated with this ascidian showed antibacterial activity on *B. subtilis* and MR *S. aureus* with IC50 values ranging from 2.5 to 9.9 μM (Mayer et al. 2021).

Antioxidant activity: The bacterial strain, *Actinomadura* sp. strain WMMB499 isolated from this tunicate yielded halomadurones A -D.(1-4) Among these compounds, compounds 3 and 4 displayed significant antioxidant activity (Chen et al. 2018).

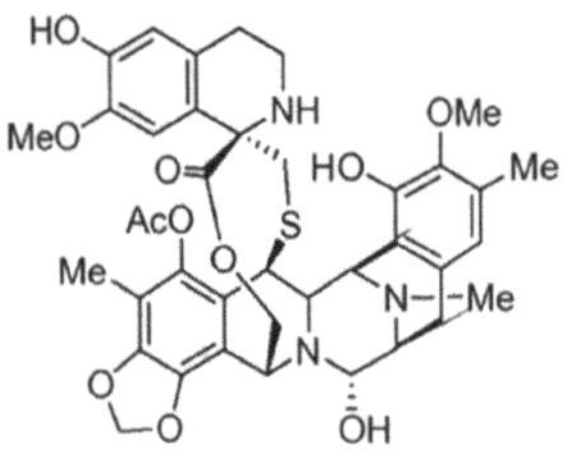

ET-743

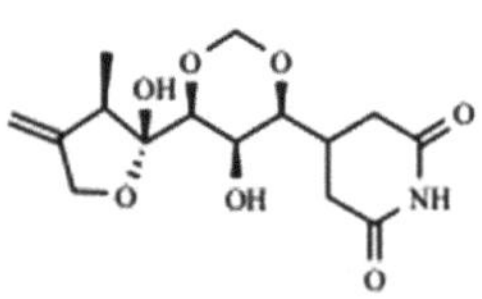

Sesbanimide A

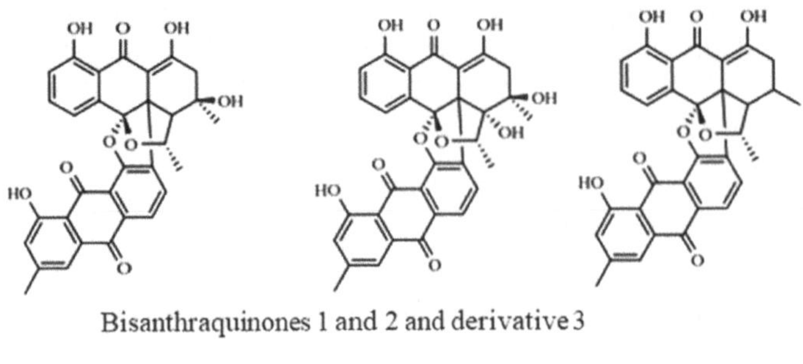

Bisanthraquinones 1 and 2 and derivative 3

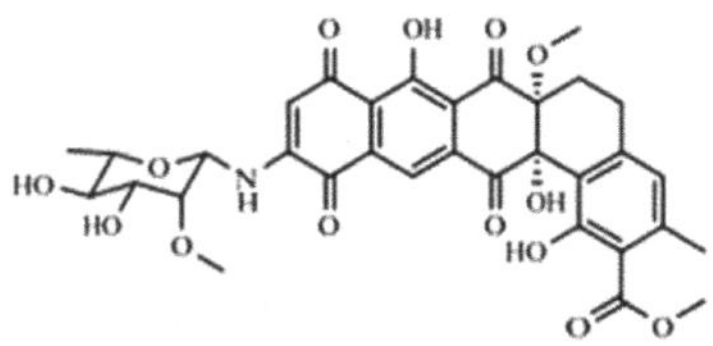

Arenimycin

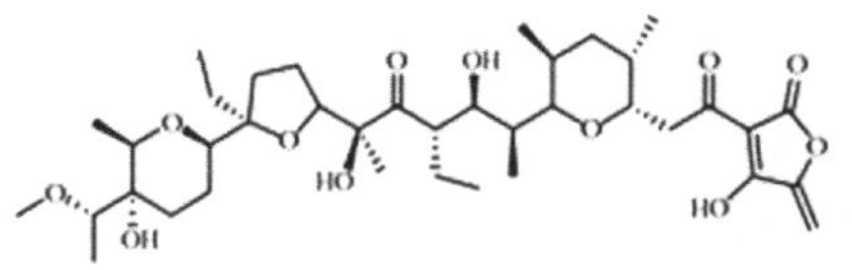

Ecteinamycin

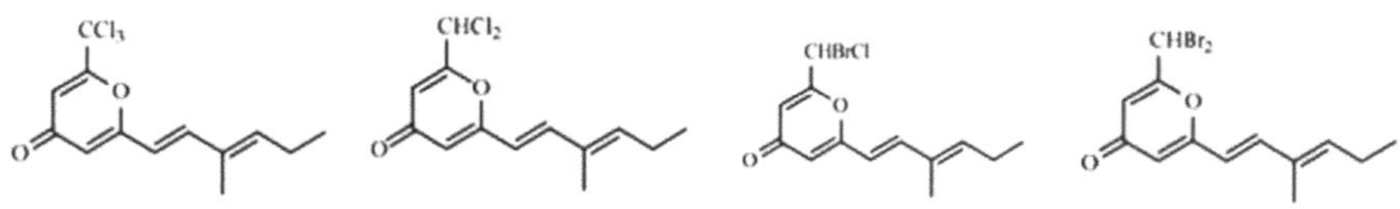

Halomadurones A–D

Order: Aplousobranchia Family: Stomozoidae

Stomozoa roseola *(= Stomozoa murrayi)*

Common name: Not reported

Global distribution: Indo-West Pacific and Atlantic Ocean: South Africa.

Habitat: Sessile.

Associated bacteria: Pharmaceutical compounds and activities

Antibacterial activity: The compounds 6-bromoindole-3-carbaldehyde and its debromo analog, indole-3-carbaldehyde have been isolated from *Acinetobacter* sp. associated with this species. Of these compounds, 6-bromoindole-3-carbaldehyde showed moderate antibacterial properties against eight bacterial strains viz. Strain SM-S1, Strain SM-S2, Strain SM-INT, Strain SM-Z, *Alteromonas haloplanktis, Abdllus marinus, Dcleya marina,* and *Vibrio campbellii* (Olguin-Uribe et al. 1997).

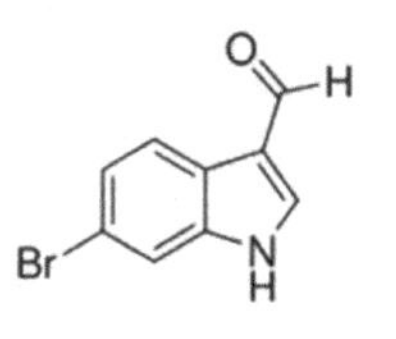

6-Bromoindole-3-Carbaldehyde

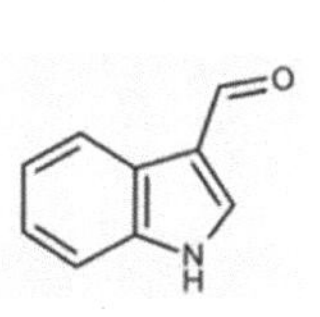

Indole-3-carbaldehyde

Order: Stolidobranchia Family: Pyuridae

Halocynthia aurantium

Image credit: A. C. Tatarinov, commons.wikimedia.

Common name: Pacific sea peach

Global distribution: Arctic and Northern Pacific. Polar and subtropical.

Habitat: Sessile; depth range 0 - 100 m

Associated bacteria: Pharmaceutical compounds and activities

The bacterial strain, *Bacillus pumilus* strain KMM 1364 isolated from this species yielded surfactin-like cyclic depsipeptides 1, 2, 6, 7, and 8 which may have biotechnological applications (Chen et al. 2018).

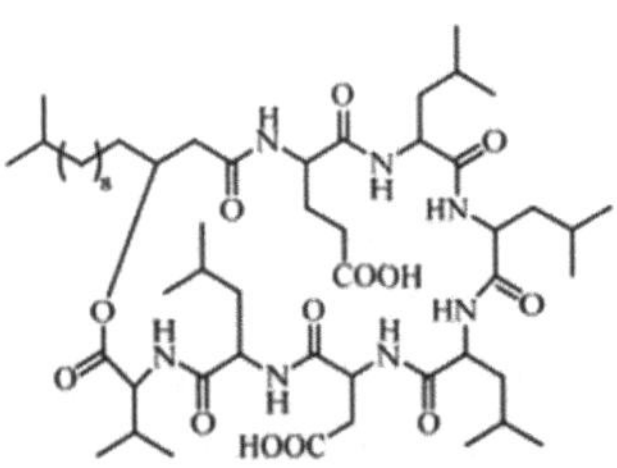

Depsipeptide 1

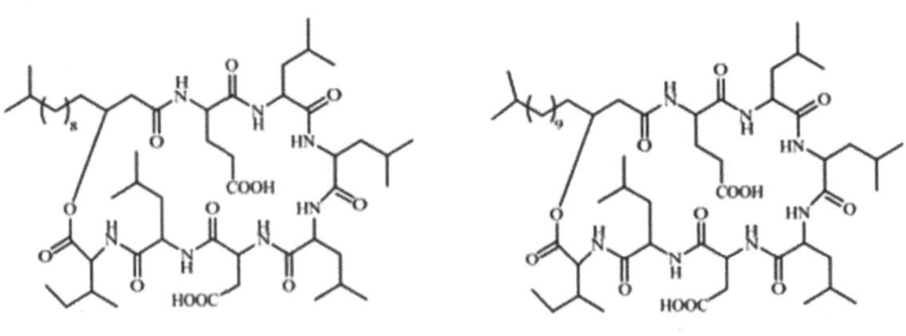

Depsipeptides 2,6

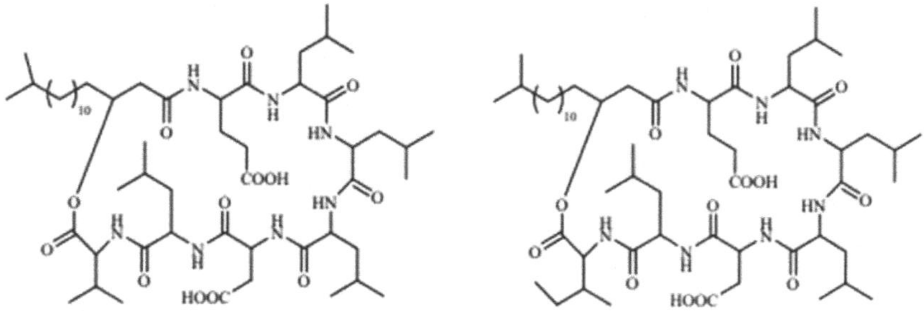

Depsipeptides 7,8

Halocynthia roretzi

Image credit: bifyu. https://creativecommons.org/licenses/by-sa/2.0/

Common name: Sea pineapple, Korean common sea squirt

Global distribution: Northwest Pacific: Japan.

Habitat: It is usually found attached to rocks in shallow coastal waters

Associated bacteria: Therapeutic compounds and their bioactivities

The associated bacterial strain, *Hasllibacter halocynthiae* strain KME 002 of this tunicate produced compounds viz. 3,12-dihydroxy-7-ketocholanic acid, 12-hydroxy-3-keto-glycocholanic acid, nutriacholic acid, deoxycholic acid, 3,3,12- trihydroxy-7-ketocholanic acid, and 3,3,12-trihydroxy-7-deoxycholanic acid, the activities of which are yet to be known (Chen et al. 2018).

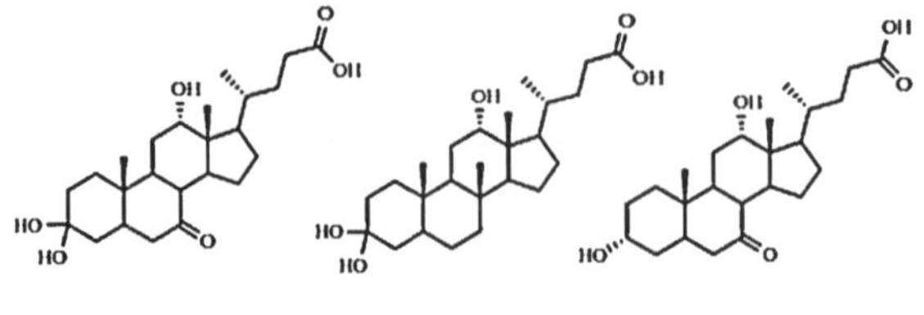

1 2 3

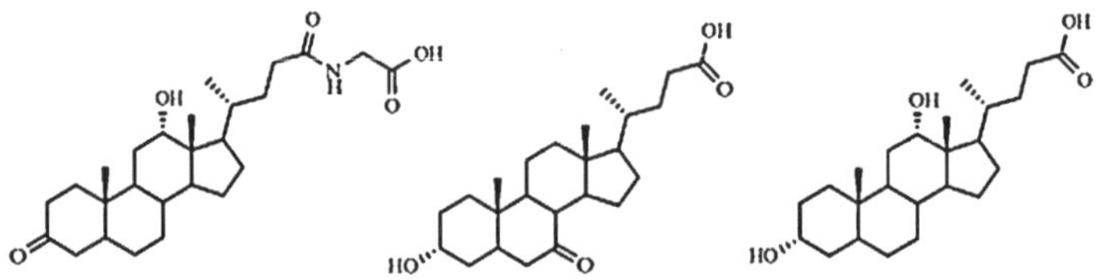

4 5 6

1: 3,3,12- trihydroxy-7-ketocholanic acid; 2: 3,3,12-trihydroxy-7-deoxycholanic acid; 3: 3α,12α-dihydroxy-7-ketocholanic acid; 4: 12-hydroxy-3-keto-glycocholanic acid; 5:, nutriacholic acid; 6: deoxycholic acid.

Chen et al. (2018) reported on the isolation of the bacterium *Aeromicrobium halocynthiae* KME 001T which produced the compound, taurocholic acid.

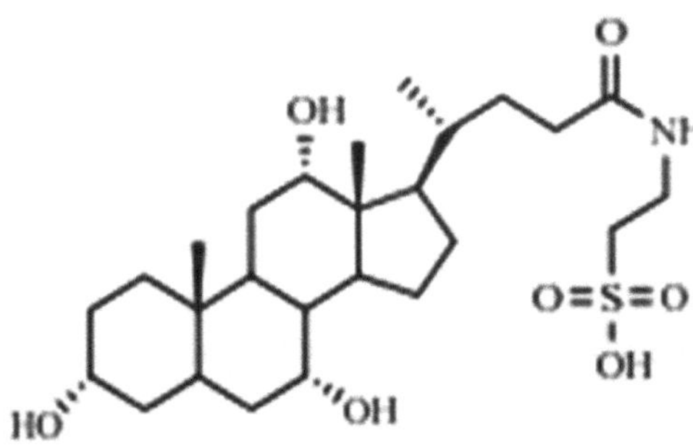

Taurocholic acid

Order: Stolidobranchia Family: Pyuridae

Herdmania momus

Common name: Red-throated ascidian,

Global distribution: Distributed across the globe and is a common inhabitant of the Great Barrier Reef.

Habitat: Commonly found attached to rocks from depths of 3–50 m.

Associated bacteria: Pharmaceutical compounds and activities

Antibacterial activity: The associated *Bacillus* sp. of this species showed antibacterial activity against pathogenic *Pseudomonas aeruginosa* (Mary et al. 2016).

Order: Stolidobranchia Family: Pyuridae

Microcosmus exasperatus

Common name: Not reported

Global distribution: Indo-West Pacific, and Atlantic Ocean

Habitat: Sessile; depth range 0–17 m

Associated bacteria: Pharmaceutical compounds and activities

Antibacterial activity: The associated *Bacillus* sp. of this species showed antibacterial activity against pathogenic *Escherichia coli* and *Staphylococcus aureus*. Further, its *Vibrio algfinolyticus* acted against *Pseudomonas aeruginosa* (Mary et al. 2016).

Order: Stolidobranchia Family: Styelidae

Botryllus schlosseri

Image credit: Massimiliano DeMartino. https://creativecommons.org/licenses/by-sa/3.0/

Common name: Golden star tunicate or star sea squirt

Global distribution: Adriatic, Black sea, Europe, and Mediterranean

Habitat: Grows on slow-moving, submerged objects, plants, and animals in nearshore saltwater environment

Associated bacteria: Pharmaceutical compounds and activities

Antibacterial and glucose-lowering activities: The bacterial strain *Nocardiopsis dassonvillei* HQA404 isolated from this species produced 1,6-dihydroxyphenazine displayed antibacterial activity against *Vibrio anguillarum* and *Vibrio parahaemolyticus*. Further, this compound serves as a oral glucose-lowering agent by inhibiting the activity of the enzyme, alpha-glucosidase. The other compound produced by the above bacterial strain is 2-(acetylamino)-phenol (Chen et al. 2018).

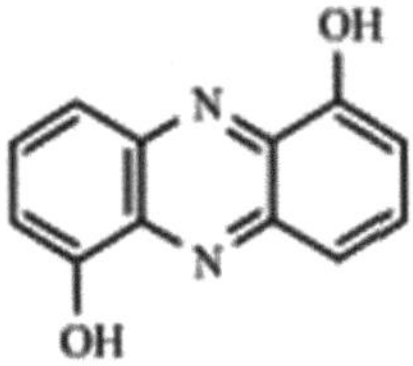

1,6-dihydroxyphenazine

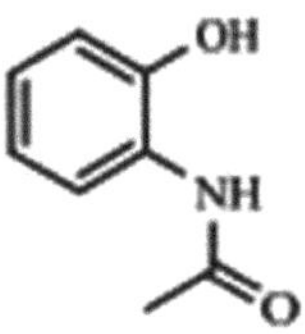

2-(acetylamino)-phenol

Order: Stolidobranchia Family: Styelidae

Styela canopus

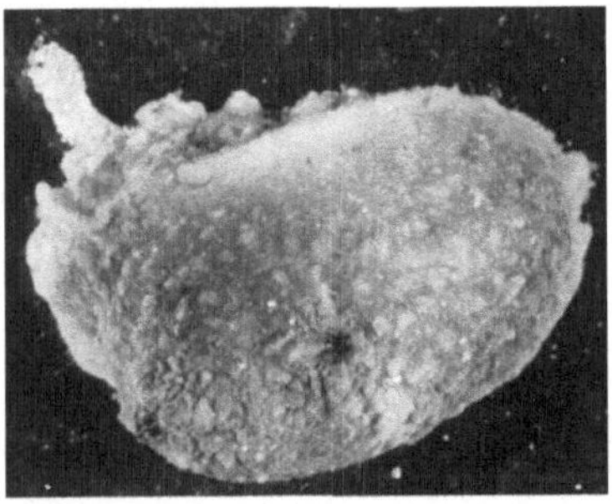

Image credit: Smithsonian Environmental Research, https://creativecommons.org/licenses/by/2.0/

Global distribution: It is widely distributed in temperate and tropical seas

Habitat: Coastal waters

Associated bacteria: Pharmaceutical compounds and activities

Antibacterial activity: The naphthoquinone derivatives granaticin, granatomycin D and dihydrogranaticin B of *Streptomyces* PTY08712 isolated from this tunicate showed antibacterial activity on *B. subtilis* and MR *S. aureus*. The IC50 values recorded for the first two compounds were 1.6 and 6.2 μg/mL, respectively (Nweze et al. 2020; Mayer et al. 2021).

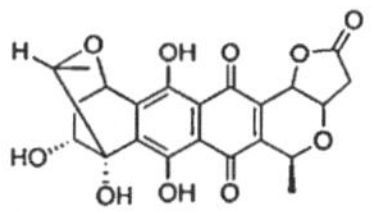

Granaticin

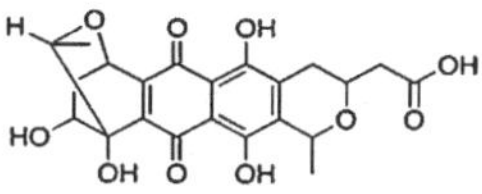

Granatomycin D

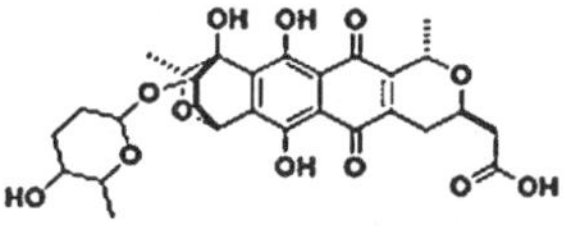

Dihydrogranaticin B

Styela clava

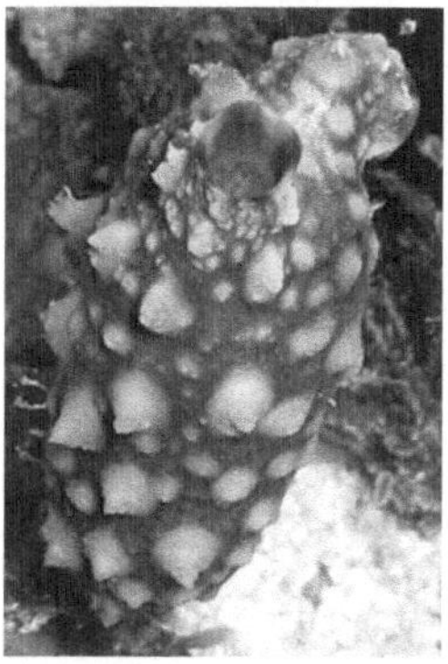

Image credit: Matthieu Sontag, https://creativecommons.org/licenses/by-sa/3.0/

Common names: Asian tunicate, leathery sea squirt, or club sea squirt,

Global distribution: It is native to the Pacific coast of Asia. It was introduced to Australia, New Zealand, North America and Europe.

Habitat: It is mainly found on hard substrates such as rocks, pilings, and shells. It is an euryhaline species surviving in both brackish water and marine conditions.

Associated bacteria: Therapeutic compounds and their bioactivities

Its associated bacterium Streptomyces coelicoflavus strain HQA809 has been reported to produce germicidin and 6-isopropyl group-3-ethyl-4-hydroxy-2-pyrone, the bioactivity of which is known (Chen et al. 2018).

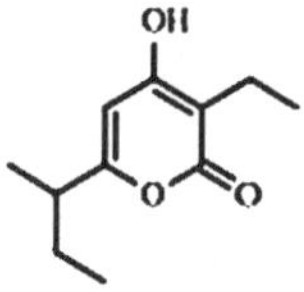

1

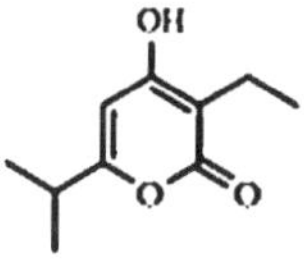

2

1: Germicidin; 2: 6-isopropyl group-3-ethyl-4-hydroxy-2-pyrone

Order: Stolidobranchia Family: Styelidae

Symplegma brakenhielmi

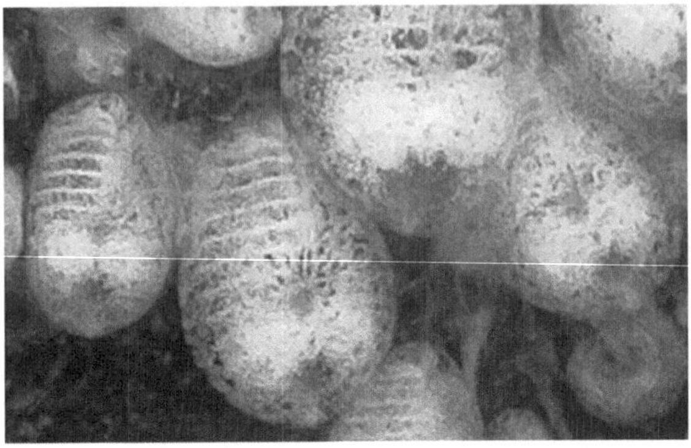

Image credit: Stephen Thorpe, https://creativecommons.org/licenses/by/4.0/

Common name: Not reported

Global distribution: Indo-Pacific, Western Atlantic and the Mediterranean Sea.

Habitat: Sessile; brackish; depth range 0–33 m

Associated bacteria: Pharmaceutical compounds and activities

Antibacterial activity: The bacterial species *Micromonospora* sp. isolated from this species yielded two halimane-type diterpenoids, micromonohalimanes A and B. Among them Micromonohalimane B demonstrated moderate antibacterial activity against methicillin-resistant *Staphylococcus aureus* (Zhang et al. 2016).

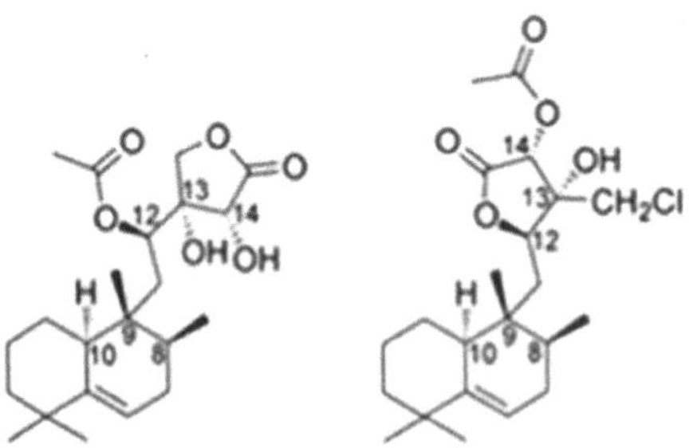

Micromonohalimanes A and B

Unidentified ascidian 1

Associated bacteria: Pharmaceutical compounds and activities

Anticancer activity: The bacterial symbiont *Saccharopolyspora* sp. strain SS081219JE-28 of this tunicate produced the compound, JBIR-66 (3Z,6E,8E)-N-(4-acetamido-3-hydroxybutyl)-2-hydroxy-4,8-dimethylundeca-3,6,8-trienamide) which exhibited fairly weak activity on human lymphoblastoid namalwa cell line with an IC50 value of 440 mM (48 h) (Takagi et al. 2010). lymphoblastoid namalwa cells with an IC50 value of 440 mM (48 h) (Takagi et al. 2010).

Unidentified ascidian 2

Associated bacteria: Pharmaceutical compounds and activities

Anti-inflammatory activity/cell-cell adhesion inhibitory activity: The bacterium *Streptomyces* sp. isolated from this species produced the compound, bohemamine which is believed to serve as an inhibitor of cell-cell adhesion and an agent of anti-inflammatory activity (Zhang et al. 2003; Chen et al. 2018).

Unidentified ascidian 3

Associated bacteria: Pharmaceutical compounds and activities

Cytotoxicity: The bacterium *Streptomyces* sp. the symbiont of this species produced S-methyl-2,4-dihydroxy-6-isopropyl-3,5-dimethylbenzothioate which is believed to possess cytotoxicity (Mahyudin et al. 2012; Chen et al. 2018).

Unidentified ascidian 4

Associated bacteria: Pharmaceutical compounds and activities

The bacterium *Saccharopolyspora* sp. JE-28 SS081219 isolated from this species produced macrolactins E) and F together with gilvocarcins M and V (Chen et al. 2018). Fischer et al. (2003) reported on the potent bactericidal, virucidal, cytotoxic, and antitumor activities of gilvocarcin V.

Unidentified ascidian 5

Associated bacteria: Pharmaceutical compounds and activities

Androgen receptor inhibitory activity: The actinobacterial strain *Verrucosispora gifhornensis* YM28-088 of this tunicate produced gifhornenolones A and B; and cyperusol C. Among these compounds, only gifhornenolone A showed significant activity against the androgen receptor(binding the androgenic hormones, including testosterone and dihydrotestosterone in the cytoplasm) (Chen et al. 2018).

Unidentified ascidian 6

Associated bacteria: Pharmaceutical compounds and activities

Hemolytic activity: *Nocardia* sp. strain KMM 3749, the symbiont of this tunicate produced ubiquinone Q9 which caused hemolysis of mouse erythrocytes (Chen et al. 2018).

Unidentified ascidian 7

Associated bacteria: Pharmaceutical compounds and activities

Antibacterial activity: The bacterial strain, *Streptomyces* sp. PTY087I2 of this tunicate produced three naphthoquinone derivatives, granaticin, granatomycin D, and dihydrogranaticin B which showed antibacterial activity against the human pathogens *Bacillus subtilis*, methicillin-sensitive *Staphylococcus aureus* (MSSA), methicillin-resistant *Staphylococcus aureus* (MRSA), and *Pseudomonas aeruginosa*) (Chen et al. 2018).

Unidentified orange ascidian 8

Associated bacteria: Pharmaceutical compounds and activities

Antimicrobial and cytotoxic activity: The bacterial strain LL-14I352, a symbiont of this species produced phenazine compounds LL-14I352 (or pelagiomicin) and which possessed several bioactivities including RNA and protein synthesis, cytotoxic activity and DNA inhibitory activity (Chen et al. 2018).

Unidentified greenish ascidian 9

Associated bacteria: Pharmaceutical compounds and activities

Cytotoxicity: The bacterial strain *Streptomyces* sp. YM14-060 isolated from this tunicate produced piericidin compounds, A1 and A2; and C7 and C8 which inhibited Neuro-2a mouse neuroblastoma cell line and it also showed cytotoxicity of rat glial cells (RG-E1A-7) (Chen et al. 2018).

Unidentified greenish ascidian 10

Associated bacteria: Pharmaceutical compounds and activities

Antitumor and immunosuppressive activity: The bacterium *Agrobacterium* PH-A034C isolated from this species produced the compound, sesbanimide C (Chen et al. 2018) which is said to possess antitumor and immunosuppressive activity (Luzzio 2019).

Unidentified ascidian 11

Associated bacteria: Pharmaceutical compounds and activities

Antibacterial activity: A naphtoquinone terpenoid compound, napyradiomycin A1which was isolated from a marine ascidian-associated *Streptomyces* sp. strain CA-271078 displayed activity against MR *S. aureus* with IC50 values ranging from 0.5 to 1 μg/mL (Mayer et al. 2021).

Unidentified tunicate 12

Antibacterial activity: The ethyl acetate extracts of a tunicate–associated Gram-negative marine bacterium, *Pseudoalteromonas rubra* TKJD 22 showed the presence of a red-orange crystalline solid, isatin, which displayed antibacterial actiivyt against MDR *Escherichia coli* and MDR-ESBL *Escherichia coli* (Nweze et al. 2020).

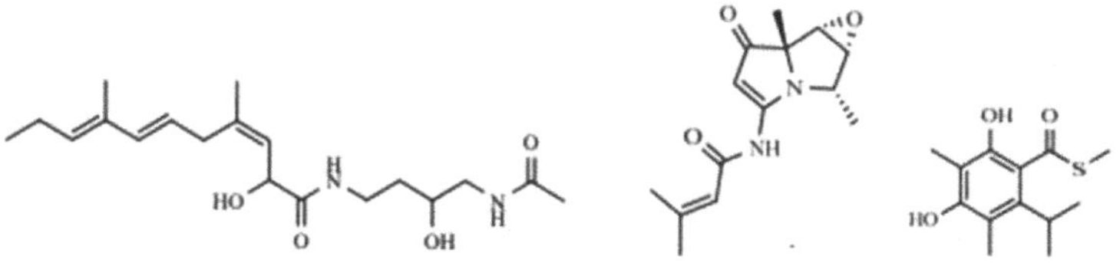

JBIR-66 Bohemamine S-methyl-2,4-dihydroxy-6-isopropyl-3,5-
dimethylbenzothioate

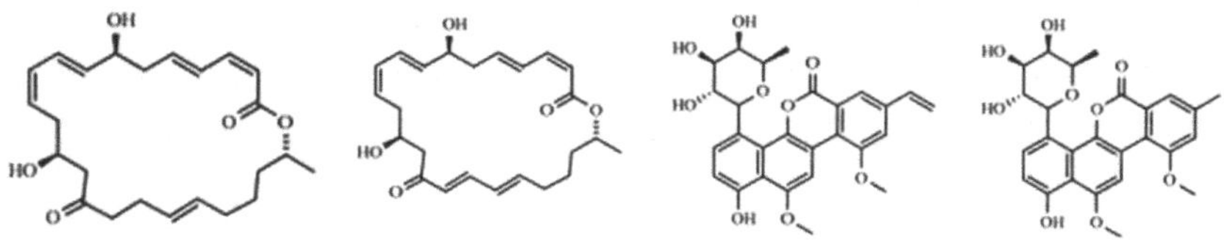

Macrolactin E Macrolactin F Gilvocarcin M Gilvocarcin V

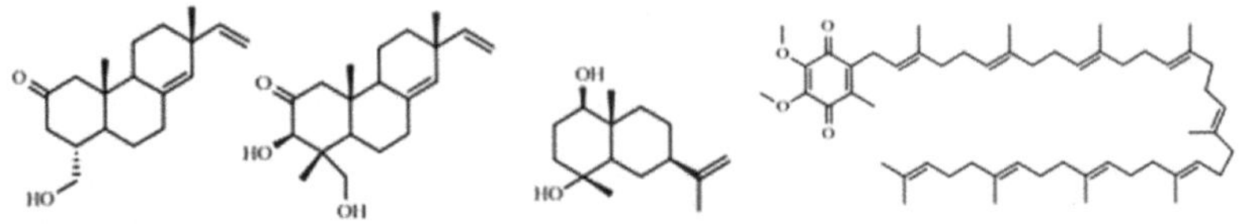

Gifhornenolones A,B   Cyperusol C   Ubiquinone Q9

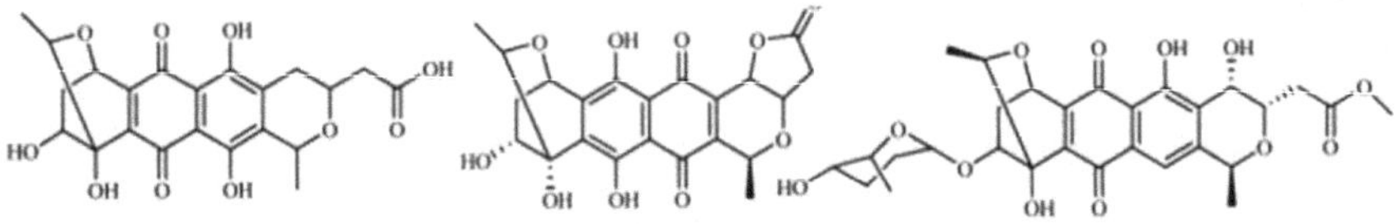

Granaticin   Granatomycin D   Dihydrogranaticin B

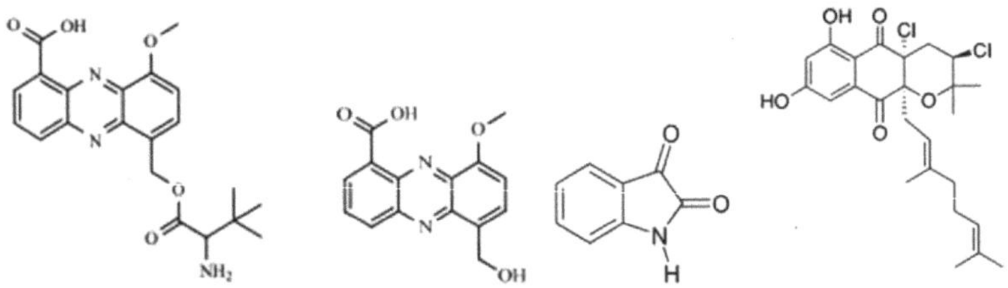

LL-14I352 α   LL-14I352 β   Istatin   Napyradiomycin A1

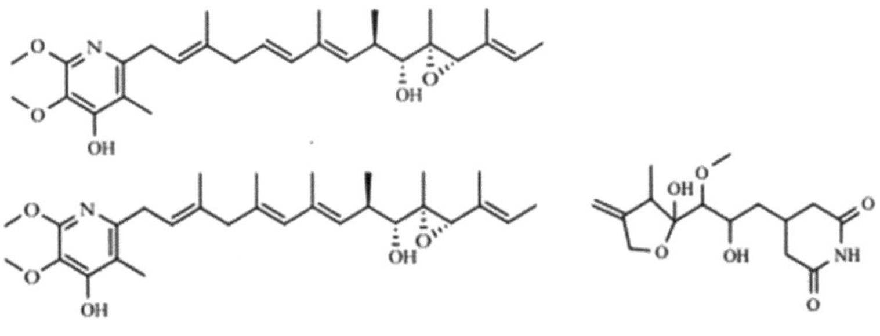

Piericidins C7, C8   Sesbanimide C

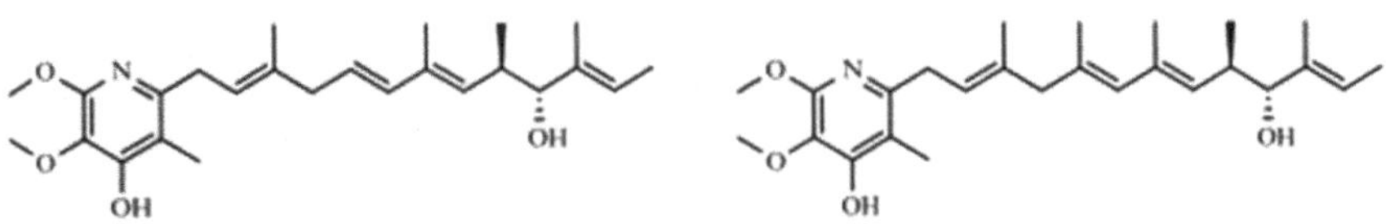

Piericidins A1, A2

FISHES

CARTILAGINOUS FISHES

Superclass: Pisces   Class: Elasmobranchii   Order: Squaliformes
Family: Etmopteridae

Centroscyllium fabricii

Image credit: Public domain.

Common name: Black dogfish

Global distribution: Both sides of the North Atlantic, off western Greenland and Iceland; also inhabits most of the waters along the West African coastline, from Guinea to South Africa

Habitat: Prefers continental shelves at depths of up to 275 m.

Associated bacteria: Pharmaceutical compounds and activities

Antimicrobial activity: *Bacillus amyloliquefaciens* (strain BTSS-3) isolated from the gut of this species showed antimicrobial activity against pathogenic bacteria including *Salmonella typhimurium, Proteus vulgaris, Clostridium perfringens, Staphylococcus aureus, Bacillus cereus, Bacillus circulans, Bacillus macerans*, and *Bacillus pumilus* (Bindiya et al. 2015).

TELEOST FISHES

Class: Actinopterygii   Order: Perciformes   Family: Carangidae

Decapterus maruadsi

Image credit: Public domain.

Common name: Japanese scad

Global distribution: Indo-West Pacific

Habitat: This reef-associated species is seen mainly in coastal waters down to 20 m. It is also observed to aggregate near buoys and fish aggregating devices.

Associated bacteria: Pharmaceutical compounds and activities.

Histamine and biogenic amines production: The bacterial species *Arthrobacter bergeri*, *Pseudomonas* sp., *Psychrobacter* sp., *Shewanella baltica*, and *Aeromonas salmonicida* isolated from this species produced hiastamines and *Aeromonas salmonicida* also produced amines like spermidine, tryptamine, tyramine, spermine, putrescine, and cadaverine. Histamine plays a key role in immune responses (Hu et al. 2014).

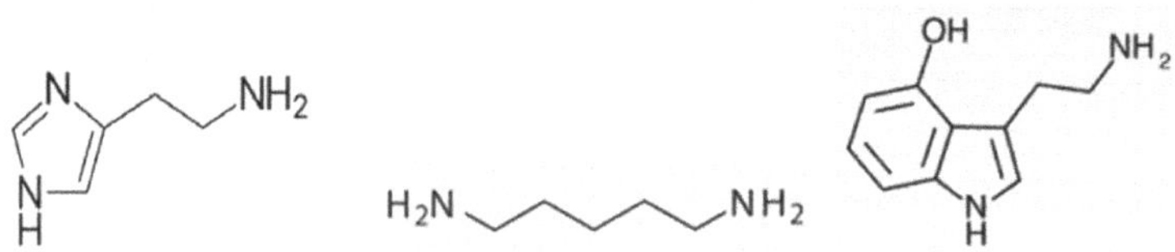

Histamine Cadaverine Tryptamine

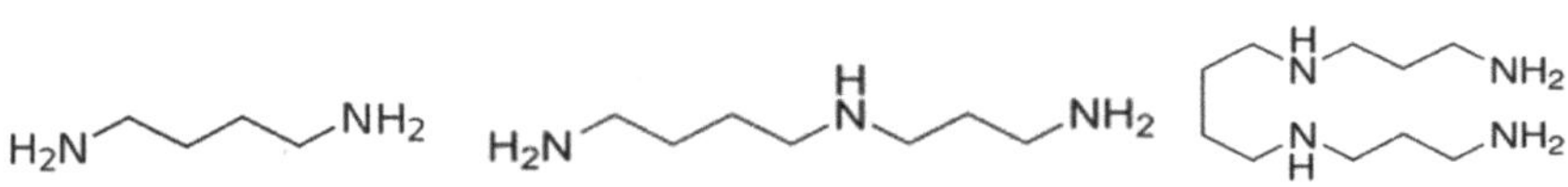

Putrescine Spermidine Spermine

Order: Perciformes Family: Carangidae

Seriola lalandi

Image credit: A. R. Mc Culloch, Public domain.

Common name: Yellowtail amberjack

Global distribution: Indo-Pacific, Eastern Pacific and Eastern Atlantic

Habitat: Marine; brackish; benthopelagic; depth range 3–825 m

Associated bacteria: Pharmaceutical compounds and activities

Antibacterial activity: In this species, the associated bacteria which were consisting of Alphaproteobacteria (2%), Bacilli (7%), Actinobacteria (8%), and Gammaproteobacteria (80%) displayed inhibitory activity against several pathogenic bacterial strains and improved immunological properties and growth (Ramírez et al. 2017).

Order: Perciformes Family: Cichlidae

Oreochromis niloticus

Image credit: Germano Roberto Schüür, https://creativecommons.org/licenses/by-sa/4.0/

Common name: Nile tilapia

Global distribution: Africa: also in coastal rivers of Israel

Habitat: This benthopelagic species is seen in coastal river basins and brackish water at depths of 0–20 m. It is also potamodromous.

Associated bacteria: Therapeutic compounds and their bioactivities

Antibacterial activity: The isolate Bacillus UBRU4 of *Oreochromis niloticus*, the symbiont of this fish has been reported to show antibacterial activity against *Aeromonas hydrophila*, a gram-negative bacteria responsible for gastroenteritis in humans (Chantharasophon et al. 2011).

Order: Perciformes Family: Latidae

Lates calcarifer

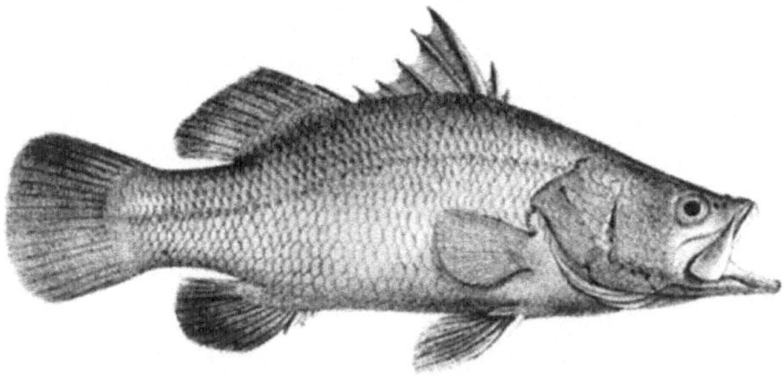

Image credit: George Henry Ford / Carl Ludolf Griesbach, Public domain.

Common name: Barramundi,

Global distribution: Indo-West Pacific

Habitat: Marine; freshwater; brackish; demersal; catadromous; depth range 10 - 40 m

Associated bacteria: Pharmaceutical compounds and activities

Antimicrobial activities: The exopolysaccharides (EPS) extracted from the bacterial species *Bacillus cereus* and *Brachybacterium* sp. of this species showed antimicrobial activity against both grams positive and negative bacteria (Orsod et al. 2012).

Order: Perciformes Family: Sparidae

Sparus aurata

Image credit: Roberto Pillon, https://creativecommons.org/licenses/by/3.0/

Common name: Gilthead seabream.

Global distribution: Eastern Atlantic

Habitat: This demersal species is seen in both marine brackish water at depths of 1 - 150 m

Associated bacteria: Therapeutic compounds and their bioactivities

Antibacterial activity: The glycolipids extracted from the associated bacterial genera of this species viz. *Pseudomonas*, *Acinetobacter*, *Sphingomonas*, and *Aeromonas* exhibited antibacterial activity (Floris et al. 2018).

Order: Scorpaeniformes Family: Scorpaenidae

Pterois volitans

Image credit: Michael Gäbler, https://creativecommons.org/licenses/by/3.0/

Common name: Red lionfish

Global distribution: Pacific Ocean: Cocos-Keeling Islands, and Western Australia

Habitat: Reef-associated; depth range 2–55 m

Associated bacteria: Pharmaceutical compounds and activities

Antibacterial activity: A total of 54 isolates belonging to *Alteromonas* spp., *Pseudoalteromonas* spp., *Bacillus* spp., *Photobacterium* spp., *Ruegeria* sp. *Vibrio* sp. and *Exiguobacterium* sp. were isolated from this fish as symbionts. These isolates inhibited the growth of *V. parahaemolyticus* strain LA-4T-1, *V. vulnificus* strains 1-FT-1 and 76-FC-1, *Photobacterium damselae* ssp. *piscicida* and *V. harveyi* (Stevens et al. 2016).

Fistularia commersonii

Image credit: Matthias Kleine, https://creativecommons.org/licenses/by-sa/3.0/

Common names: Smooth cornetfish or bluespotted cornetfish,

Global distribution: Tropical and subtropical Indo-Pacific

Habitat: Reef-associated; depth range 0–132 m

Associated bacteria: Pharmaceutical compounds and activities

The bacterial strain *Paracoccus zeaxanthinifaciens* 22-5T (5KCTC 22803T 5CCUG 58401T) isolated from this fish was found to have ubiquinone-10 (Q-10) as the predominant isoprenoid quinone (Kim et al., 2010). This compound may slow the progression of age-related diseases, including hearing loss by enhancing mitochondrial activity and mitochondrial antioxidant defense system activity and increasing ATP production (Tian et al. 2014).

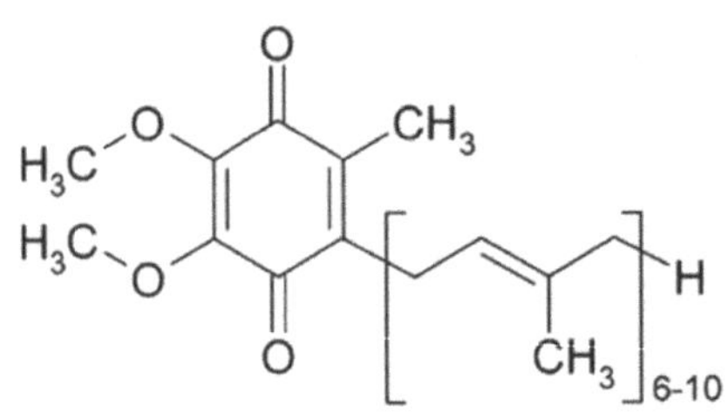

Ubiquinone-10

Order: Syngnathiformes Family: Syngnathidae

Hippocampus kuda

Image credit: Emőke Dénes, https://creativecommons.org/licenses/by-sa/4.0/

Common name: Spotted seahorse

Global distribution: Indo-Pacific

Habitat: This reef-associated species is seen in both marine and brackish water at depths of 0–70 m

Associated bacteria: Pharmaceutical compounds and activities

Lipase activity: The predominant microbial species associated with this fish viz. *Burkholderia cenocepacia* which possessed the ability to degrade lipids, cellulose, xylan, starch and proteins. Lipase activity relates to the digestion of foods rich in lipids (Tanu et al. 2012).

Order: Tetraodontiformes Family: Triacanthidae

Pseudotriacanthus strigilifer *(= Triacanthus strigilifer)*

Image credit: Francis Day, Public domain.

Common name: Long-spined tripodfish

Global distribution: Indo-Pacific: Gulf of Oman to Indonesia and the Philippines.

Habitat: Marine; brackish; demersal; depth range 2–110 m

Associated bacteria: Pharmaceutical compounds and activities

Antibacterial activity: Five isolates (1-5) of the gram-positive, *Lysinibacillus fusiformis* deerived from this species displayed antibacterial activity against several pathogenic bacterial species including *Proteus* sp, *Serratia merche*, *E. coli*, *Streptococcus viridae*, *Streptococcus epidermis*, *Staphylococcus aureus*, *Pseudomonas aeruginosa*, *Bacillus aerogenosa*, *Bacillus megaterium*, *Solmonella typhi*, *Streptococcus pneumoniae* and *Klebsiella pneumoniae* (Abideen and Babuselvam 2014).

Unidentified Puffer Fish

Associated bacteria: Pharmaceutical compounds and activities

Antimicrobial activity: The *Enterococcus faecium* AD1 isolated from this fish showed antimicrobial activity against *Candida albicans, Salmonella typhi, Pseudomonas aeruginosa, Staphylococcus sciuri, Staphylococcus aureus,* and *Micrococcus luteus* (Nguyen et al. 2015).

Antioxidant activity: The exopolysaccharide (EPS) extracted from the bacterium *Enterococcus faecium* isolated from this species showed the highest amount of antioxidant activity with an IC50 value of 5 mg/mL (Nguyen et al. 2015).

Anticancer activity: The exopolysaccharide (EPS) extracted from the *Enterococcus faecium* isolated from this species showed activity against Hela and Hep G2 cancer cells. The percent values of cytotoxicity for crude EPS were 90.4 and 70.3 for Hela and Hep G2, respectively (Nguyen et al. 2015).

TTX producing Pufferfishes and Gobiid fish

The endosymbiotic bacteria of certain species of puffer fish genera *Chelonodon, Lagocephalus,* and *Takifugu* (Tetraodontiformes, Tetraodontidae) and gobiid fish, *Yongeichthys* (Perciformes, Gobiidae) have been reported to produce Tetrodotoxin (TTX) (Lago et al. 2015; Magarlamov et al. 2017).

Microbes of Shark Mouth

Associated bacteria: Pharmaceutical compounds and activities

Cytoxicity: The ethyl acetate extract of *Bacillus cereus* and *Vibrio alginolyticus* isolated from the mouth of unidentified sharks showed cytoxicity against human liver hepatocellular carcinoma (Hep G2) cells. The concentrations that caused 50% cell death (IC50) in the cell viability assay on Hep G2 cells were 0.764 μg/ml and 0.918 μg/ml for *Bacillus cereus* and *Vibrio alginolyticus*, respectively (Ramlakhan 2015).

Mucus from sharks

Associated bacteria: Pharmaceutical compounds and activities

Antibacterial activity: The culturable bacterial isolates from the mucus of white sharks, *Carcharodon carcharias* have been reported to possess antibacterial activity (Luer and Walsh 2018).

CHAPTER 5

Medicinal Fungi Associated with Marine Plants and Animals: Their Compounds and Bioactivities

FUNGI ASSOCIATED WITH MARINE PLANTS

SEAWEEDS (MARINE MACROALGAE)

Phylum (Division): Chlorophyta Class: Ulvophyceae
Order: Bryopsidales Family: Caulerpaceae

Caulerpa sp.

Image credit: Nhobgood Nick Hobgood, https://creativecommons.org/licenses/by-sa/3.0/

Common name (s): Not applicable
Global distribution: Tropical and temperate seas

Medicinal Marine Microbiota. Santhanam Ramesh, Ramasamy Santhanam, Arumugam Uma & Raju Mohanraju (Authors)

Habitat: The species of this genus occur largely in shallow intertidal zones up to depths of about 100 m.

Associated fungi and their pharmaceutical compounds and activities

Antimycobacterial, antimicrobial, cytotoxic, antiviral, immunodepressant activities: The crude extracts of the fungi *Beauveria felina* strains AcSS8 and AcSS13 associated with this green alga yielded cyclodepsipeptide destruxins viz. destruxin B, destruxin E chlorohydrin, Me-Pro destruxin E chlorohydrin, and pseudodestruxin C, which have been reported to possess activities such as antimycobacterial, antibacterial, antifungal, and cytotoxic (de Vita-Marques et al., 2008; Suryanarayanan 2012).

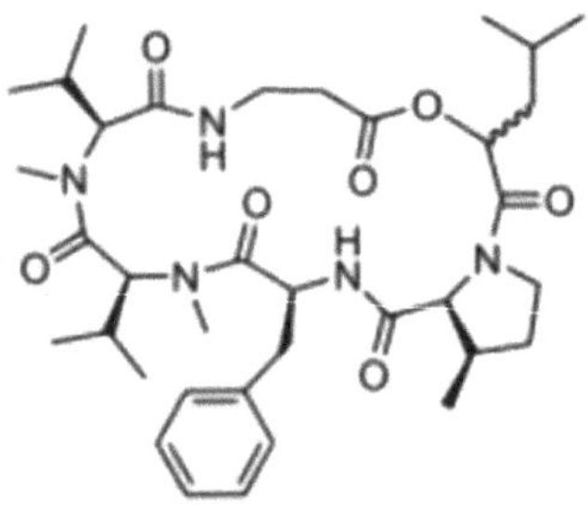

Pseudodestruxin C

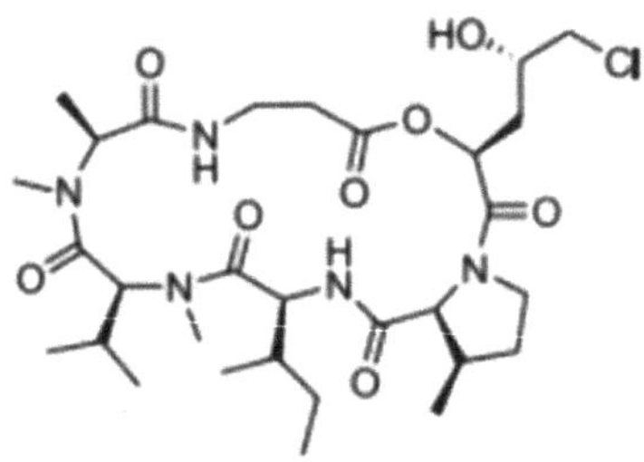

β-Me-Pro destruxin E chlorohydrin

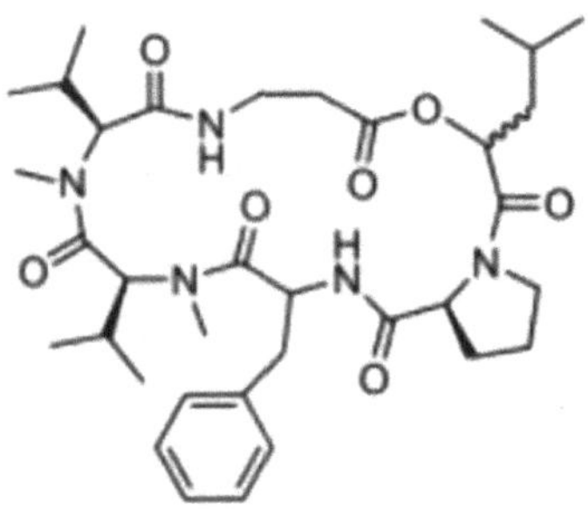

Destruxin B

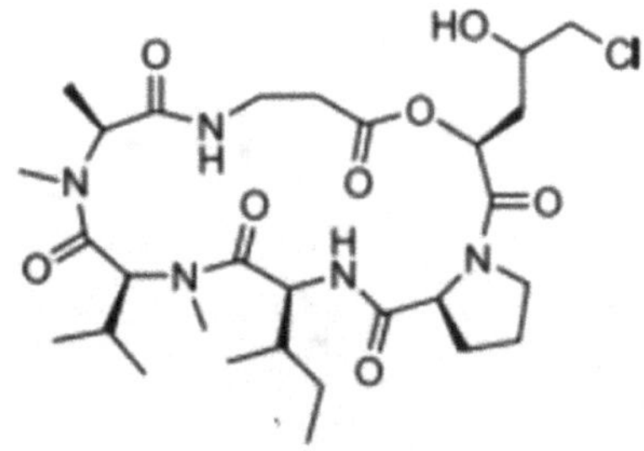

Destruxin E chlorohydrin

Order: Bryopsidales Family: Halimedaceae

Flabellia petiolata

Image credit: Esculapio, https://creativecommons.org/licenses/by/3.0/

Common name (s): No common names

Global distribution: Eastern Atlantic Ocean and the Mediterranean Sea

Habitat: It grows in shallow coastal waters with soft sediment.

Associated fungi and their pharmaceutical compounds and activities

Antibacterial activity: The extracts of the fungal strains *Microascacea* sp. strain MUT 4861 and *Beauveria bassiana* strain MUT 4865 associated with this green alga have been reported to possess antibacterial activity (Nweze et al. 2020).

Order: Bryopsidales Family: Udoteaceae

Penicillus capitatus

Image credit: W.Strickling. https://creativecommons.org/licenses/by-sa/4.0/

Common name (s): Shaving brush alga

Global distribution: It is more common in Caribbean, South Florida and Bahamas

Habitat: It forms large colonies in shallow, lagoons, and bays. It may also be found mixed with sea grasses.

Associated fungi and their pharmaceutical compounds and activities

Anti-tumor and cytotoxic activities: The crude extract of the fungus *Aspergillus versicolor* associated with this alga yielded sesquiterpenoid nitrobenzoyl esters. Among them, 9, 14-Dihydroxy-6-p-nitrobenzoylcinnamolide (= insulicolide) has been reported to possess antitumor activity and moderate cytotoxicity human colon carcinoma cell line (HCT-116) with an LC50 value of 1.1 g mL^{-1} (Belofsky et al., 1998).

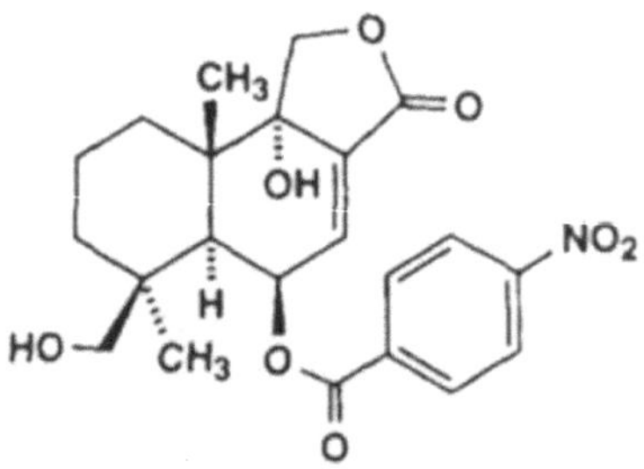

9α, 14-Dihydroxy-6β-p-nitrobenzoylcinnamolide

Order: Cladophorales Family: Siphonocladaceae

Dictyosphaeria versluyii

Image credit: JulsCanja, Public domain.

Common name (s): Not known

Global distribution: Indian and Pacific Oceans, Eastern Atlantic and Caribbean Sea

Habitat: This benthic species occurs both in calm habitats and with strong wave action or currents. It is commonly found attached to coral rubble and rocks and coral rubble in tidal pools or subtidal locations up to a depth of about 70 m.

Associated fungi; and their pharmaceutical compounds and activities

Antibacterial activity: The *Penicillium* sp. associated with this alga yielded two polyketide compounds viz. dictyosphaeric acid A and B which possessed antibacterial activity (Bugni 2003).

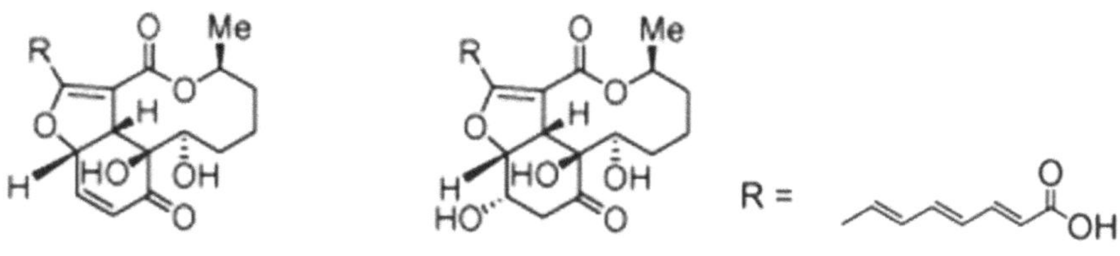

Dictyosphaeric acid A, B

Order: Ulotrichales Family: Monostromataceae

Monostroma hariotii

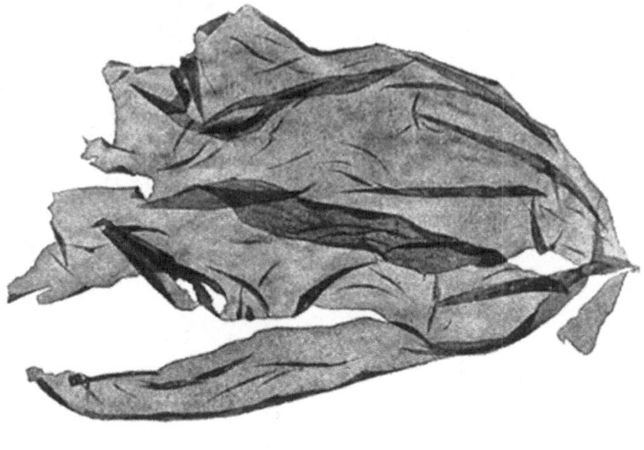

Image credit: Public domain.

Common name (s): Green laver

Global distribution: it is distributed widely from temperate to polar areas

Habitat: This benthic species occurs on pools and rocks in the intertidal zone

Associated fungi; and their pharmaceutical compounds and activities

Antifungal and trypanocidal activities: The extract of the fungus *Penicillium* sp. UFMGCB 6120 which is associated with this alga has been reported to possess antifungal and trypanocidal activities. This extract showed antifungal activity against the pathogenic fungus *Cladosporium sphaerospermum*, with 96% inhibition and an MIC value of 250 mg/mL. It is worth mentioning here that this filamentous and harmful alga causes allergic reactions, asthma and other respiratory diseases and skin infections (https://library.bustmold.com/cladosporium/cladosporium-sphaerospermum/) The extract of UFMGCB 6120 also displayed 100% inhibition of trypomastigote forms of the protozoan parasite, *T. cruzi*, for which an IC50 value of 0.45 gm L^{-1} was recorded (Godinho et al. 2013).

Order: Ulvales Family: Ulvaceae

Ulva australis *(= Ulva pertusa)*

Common name (s): Sea lettuce

Global distribution: Indo-Pacific Ocean

Habitat: It is found growing in the upper subtidal and lower littoral zones and is found attached to rocks or other marine organisms.,

Associated fungi; and their pharmaceutical compounds and activities

Cytotoxic activity: The fungal strain, *Chaetomium globosum* QEN-14 associated with this alga has been reported to produce ytochalasan derivative *viz.* cytoglobosins A–G; and two related compounds, isochaetoglobosin D and chaetoglobosin F. Among these compounds, cytoglobosin C and D exhibited cytotoxic activity against the A-549 tumor cell line (Cui et al. 2010).

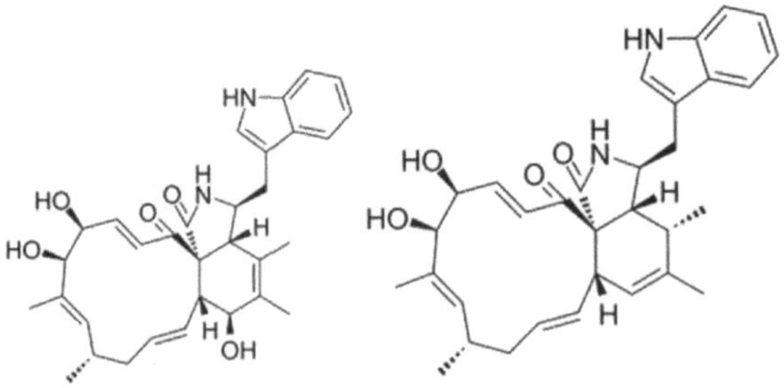

Cytoglobosin C, D

Ulva compressa *(= Enteromorpha compressa)*

Image credit: Mirgolth, https://creativecommons.org/licenses/by-sa/3.0/

Common name (s): Not known

Global distribution: This cosmopolitan species is commonly found in the Pacific Ocean and North and South Atlantic.

Habitat: It is commonly seen in the intertidal zone, rock pools and tide pools

Associated fungi and their pharmaceutical compounds and activities

Tyrosinase inhibitory activity: The fungus *Myrothecium* sp. associated with this alga has been reported to produce cyclopentenone derivatives *viz*. 6-n-pentyl-alpha-pyrone, myrothenone A and B, trichodenone A and cyclonerodiol. Among these compounds, 6-n-pentyl-alpha-pyrone and myrothenone displayed significant tyrosinase inhibitory activity with IC50 values of 0.8 and 6.6 μM, respectively. As these compounds are capable of controlling the biosynthesis of melanin, they offer vast scope in the production of cosmetic and hyperpigmentation agents (Li et al. 2005).

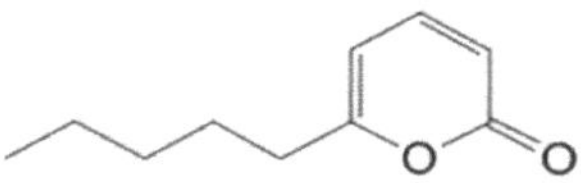

6-n-pentyl-alpha-pyrone

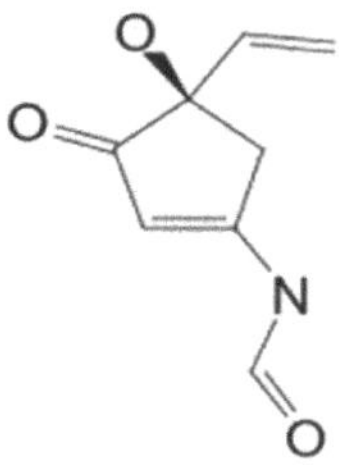

Myrothenone A

Ulva flexuosa *(= Enteromorpha tubulosa)*

Image credit: By Pierre-Louis Crouan, Public domain.

Common name (s): Grass kelp or green alga

Global distribution: It has worldwide distribution

Habitat: This species prefers littoral zones with some wave action. This euryhaline alga is also known to tolerate freshwater conditions.

Associated fungi; and their pharmaceutical compounds and activities

Immune regulatory effect: The fungus, *Aspergillus flavus* associated with this alga has been reported to produce 2-Pyrone derivatives viz. 4-(hydroxymethyl)-5- hydroxy-2H-pyran-2-one and (5-hydroxy-2-oxo-2Hpyran-4-yl) methyl acetate. Among these compounds, 4-(hydroxymethyl)-5- hydroxy-2H-pyran-2-one showed immune regulatory effect by significantly increasing the cyclic adenosine monophosphate (cAMP) in Chinese hamster ovary cells (CHO) and human embryonic kidney cells (HEK293). cAMP has been reported to be a promising regulator of innate immune cell functions (Lin et al. 2008).

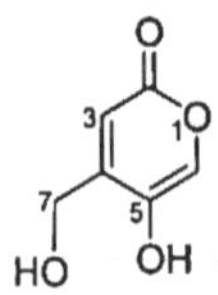

1

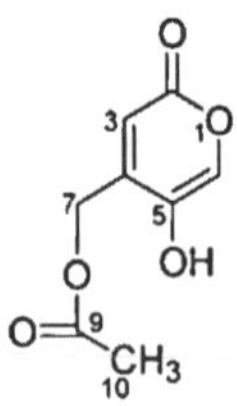

2

1: 4-(hydroxymethyl)-5- hydroxy-2H-pyran-2-one; 2: (5-hydroxy-2-oxo-2Hpyran-4-yl) methyl acetate

***Ulva* sp.1** (= *Enteromorpha* sp.)

Associated fungi; and their pharmaceutical compounds and activities

Fat-accumulation inhibitory activity: The fungus *Cadophora malorum* which is associated with this unidentified species of *Ulva* yielded (+)-sclerosporin and hydroxylated sclerosporin derivatives viz. 15-hydroxysclerosporin, 12-hydroxysclerosporin, 11-hydroxysclerosporin, and 8-hydroxysclerosporin. Among these compounds, 8-hydroxysclerosporin has been reported to possess inhibitory effects on 3T3-L1 murine adipocytes resulting controlled fat accumulation (Almeida et al.,2010a).

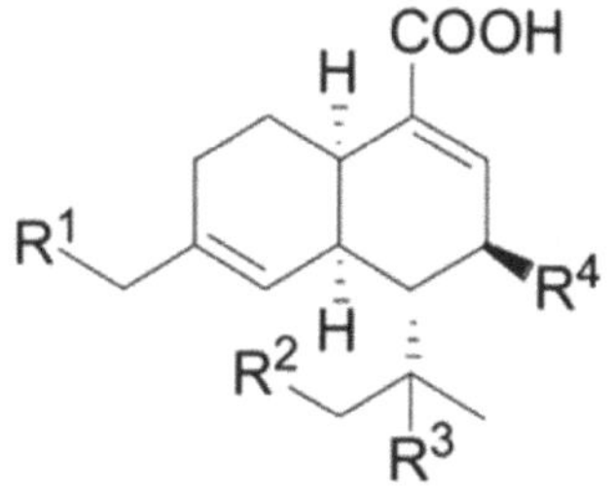

R1=H, R2=H, R3=H, R=OH

8-hydroxysclerosporin

***Ulva* sp.2**

Associated fungi; and their pharmaceutical compounds and activities

Antibacterial, antimycobacterial and HILE inhibitory activities: The fungus, *Coniothyrium cereale* associated with this alga yielded phenalenone derivatives viz. coniosclerodin, (Z)-coniosclerodinol, sclerodinol, conioscleroderolide, coniolactone, (-) sclerodin, and scleroderolide. Among these compounds, sclerodinol and scleroderolide displayed antibacterial properties on *Staphylococcus aureus* SG 511 for which an 24 mM was recorded as MIC. Compounds (Z)-coniosclerodinol, sclerodinol and coniolactone exhibited antitubercular activity against *Mycobacterium phlei*; and compounds coniosclerodin, conioscleroderolide, and scleroderolide displayed promising HLE (human leukocyte elastase or human neutrophil elastase) inhibitory activity with IC50 values of 7, 13 and 11 μM, respectively (Elsebai et al. 2011). The HLE which is a serine proteinase largely involved in tissue degradation and inflammation (Elsebai et al. 2011).

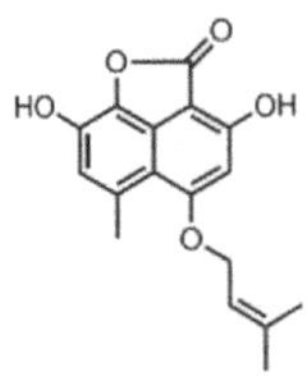 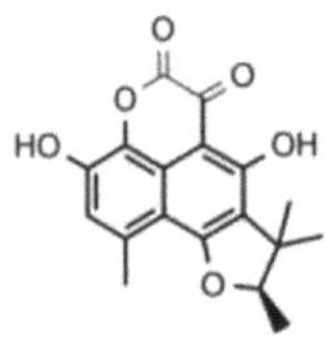 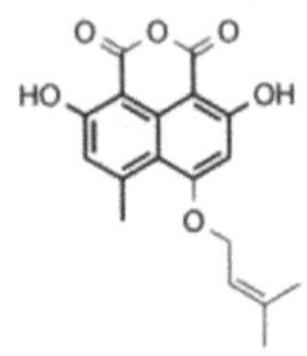

Coniolactone Sclerodertolide Conioscerodin

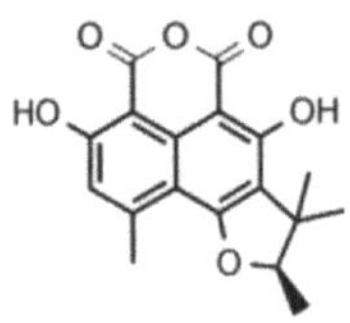 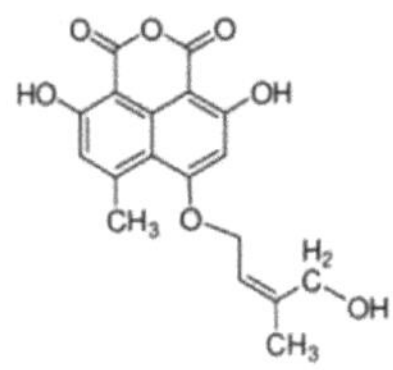 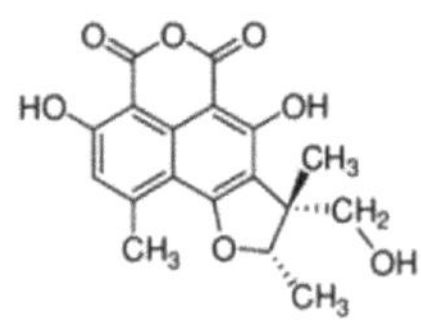 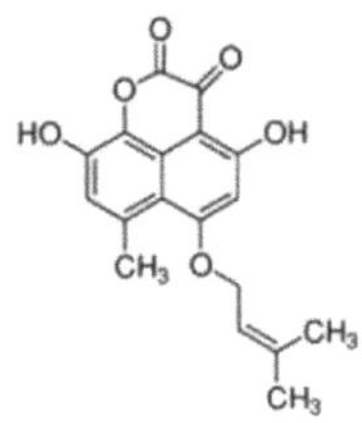

Sclerodin (Z)-coniosclerodinol Sclerodinol Coniosclerodrolide

Unidentified green alga

Associated fungi and their pharmaceutical compounds and activities

Antitumor activity: The fungus *Emericella nidulans* var. *acristata* which is associated with this alga has been reported to produce an indole alkaloid compound viz. emindole DA which displayed antitumor activity against a total of 36 human tumor cell lines with a mean IC50 value of 5.5 μg/mL (Kraj et al. 2006).

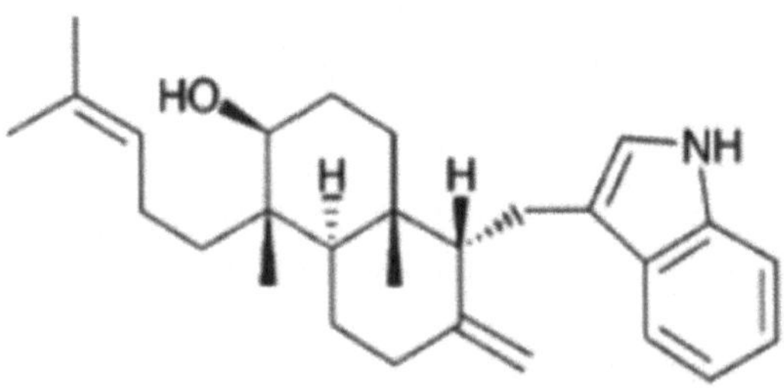

Emindole DA

Phylum (Division): Phaeophyta Class: Phaeophyceae

Order: Ectocarpales Family: Scytosiphonaceae

Colpomenia sinuosa

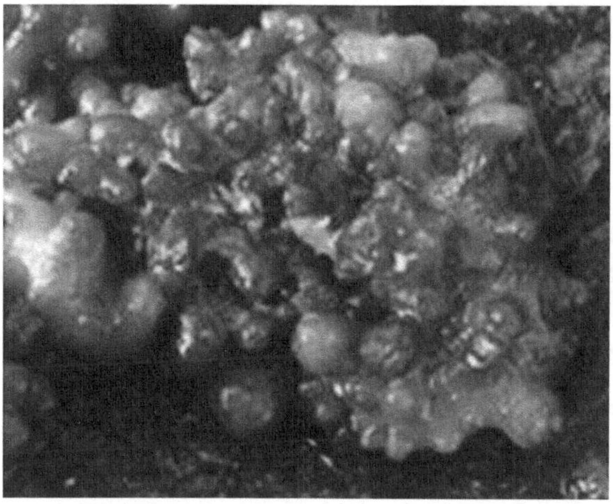

Image credit: J. McKenna, Public domain.

Common name (s): Sinuous ballweed or oyster thief

Global distribution: It is commonly distributed in tropical to temperate seas

Habitat: It is found attached to rocks or other algae in the mid intertidal zone

Associated fungi; and their pharmaceutical compounds and activities

Antifungal activity: The fungal strain *Aspergillus niger* EN-13 associated from this alga produced 5,7-dihydroxy-2-[1-(4-methoxy-6-oxo-6H-pyran-2-yl)-2-phenylethylamino]-[1,4]naphthoquinone which exhibited moderate antifungal activity. Further, the above fungal strain also produced the compound, asperamide A, which exhibited moderate antifungal activity against *Candida albicans* (Suryanarayanan, 2012).

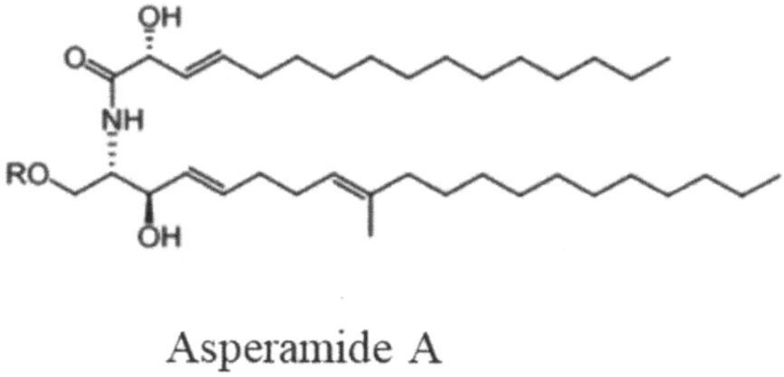

R=H

Asperamide A

Order: Sphacelariales Family Cladostephaceae

Cladostephus spongiosus *(= Cladostephus verticillatus.)*

Image credit: Augusta Foote Arnold, Public domain.

Common name (s): Wolliger Seequirl

Global distribution: it has worldwide distribution

Habitat: It is commonly seen in shallow subtidal and inertidal areas where it is found attached to stones and rocks.

Associated fungi; and their pharmaceutical compounds and activities

Antioxidant activity: The fungus *Acremonium* sp. associated from this alga has been reported to produce a hydroquinone compound viz. acremonin A which possessed significant antioxidant activity (Abdel-Lateff et al. 2002).

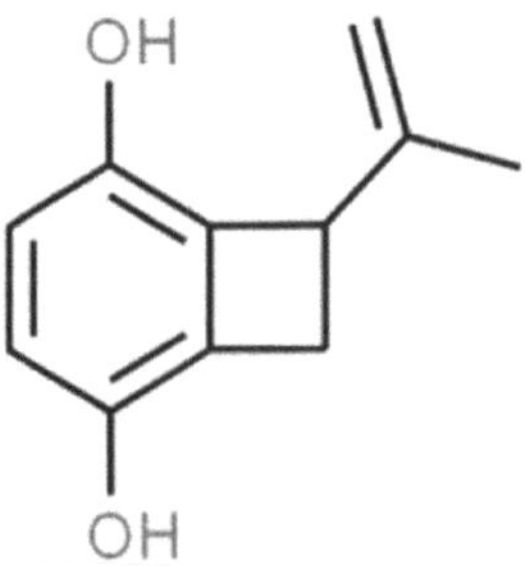

Acremonin A

Order: Sphacelariales Family: Stypocaulaceae

Halopteris scoparia *(= Stypocaulon scoparium)*

Image credit: Pierre-Louis Crouan, Public domain.

Common name (s): Sea broom

Global distribution: It is commonly distributed in parts of the Atlantic Ocean and the Mediterranean and.

Habitat: It is abundant in the subtidal zone and is found attached to rocks facing the open sea.

Associated fungi: Therapeutic compounds and their bioactivities

Antioxidant and antimicrobial activity: The extracts fungus *Aspergillus chevalieri* associated with this alga have been reported to display the highest antioxidant activity by scavenging DPPH free radicals with an EC50 value of 35.5 µg/mL. Further these extracts have also shown promising antifungal activity against *Cutibacterium acnes* (Calado et al. 2021).

Phylum (Division): Rhodophyta Class: Florideophyceae

Order: Ceramiales Family: Dasyaceae

Dasysiphonia japonica *(= Heterosiphonia japonica)*

Common name (s): Not reported

Global distribution: It is commonly distributed in Norway. Netherlands, Spain, and France

Habitat: This drift alga is seen attached to hard substrata and other algae in subtidal areas of depths of 1-20 m.

Associated fungi; and their pharmaceutical compounds and activities

Antibacterial activity: The fungus *Aspergillus oryzae* associated with this alga yielded indoloditerpene derivatives viz. asporyzin A-C, and related indoloditerpenes viz. JBIR-03, emindole SB and emeniveol. Among these compounds, asporyzin C has been reported to possess antibacterial activity against *E. coli* (Qiao et al. 2010).

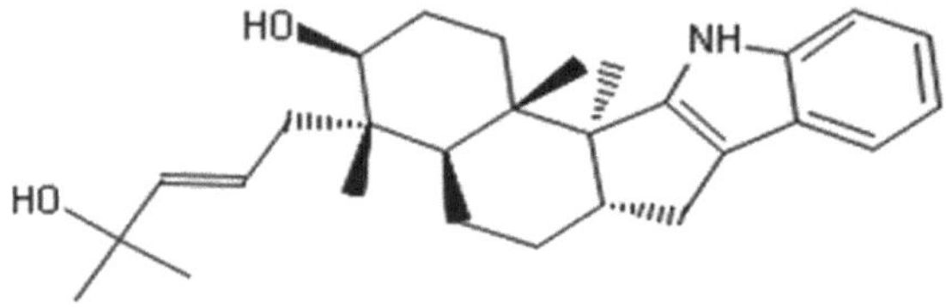

Asporyzin C

Order: Ceramiales Family: Rhodomelaceae

Vertebrata fucoides *(= Polysiphonia violacea)*

Image credit: Pierre-Louis Crouan, Public domain.

Common name (s): No common name

Global distribution: It is commonly distributed around Ireland, Isle of Man and Great Britain,

Habitat: It occurs in pools in the littoral zone and on open rock

Associated fungi: and their pharmaceutical compounds and activities

Cytotoxicity: The fungus *Apiospora montagnei* associated with this alga yielded compounds viz. polyketide apiosporic acid, diterpene myrocin, monomethyl ester 9-hydroxyhexylitaconic acid, enantiomer, (+)-hexylitaconic acid, and (+)-epiepoxydon. Among these compounds, (+)- epiepoxydon has been reported to display potent cytotoxicity on a number of human cancer cells (Klemke et al. 2004).

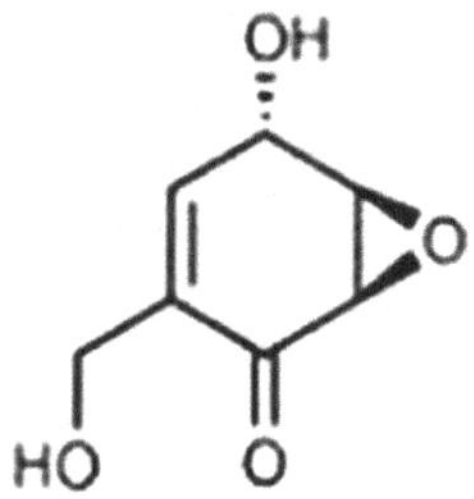

(+)- epiepoxyodon

Polysiphonia stricta *(= Polysiphonia urceolata)*

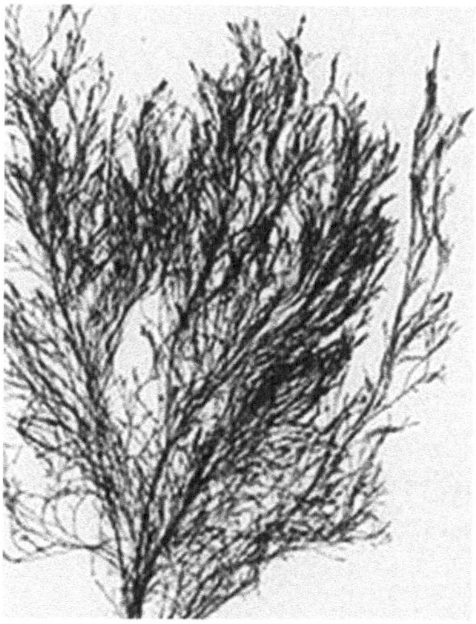

Image credit: Pierre-Louis Crouan, Public domain.

Common name (s): No common name

Global distribution: It is seen around West Atlantic, American Atlantic and British Isles

Habitat: This epiphytic and epizoic species is seen in the low littoral areas down to to a depth of 20 m.

Associated fungi; and their pharmaceutical compounds and activities

Antioxidant and anti-tumor activities: The fungus *Chaetomium globosum* which is associated with this alga yielded benzaldehyde derivatives *viz.* 2-(2',3-epoxy-1',3'-heptadienyl)-6-hydroxy-5-(3-methyl-2-butenyl) benzaldehyde, chaetopyranin, isotetrahydroauroglaucin and, erythroglaucin. All these compounds have been reported to show moderate DPPH (1,1- diphenyl-2-picrylhydrazyl) radical-scavenging activity with IC50 values of 88, 35, 26, and 62 μg/mL, respectively. Chaetopyranin also exhibited weak or moderate cytotoxicity against three tumor cell lines viz. hepatocellular carcinoma cells, human microvascular endothelial cells and human lung epithelial cells with IC50 values of 28.5, 15.4, and 39.1 μg/mL, respectively (Wang et al. 2006).

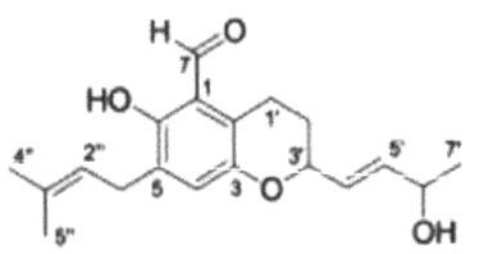

Chaetopyranin

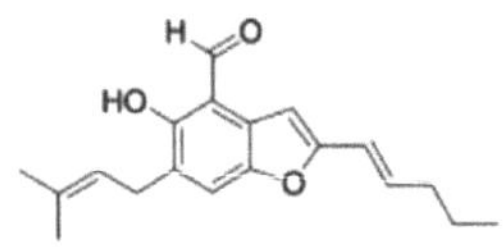

2-(2',3-epoxy-1',3'-heptadienyl)-6-hydroxy-5-(3-methyl-2-butenyl)benzaldehyde

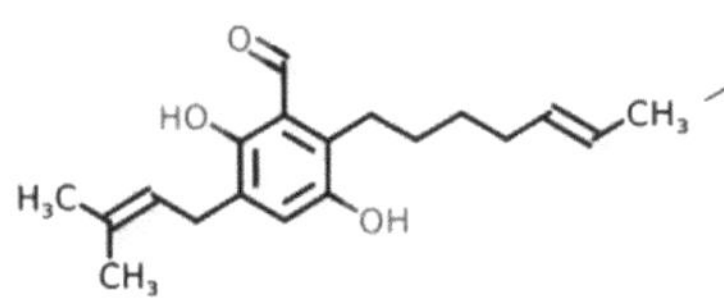

Isotetrahydroauroglaucin

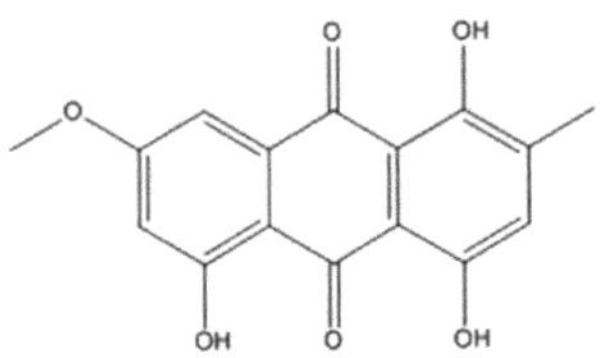

Erythroglaucin

Polysiphonia sp.

Associated fungi;; and their pharmaceutical compounds and activities

Antimicrobial activity: The fungus, *Geniculosporium* sp. associated with this alga has been reported to produce 11 tricyclic sesquiterpenes which showed antimicrobial activity against pathogenic bacteria and fungi (Krohn et al., 2005; Suryanarayanan, 2012).

Laurencia similis

Common name (s): Not designated

Global distribution: It is a new species form Queensland, Australia

Habitat: Not reported

Associated fungi; and their pharmaceutical compounds and activities

The fungus, *Exophiala oligosperma* (EN-21) associated with this alga has been reported to produce compounds *viz.* 2-phenoxynaphthalene, (2S, 3R, 4E, 8E)-1-O-β-D-glucopyranosyl-3-hydroxy-2-[(R)-2′-hydroxyoctadecanoyl] amino-9-methyl-4, 8-octadeca-diene,, (22E,24R)-ergosta-7,22-dien-3β,5α,6β-triol, (22E, 24R)-3β, 5α, 9α-trihydroxy-ergosta-7, 22-dien-6-one,, (22E, 24R)-5α, 6α-epoxy-ergosta-8, 22-dien-3β, 7α-diol, (22E, 24R)-ergosta-4, 6, 8(14), 22-tetraen-3-one, and euphorbol the activities of which are yet to be known (Li et al. 2011).

***Laurencia* sp.**

Image credit: Pierre-Louis Crouan, Public domain.

Associated fungi and their pharmaceutical compounds and activities

Anticancer activity: The endophytic fungal strain, *Penicillium chrysogenum* QEN-24S derived from this unidentified algal species yielded polyketide derivatives *viz.* penicitides A and B; glycerol derivatives *viz.* 2-(2,4-dihydroxy-6-methylbenzoyl)-glycerol and 1-(2,4-dihydroxy-6-methylbenzoyl)- glycerol; and a monoterpene derivative, penicimonoterpene. Of these compounds, penicitide A displayed moderate cytotoxic activity against human hepatocellular liver carcinoma cell line (Gao et al. 2011).

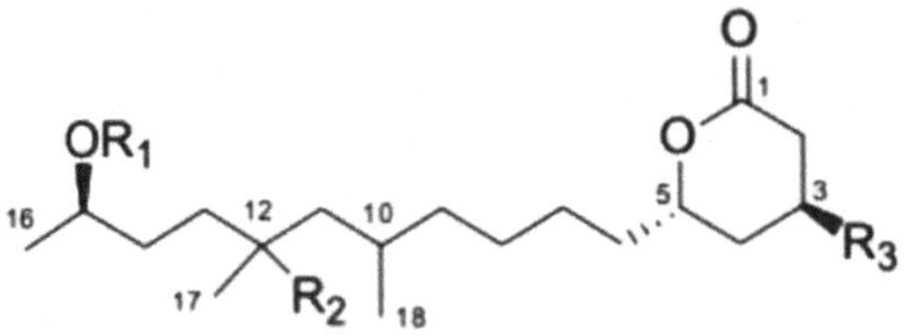

R1=R2=H, R3=OH

Penicitide A

Order: Gigartinales Family: Dumontiaceae

Hyalosiphonia caespitosa

***Epipolasis* sp.**

Image credit: Sponges of Plolynesia, https://creativecommons.org/licenses/by/4.0/

Common name (s): Not reported

Global distribution: Japan Sea, Korea, China and Pacific coast of Honshu

Habitat: It grows on rocks slightly below the low tide level

Associated fungi and their pharmaceutical compounds and activities

Anti-tyrosinase activity: The fungal strain *Botrytis* sp.(Family: Sclerotiniaceae) associated with this alga has been reported to produce pyrone derivatives viz. 6- [(E)-hept-1-enyl]–pyrone and 6-[(E)-pent-1-enyl]–pyrone. Of the above compounds, 6- [(E)-hept-1-enyl]--pyrone exhibited anti-tyrosinase effects for which 4.5 M was recorded as IC50 value (Zhang et al. 2007).

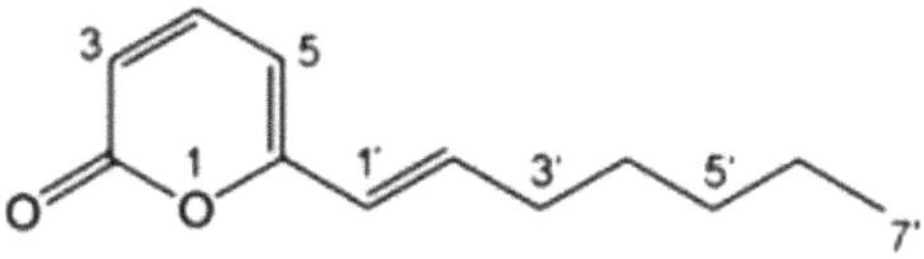

6- [(E)-hept-1-enyl]-α-pyrone

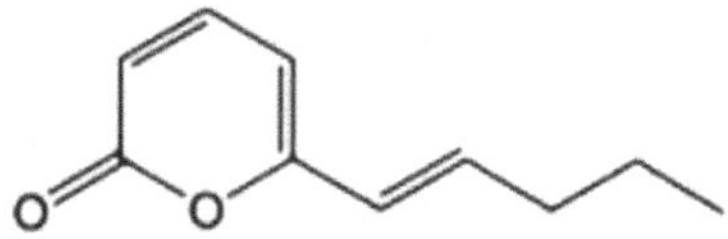

6-[(E)-pent-1-enyl]-α-pyrone

Order: Gigartinales Family: Polyidaceae

Polyides rotundus

Image credit: Gabriele Kothe-Heinrich, Public domain.

Common name (s): Discoid fork weed

Global distribution: Europe and North America

Habitat: It is an epilithic and salt-tolerant species and is commonly seen attached to rock and stones in the upper sublittoral and littoral areas

Associated fungi; and their pharmaceutical compounds and activities

Antimicrobial activity: The fungus, *Pseudoanguillospora* sp. associated with this alga has been reported to yield isochromans *viz.* pseudoanguillosporin A, B, and cephalochromin which have shown broad antimicrobial activities (Kock et al. 2009).

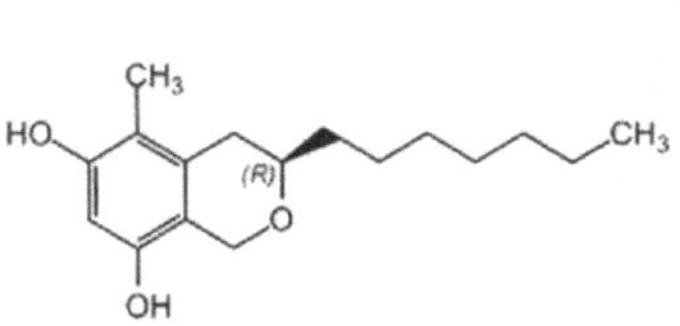

Pseudoanguillosporin A

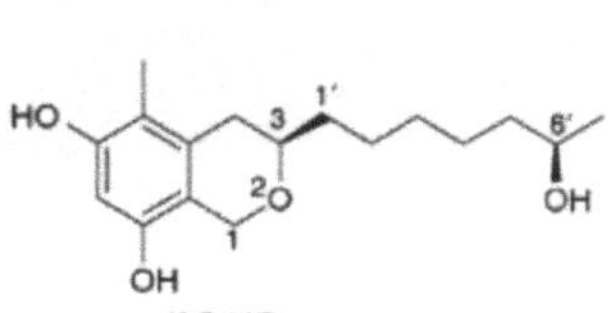

Pseudoanguillosporin B

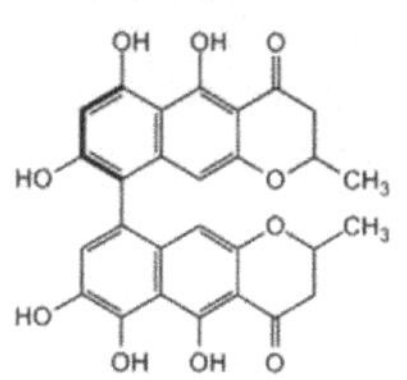

Cephalochromin

Order: Gracilariales Family: Gracilariaceae

Gracilaria folifera

Common name (s): Graceful red weed

Global distribution: Atlantic Ocean, the Mediterranean, and Western Indian Ocean

Habitat: It is a sessile species

Associated fungi; and their pharmaceutical compounds and activities

Antibacterial activity: The fungus *Curvularia* sp. associated with this algal species produced the macrolide, curvulone A which exhibited antibacterial activity against the pathogenic bacterium *Bacillus subtilis* (Zhang et al. 2021).

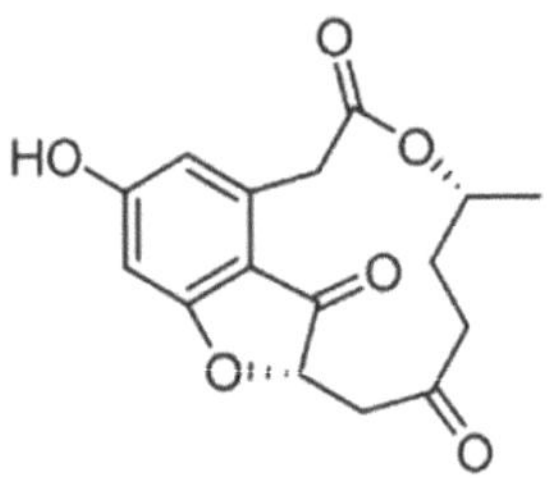

Curvulone A

Order: Nemaliales Family: Galaxauraceae

Actinotrichia fragilis

Image credit: Philippe Bourjon, https://creativecommons.org/licenses/by/4.0/

Common name (s): Spikeweed

Global distribution: Western Central Pacific Ocean: Palau, Guam and Philippines.

Habitat: This epiphytic species is seen at depths of 0–50 m.

Associated fungi; and their pharmaceutical compounds and activities

Anticancer activity: The fungus, *Penicillium citrinum*, associated with this alga yielded a pentacyclic alkaloid, citrinadin A which has been reported to display anticancer activity (Suryanarayanan,2012)

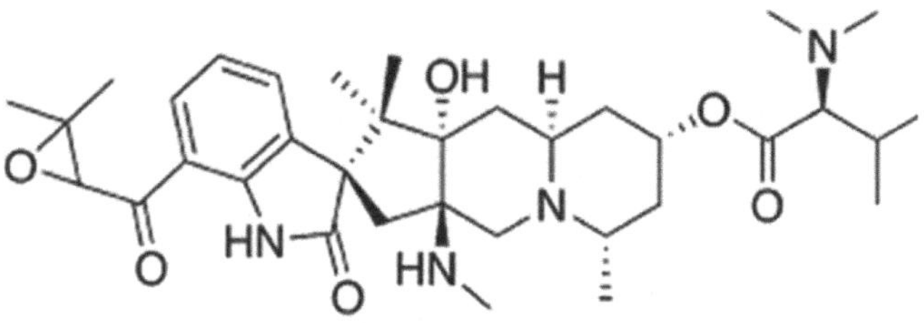

Citrinadin A

Order: Nemaliales Family: Liagoraceae

Liagora viscida

Image credit: Diego Delso, https://creativecommons.org/licenses/by-sa/4.0/

Common name (s): Not designated

Global distribution: It is a common species in NW spain

Habitat: It is often found attached to reefs or on small boulders in rockpools at depths of 0-20m

Associated fungi and their pharmaceutical compounds and activities

Antiplasmodial activity: The fungus *Drechslera dematioidea* associated with this alga has been reported to produce 10 sesquiterpenoids *viz.* isosativenetriol, drechslerines A and B, 9-hydroxyhelminthosporol, and drechslerines C–G. Among these compounds, drechslerines E,, F and G have shown antiplasmodial activity by inhibiting the growth of *Plasmodium falciparum* (Osterhage et al., 2002).

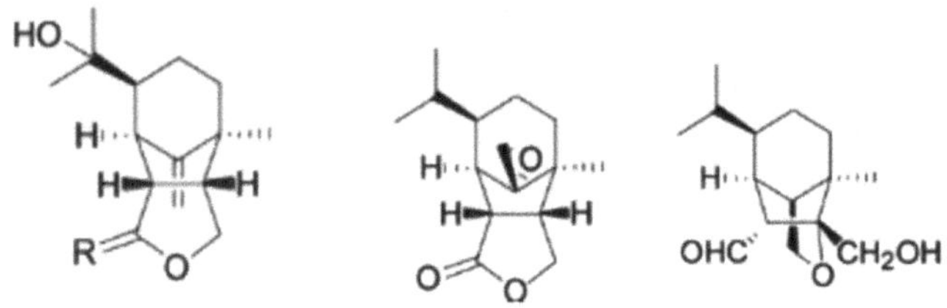

R= α-H, β-OCH3

Drechslerine E Drechslerine F, G

Order: Palmariales Family: Palmariaceae

Palmaria decipiens

Common name (s): Austral dulse

Global distribution: This polar seas have been reported from the parts of the Southern Ocean *viz.* Antarctica and sub-Antarctic islands.

Habitat: It is a sessile species

Associated fungi; and their pharmaceutical compounds and activities

Trypanocidal activity: The extracts of the fungal strain, *Penicillium* sp. UFMGCB 6034 associated with this alga have shown trypanocidal properties on the trypomastigotes of *Trypanosoma cruzi* with 100% inhibition and an IC50 value of 1.28 gm L^{-1} (Godinho et al. 2013).

Unidentified marine alga

Associated fungi and their pharmaceutical compounds and activities

UV-A absorbing activity: The fungus *Aspergillus terreus* associated with this unidentified marine alga has been reported to yield the chiral dipyrrolobenzoquinone derivative viz. terreusinone which displayed UV-A absorbing activity with an ED50 value of 70 µg/mL. As this compound could serve as a promising UV-A protectant, it may have wide applications in cosmetic industries (Lee et al. 2004; Suryanarayanan, 2012).

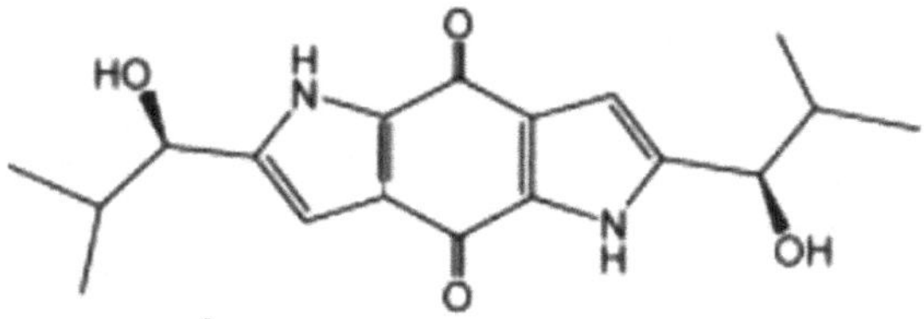

Terreusinone

SEAGRASSES

Kingdom: Plantae Phylum (Division)): Tracheophyta Class: Magnoliopsida
Order: Alismatales Family: Cymodoceaceae

Cymodocea serrulate

Associated fungi; and their pharmaceutical compounds and activities

Antimicrobial activity: The extracts of the fungus, *Hypocreales* sp. associated with this plant showed antimicrobial activity against *Candida albicans* ATCC 90028, *C. albicans* NCPF 3153, *Cryptococcus neoformans* ATCC 90112, *C. neoformans* ATCC 90113 and *Microsporum gypseum* (clinical isolate) with 8, 8, 16, 8, and 2 µg/ml respectively as MIC values (Supaphon et al. 2013).

Syringodium isoetifolium *(Ascherson) Dandy, 1939*

Order: Alismatales Family: Cymodoceaceae

Associated fungi: Therapeutic compounds and their bioactivities

Antibacterial activity: The fungal species associated with this plant species include *Aspergillus* sp., *Paecilomyces* sp., *Penicillium* sp., and *Sclerotinia* sp. Among these fungal species, the crude ethyl acetate of *Sclerotinia* sp. showed antibacterial activity against *Pseudomonas aeruginosa* and *Staphylococcus aureus* with inhibition zone values of 9 and 9 mm respectively (Notarte et al., 2018).

Order: Alismatales Family: Hydrocharitaceae

Enhalus acoroides

Associated fungi and their pharmaceutical compounds and activities

Antibacterial activity: The crude ethyl acetate of the associated fungal species of this plant showed antibacterial activity. *Aspergillus* sp. showed activity against *Pseudomonas aeruginosa* and *Staphylococcus aureus* with inhibition zone values of 9 and 7 mm, respectively; *Aspergillus tubingensis* showed activity with 17 and 19 mm; and *Mycelia sterilia* against *Pseudomonas aeruginosa* and *Staphylococcus aureus* with 9 and 7 mm respectively (Notarte et al. 2018).

Trypanocidal and cytotoxic activities: A cyclic pentapeptide compound viz. malformin A was derived from the fungus *Aspergillus tubingensis* IFM 63452 associated with the leaves of this plant. Malformin A1 displayed trypanocidal effects on *Trypanosoma congolense* with an IC50: value of 15.08 ng/mL. Further, this compound also exhibited cytotoxic activity on the cell lne HeLa with an IC50: value of 50.15 ng/mL; and on P388 cell line with an IC50 value of 70.38 ng/mL (Notarte et al. 2017).

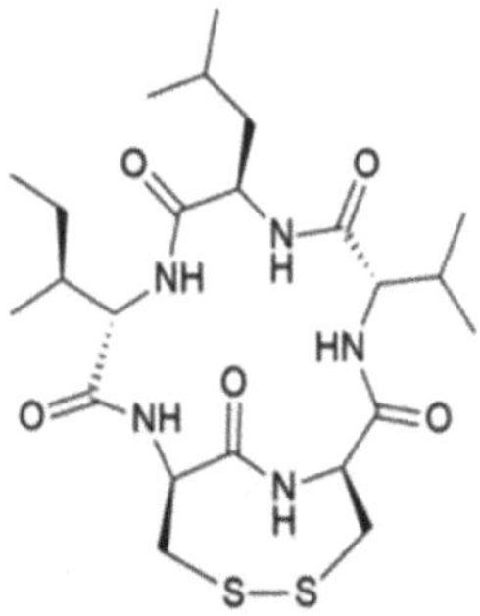

Malformin A1

Halophila ovalis

Common name (s): Spoon seagrass

Global distribution: Throughout the tropical Indo-West Pacific region and in some parts of temperate Australia

Habitat: Reef sand estuaries; depth range 2–5 m

Associated fungi and their pharmaceutical compounds and activities

Antimicrobial activity: The crude extract of the fungal species, *Trichoderma* sp 1 associated with this plant showed antimicrobial activity against

Staphylococcus aureus ATCC 25923, *Candida albicans* ATCC 90028, and *C. albicans* NCPF 3153, with MIC values of 64, 128 and 2 μg/ml respectively. Further, its *Trichoderma* sp 2 showed antimicrobial activity against *Staphylococcus aureus* ATCC 25923, and *Candida albicans* NCPF 3153 with MIC values of 64 and 1 μg/ml, respectively (Supaphon et al. 2013).

Thalassia hemprichii

Associated fungi and their pharmaceutical compounds and activities

Antimicrobial activity: The crude extract of the fungal species *Penicillium* sp. which is associated with this plant species displayed antifungal activity against *Cryptococcus neoformans* ATCC 90112 and *C. neoformans* ATCC 90113 with MIC values of 8 and 32 μg/ml respectively. Its *Fusarium* sp. showed antimicrobial activity against *Staphylococcus aureus* ATCC 25923, MRSA and *Cryptococcus neoformans* ATCC 90113 with MIC values of 8,8 and 4 μg/ml respectively; its *Stephanonectria* sp. 1 showed antibacterial activity against *Staphylococcus aureus* ATCC 25923 with an MIC value of 8 μg/ml; its *Stephanonectria* sp. 2 showed antibacterial activity against *Staphylococcus aureus* ATCC 25923 with an MIC value of 8 μg/ml; and its sterile fungi (mycelia sterilia) showed antimicrobial activity against *Cryptococcus neoformans* ATCC 90112 and *C. neoformans* ATCC 90113 with MIC values of 4 and 32 μg/ml, respectively (Supaphon et al. 2013).

Cytotoxicity: The metabolites, polyporapyranones A–H were isolated from two olyporales fungal species (division Basidiomycota) associated with this plant. Among these compounds, polyporapyranones A and D exhibited moderate and weak inhibition of the vero cell line (derived from the kidney of an African green monkey), respectively (Blunt et al. 2015).

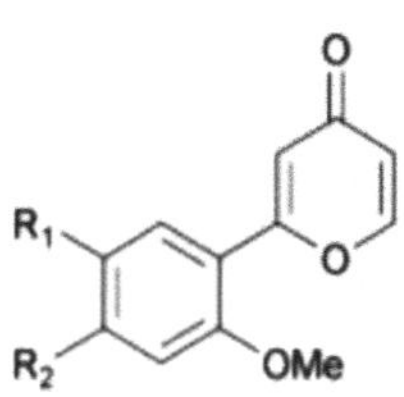

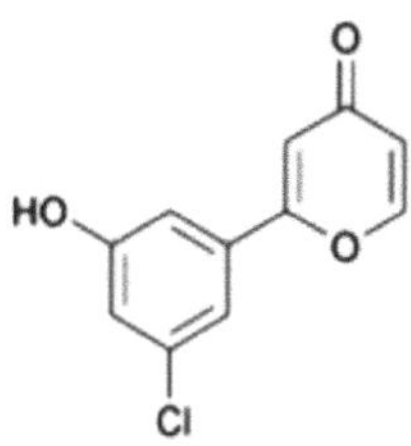

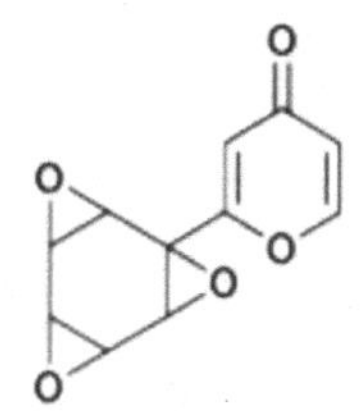

R1=Ome, R2=H; R2= H, R2= Ome; R1= R2=Ome

Polyporapyranones A-C Polyporapyranone D Polyporapyranone E

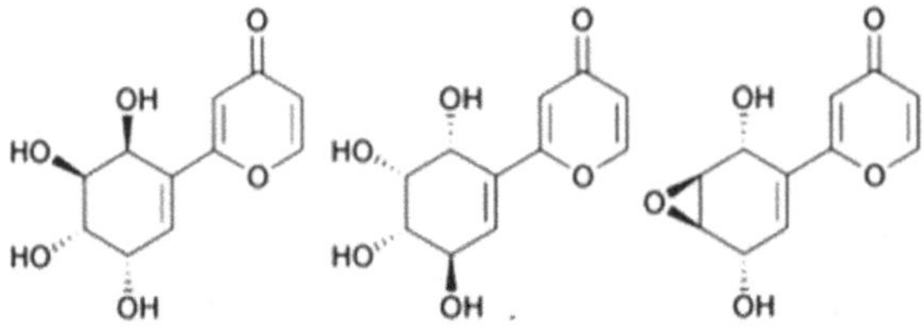

Polyporapyranones F-H

Others: Maharaj (2015) reported on the isolation of β-resorcyclic macrolide 5′hydroxyzearalenone from the fungus *Fusarium* sp. associated with this plant species.

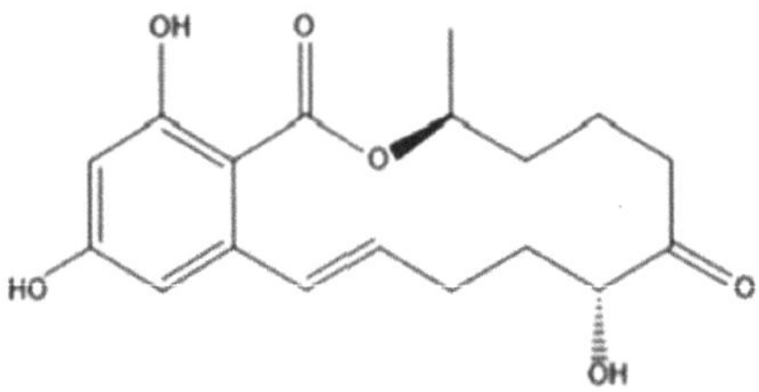

5′hydroxyzearalenone

Order: Alismatales Family: Zosteraceae

Zostera marina Linnaeus

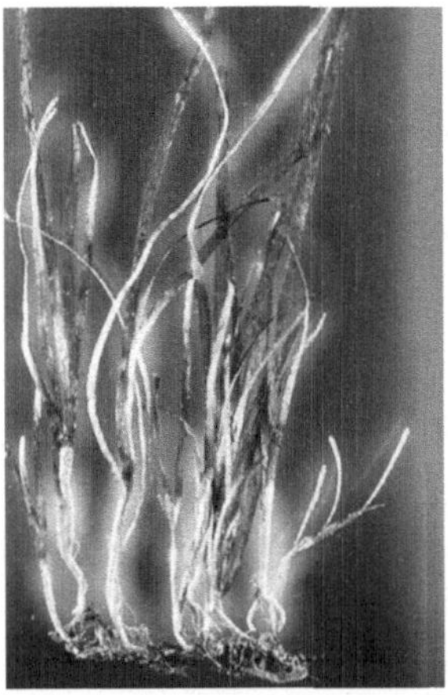

Image credit: https://creativecommons.org/publicdomain/zero/1.0/

Common name (s): Eel grass

Global distribution: It is widespread through Atlantic and Pacific; and is the only seagrass species which is found extending into the Arctic Circle.

Habitat: It is living mainly on sand to fine gravel bottom of the subtidal zone at depths of 0-4m It may thrive in a variety of habitats like estuaries, lagoons and bays.

Associated fungi; and their pharmaceutical compounds and activities

Immune function: The fungi *Penicillium glabrum* and *Penicillium implicatum* associated with this plant yielded the compounds sulochrin and 4-Methoxy-3-methylgoniothalamin. Among these compounds sulochrin has been reported to have an inhibitory activity to eosinophil degranulation which is the principal pathogenic condition in chronic eosinophilic diseases (Ohashi et al. 1999; Nicoletti and Trincone 2016).

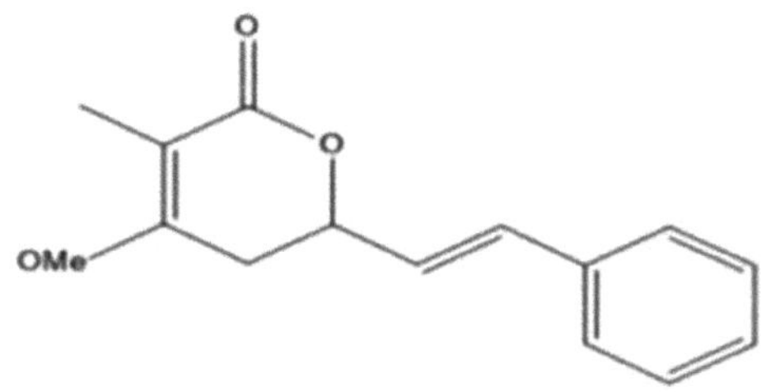

4-Methoxy-3-methylgoniothalamin

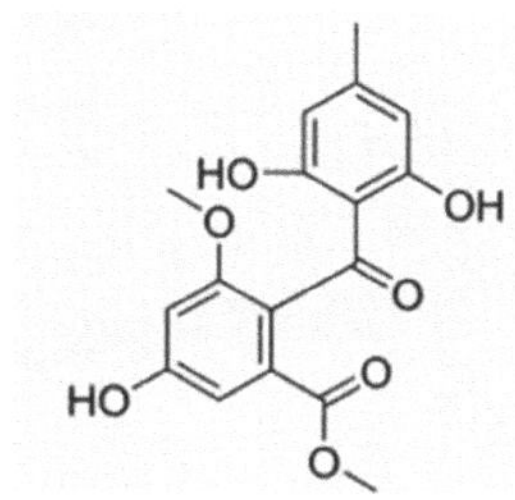

Sulochrin

Antimicrobial activities: The phthalide derivative 3-[2®-hydroxybutyl]-7-hydroxyphthalide (2) and other compounds (–)-3-butyl-7-hydroxyphthalide (1), isopatulin (3), m-hydroxybenzyl alcohol (4), cyclopenin (5), and cyclopeptine (6) were isolated from the mycelia fungus *Penicillium claviforme* associated with this seagrass species. These compounds have shown antimicrobial activities (Nalini and Poonam 2012; Afiyatullov et al. 2015; Kinghorn et al. 2017).

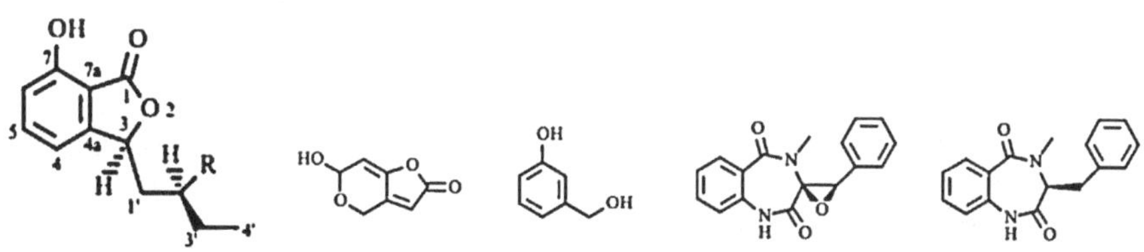

1: R=H; 2: R=OH 3 4 5 6

1,2

1: (–)-3-butyl-7-hydroxyphthalide; 2: 3-[2(R)-hydroxybutyl]-7-hydroxyphthalide; 3: isopatulin;4: m-hydroxybenzyl alcohol; 5: cyclopenin; 6: cyclopeptine

Antibacterial activity: The fungi *Penicillium antarcticum*, *Penicillium atrovenetum*, *Penicillium atramentosum*, *Trichoderma harzianum*, *Phoma macrostoma*, and *Cladosporium langeronii*, have been found to be associated with this plant species. Ethyl acetate extracts of all these fungal strains displayed antibacterial activity against the Gram-negative bacterium (Petersen et al. 2019).

Anticancer activity: The polyketides, zosteropenillines A–L (1–12) and pallidopenilline A have been isolated from the ethylacetate extract of the fungus *Penicillium thomii* associated with this plant species. Among these compounds, compounds 2, 8, and 10 induced a moderate downregulation of NO production in LPS- stimulated macrophages at non-cytotoxic concentration of 10.0 μM. Further, the most effective compound was zosteropenilline B which exhibited the maximal pronounced inhibition of NO formation in lipopolysaccharide-stimulated RAW 264.7 cells (Afiyatullov et al. 2017).

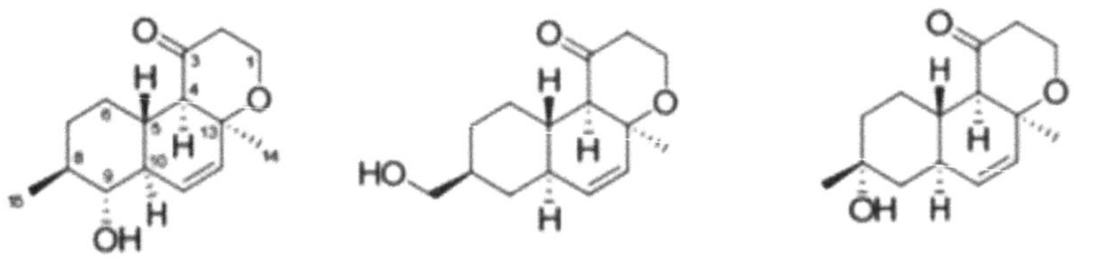

Zosteropenillines A-C

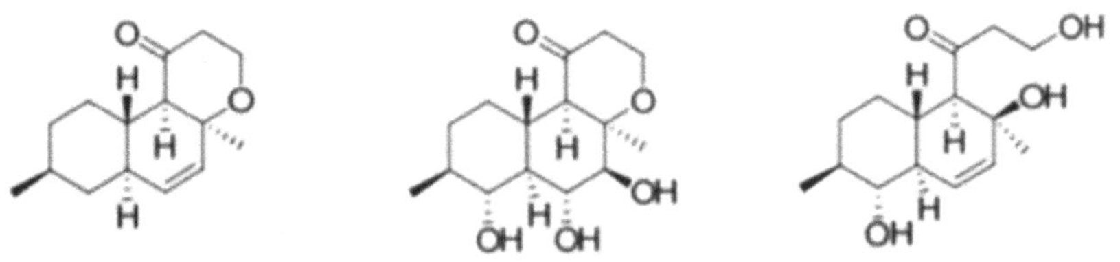

Zosteropenillines D-F

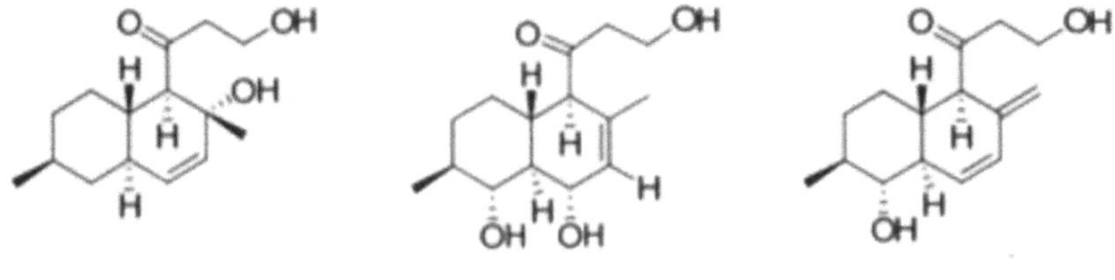

Zosteropenillines G-I

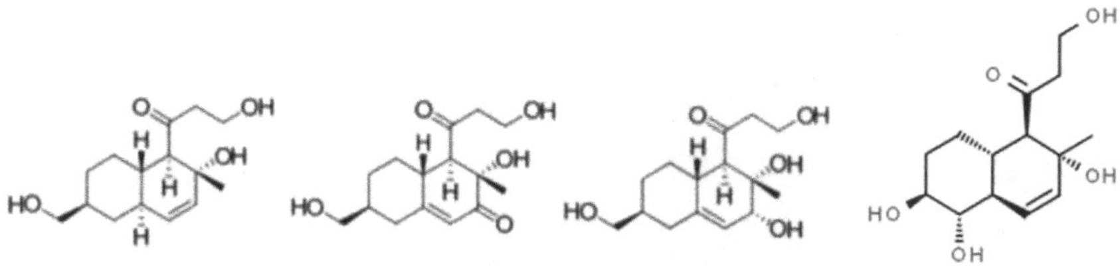

Zosteropenillines J-L Pallidopenilline A

Unidentified seagrass

Associated fungi: Therapeutic compounds and their bioactivities

Antitrypanosomal activity: The fungus *Aspergillus tubingensis* IFM 63452 associated with an unidentified seagrass species produced a pentapeptide-compound, malformin A1 which exhibited significant antitrypanosomal activity against the protozoan parasite *Trypanosoma congolense* with an IC50 value of 0.015 μg/mL (Mayer et al. 2021).

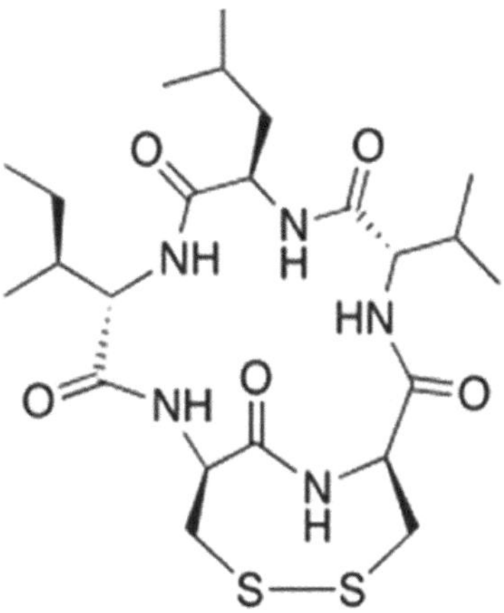

Malformin A1

MANGROVE PLANTS AND HALOPHYTES

(a) Mangrove plants

Phylum/Division: Tracheophyta Class: Magnoliopsida

Order:Lamiales Family: Acanthaceae

Acanthus ilicifolius

Image credit: Shagil Kannur. https://creativecommons.org/licenses/by-sa/4.0/

Common name (s): Holly-leaved acanthus, holy mangrove or sea holly

Global distribution: It is common in Australia, Australasia, and Southeast Asia

Habitat: It is seen in mangrove habitats.

Associated fungi and their pharmaceutical compounds and activities

Antidiabetic activity: The endophytic fungus *Lasiodiplodia theobromae* ZJ-HQ1 associated with this mangrove plant has been reported to produce a polyketide, lactone compound, lasiodiplactone which displayed α-glucosidase inhibitory activity with an IC50 value of 29.4 μM (Mayer et al. 2021).

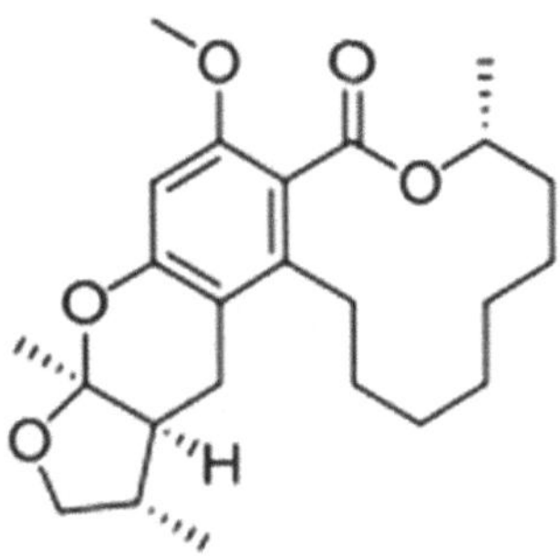

Lasiodiplactone

Order:Lamiales Family: Lamiaceae

Clerodendrum inerme

Image credit: Vengolis, https://creativecommons.org/licenses/by-sa/3.0/

Common name (s): Wild jasmine, seaside cleodendrum

Global distribution: Asia, Australia, Malesia and the Pacific islands

Habitat: This evergreen mangrove shrub occurs in high sand islands of coastal regions

Associated fungi and their pharmaceutical compounds and activities

Antidiabetic activity: The endophytic fungus *Trichoderma* sp. 307 associated with this mangrove plant has been reported to produce a polyketide compound, de-O-methyllasiodiplodin which displayed anti-diabetic activity by inhibiting α-glucosidase with an IC50 value of 25.8 μM (Mayer et al. 2021).

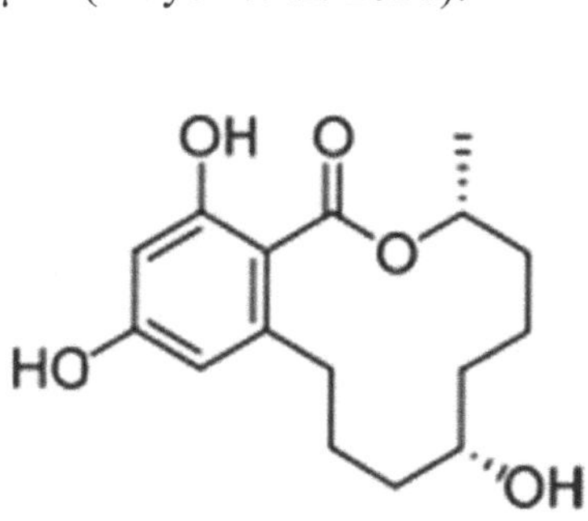

de-O-methyllasiodiplodin

Order: Malpighiales Family: Rhizophoraceae

Brguiera sexangula* var. *rhynchopetala

Image credit: Psumuseum. https://creativecommons.org/licenses/by-sa/3.0/

Common name (s): Mangrove plant

Global distribution: Indo-West Pacific; common in South China Sea

Habitat: It is a mangrove tree species tolerating fresh, brackish, and marine waters

Associated fungi and their pharmaceutical compounds and activities

Antibacterial activity: The fungus *Daldinia eschscholtzii* HJ001, (Hypoxylaceae) associated with this mangrove plant yielded a sorbicillin derivative *viz.* [11]- cytochalasa-5(6),13-diene-1,21-dione-7,18-dihydroxy-16,18- dimethyl 10-phenyl-(7S*,13E,16S*,18R*), which displayed weak antibacterial activity against MRSA ATCC with an MIC value of 80 μg/mL (Nweze et al. 2020).

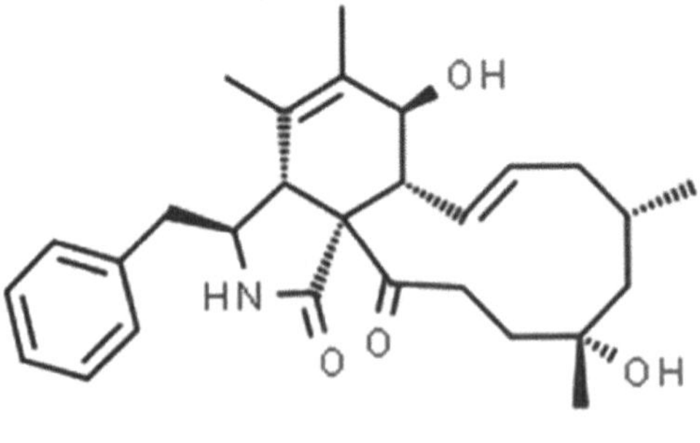

[11]-cytochalasa-5(6),13-diene-1,21-dione-7,18-dihydroxy-16,18- dimethyl-10-phenyl-(7S*,13E,16S*,18R*)

Kandelia obovata

Image credit: Itsmine, https://creativecommons.org/licenses/by-sa/3.0/

Common name (s): Not known

Global distribution: Eastern Asia: Vietnam, I Hong Kong, Taiwan, Indonesia, Southern China,, and Japan.

Habitat: This evergreen, small mangrove tree is seen on banks of r tidal rivers

Associated fungi and their pharmaceutical compounds and activities

Antidiabetic activity: The endophytic fungus *Talaromyces amesolkiae* associated with this mangrove plants produced a polyketide compound, sescandelin B which exhibited moderate antidiabetic activity by inhibiting α-glucosidase with an IC50 value of 17.2 μM (Mayer et al. 2021).

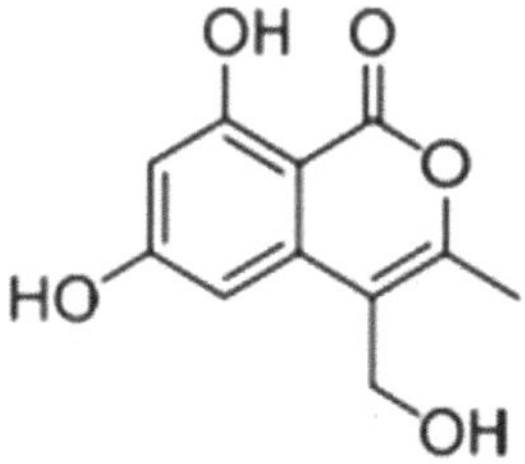

Sescandelin B

Order: Malvales Family: Malvaceae

Heritiera fomes

Common name (s): Sunder, sundri,

Global distribution: Indo-Pacific; it is the commonest mangrove tree species of the Sundarbans of Bangladesh and India,

Habitat: Coastal region

Associated fungi; and their pharmaceutical compounds and activities

Antibacterial activity: The ethyl acetate extract and methanol extract of the fungus, *Pestalotia* sp.(Amphisphaeriaceae) associated with this mangrove plant yielded the compounds xylitol and oxysporone, respectively. These extracts have been reported to exhibit antibacterial activity against strains of MRSA wi th the MIC values ranging from 32 to 128 μg/ml (Nweze et al. 2020).

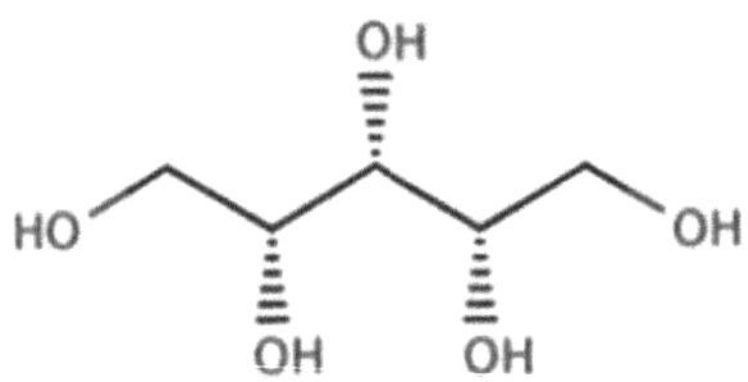

Xylitol

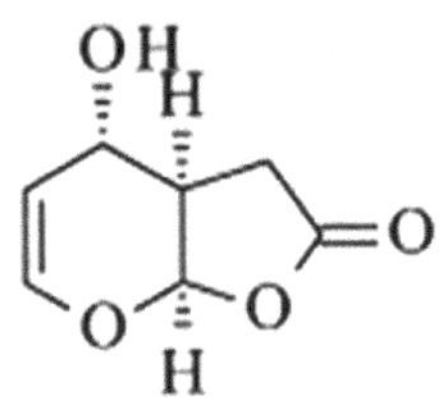

Oxysporone

Order: Myrtales Family: Lythraceae

Sonneratia alba

Image credit: Ton Rulkens. https://creativecommons.org/licenses/by-sa/2.0/

Common name (s): Not reported

Global distribution: Tropical and subtropical seas

Habitat: This mangrove tree is seen in tidal creeks and sheltered, sandy seashores

Associated fungi and their pharmaceutical compounds and activities

Immune system activity: The endophytic fungus *Penicillium* sp. ZJ-SY2 associated with this mangrove tree has been reported to produce a benzophenone polyketide compound, peniphenone which displayed lymphocyte immune suppression with IC50 values ranging from 8.1 to 9.3 μg/mL (Mayer et al. 2021).

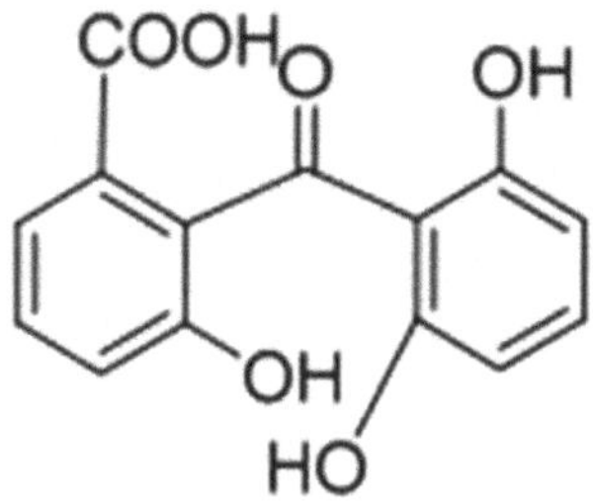

Peniphenone

Unidentified mangrove plant 1

The algae-like fungi *viz.* thraustochytrids (Class, Phycomycota; Family, Thraustochytriaceae) otherwise called as heterotrophic microalgae constitute a major component of the microbiota in mangrove ecosystems. The medicinal values of the thraustochytrids *viz. Schizochytrium* spp. which are associated with the mangrove plants is detailed below:

Antioxidant activity: The water extracts of *Schizochytrium* sp. has been reported to show the highest antioxidant activity with a maximum scavenging effect of 81.6% (Yu et al., 2017).

Park et al. (2018) reported that the strain, *Schizochytrium* sp. SHG104 showed significant antioxidant activity owing to its high levels of DHA (32.1%), palmitic acid and carotenoid-based antioxidant pigment viz. astaxanthin.

Luo et al. (2015) reported that the compounds campesterol, 24-Methylene cholesterol, 24-Methyl-colest-7-en-3β-ol, ergosterol, stigmasterol and other lipids isolated from *Schizochytrium aggregatum* displayed antioxidant activity and the IC50 value in DPPH radical scavenging study was 5.76 mg/mL.

Immunomodulatory effects: Due to the presence of several immunomodulatory compounds such as polysaccharides, nucleotides, peptides, flavonoids, β-glucans, β-carotene, *Schizochytrium* sp. has been reported to be safe for oral consumption and it has been used recently for the production of a highly immunoprotective influenza vaccine (Ramos-Vega et al. 2018).

Cholesterol-lowering activity: The compounds, lathosterol, ergosterol, stigmasterol, and 24-Ethylcholesta-5,7,22-trienol isolated from *Schizochytrium* sp. displayed cholesterol-lowering activity. Diet with this species at a concentration of 0.060 g/kg diet has been reported to decrease blood cholesterol by about 20% (Luo et al. 2015).

Antinociceptive and anticancer effects: Ramos-Vega et al. (2018) reported that its high DHA and EPA have produced antinociceptive and anticancer effects. As biofactory of antigens: Ramos-Vega et al.(2018) reported that *Schizochytrium* sp. has the potential for use as the biofactory of antigens and as a natural microcapsule (9–14 μm), which would be cheaper than the synthetic microparticles.

Unidentified mangrove plant 2

Associated fungi and their pharmaceutical compounds and activities

Nervous system activity: A mangrove plant–associated endophytic fungus *Penicillium* sp. SK5GW1L produced a meroterpenoid compound, terreulactone C which exhibited nervous system activity by inhibiting AChE with an IC50 value of 28 nM (Mayer et al. 2021).

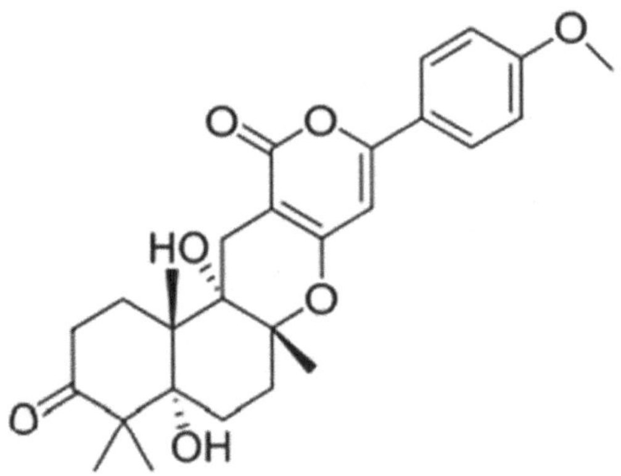

Terreulactone C

(b) Halophytes

Class: Magnoliopsida Order: Caryophyllales Family: Amaranthaceae

Salsola oppositifolia

Image credit: https://creativecommons.org/licenses/by-sa/3.0/

Common name (s): Pink saltbush

Global distribution: It is native to Asia, Europe, and Africa

Habitat: Grows in salty areas or in areas with alkaline salts

Associated fungi and their pharmaceutical compounds and activities

Antimicrobial activities: The endophytic fungi *Seimatosporium* sp. and *Microsphaeropsis* sp. which were associated with this plant species yielded a naphthalene derivative named seimatorone (1) and A polychlorinated triphenyl diether named microsphaerol (2). Compound 2 displayed potent antibacterial activity on *Escherichia coli* and *Bacillus megaterium* and antifungal activity on *Microbotryum violaceum* and *Chlorella fusca*. On the other hand, compound 1 exhibited moderate antibacterial and antifungal activities (Jin et al. 2016).

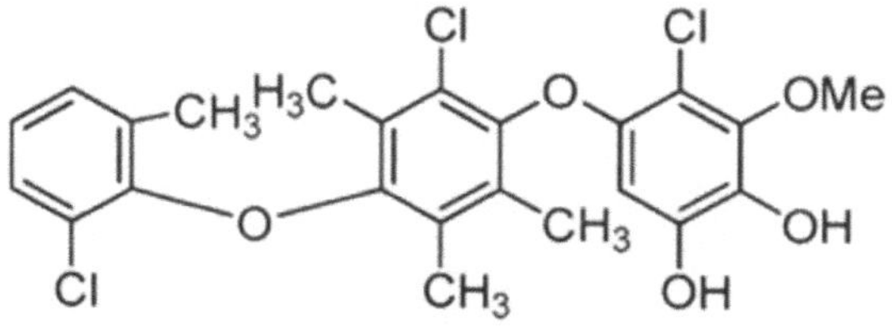

Microsphaerol

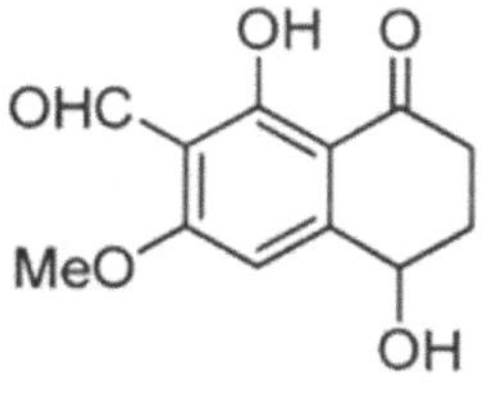

Seimatorone

FUNGI ASSOCIATED WITH MARINE INVERTEBRATE ANIMALS

SPONGES

Phylum: Porifera Class: Demospongiae

Order: Agelasida Family: Agelasidae

Agelas sp.

Image credit: Esculapio, https://creativecommons.org/licenses/by/3.0/

Associated fungi: Therapeutic compounds and their bioactivities

Antioxidant activity: The supernatant extracts of the unidentified fungus FS1 associated with this sponge showed the highest free radical scavenging activity (FRSA) against DPPH radical (El-Hady et al., 2014a).

Immunological activity: Holler (1999) reported on the isolation of *Microspaeropsis olivacea* from this sponge and this produced cerebroside and 10-Methyl-(9Z)- octadecenoic acid, 1-O-(10-methyl- (9Z)-octadecenoyl)-glycerol. Among these compounds, the former showed immunological activity.

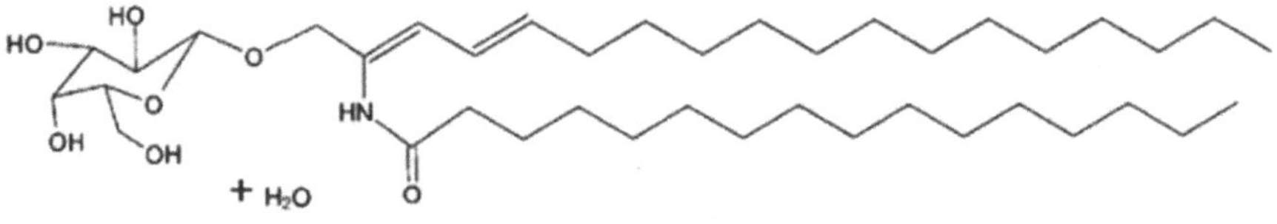

Cerebroside

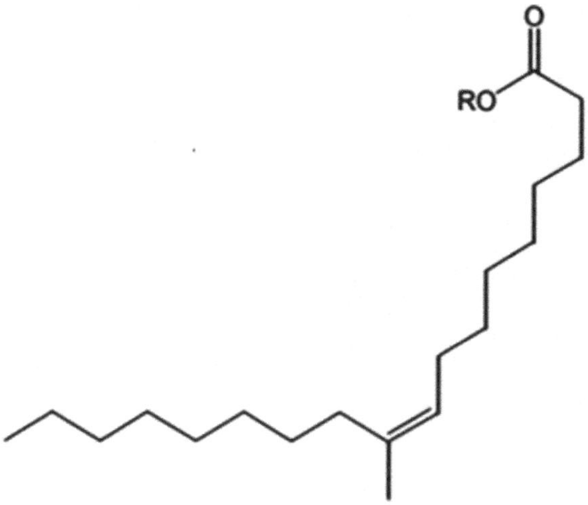

10-Methyl-(9Z)- octadecenoic acid, 1-O-(10-methyl- (9Z)-octadecenoyl)-glycerol

Order: Axinellida Family: Axinellidae

Axinella corrugata

Common name (s): Tropical sponge

Global distribution: Western Central Atlantic: Caribbean and USA.

Habitat: Sessile; depth range 5 - 71 m

Associated fungi and their pharmaceutical compounds and activities

Antimicrobial and cytotoxic activities: The fungus *Penicilium raistrikii* isolated from this species produced norliquexanthone (1,3,6- trihydroxy-8-methyl-9H-xanthen-9-one) which served as antimicrobial and/or cytotoxic agents (Kossuga et al. 2012).

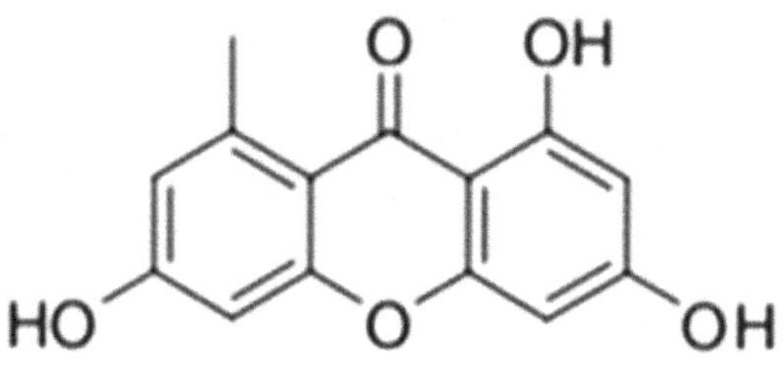

Norliquexanthone

Axinella damicornis

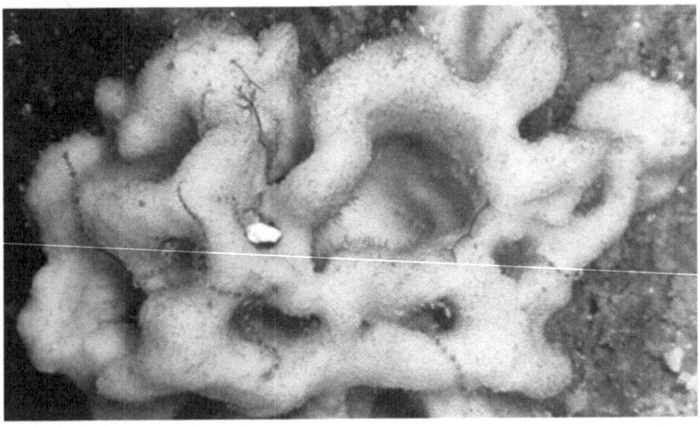

Image credit: Matthieu Sontag, https://creativecommons.org/licenses/by-sa/3.0/

Common name (s): Mediterranean sponge

Global distribution: Mediterranean and Northeast Atlantic

Habitat: Sessile; depth,15 m; often found partly hidden in hydroid/bryozoan

Associated fungi and their pharmaceutical compounds and activities

Cytotoxic and neuroprotective effects: The compounds 3,3-bicoumarin (bicoumanigrin A) (1)and aspernigrin B (2) were obtained from the fungus *Aspergillus niger* associated with this this sponge species. While the first compound exhibited moderate cytotoxic activity on human cancer cells, the 2nd compound showed a potent neuroprotective activity (Thomas et al. 2010).

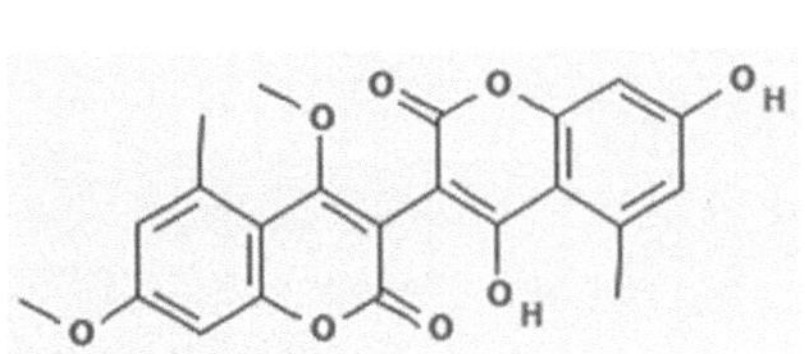

3,3′-bicoumarin

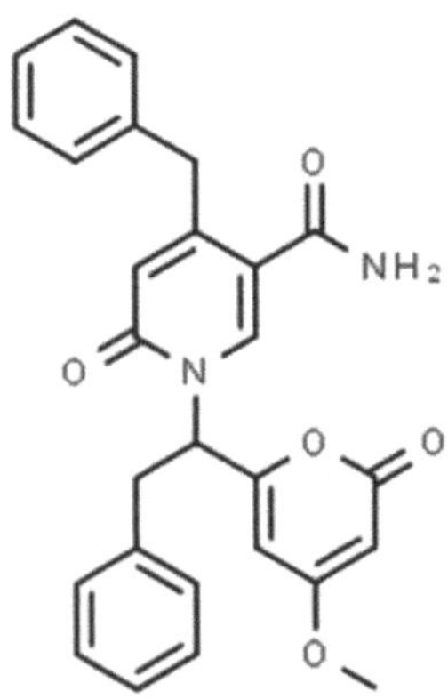

Aspernigrin B

Axinella verrucosa

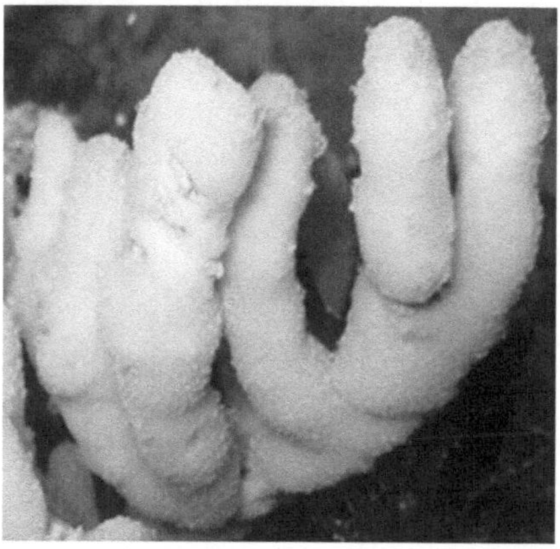

Common name (s): Not reported

Global distribution: Mediterranean.

Habitat: Sessile on cliffs and on pebble bottom, 10-74 m

Associated fungi and their pharmaceutical compounds and activities

Antiproliferative and antifungal activities: Ethyl acetate extract of *Penicillium* sp. associated with this sponge species produced compounds such as communesin B-D, dechlorogriseofulvi, oxaline, and griseofulvin. Among these compounds, the communesins and oxaline showed noderate antiproliferative efffects and griseofulvin displayed antifungal activtiy (Thomas et al. 2010).

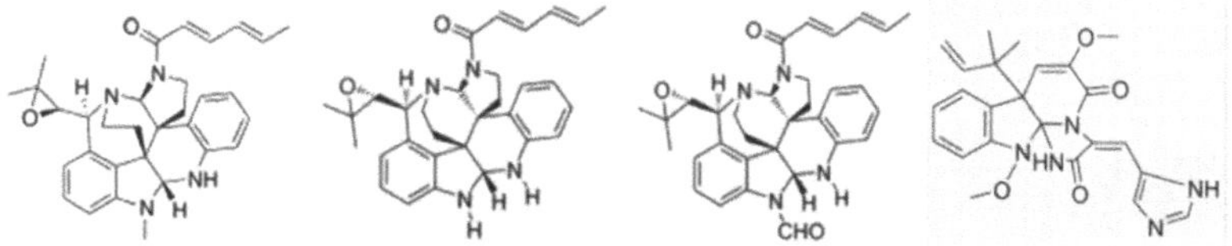

Communesin B -D Oxaline

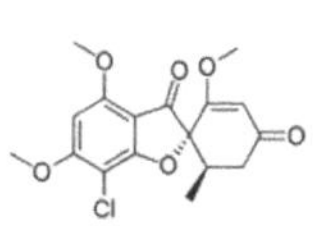

Griseofulvin

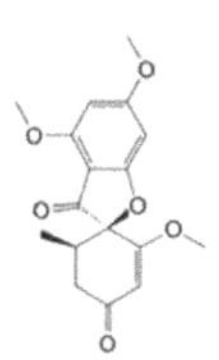

Dechlorogriseofulvin

Antibacterial activity: Its associated fungus, *Talaromyces* sp. strain LF458 has been reported to produce the compounds, talaromycesone A, B which showed antibacterial activity against MRSA with an IC50 value of 4.6 μM (Nweze et al. 2020).

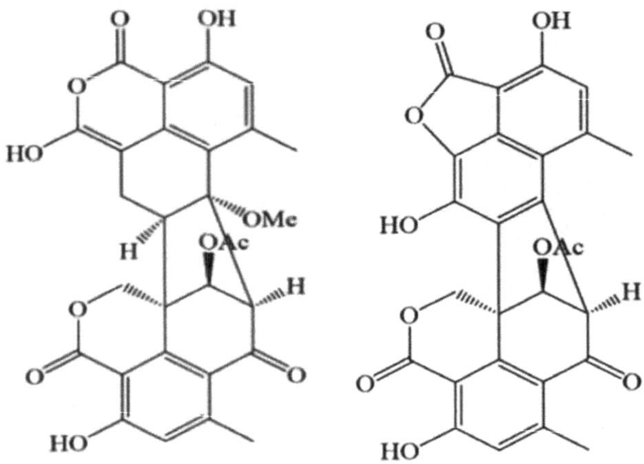

Talaromycesone A, B

Axinella sp.

Associated fungi and their pharmaceutical compounds and activities

Antifungal activity: *Myrothecium* sp. JS9 which is a symbiotic fungal strain isolated from this unidentified species produced compounds viz.roridins A and D (macrocyclic trichothecenes) which displayed antifungal activity (Thomas et al. 2010).

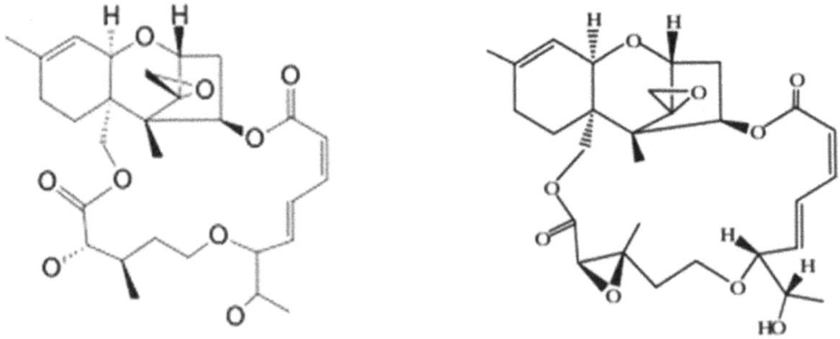

Roridin A, D

Antibacterial activity: Fungus *Penicillium citrinum*, harboured by this sponge produced steroids, isocyclocitrinol A and 22-acetylisocyclocitrinol A which displayed modest antibacterial activity against *Staphylococcus epidermidis* and *Enterococcus durans* with an MIC value of 100µg/mL (Thomas et al., 2010).

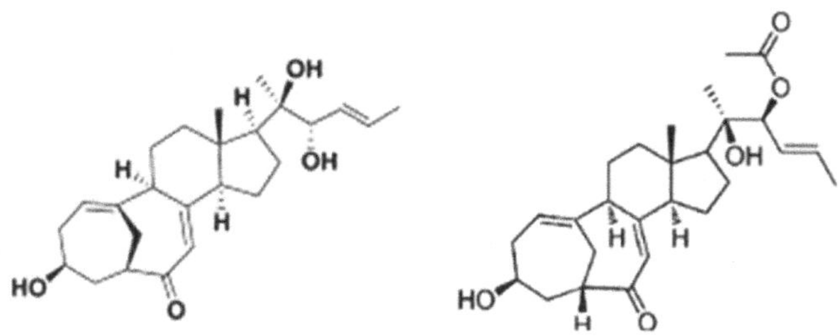

Isocyclocitrinol A 22-acetylisocyclocitrinol A

Panizel et al. (2013) reported that the compound peptaibol produced by the associated fungus *Trichoderma atrovorode* of this sponge displayed antibacterial activity on *Shewanella* sp. (III.07), *Microbacterium* sp. (PII.14), *Sporosarcina* sp. (NB90), and *Bacillus* sp. (NB36).

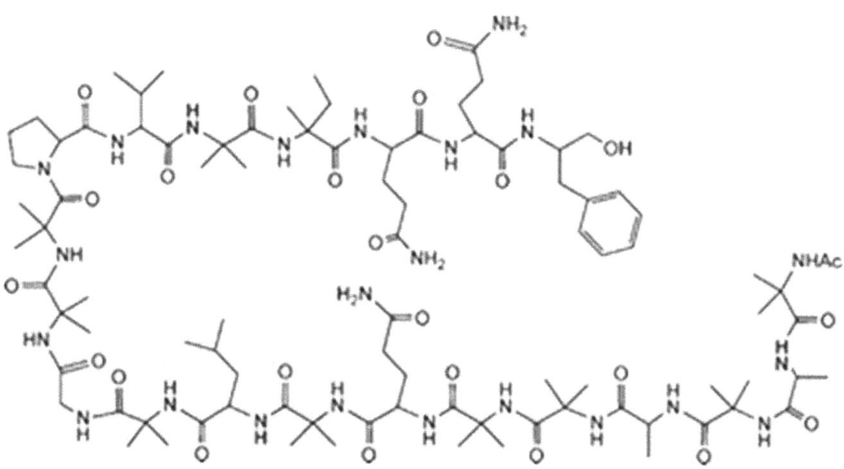

Peptaibol

Cytotoxicity and antibacterial activity: The fungal strain *Acremonium* sp. 021172C associated with this sponge species produced compounds such as peptaibiotic efrapeptins E, F, G; efrapeptins E and H; linear octapeptides,

RHM1-4; and N-methylated scytalidamides A and B. While RHM1 and RHM2 displayed weak cytotoxic activity on murine cancer cell lines,but RHM1 showed antibacterial activity. Efrapeptins showed both potent cytotoxic activity on murine and human cancer cells and antibacterial activity (Boot et al. 2006). Misra (2019) reported on the cytotoxic activity of scytalidamides.

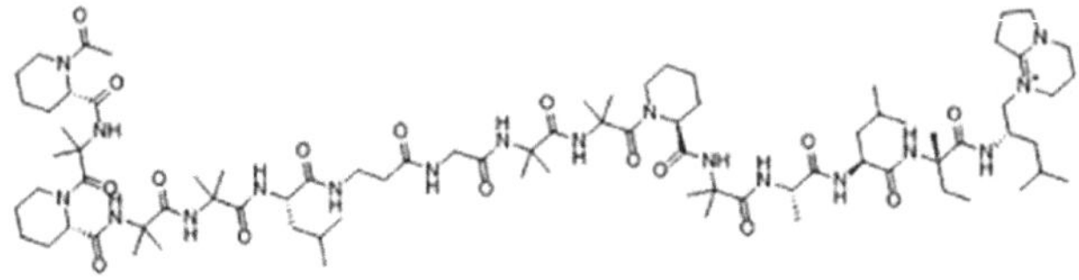

Efrapeptin F

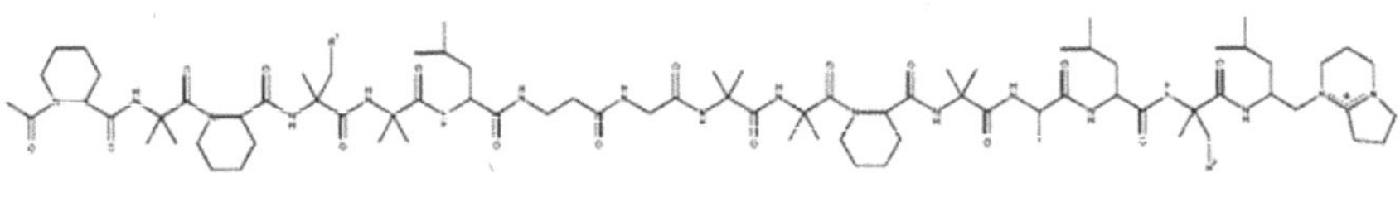

R1=R2=Me

Efrapeptin G

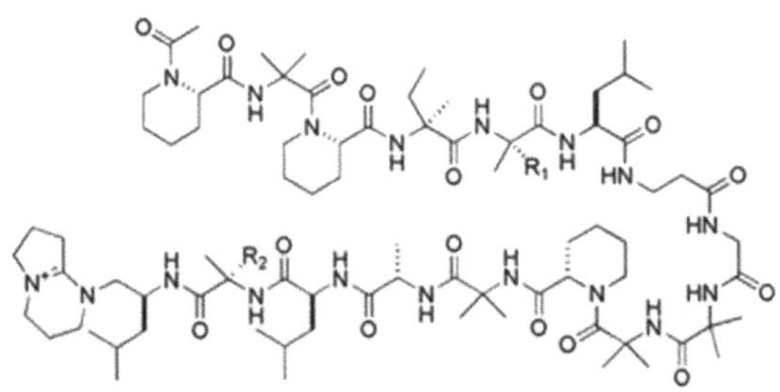

Efrapeptin Eα: R1=CH3; R2=CH3 Efrapeptin H: R1=CH2CH3; R2=CH2CH3

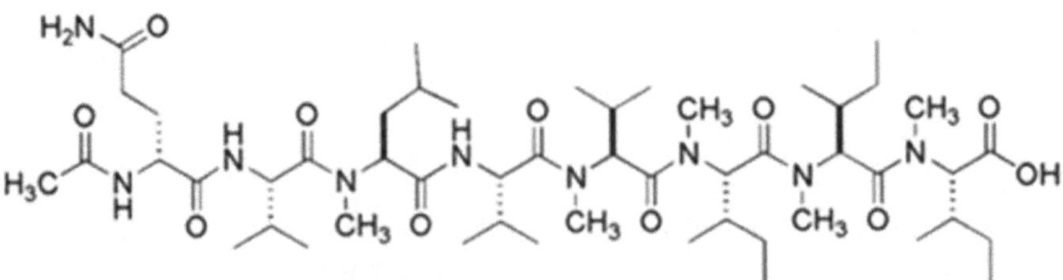

RHM 3

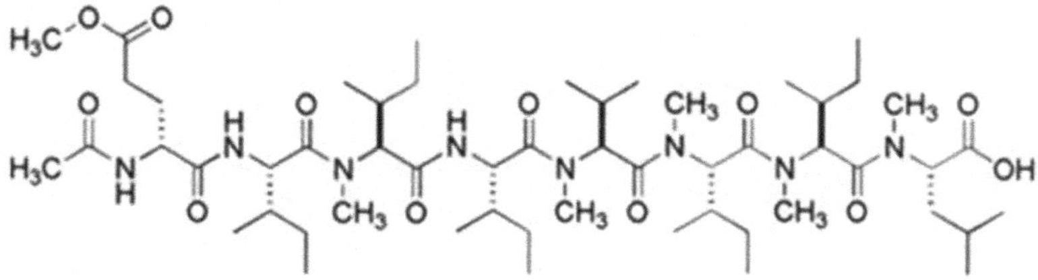

RHM 4

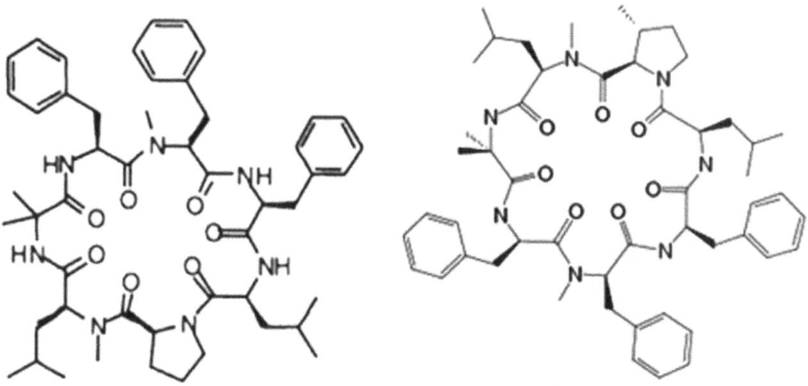

Scytalidamides A,B

Axinella sp. *(= Teichaxinella* sp.*)*

Common name (s): Orange ruffled sponge

Associated fungi and their pharmaceutical compounds and activities

Cytotoxicity: The N-methylated linear octapeptides, RHM1 and RHM2 derived from the associated fungus *viz. Acremonium* sp. of this sponge exhibited poor cytotoxic activity on murine L1210 cells (Zhang et al. 2017).

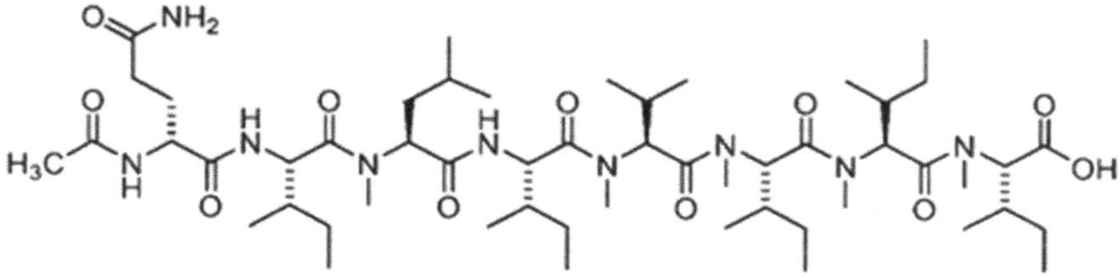

RHM1

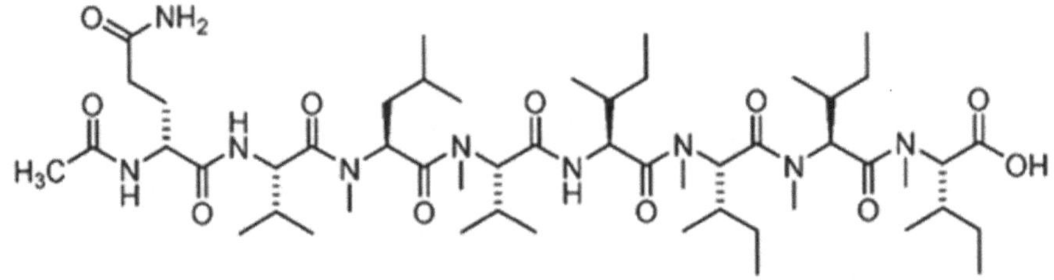

RHM 2

Phakellia fusca

Phakellia sp.

Image credit: Public domain.

Common name (s): Not reported

Global distribution: Western Central Pacific: South China Sea.

Habitat: Sessile

Associated fungi and their pharmaceutical compounds and activities

Antibacterial and cytotoxic activities: Compounds such as pestaloisocoumarins A and B (1, 2), gamahorin (3), isopolisin B (4), pestalotiol A (5), pestalachloride B (6) and pestalachloride E (7), and a mixture of pestalalactone atropisomers (8, 9) have been isolated from the fungus *Pestalotiopsis heterocornis* associated with this sponge species. Compounds 13, showed antibacterial activities against *Bacillus subtilis* and *Staphylococcus aureus* with a range of MIC values from 25 to 100 μg/mL and weak antifungal activities were observed with such compounds. The chlorinated benzophenone compounds 69 displayed antibacterial activities against *Staphylococcus aureus* and *Bacillus subtilis* with MIC values ranging from 3 to 50 μg/mL and cytotoxicity on 4 human cancer cells *viz.* BCG-823, H-460, PC-3, and SMMC-7721 with IC50 values ranging from 7 to 88 μM (Lei et al. 2017).

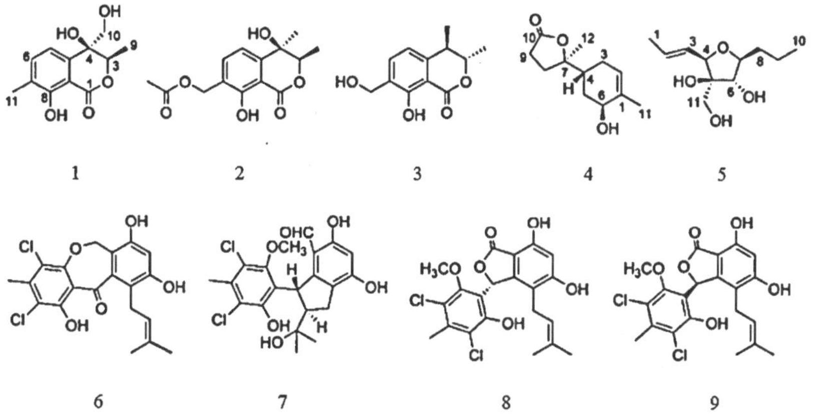

1,2: pestaloisocoumarins A, B; 3: gamahorin; 4: isopolisin B; 5: pestalotiol A;6: pestalachloride B; 7: pestalachloride E; 8,9: pestalalactone atropisomers

Order: Axinellida Family: Raspailiidae

Ectyoplasia ferox

Image credit: iNaturalist user: thibaudaronson, https://creativecommons.org/licenses/by-sa/4.0/

Common name (s): Brown encrusting octopus sponge

Global distribution: Occasional Florida Keys; common in the Bahamas and Caribbean.

Habitat: Inhabits reefs and adjacent areas. Often grow over coral rubble and dead areas of the reef.

Associated fungi and their pharmaceutical compounds and activities

Antimicrobial activity and cytotoxicity: The fungus *Coniothyrium* sp. 193477 associated with this sponge species produced 2-(1(E)-propenyl)-octa-4(E),6(Z)-diene-1,2-diol, (3R)-6-methoxymellein, (3S)-(3,5-dihydroxyphenyl)butan-2-one, (3R)-6-methoxy-7-chloromellein and cryptosporiopsinol. Of these compounds, only cryptosporiopsinol showed potent antimicrobial activity. Further, the fungus Phoma sp. derived from this sponge produced compounds *viz.* epoxyphomalin A and B. Among these, compound A displayed strong antitumor activity. Furthermore, the associated fungus *Spicellum roseum* 193H15 yielded trichothecenes such as trichodermol and 8-deoxytrichothecin which inhibited the enzyme LacCer synthase. The latter has been reported to play an important role in cell proliferaiton and oncogene expression. *S. roseum* also produced cyclohexadepsipeptide compounds *viz.*, spicellamides A and B, the bioactivities of which are yet to be known (Thomas et al. 2010).

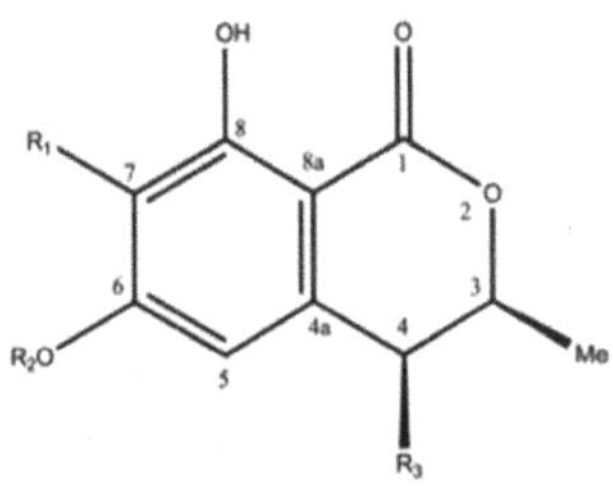

R1=R3=H, R2=Me

(3R)-6-methoxymellein

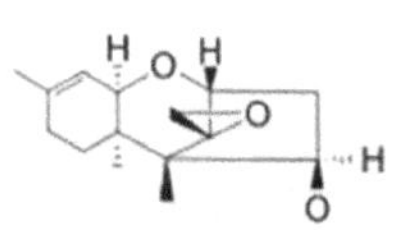

Trichodermol

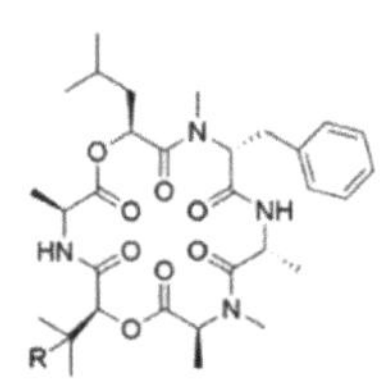

Spicellamide A: R= H

Spicellamide B: R=OH

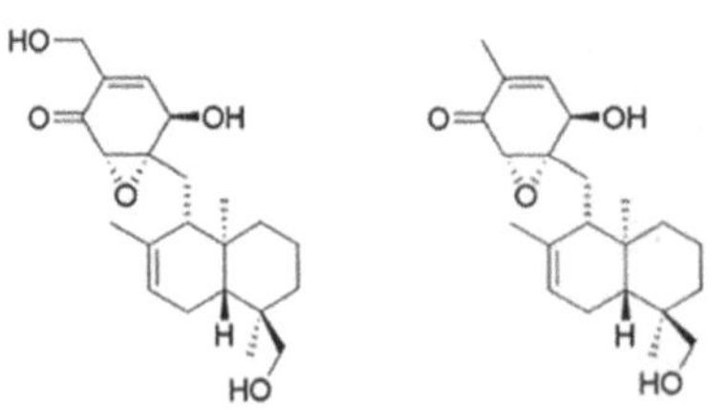

Epoxyphomalins A, B

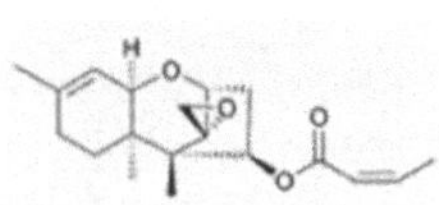

8-deoxytrichothecin

Order: Bubarida Family: Dictyonellidae

Dictyonella incisa

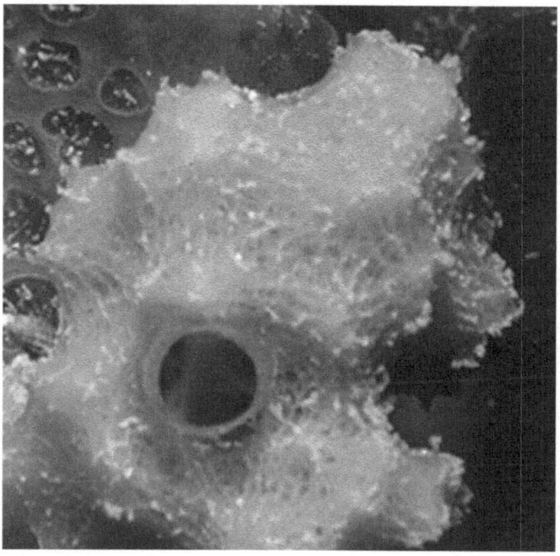

Image credit: Géry Parent. Public domain.

Common name (s): It has no common name

Global distribution: Subtropical Northeast Atlantic.

Habitat: It is a sessile species

Associated fungi and their pharmaceutical compounds and activities

Antibacterial activity: The fungal strain *Trichoderma saturnisporum* DIIA harbored by this sponge produced the sorbicillinoid compounds *viz.* saturnispol F and saturnispol H which displayed antibacterial activity against the strain *Escherichia faecalis* A4 (VRE) with MIC values of 1.63 and 12.9 g/mL, respectively (Nweze et al. 2020).

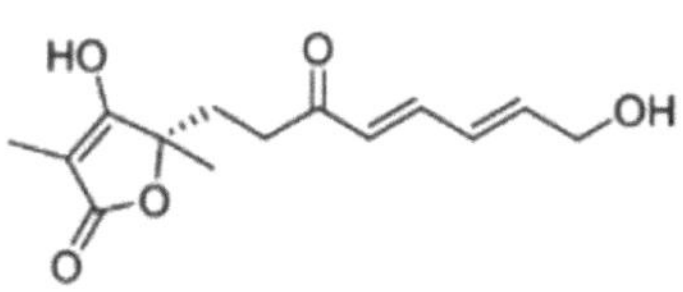

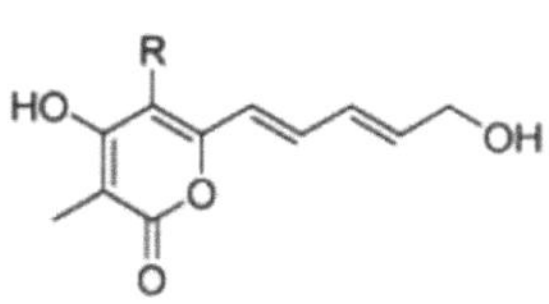

R=H

Saturnispol F Saturnispol H

Order: Chondrosiida Family: Chondrosiidae

Chondrosia reniformis

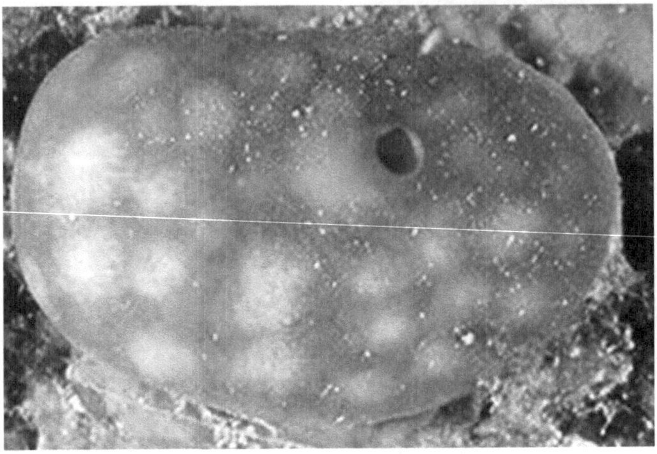

Image credit: Guido Picchetti https://creativecommons.org/licenses/by-sa/3.0/

Common name (s): Not reported

Global distribution: Mediterranean; Portugal;

Habitat: Sessile; shallow water, exposed habitats.

Associated bacteria: Pharmaceutical compounds and activities

Anti-infective activity: The fungal species *Penicillium rugulosum* (Ascomycota, fungus) associated with this sponge species produced compounds viz. prugosenes A1, A2, A3, B1, B2, C1, C2 which have been reported to have the potential to act as template for anti-infective compounds (Thomas et al. 2010; Pallela and Ehrlich, 2016).

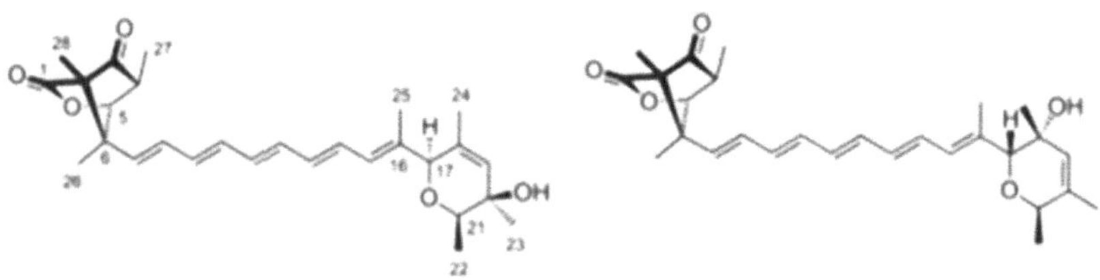

1,2

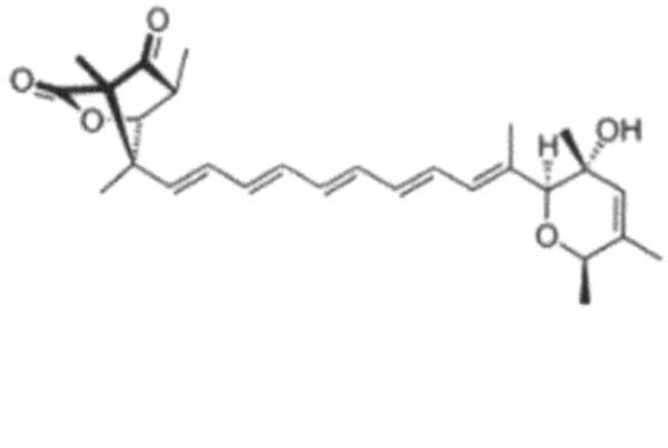

3

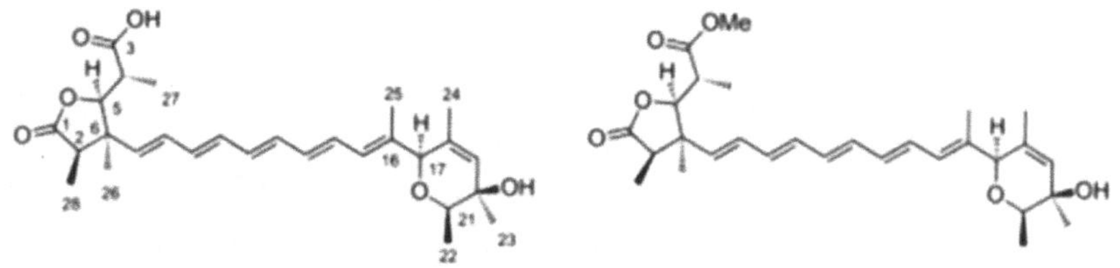

4,5

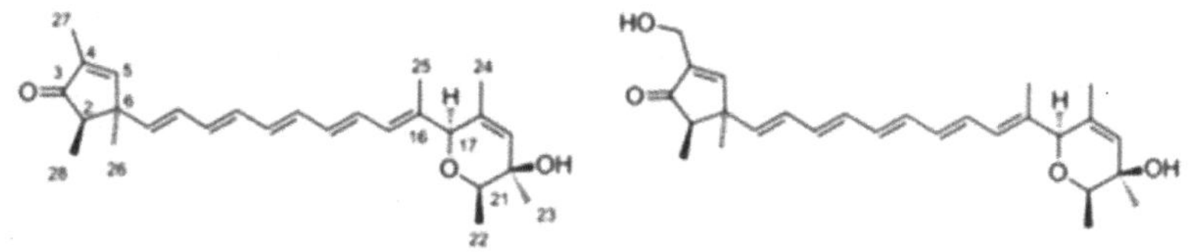

6,7

1-3: Prugosenes A1-A3; 4,5: Prugosenes B1,B2; 6,7: Prugosenes C1,C2

Order: Clionaida Family: Clionaidae

Cliona celata

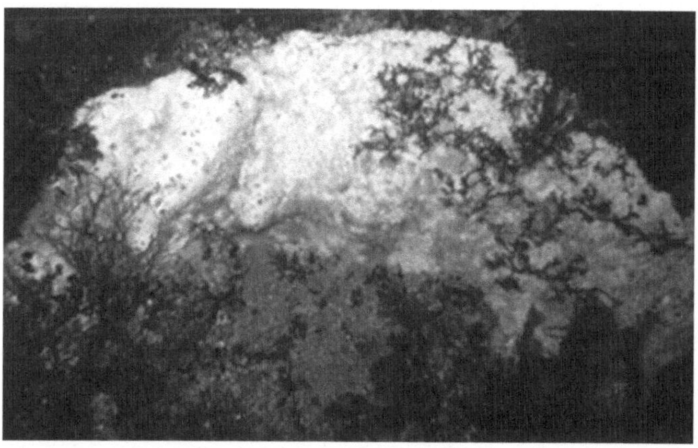

Image credit: https://creativecommons.org/licenses/by-sa/3.0/

Common name (s): Red boring sponge

Global distribution: Widely distributed along all Western European coastlines.

Habitat: This massive species occurs mainly on rocky surface. This species has been reported to begin its life by boring into substrates like molluscan shells, limestone, or calcareous red algae.

Associated fungi; and their pharmaceutical compounds and activities

Antimicrobial activity: The fungal species *Cladosporium cladosporioides* and *Aspergillus oryzae* associated from this sponge displayed antimicrobial activity against pathogenic bacteria and fungi viz. *Escherichia coli*, *Pseudomonas aeruginosa*, *Staphylococcus aureus*, *Staphylococcus fecalis*, *Bacillus subtilis*, *N. gonorroeae*, and *Candida albicans* with inhibition zone diameter values ranging from 8.7 to 13.3 mm (*Aspergillus oryzae*); and from 0 to 9.3 mm (*Cladosporium cladosporioides*) (Sayed et al. 2016).

Spheciospongia vagabunda *(= Spirastrella vagabunda)*

Associated fungi; and their pharmaceutical compounds and activities

Antibiotic activity: The polyketide compound, 14,15-secocurvularin derived from its associated unidentified fungus was found to be mildly antibiotic against *Bacillus subtilis* (Thomas et al. 2010).

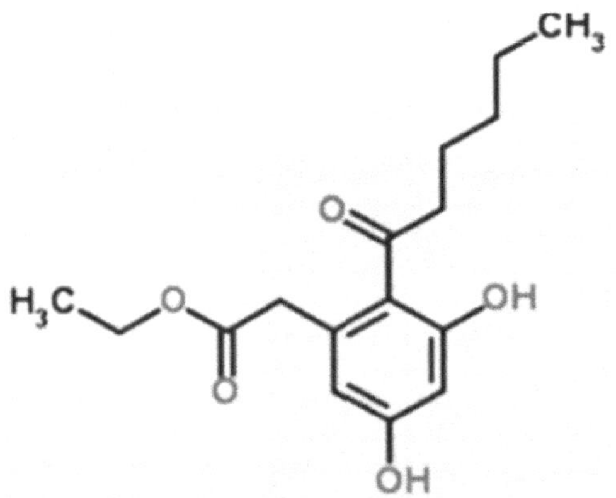

Order: Dictyoceratida Family: Irciniidae

***Psammocinia* sp.**

Associated fungi; and their pharmaceutical compounds and activities

Antifungal activity and cytotoxicity: Three meroterpenoids, insuetolides AC have been obtained from the fungus *Aspergillus insuetus* (OY-207) isolated from this Mediterranean sponge. While the compound insuetolide C displayed weak cytotoxicity on MOLT-4 cancer cell line, the compound insuetolide A showed antifungal activity against the fungus *Neurospora crassa*, (Elissawy et al. 2015).

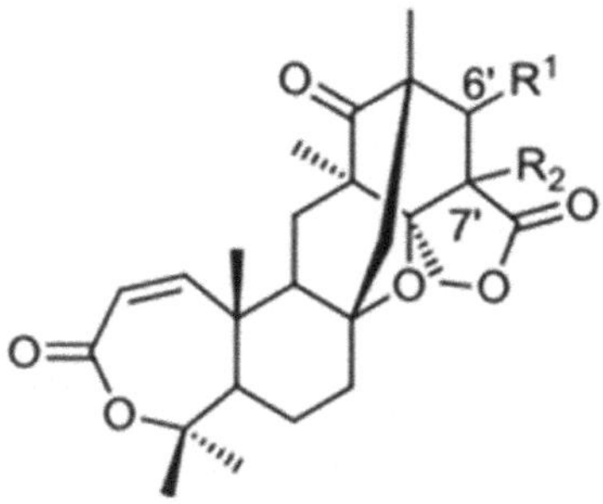

Insuetolide A: R1=R2=H; Insuetolide B: R1=αOH, R2=H; Insuetolide C: R1,R2= didehydro

Elissawy et al.(2015) reported on the isolation of four drimane-type sesquiterpenoid lactones viz. (E)-6-(4' hydroxy-2'-butenoyl)-strobilactone A (1), 2,9,11-trihydroxy-6oxodrim-7-ene (2), strobilactone A (3) and (E,E)-6-(6',7'-dihydroxy-2',4'-octadienoyl)-strobilactone A (4) were isolated from the fungus *Aspergillus insuetus* (OY-207) isolated from this sponge species. While compounds 1 and 4 exhibited weak cytotoxic activity against MOLT-4 cancer cell lines, compounds 3 and 4 showed mild antifungal activity against the fungus *Neurospora crassa*.

Sarcotragus fasciculatus *(= Ircinia fasciculata)*

Associated fungi; and their pharmaceutical compounds and activities

Cytostatic activity: The compound sorbicillactone A isolated from the fungus *Penicillium chrysogenum* associated with this sponge displayed selective cytostatic activity against L5178y murine leukemic lymphoblasts and protected human T cells by inhibiting the cytopathic effects of HIV-1 (Thomas et al. 2010).

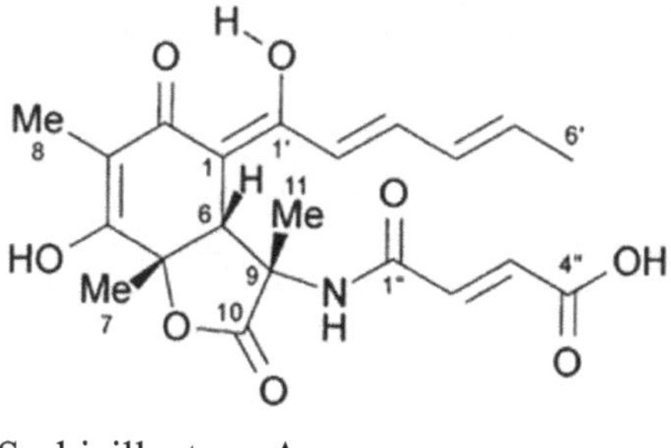

Sorbicillactone A

Order: Dictyoceratida Family: Spongiidae

Hyattella cribriformis

Common name (s): No reported

Global distribution: Caribbean; Australia and New Zealand

Habitat: Tropical reefs and even the deep sea.

Associated fungi; Therapeutic compounds and their bioactivities.

Antibacterial activity: The fungi associated with this sponge viz. *Aspergillus flavus*, *A. terreus*, *A. fumigatus*, *A. niger*, *Trichoderma* sp. and *P. citrinum* displayed antibacterial activity against pathogenic bacteria such as *S. aureus*, *S. typhi*, *S. paratyphi*, *K. oxytoca*, *E. coli*, *L. vulgari*, *K. pneumoniae*, *V. cholerae*, and *A. tumefaciens* with values of zone of clearance ranging from 1 to 11 mm (Vasanthabharathi and Jayalakshmi 2012)

Spongia (Spongia) ceylonensis *(= Spongia officinalis var. ceylonensis)*

Associated fungi; and their pharmaceutical compounds and activities

Antibacterial activity: The associated fungi viz. *Aspergillus flavus, A. fumigatus, A. niger, A. terreus, Penicillium spp.* and *Trichoderma viride* of this sponge species showed antibacterial activity against pathogenic bacteria such as *S. aureus, S. typhi, S. paratyphi, K. oxytoca, E. coli, L. vulgari, K. pneumoniae*, *V. cholerae* and *A. tumefaciens* with values of zone of clearance ranging from 1 to 11 mm (Vasanthabharathi and Jayalakshmi 2012).

***Spongia* sp.**

Associated fungi and their pharmaceutical compounds and activities

Cytotoxicity: The fungal strain *Myrothecium verrucaria* 973023 derived from this sponge yielded compounds viz. isororidin A, epiroridin E, verrucarin A, 13-acetyltrichoverrin B, miophytocen C, 3-hydroxyroridin E, roridin A, L, M, and M, and trichoverrin A and B. Except miophytocen C, all other compounds displayed potent cytotoxic activity on murine and human tumor cells (Thomas et al. 2010).

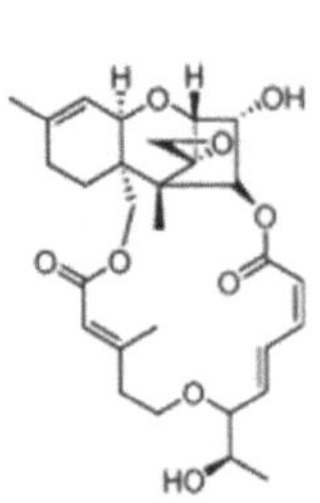

3-hydroxyroridin E

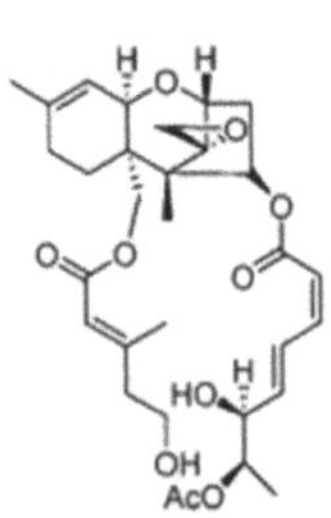

Miophytocen C

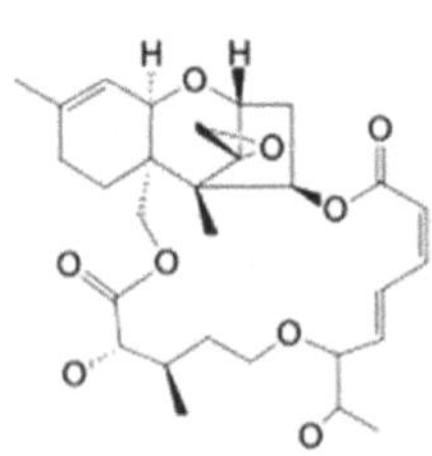

Roridin A

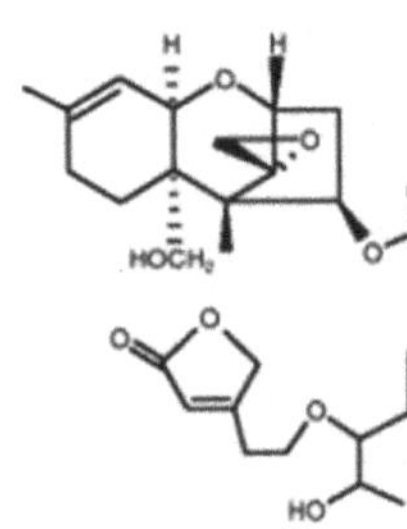

Roridin L

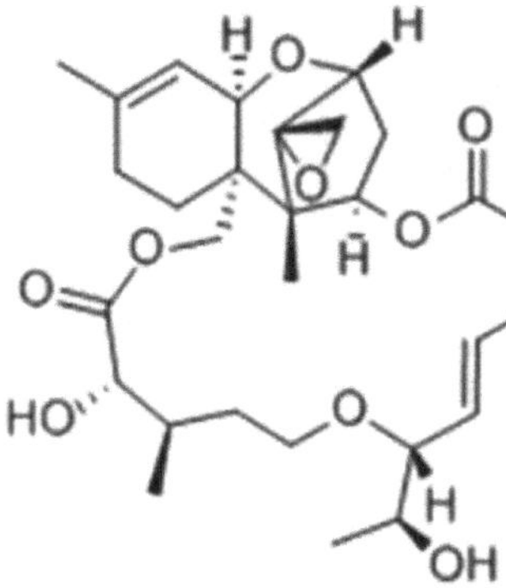

Isororidin A

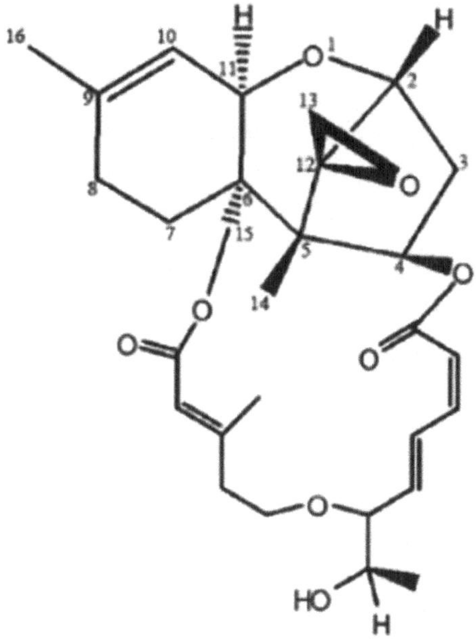

Epiroridin E

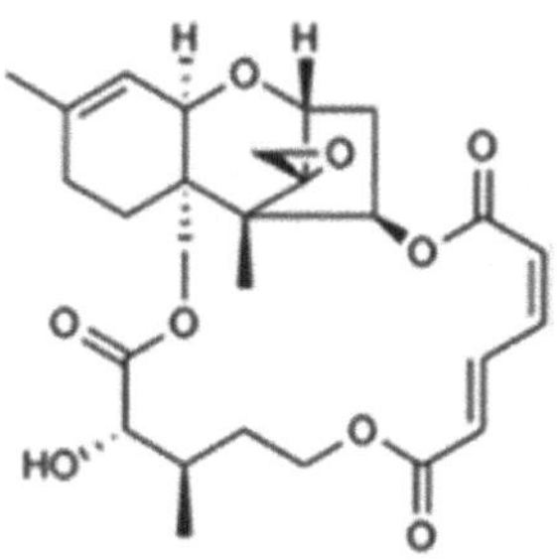

Verrucarin A

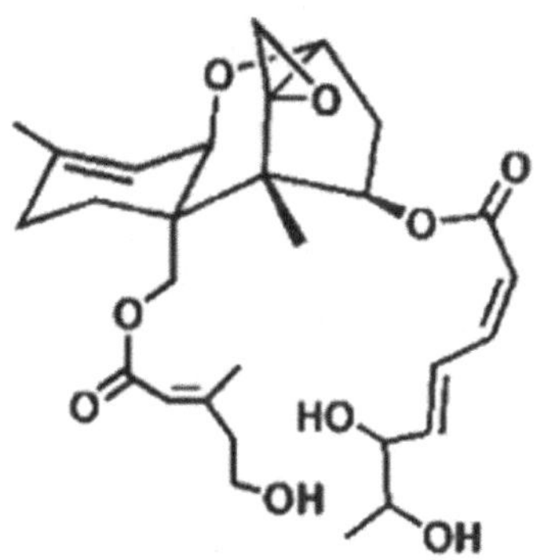

Trichoverrins

Order: Dictyoceratida   Family: Thorectidae

Hyrtios erectus *(= Heteronema erectus)*

Common name: No informatin

Global distribution: Indo-Pacific: Palau and Hawaii.

Habitat: Sessile; attached to dead coral at base of pinnacles; depth range 0 – 100 m

Associated fungi and their pharmaceutical compounds and activities

The fungus *Emericellopsis minima* associated with this sponge yielded a bicyclic sesquiterpene, (5E)-2-methyl-5[(1'R*,5'R*)-2-methylidene-7-oxo-bicyclo[3.2.1]oct-6-ylidene]-4-oxopentanoic acid which showed neither antimicrobial nor cytotoxic effects (Elissawy et al. 2015).

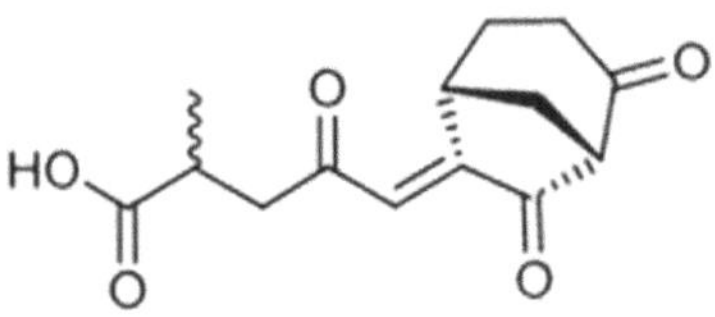

(5E)-2-methyl-5[(1'R*,5'R*)-2-methylidene-7-oxobicyclo[3.2.1]oct-6-ylidene]-4-oxopentanoic acid

Hyrtios proteus

Duchassaing and Michelotti, 1864

Common name (s): Not reported

Global distribution: Western Central Atlantic

Habitat: Only known from reef localities.; sessile; brackish; depth range 0 - 5 m

Associated fungi and their pharmaceutical compounds and activities

Antitumor activity: The culture extract of the fungal strain of *Aspergillus niger* from this sponge yielded compounds *viz.* asperazine and malformin C which exhibited leukemia and tumor-selective bioactivity (Thomas et al. 2010).

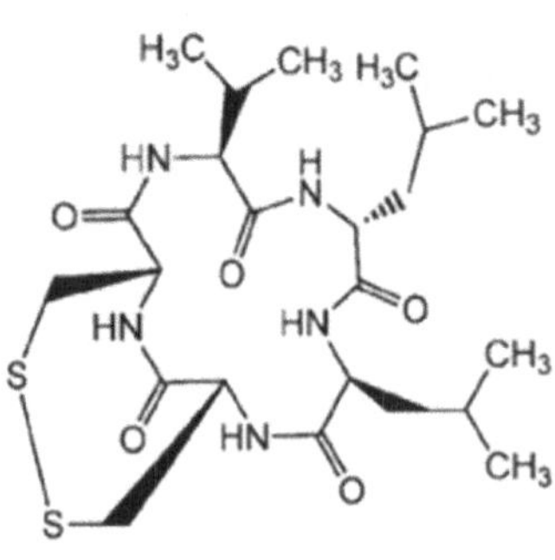

Malformin C

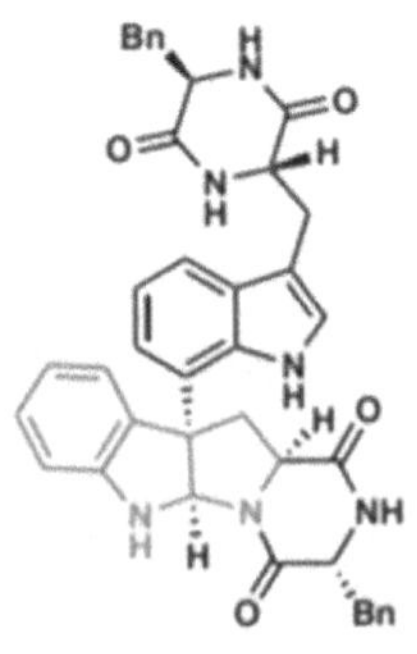

Asperazine

***Hyrtios* sp.**

Associated fungi; and their pharmaceutical compounds and activities

Cytotoxicity: The compound asperazine (A diketopiperazine alkaloid) derived from the culture of the fungus *Aspergillus niger* obtained from this sponge displayed remarkable cytotoxicity and leukemia selectivity (Thomas et al. 2010).

***Scalarispongia scalaris* (= *Cacospinga scalaris*)**

Image credit: Diego Delso, https://creativecommons.org/licenses/by-sa/4.0/

Common name (s): Not designated

Global distribution: NE Atlantic; Mediterranean.

Habitat: It occurs in shallow-waters down to a depth of 250 m.

Associated fungi; and their pharmaceutical compounds and activities

Antibacterial activity: The fungus, *Engyodontium album* strain LF069 associated with this sponge, has been reported to synthesize the polyketide compounds, engyodontochones which are active against MRSA (DSM 18827). The activity of these compounds is believed to be even ten times stronger than the control antibiotic chloramphenicol (Nweze et al. 2020).

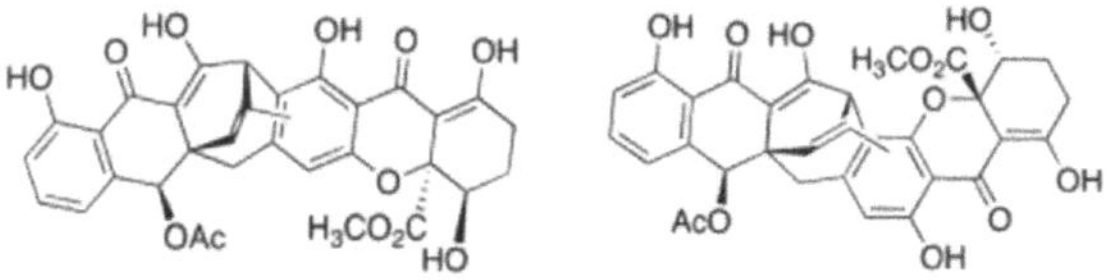

Engyodontochone A,B

Order: Halichondrida Family: Halichondridae

***Epipolasis* sp.**

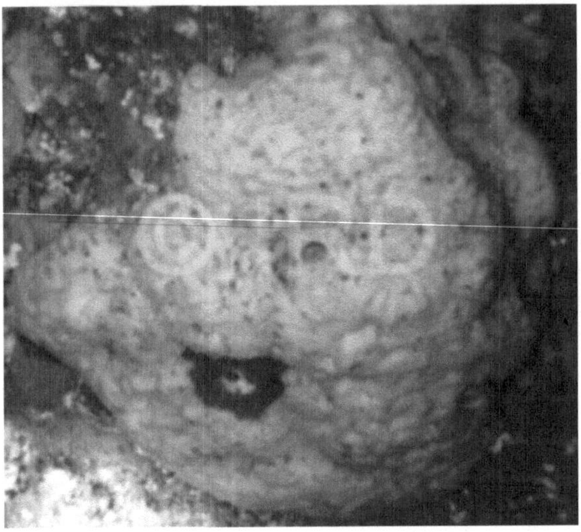

Image credit: Sponges of Polynesia Queensland Museum (2014). https://creativecommons.org/licenses/by/4.0/

Common name: Not known

Global distribution: French Polynesia: Marquesas islands

Habitat: It occurs on rocky slope/wall.

Associated fungi; and their pharmaceutical compounds and activities

Antibacterial activity: The fungus *Aspergillus candidus* KUFA0062 associated with this sponge has been reported to synthesize the hydroxypyrrolidine alkaloid compound *viz.* preussin which showed significant antibacterial activity against MRSA 66/1, and VR *Escherichia faecalis* B3/101 with an MIC value of 32 μg/mL (Nweze et al. 2020).

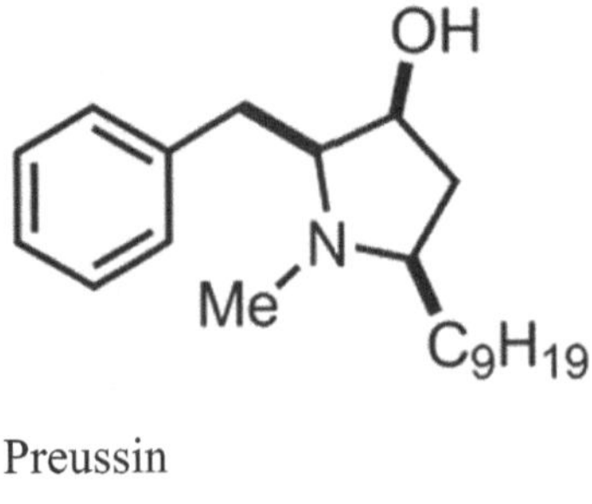

Preussin

Order: Haplosclerida Family: Callyspongiidae

Callyspongia (Cladochalina) aerizusa

Image credit: Bernard Dupont from France. https://creativecommons.org/licenses/by-sa/2.0/

Common name (s): Not reported

Global distribution: Indo-West Pacific.

Habitat: Sessile; depth range 0 - 100 m;

Associated fungi; Therapeutic compounds and their bioactivities

Antibacterial activity: The fungus *Cladosporium herbarum* derived from this sponge yielded acetyl Sumiki's acid which showed antibacterial activity against *Staphylococcus aureus* and *Bacillus subtilis* at 5 g/disc (Thomas et al. 2010).

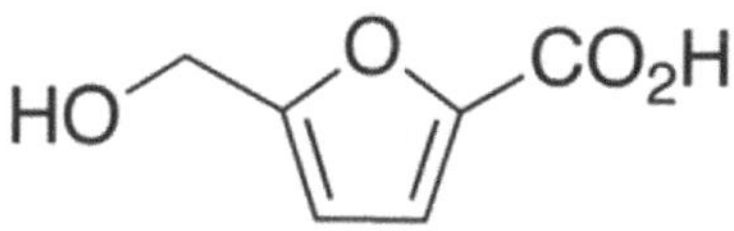

Sumiki's acid

Callyspongia (Cladochalina) diffusa

Common name (s): Champagne sponge

Global distribution: Indo-West Pacific.

Habitat: Sessile; depth range 0 - 100 m

Associated fungi; and their pharmaceutical compounds and activities

Antibacterial activity: The associated fungi *viz. Aspergillus flavus, A. fumigatus, A. niger, A. terreus, P. citrinum* and *Trichoderma* sp. of this sponge species showed antibacterial activity against pathogenic bacteria such as *S. aureus, S. typhi, S. paratyphi, K. oxytoca, E. coli, L.vulgari, K. pneumoniae, V. cholerae* and *A. tumefaciens* with values of zone of clearance ranging from 1 to 18 mm (Vasanthabharathi and Jayalakshmi 2012).

Antifungal activity: The associated fungi *viz. Aspergillus flavus, A. fumigatus, A. niger, A. terreus, P. citrinum* and *Trichoderma* sp. of this sponge species showed antifungal activity against *F. oxysporum, B.cinerea, A. alternata, R. solani, F. udam, M. phaseolina, P. hypolateritia, C. theae,* and *P. thae* with values of zone of clearance ranging from 1 to 11 mm (Vasanthabharathi and Jayalakshmi 2012).

Callyspongia (Cladochalina) vaginalis

Image credit: Nhobgood (talk) Nick Hobgood, https://creativecommons.org/licenses/by-sa/3.0/

Common name (s): Branching vase sponge

Global distribution: Western Central Pacific and Western Atlantic

Habitat: Reef-associated; brackish; depth range 2–70 m

Associated fungi and their pharmaceutical compounds and activities

Antimicrobial activity: The fungus *Ulocladium botrylis* isolated form this sponge species produced 1-hydroxy-6-methyl-8-(hydroxyl-methyl) xanthone and ulocladol. Both the compounds displayed antimicrobial activity (Thomas et al. 2010).

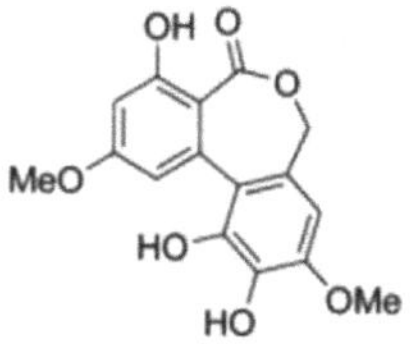

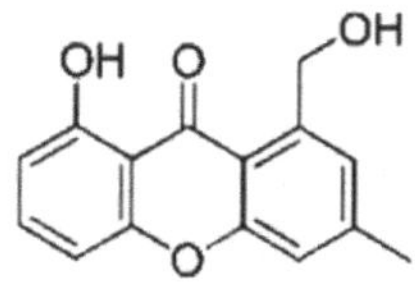

Ulocladol 1-hydroxy-6-methyl-8-(hydroxyl-methyl)xanthone.

***Callyspongia* sp.**

Associated fungi; and their pharmaceutical compounds and activities

The fungus *Stachylidium* sp isolated from this sponge produced compounds such as o-prenylated natural product stachyline A-D (112-115) and phthalide derivatives marilone A-C. Among thee compounds, marilone A was active against *Plasmodium berghei* liver stages while marilone B was a selective antagonist of serotonin receptor 5-HT2B. However, both the compounds displayed weak antiproliferative activity on 3 HTCLs (Maharaj,2015).

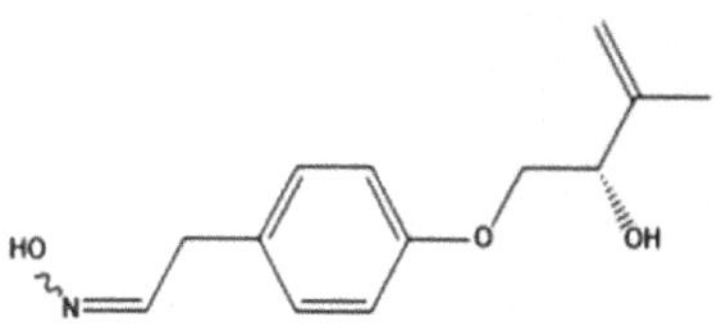

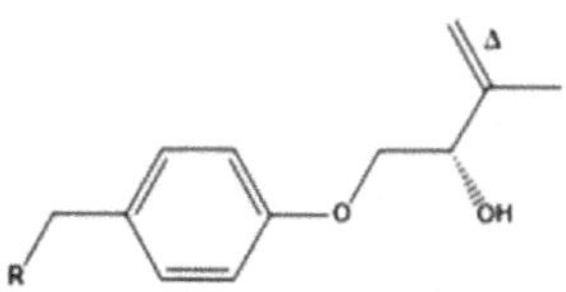

Stachyline A Stachyline B, R= COOH; Stachyline C, R= CH2OH;

Stachyline D, R= COOH,Δ saturated

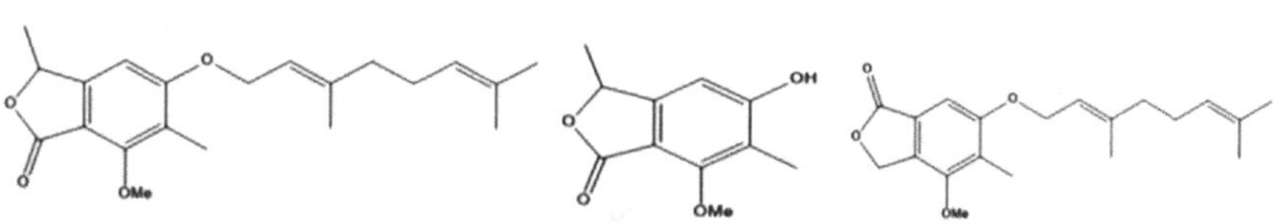

Marilone A-C

Antiviral activity: The fungus *Aspergillus* sp. SCSIO XWS02F40 associated with this sponge has been reported to produce a polyketide compound, asteltoxin E which displayed moderate antiviral activity against H1N1 and H3N2 influenza virus with IC50 values of 3.5 and 6.2 μM, respectively (Mayer et al. 2021).

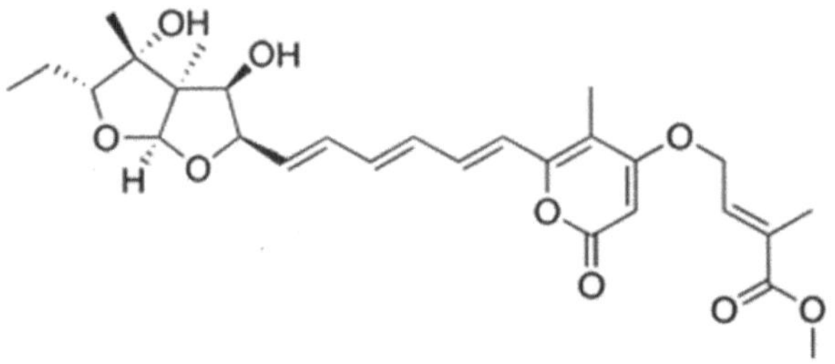

Asteltoxin E

Order: Haplosclerida Family: Niphatidae

Amphimedon viridis

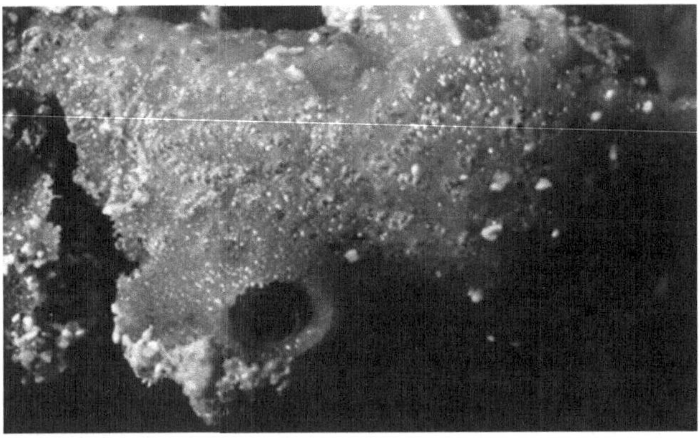

Image credit: Soledade, G.O. et al., https://creativecommons.org/licenses/by/4.0/

Common name (s): Green sponge

Global distribution: Indo-Pacific and Western Central Atlantic; Florida, Bahamas, and Caribbean.

Habitat: Sessile; brackish; depth range 1–14 m; favors lagoons, but is also found in mangroves, seagrass fields and on rocks and in reefs.

Associated fungi and their pharmaceutical compounds and activities

Antioxidant activity: The supernatant extracts of the unidentified fungus FS1 associated with this sponge species showed the highest free radical scavenging activity (FRSA) against DPPH radical (El-Hady et al., 2014a).

Antimicrobial activity: The supernatant from the static culture of the fungus FS1 isolated from the sponge *Amphimedon viridis* was found to be highly effective against *Staphylococcus aureus* and *Candida albicans* (El-Hady et al., 2014a).

Gelliodes carnosa *(= Sigmadocia carnosa)*

Gelliodes sp.

Image credit: Adapted from M.S. Khakshoor et al., 2014. https://jifro.ir/article-1-1733-fa.pdf

Common name (s):Not reported

Global distribution: Indo-West Pacific: India and New Caledonia

Habitat: Sessile; depth range 0–100 m

Associated fungi; and their pharmaceutical compounds and activities

Anti-inflammatory activity: The fungal strain *Hypocreales* sp. HLS-104 derived from this marine sponge yielded two sesquiterpenoid compounds *viz.* hypocreaterpenes A and B; and 1R,6R,7R,10S-10-hydroxy-4(5)-cadinen-3-one. Among these compounds, 1R,6R,7R,10S-10-hydroxy-4(5)-cadinen-3-one displayed poor inhibitory effect on NO production in the cell assay of RAW 264.7. Overproduction of NO induces tissue damage associated with acute and chronic inflammations (Elissawy et al. 2015).

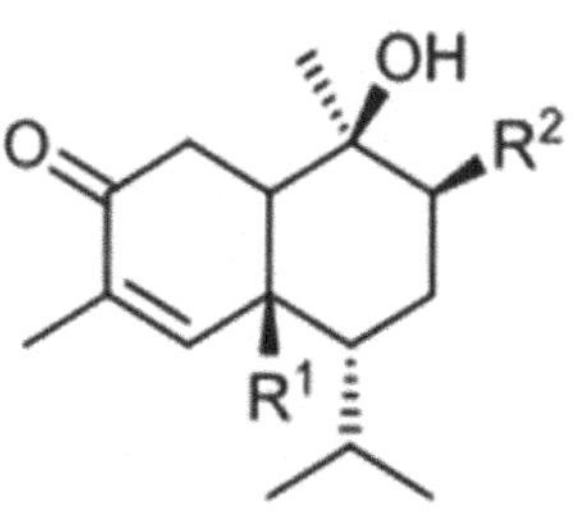

Compound1: R1=OH, R2=H; Compound 2: R1=H,R2=OH; Compound 3: R1,R2=H

Antibacterial activity: The associated fungi *viz. Aspergillus flavus, A. niger,* and *Fusarium* sp. of this sponge species showed antibacterial activity against pathogenic bacteria such as *S. aureus, S. typhi, S. paratyphi, K. oxytoca, E. coli, P. mirabilis, L. vulgaris, K. pneumoniae, V. cholerae* and *A. tumefaciens* with values of zone of clearance ranging from 1 to 6 mm (Vasanthabharathi and Jayalakshmi 2012).

Niphates olemda

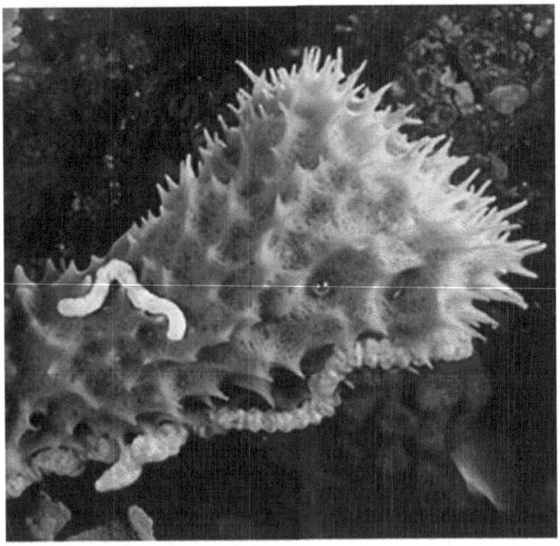

Niphates sp.

Common name (s): Tube sponge

Global distribution: Western Central Pacific: Indonesia, New Caledonia, Palau and Philippines.

Habitat: Sessile; depth range 8–25 m.

Associated fungi; and their pharmaceutical compounds and activities

Antibacterial activity: Fungus *Curvularia lunata* derived from this sponge produced three compounds viz. bisanthraquinone cytoskyrin A, plant hormone (+)-abscisic acid and lunatin (1,3,8-trihydroxy-6-methoxyanthraquinone. Among these compounds, the anthraquinones displayed potent antibacterial activity against *Escherichia coli, Bacillus subtilis,* and *Staphylococcus aureus* (Jadulco et al. 2002; Thomas et al. 2010).

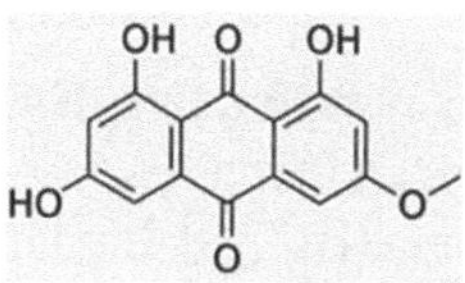

Lunatin

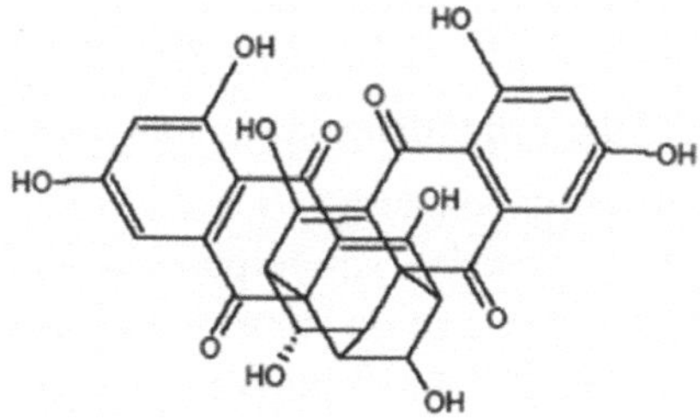

Cytoskyrin A.

Niphates recondite

Common name (s): Not reported

Global distribution: Bahamas and Curacao.

Habitat: In reef environments, on dead corals and other hard substrata; depth range from 6 to 52 m.

Associated fungi and their pharmaceutical compounds and activities

Lipid-lowering effects: The endophytic fungus *Stachybotrys chartarum* isolated from this sponge afforded phenylspirodrimanes, named chartarlactams which displayed strong lipid-lowering activity on HepG2 cell line at a concentration of 10 M (Jin et al. 2016).

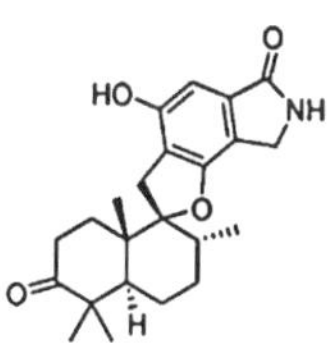

1

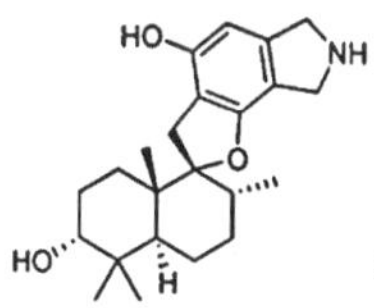

2

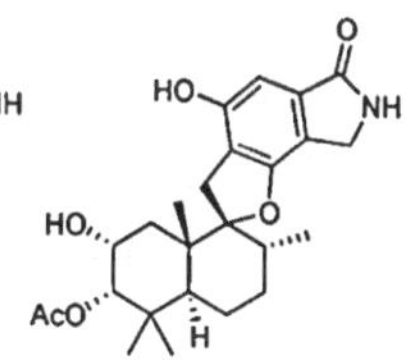

3

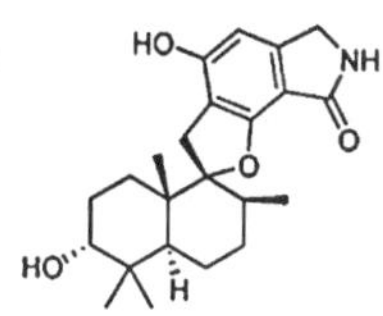

4

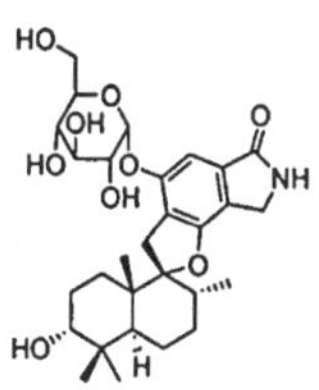

5

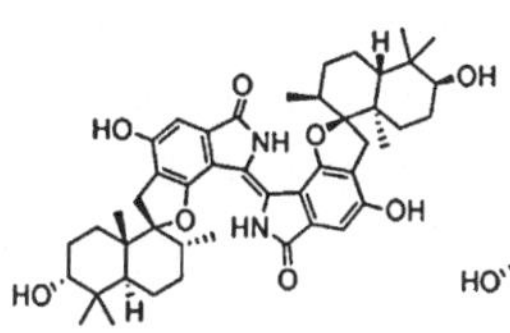

6

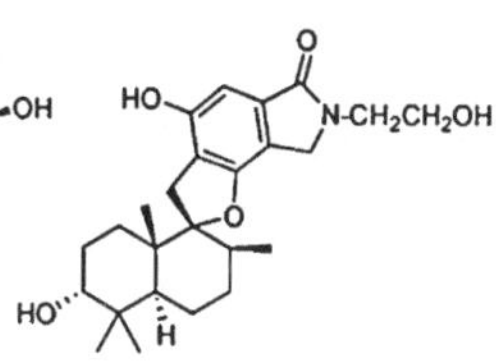

7

1-7: Chartarlactams

Niphates sp.

Associated fungi; and their pharmaceutical compounds and activities

Cytotoxicities: The associated fungus *Hansfordia sinuosae* of this sponge produced caryophyllene-based sesquiterpenoids *viz.* punctaporonins HM which showed poor cytotoxicity on human lung adenocarcinoma (A549), human ovarian carcinoma (A2780) human colon carcinoma (HCT-8), human hepatoma (Bel7402), and human gastric carcinoma (BGC823), with IC50 values >10 μM (Zhang et al. 2017).

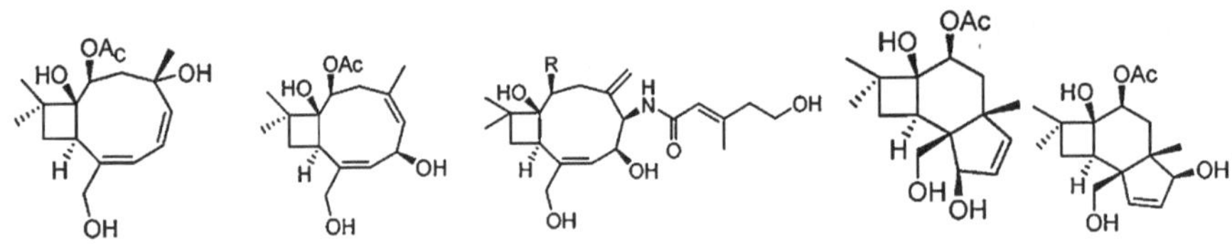

H I J,K L M

J; R=OAc; K: R=H)

Punctaporonins H-L

Order: Haplosclerida Family: Petrosiidae

Petrosia (Petrosia) ficiformis

Associated fungi; and their pharmaceutical compounds and activities

Immunosuppressive activity: The fungal strain, *Penicillium brevicompactum* associated with this sponge species provided two peptides, *viz.* petrosifungins A and B, and other fungal metabolites such as mycophenolic acid, asperphenamate and brevianamide A. Among these compounds, mycophenolic acid is an immunosuppressive agent; and petrosifungins A and B are known to serve as potent compounds for other pharmacologically and toxicologically important derivatives. Further, another strain of *Aspergillus insuetus* derived from this sponge yielded two compounds *viz.* terretonins E and F which are considered to be importang inhibitors of human mitochondrial respiratory chain (The mitochondrial respiratory chain supplies the cell with energy in the form of ATP. Mutations affecting this chain can cause diseases known as mitochondrial encephalomyopathies) (Mauro,2003; Thomas et al. 2010).

***Petrosia* sp.**

Associated fungi and their pharmaceutical compounds and activities

Cyctotoxicity: The fungal species *Aspergillus versicolor* derived from this unidentified species yielded fellutamide F which displayed cytotoxicity (Lee et al. 2011).

Cytotoxicity and antibacterial activity: Lee et al.(2007) reported that the fractionation of the cultured filtrates of *Aspergillus versicolor*, which was derived from this marine sponge yielded three polyketides: decumbenones A and B which displayed cytotoxic activity on 5 human tumor cell lines and antibacterial activity against twenty clinically isolated methicillin-resistant strains.

Cytotoxicity and melanin inhibition: Thomas et al. (2010) reported that the fungus *Aspergillus versicolor* isolated from this sponge yielded three polyketide compounds viz. versiol and decumbenones A, B; and lipopeptide, fellutamide C. While fellutamide C possessed cytotoxic properties and decumbenone A is known to be an ideal melani inhibitor.

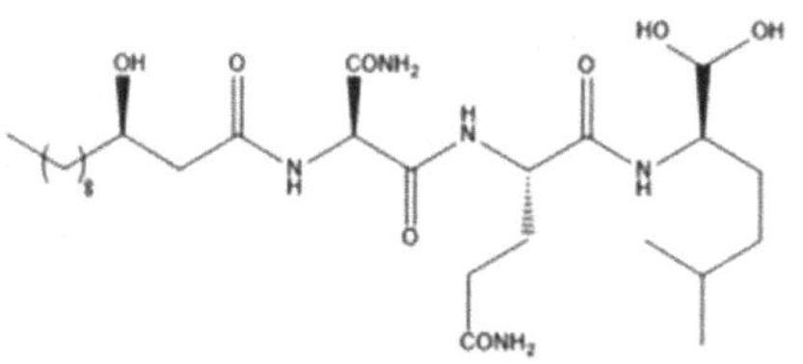

Fellutamide F

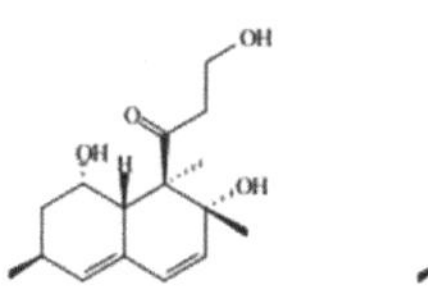

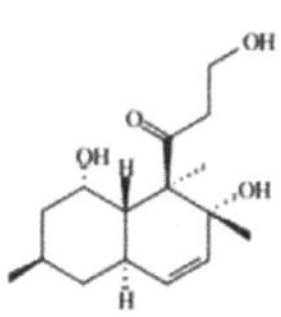

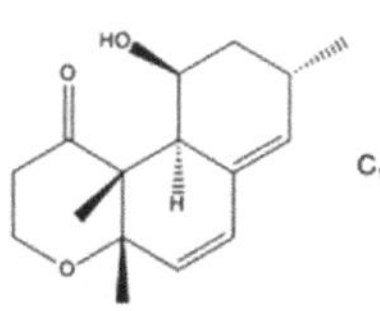

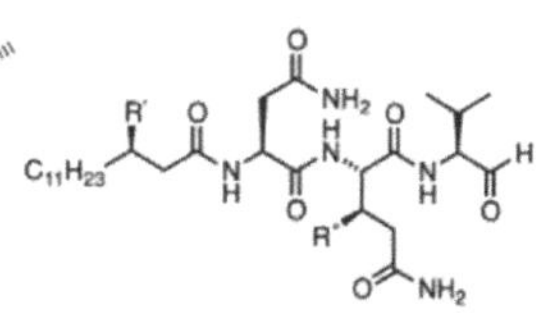

Decumbenones A, B Versiol Fellutamide C

R1=R2=H

Cytotoxicity: The fungal strain *Aspergillus versicolor* derived from this marine sponge afforded a new xanthone which displayed potent ytotoxicity on several human tumor cells viz. XF-498, and HCT-15, A-549, SK-OV-3, and SK-MEL-2 with IC50 values ranging from 1.2 to 4.6 μg/mL (Zhang et al., 2017).

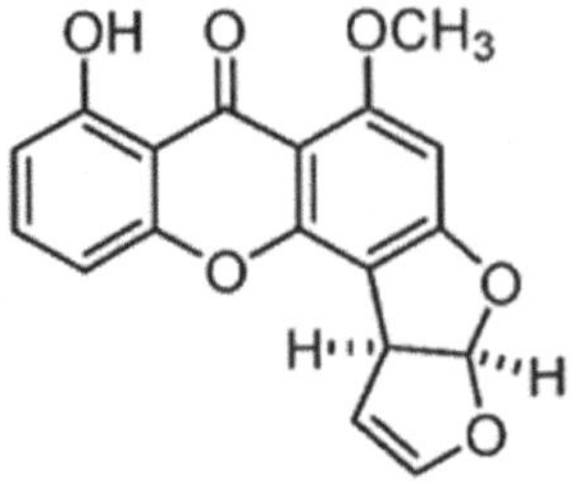

Xanthone

Neopetrosia chaliniformis *(= Neopetrosia exigua; Xestospongia exigua)*

Global distribution: Western Central Pacific: Indonesia.

Habitat: Sessile

Associated bacteria: Pharmaceutical compounds and activities

Cytotoxic and antibacterial activities: The ethyl acetate extracts of the symbiotic fungi isolates NC 1,3,4,and 7-11 associated with this sponge species showed antibacterial activity against human pathogenic bacteria of Gram positive *Bacillus subtilis* and *Staphylococcus epidermidis;* and Gram negative *Escherichia coli* and *Salmonella typhosa* with values of zone of inhibition dia. ranging from 7 to 21 mm. Further, these fungal (NC 1-13) extracts also showed cytotoxic activity of cell lines with LC50 values ranging from 16.8 to >1000 ppm (Handayani and Artasasta 2017).

Antimicrobial activity: The strains of *Penicillium* cf. *montanense* derived from this sponge yielded three compounds viz. xestodecalactones A-C. Among these compounds, only xestodecalactone B displayed antifungal activity on *Candida albicans*. Further, the compound aspergillitine of the fungus *Aspergillus versicolor* which is associated with this sponge displaued poor antibacterial activity on *Bacillus subtilis* (Thomas et al. 2010).

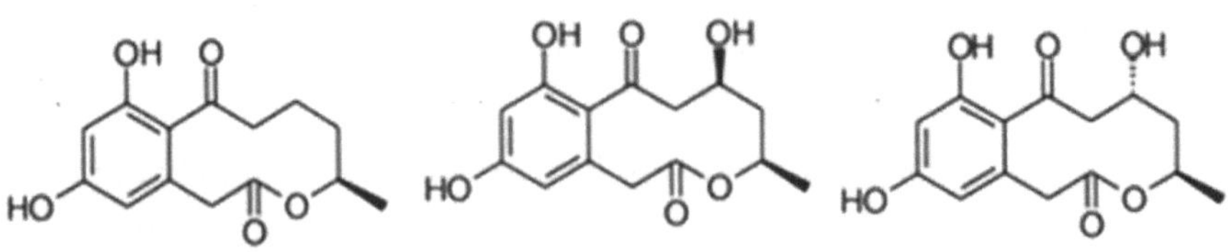

Xestodecalactone A-C

Xestospongia testudinaria

Associated fungi and their pharmaceutical compounds and activities

Cytotoxicity: Three dimeric bisabolane sesquiterpenoids, disydonols AC together with the precursor (+)-S-sydonol were isolated from the fungus *Aspergillus* sp., associated with this sponge species. Among these compounds, disydonol C showed selective cytotoxicity on Caski and HepG-2 cell lines with IC50 values of 21.0 and 6 μM, respectively. On the other hand, disydonol A showed moderate cytotoxic activity on these two cell lines with IC50 values of 25.5 and 19.1 μM, respectively (Elissawy et al. 2015).

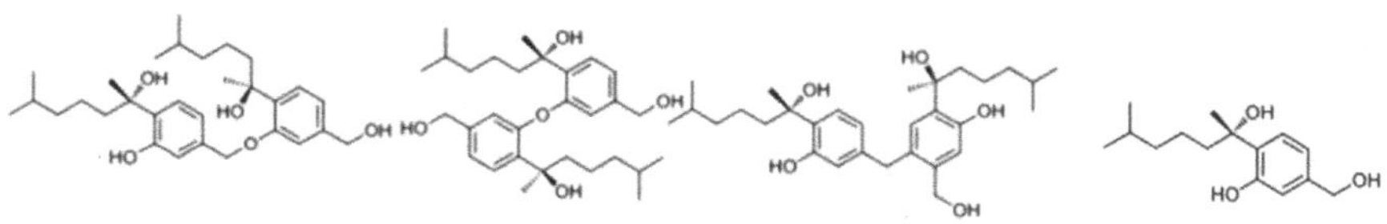

Disydonols A–C (+)-S-sydonol

Antiproliferative activity: The compound, asperaculin A, a sesquiterpenoid, was isolated from the fungal strain *Aspergillus aculeatus* CRI323-04 associated with this sponge species. Asperaculin A displayed antiproliferative activity on HuCCA-1 HepG2, MOLT-3, and A549 cancer cell lines (Elissawy et al. 2015).

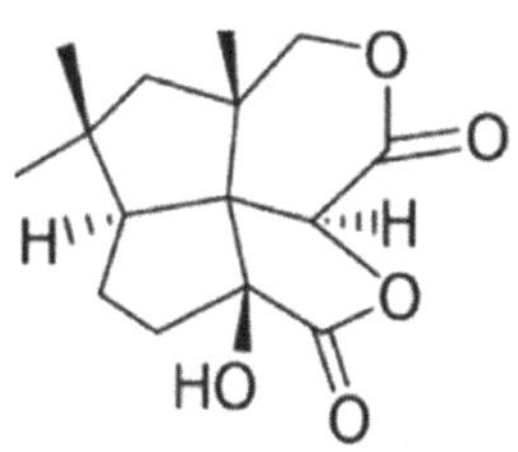

Asperaculin A

Order: Poecilosclerida Family: Acarnidae

Zyzzya sp.

Associated fungi and their pharmaceutical compounds and activities

Antiproliferative activity: The compound, asperaculin A, a sesquiterpenoid, was isolated from the fungal strain *Aspergillus aculeatus* CRI323-04 associated with this sponge species. Asperaculin A displayed antiproliferative activity on HuCCA-1 HepG2, MOLT-3, and A549 cancer cell lines (Elissawy et al. 2015).

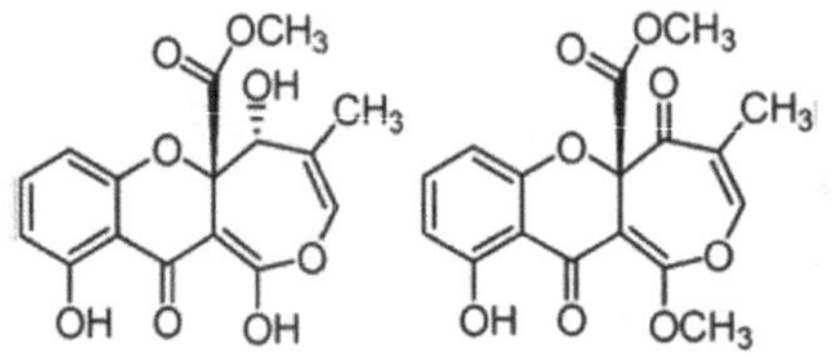

Brocaenol A Brocaenol (B, R= CH3); C, R=H)

Order: Poecilosclerida Family: Microcionidae

Clathria (Thalysias) reinwardti

Image credit: Nhobgood Nick Hobgood, https://creativecommons.org/licenses/by-sa/3.0/

Common name (s): Orange sprawling sponge

Global distribution: Indo-West Pacific

Habitat: Sessile

Associated fungi and their pharmaceutical compounds and activities

Antibacterial and antitumor peoperties: Chemical investigation of the crude ethyl acetate extract of the associated fungus *Neosartorya fennelliae* KUFA

0811 of this sponge resulted in the isolation of sitostenone, ergosta-4,6,8(14), 22-tetraen-3-one, dehydromevalonic lactone, byssochlamic acid,cyathisterone, chevalone B, aszonalenin, secalonic acid A, helvolic acid and fellutanine A. Among these compounds, sitostenone, byssochlamic acid, chevalone B, aszonalenin, fellutanine, and helvolic acid displayed antimicrobial activity. Helvolic acid exhibited strong antibacterial activities against the Gram-positive *B. subtilis* (MIC: 2 g mL1) and MRSA (MIC: 4 g mL1). Ergosta-4,6,8(14), 22-tetraen-3-one has cytotoxic activity against human gastric cancer cell and it can induce apoptosis of HepG2 cell line. Secalonic acid A possessed anti-tumor activity and reduced colchicines toxicity; and it is also known to protect against dneuron death in the mouse model of Parkinson's disease (Gomes 2014).

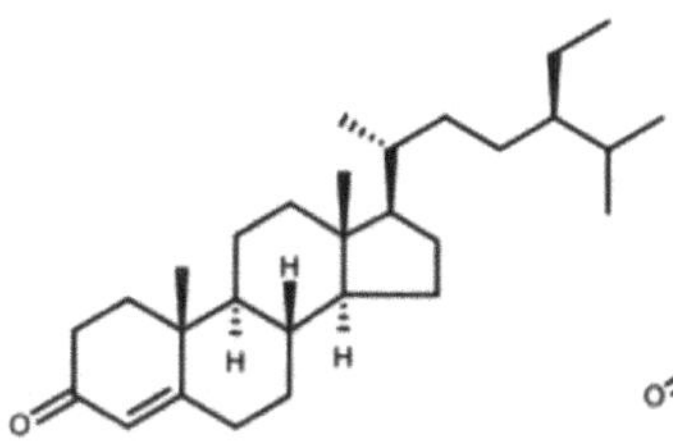

1

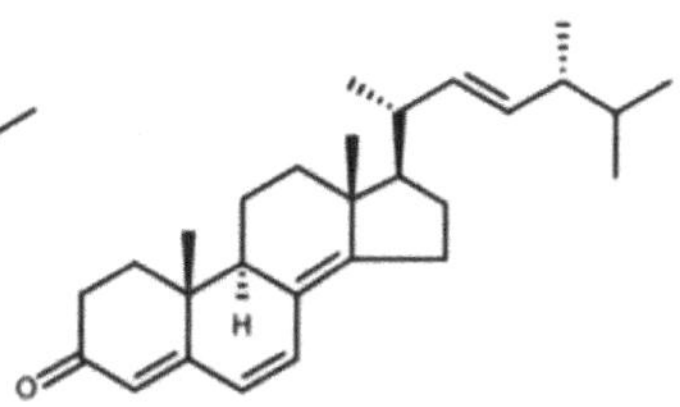

2

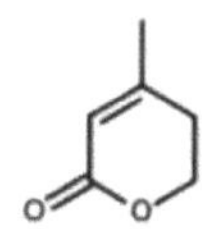

3

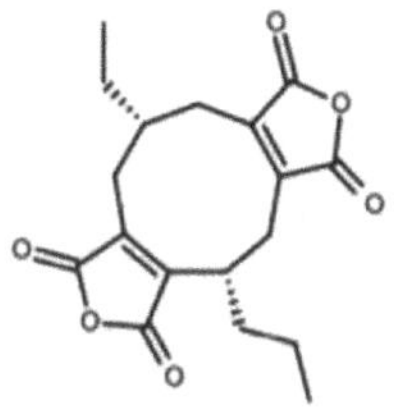

4

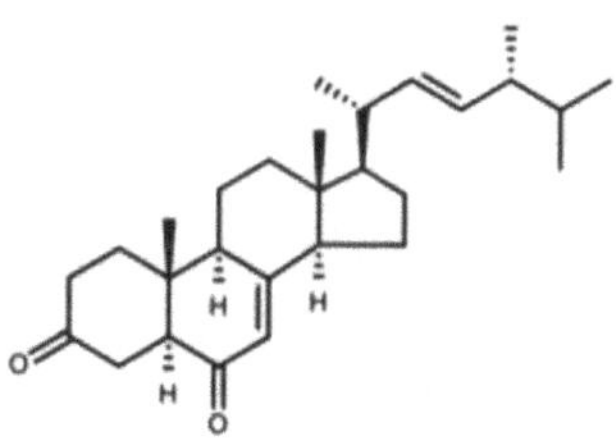

5

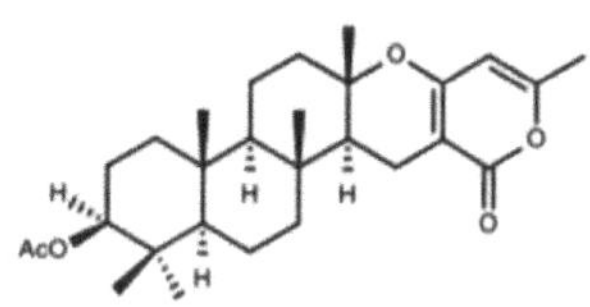

6

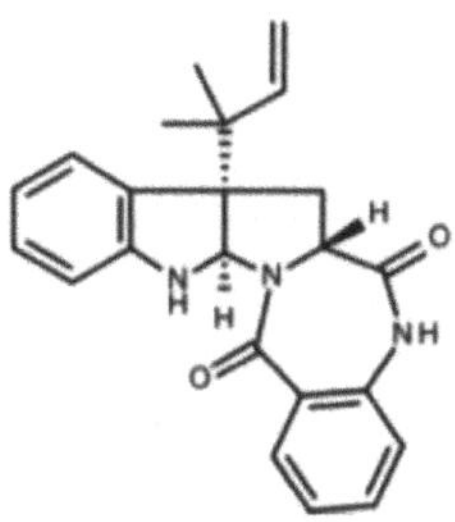

7

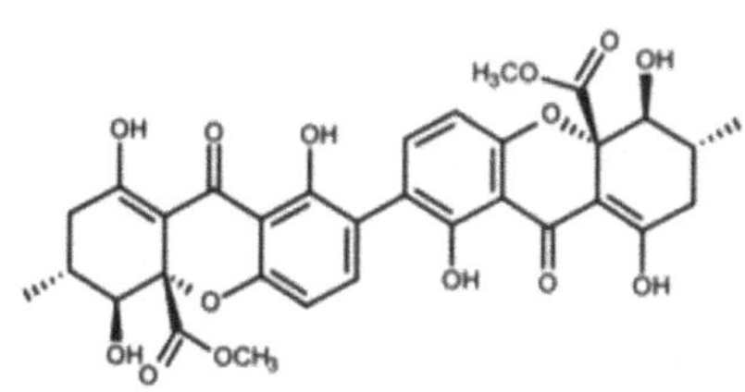

8

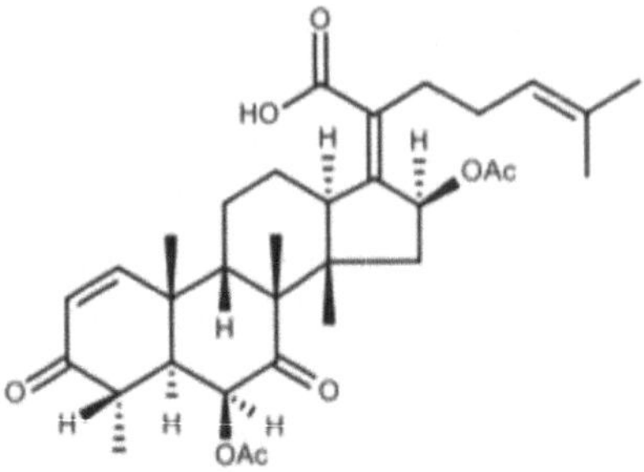

9

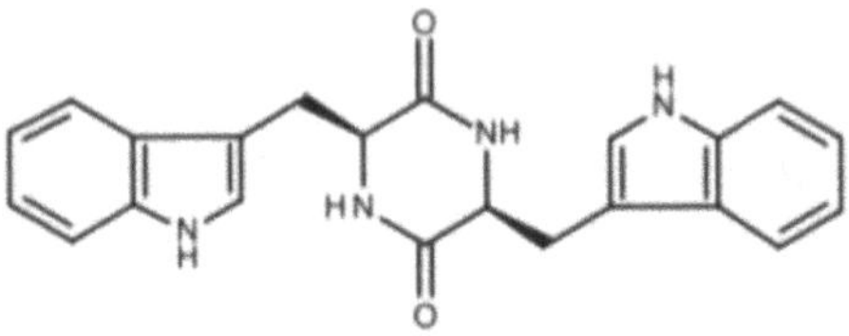

10

1: Sitostenone, 2: ergosta-4,6,8(14),22-tetraen-3-one, 3: dehydromevalonic lactone, 4: byssochlamic acid, 5: cyathisterone, 6: chevalone B, 7: aszonalenin, 8: secalonic acid A, 9: helvolic acid, 10: fellutanine A

Order: Poecilosclerida Family: Mycalidae

Mycale (Aegogropila) adhaerens

Associated fungi and their pharmaceutical compounds and activities

Antibacterial activity: The fungal strain *Exophiala pisciphila* NI10102, isolated from this sponge yielded the compound, exophilin A which showed antibacterial activity against Gram+ bacteria (Doshida et al. 1996; Thomas et al. 2010).

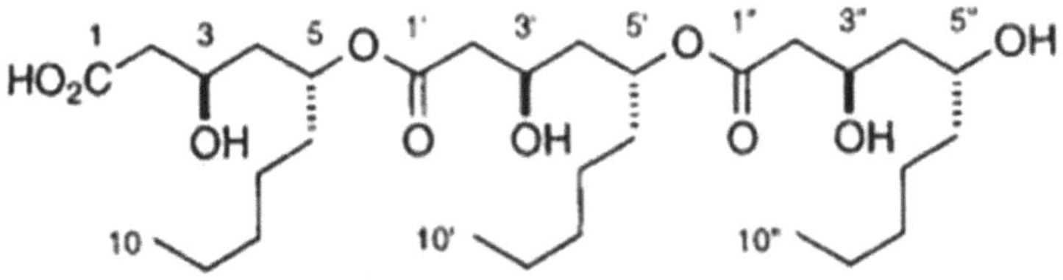

Exophilin A

Mycale (Zygomycale) angulosa

Common name (s): Not reported

Global distribution: Western Central Atlantic.

Habitat: Reef-associated; brackish; depth range 17 - 58 m

Associated fungi; and their pharmaceutical compounds and activities

Cytotoxic and antimicrobial activities: Fungal strain *Penicillium paxilli* Ma(G)K, derived from this sponge yielded compounds 5-(3-hydroxybutanoyl)- 4-methoxy-6-methyl-2H-pyran-2-one (pyrenocine B), (E)-5-but-2-enoyl-4-methoxy-6-methyl2H-pyran-2-one (pyrenocine A) and pyrenocine J ((E)-4- methoxy-5-(3-methoxybut-1-enyl)-6-methyl-2H-pyran2-one) which showed antimicrobial and cytotoxic activities against leukemia (HL60), human mammalian cancer cells (MDA-MB435), human colon cells (HCT8), and CNS cancer cells (CNS 295) (Kossuga et al. 2012).

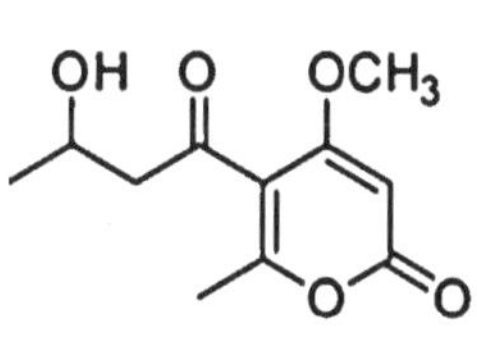

1

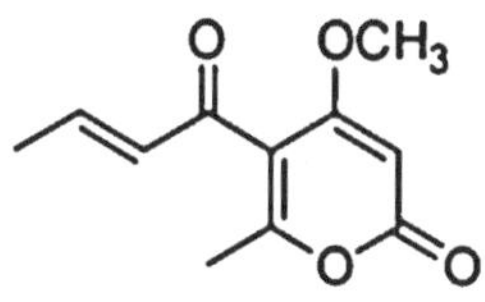

2

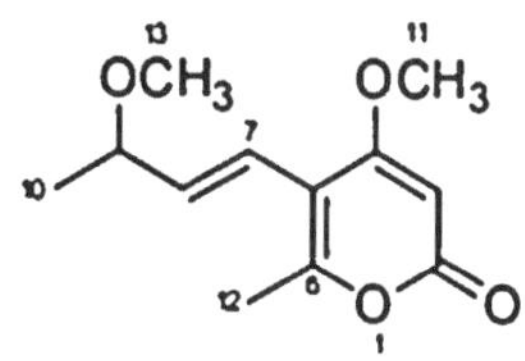

3

1: Pyrenocine B; 2: Pyrenocine; 3: A Pyrenocine J

Kossuga et al. (2012) reported that the fungal strain *Trichoderma* sp. Ma(T) H of this sponge also showed antimicrobial and cytotoxic activities with its compound roridin A.

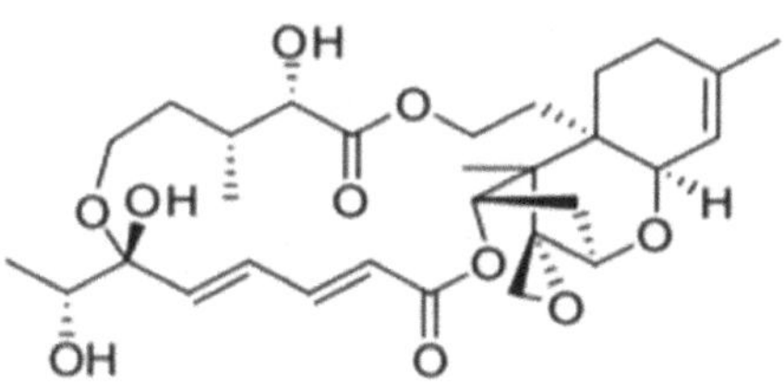

Roridin A

Mycale cecilia

Common name (s): Not reported

Global distribution: West Indian species; Galápagos Islands: Isabel

Habitat: Very common in shallow water locations such as Honolulu and Pearl Harbors.

Associated fungi and their pharmaceutical compounds and activities

Antimicrobial activity: The fungus *Trichoderma harzianum* isolated from this sponge species yielded trichoharzin which showed antimicrobial activity (Kobayashi et al. 1993).

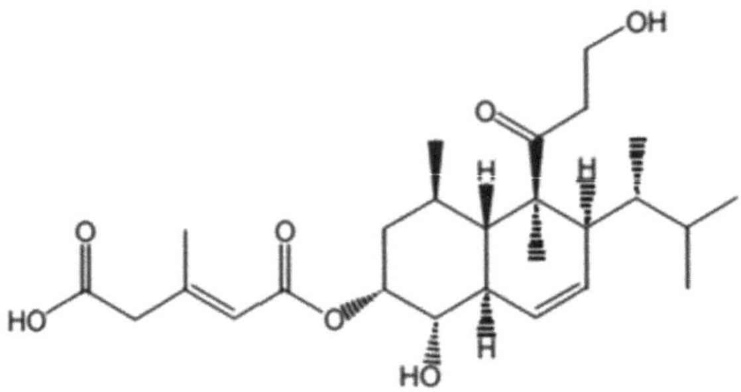

Trichoharzin

Mycale (Aegogropila) plumosa

Common name (s): Not reported

Global distribution: Indo-West Pacific

Habitat: Sessile

Associated fungi; and their pharmaceutical compounds and activities

Antitumor activities: The fungus *Penicillium auratiogriseum* isolated from this sponge species yielded compounds viz. fructigenine A and (S)-2,4-dihydroxy-1-butyl-(4-hydroxy) benzoate and which showed anti-tumor activity and exhibited potent cytotoxic effects. Further, two alkaloid compounds viz. aurantiomides B,C of another strain of this fungal species showed moderate cytotoxic activities (Thomas et al. 2010).

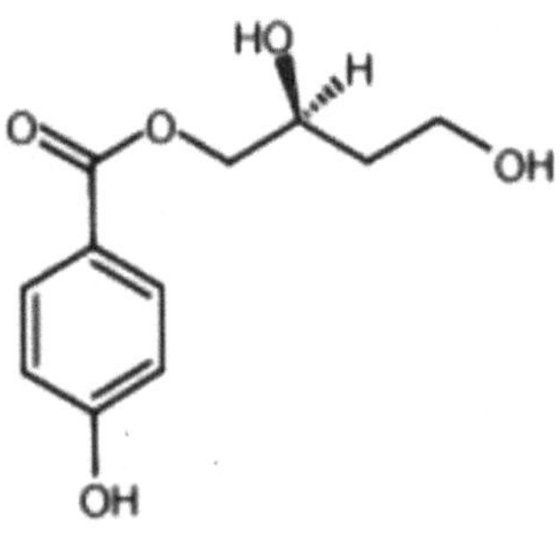

(S)-2,4-dihydroxy-1-butyl-(4-hydroxy) benzoate

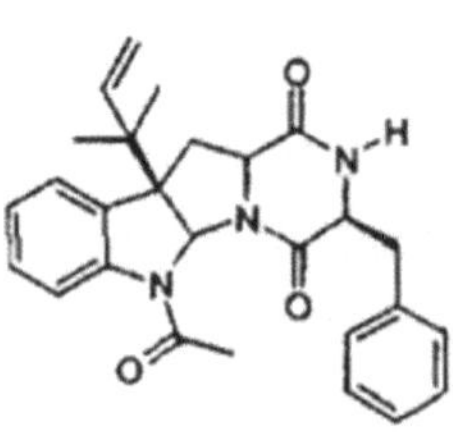

Fructigenine A

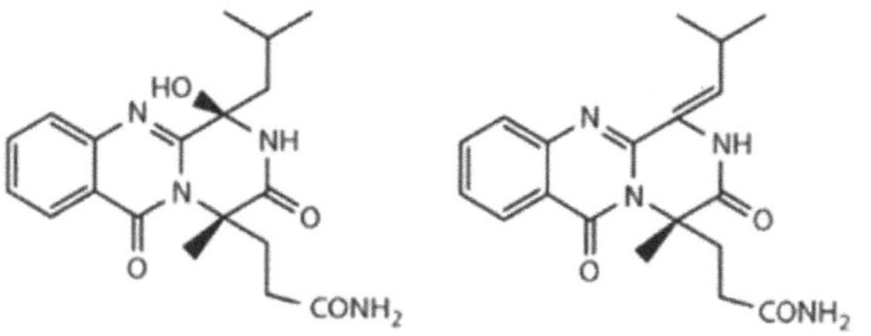

Aurantiomides B, C

Mycale sp.

Associated fungi and their pharmaceutical compounds and activities

Anti-tumor activity: The ethyl acetate fraction of the fungal species Eurotium cristatum (ECE),associated with ithis sponge species yielded 1,8-dihydroxy-6-methoxy-3-methyl-9,10-anthracenedione (physcion, (1), 2-(2', 3-epoxy-1',3'-heptadienyl)-6-hydroxy-5-(3-methyl-2-butenyl) benzaldehyde (2), and the dioxopiperazine alkaloid echinulin (3). The ECE and its metabolites showed growth inhibitory activity on three human tumor cell lines: breast adenocarcinoma (MCF-7), non-small lung cancer (NCI-H460), and melanoma (A375-C5) with the average GI50 values of 44.3, 45.5, and 71.3 μg/ml respectively. Compound 2 also displayed moderate activity against all the aforesaid three cell lines (GI50 = 58.3, 46.0, and 116.7 M for MCF-7, NCI-H460, and A375-C5, respectively), whereas compound 3 displayed poor activity on MCF-7 (GI50 = 109.7 M) and NCI-H460 (GI50 = 96.7 M) (Almeida et al. 2010b).

Order: Poecilosclerida Family: Myxillidae

Myxilla (Myxilla) incrustans

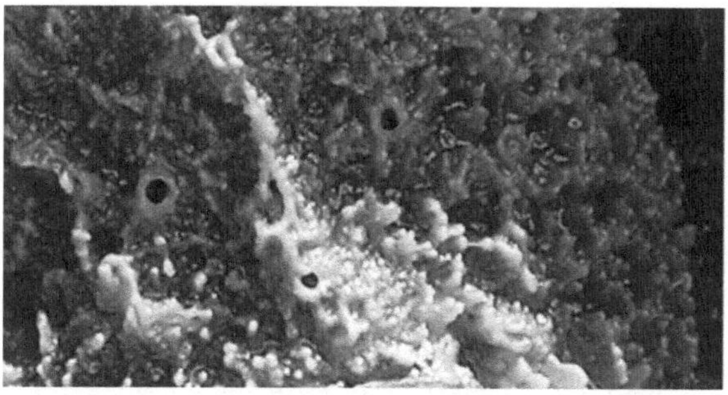

Image credit: Cwmhiraeth, https://creativecommons.org/licenses/by-sa/3.0/

Common name (s): Rough scallop sponge

Global distribution: Arctic, Bear Isl., Faroes, Norway, Atlantic coasts of Europe south to Gibraltar, Mediterranean. Recently known from many sites around the British Isles.

Habitat: In clear water. Low water mark to 400 m on stones, shells, sand and mud; lives on scallops such as *Chlamys hastata* and *Chlamys rubida*.

Associated fungi and their pharmaceutical compounds and activities

Antitumor activity: The fungus *Beauveria bassiana* from this sponge afforded a tetramic acid derivative viz. beauversetin which displayed moderate activity on an unknown human tumor cell line (Zhang et al. 2017).

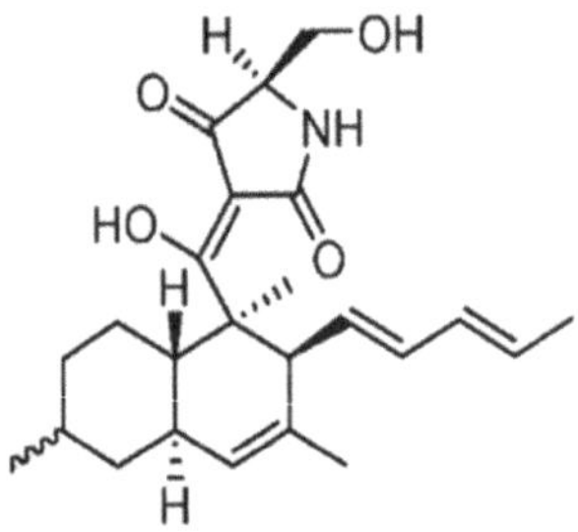

Beauversetin

Antifungal activity: The fungus *Microsphaeropsis sp.* H5-50 strain derived from this sponge yielded compounds *viz.* (R)-mellein, (3R,4S)-hydroxymellein, 4,8-dihydroxy-3,4-dihydro-2H-naphthalen-1-one, microsphaeropsisin and (3R,4R)-hydroxymellein. Among these compounds, microsphaeropsin, displayed antifungal activity (Thomas et al., 2010), an eremophilane derivative showed antifungal activity at the 50 μg level (Thomas et al. 2010).

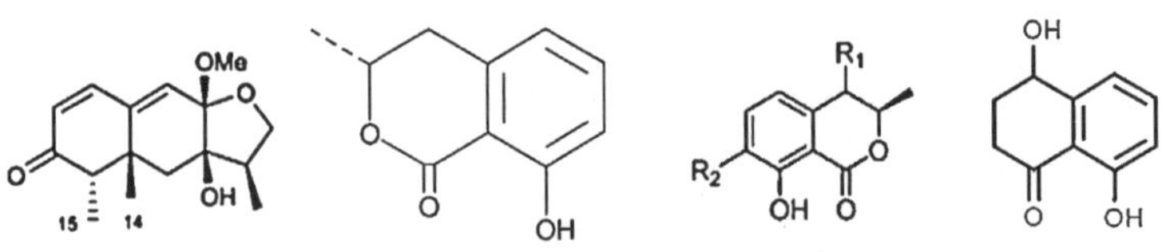

R1= αOH, R2=H

R1= βOH, R2=H

1 2 3,4 5

1: Microsphaeropsisin; 2: (R)-mellein; 3: (3R,4S)-hydroxymellein; 4:(3R,4R)-hydroxymellein; 5: 4,8-dihydroxy-3,4-dihydro-2H-naphthalen-1-one

Order: Poecilosclerida Family: Rhabderemiidae

***Rhabderemia* sp.**

Associated fungi; and their pharmaceutical compounds and activities

Anticancer ativity: The fungal strain, *Aspergillus similanensis* KUFA0013, derived from this sponge was found to produce a cyclohexapeptide, similanamide which displayed poor inhibitory activity on MCF-7 (breast adenocarcinoma), A373 (melanoma) and NCI-H460 (non-small cell lung cancer) cell lines (Zhang et al. 2017).

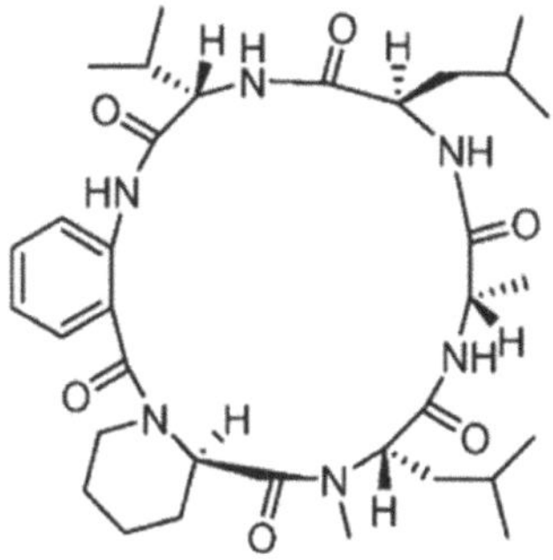

Similanamide

Order: Scopalinida   Family: Scopalinidae

Stylissa flabelliformis

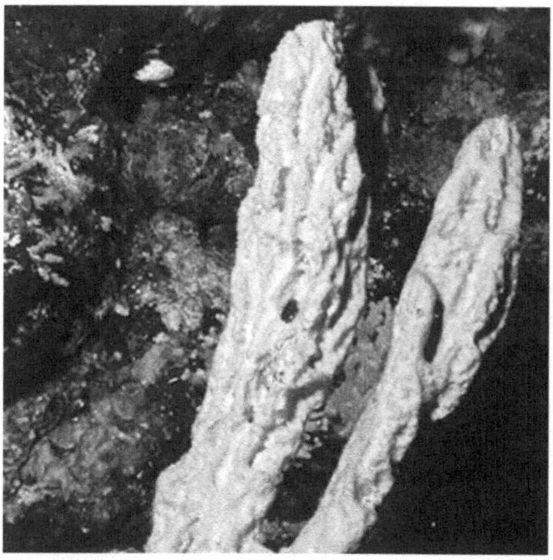

Stylissa sp.

Common name (s): Orange fan sponge

Global distribution: Indo-West Pacific.

Habitat: Sessile; depth range 3 - 76 m

Associated fungi: Therapeutic compounds and their bioactivities

Antifungal activity: The fungus *Aspergillus aculeatus* associated with this sponge species yielded pre-aurantiamine,(-)-9-hydroxyhexylitaconic acid and (-)-9-hydroxyhexylitaconic acid-4-methyl ester. Among these compounds, pre-aurantiamine showed antifungal activity against *Neurospora crassa* with an MIV value of 140 μM (Antia et al. 2011).

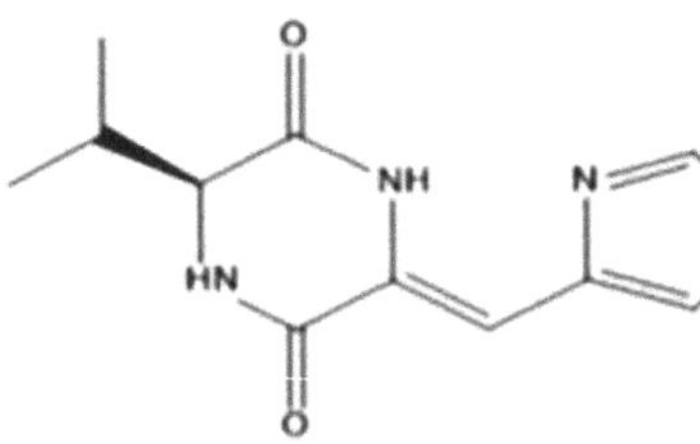

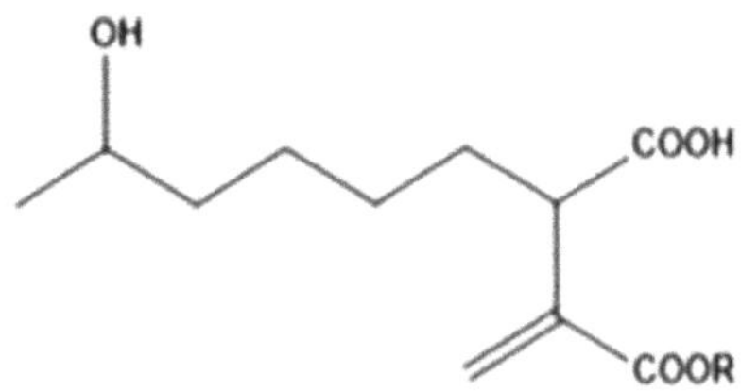

Pre-aurantiamine (-)-9- hydroxyhexylitaconic acid ((R=H)

(-)-9-hydroxyhexylitaconic acid-4-methyl ester((R=Me)

Order: Suberitida   Family: Halichondriidae

Hymeniacidon perlevis

Associated fungi and their pharmaceutical compounds and activities

Cytotoxicity: Two new prenylated diphenyl ethers obtained from the fungus *Aspergillus versicolor* Hmp-F48 which is derived from this sponge species showed moderate inhibitory activities against HL-60 (human promyelocytic leukemia) cell line with IC50 values of 6.4 and 20 μM, respectively. Further, the compound dibenzo(1,4)dioxin produced by the above strain exhibited similar inhibitory activity against HL-60 cells with IC50 3.62 μM (Zhang et al. 2017).

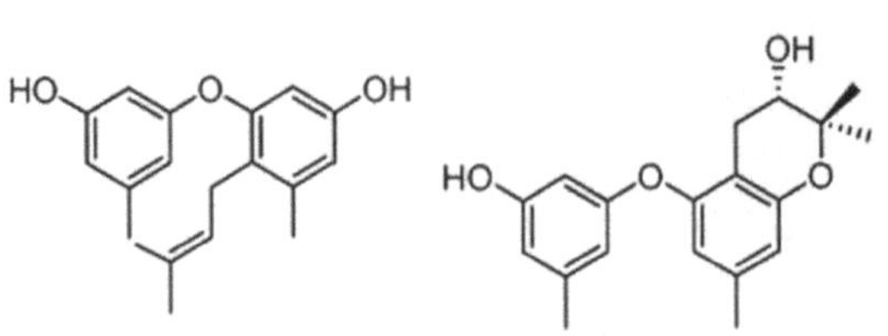

Prenylated diphenyl ethers

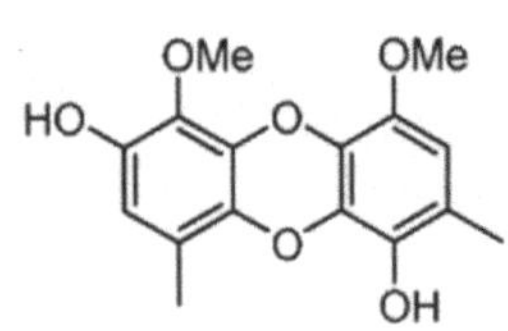

Dibenzo(1,4)dioxin

Antibacterial activity: The extract of the fungal strain of *Aspergillus versicolor* MF359 isolated from this sponge species yielded 5-methoxydihydrosterigmatocystin which displayed strong activity on *Bacillus subtilis* and *Staphylococcus aureus* with MIC values of 3 and 13 μg/mL respectively (Jin et al. 2016).

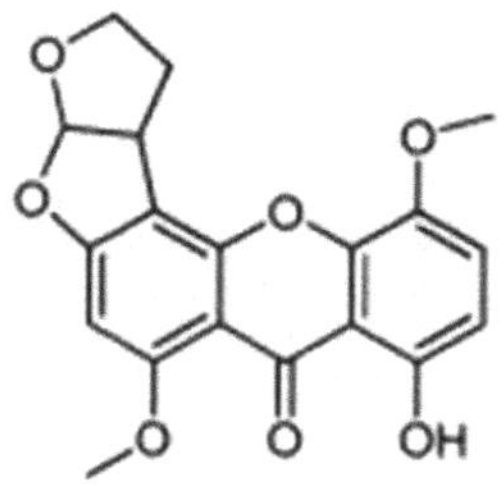

5-methoxydihydrosterigmatocystin

Nervous system activity: A polyketide, cyclopentenone compound *viz.* 5-hydroxycyclopenicillone derived from its associated fungus *Trichoderma* sp. HPQJ-34 showed nervous system activity by serving as a moderate free radical scavenger with anti-Aβ fibrillization and neuroprotective properties. The IC50 value recorded with the DHHP free radical inhibition was found to be 30 μM ((Mayer et al. 2021).

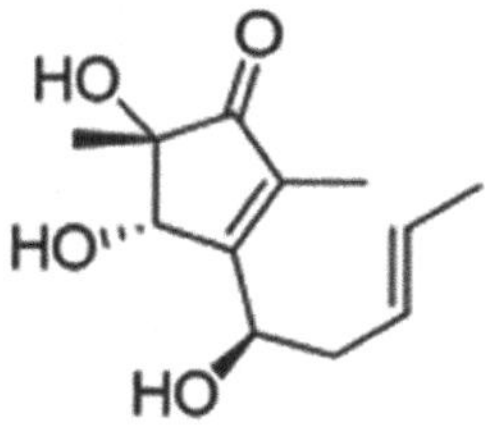

5-hydroxycyclopenicillone

***Hymeniacidon* sp.** (= *Stylotella* sp.)

Associated fungi; Therapeutic compounds and their bioactivities

Cytotoxic activity: An unidentified fungal species isolated form this sponge produced compounds Nectriapyrone A-B which have been reported to possess cytotoxicity against both human T leukemia and melanoma tumor cell lines (Guimarães et al. 2008).

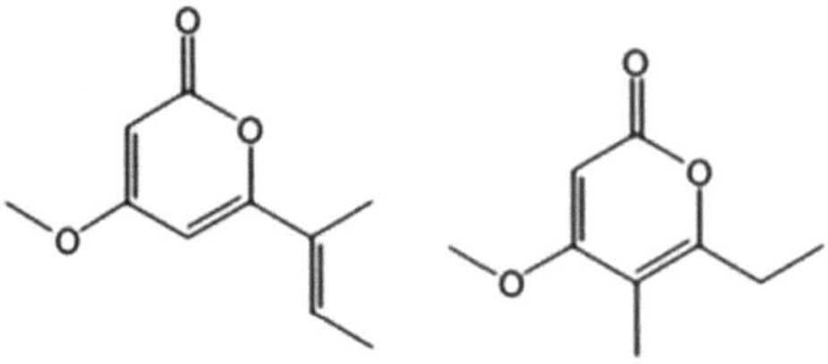

Nectriapyrones A,B

One bacterial strain, *Streptomyces carnosus*, obtained from marine sponge *Hymeniacidon* sp. has been reported tonproduce two lactones: *viz.* lobophorin C and D. While the compound, lobophorin C showed strong cytotoxicity on hepatoma 7402 cell proliferation with IC50 value of 0.6 μg/mL, compound D exhibited potent inhibitory activity on the human breast cancer cells (MDA-MB 435) with an IC50 value of 7.5 μM (Zhang et al. 2017).

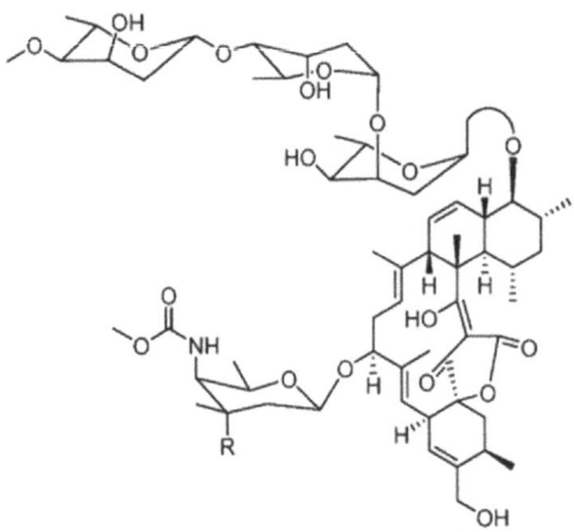

Lobophorin C (R-NO2); Lobophorin D (R=NH2)

Halichondria (Halichondria) japonica

Common name (s): Not reported

Global distribution: Japan, Korea

Habitat: Intertidal rocky shores

Associated fungi; and their pharmaceutical compounds and activities

The ascomycete fungus *Gymnascella dankaliensis* OUPS-N134 associated with this sponge species yielded the cytotoxic compounds viz. dankasterones A and B, and dankastatins A and B; and gymnastatins A-H, Q and R. Among these compounds gymnastatin Q and dankasterone A showed inhibition against breast and human cancer cells. Gymnastatins (except D,E,H) and dankastatins A and B displayed strong cytotoxicity on lymphocytic leukemia (P388). (Thomas et al. 2010).

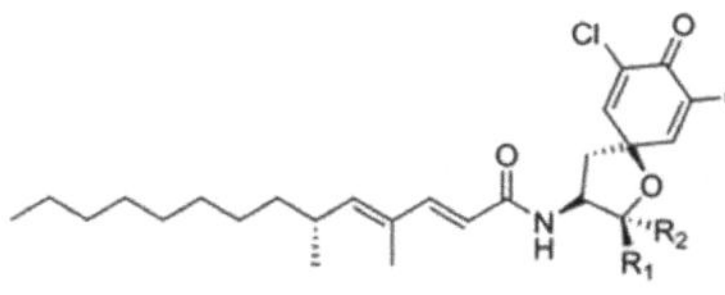

Gymnastatin A

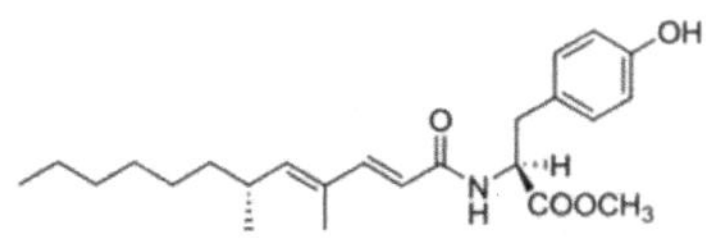

Gymnastatin H

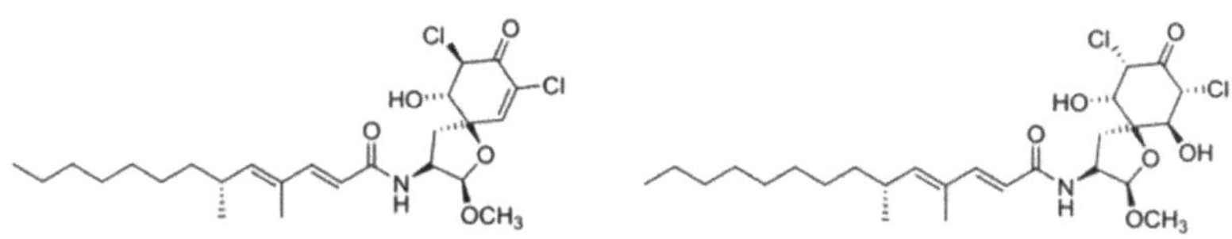

Gymnastatins B,C

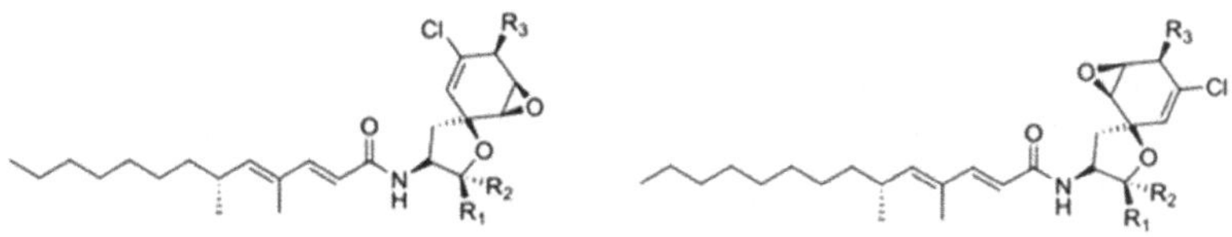

Gymnastatins D,E

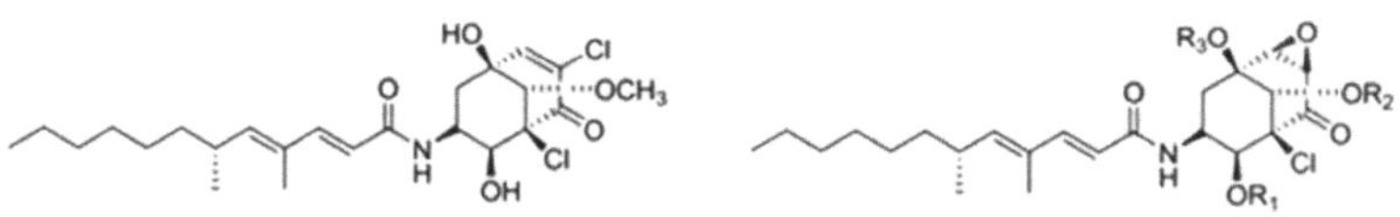

Gymnastatins F,G

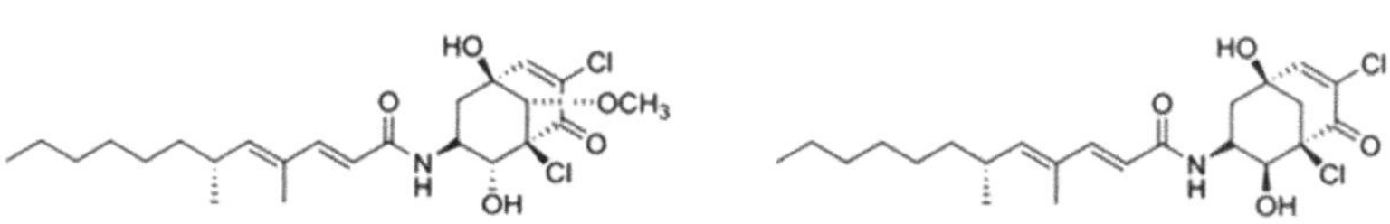

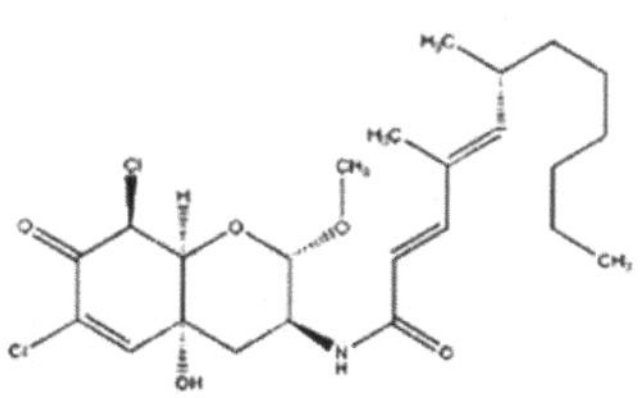

Gymnastatins Q,R

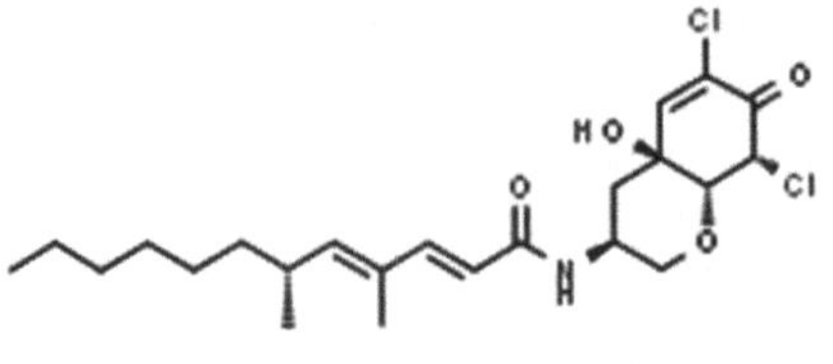

Dankastatins A,B

Antimicrobial activity: The fungal strain *Phoma* sp. Q60596 isolated from this sponge species gave rise to an antifungal antibiotic, YM-202204 which t exhibited strong antifungal activities on *Aspergillus fumigatus*, *Candida albicans,* and *Cryptococcus neoformans*. The strain *Bacillus cereus* QN03323 of this sponge yielded antibiotics, YM-266183 and YM-266184, which exhibited strong antibacterial activity on enterococci and staphylococci including MDR strains (Zhang et al. (2017).

Zhang et al. (2017) reported that unusual steroids metabolized by the fungal strain of *Gymnasella dankaliensis* isolated from this sponge exhibited potent inhibitory activity against the lymphocytic leukemia (P388) cell line with ED50 values ranging from 0.9 to 2.8 μg/mL.

Nweze et al.(2020) reported that its associated fungal strains LF327 and KF970 produced two polyketides viz. lindgomycin and ascosetin which displayed antimicrobial activity against MRSA with MIC values of 5.1 and 3.2 μM, respectively.

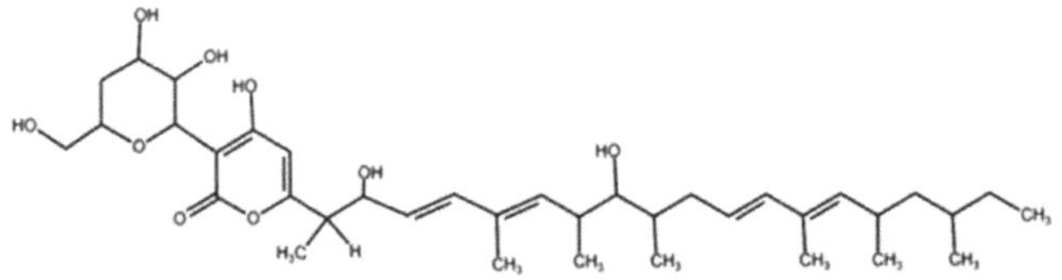

YM-202204

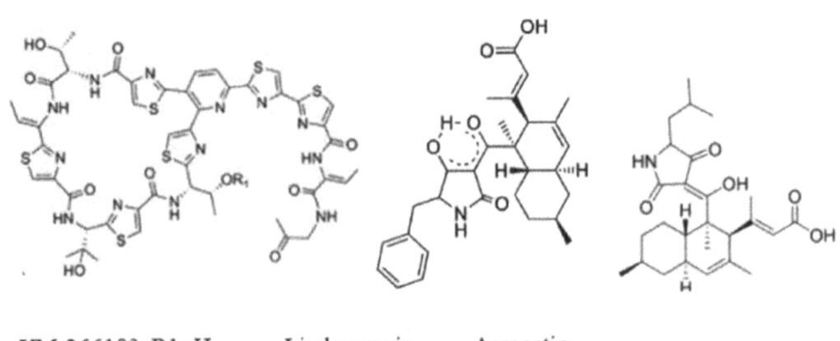

YM-266183, R1=H Lindgomycin Ascosetin

YM-266184, R1=CH3

Halichondria (Halichondria) okadai

Common name (s): Not reported

Global distribution: Pacific coast of Japan

Habitat: Intertidal areas; provides habitat for several crustaceans.

Associated fungi; and their pharmaceutical compounds and activities

Cytotoxicity: A fungal strain, *Trichoderma harzianum* OUPS-N115, isolated from this sponge species yielded metabolites *viz.* trichodenone A, B, and C; and harzialactones A, B, and R-Mevalonolactone. Among these compounds, as trichodenone A, B, and C exhibited significant cytotoxicity against leukemia P388 cell line (Thomas et al. 2010).

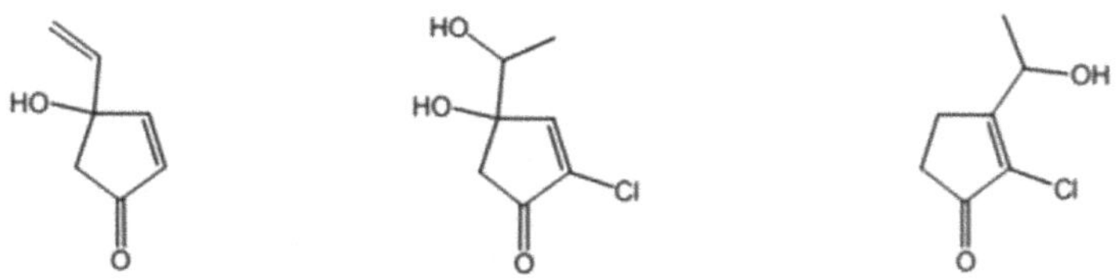

Trichodenone A-C

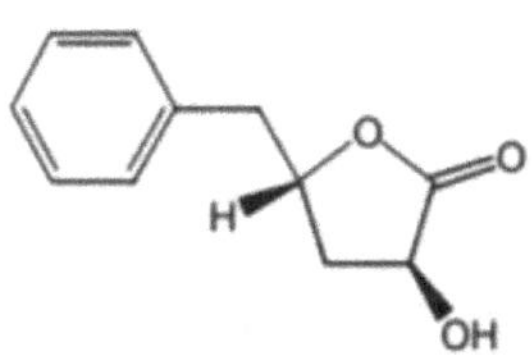

Harzialactone A,B

R-Mevalonolactone

Haliclona (Soestella) valliculata

Associated fungi and their pharmaceutical compounds and activities

Antiproliferative activity: The fungal strain *Emericella variecolor* isolated from this sponge species yielded isoemericellin (1) and evariquinone. (2). Compound 2 exhibited potent antiproliferative activity on NCI 503473, non-small cell lung cancer) cells (NCl-H460) and ATCC CCL17, human cervix carcinoma (KB) cell lines at a concentration of 3.16 μg/mL (Thomas et al. 2010; Gomes 2014).

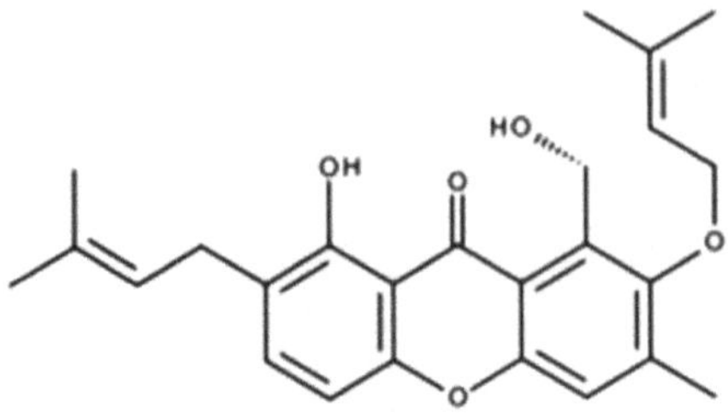

Isoemericellin

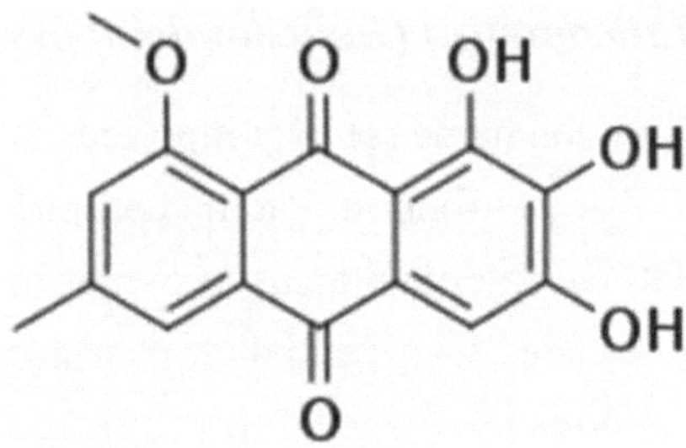

Evariquinone

Haliclona sp. 1

Associated fungi; and their pharmaceutical compounds and activities

Antibiotic activity: The saltwater culture of an unidentified fungus obtained from this sponge was found to yield hirsutane sesquiterpenes *viz.* hirsutanols AC and ent-gloeosteretriol. Among these compounds, hirsutanol A and ent-gloeosteretriol displayed poor antibiotic activity on *Bacillus subtilis* (Thomas et al. 2010).

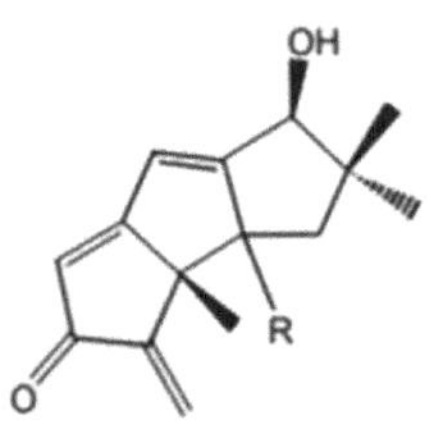

Hirsutanol A (α-OH)

Hirsutanol B (β-OH)

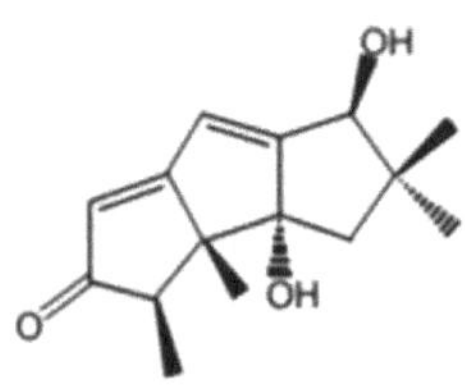

Hirsutanol C

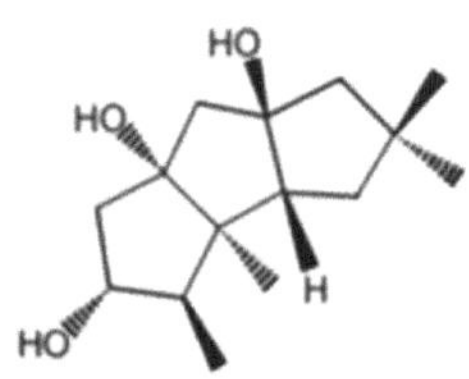

ent-Gloeosteretriol

Haliclona sp.2

Associated fungi and their pharmaceutical compounds and activities

Antimicrobial activity: The fungus *Aspergillus niger* associated with this sponge yielded 4-hydroxy- α-pyrones viz. nipyrones A–C and an analogue, germicidin C which displayed weak to moderate antimicrobial activity against MRSA (Nweze et al. 2020).

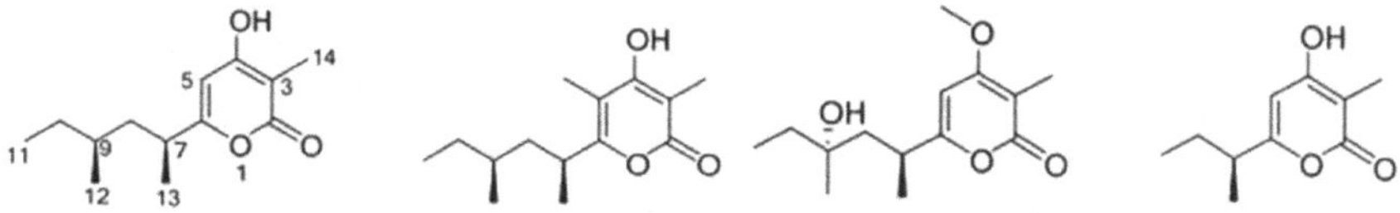

Nipyrones A,B Nipyrone C Germicidin C

Order: Suberitida Family: Suberitidae

***Homaxinella* sp.**

Associated fungi; Therapeutic compounds and their bioactivities

Cytotoxic activity: Ten new alkaloid compounds (1–10) produced by the fungal strain *Gymnasella dankaliensis*, a symbiont this sponge exhibited cytotoxic activities against the lymphocytic leukemia P388 cell line with ED50 values ranging from 0.03 to 18.0 μg/mL. Further, the 7th compound showed inhibitory activity on MKN7 (stomach) and BSY-1 (breast) tumor cell lines with GI values of 5.2 and 5.5, respectively. Further, the alkaloid, dankastatin C of this fungal strain acted against P388 (murine lymphocytic leukemia) cell line with an ED50 value of 57 ng/mL (Zhang et al. 2017).

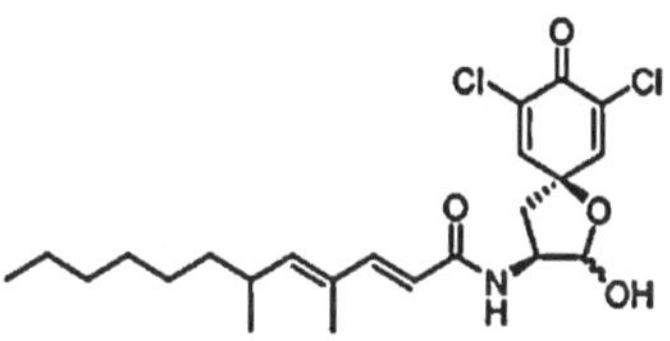

1

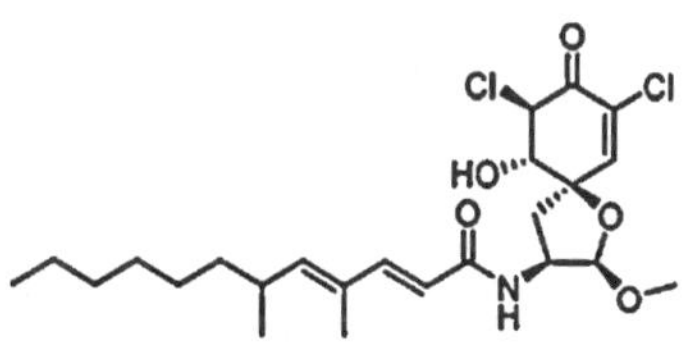

2

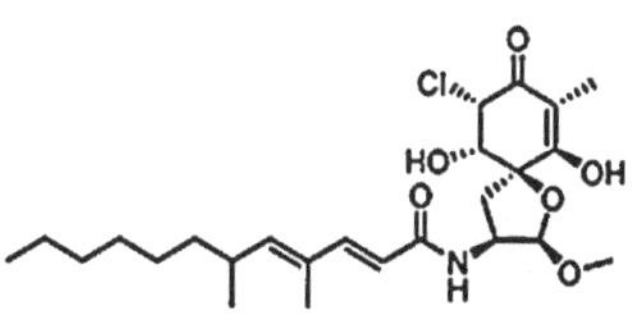

3

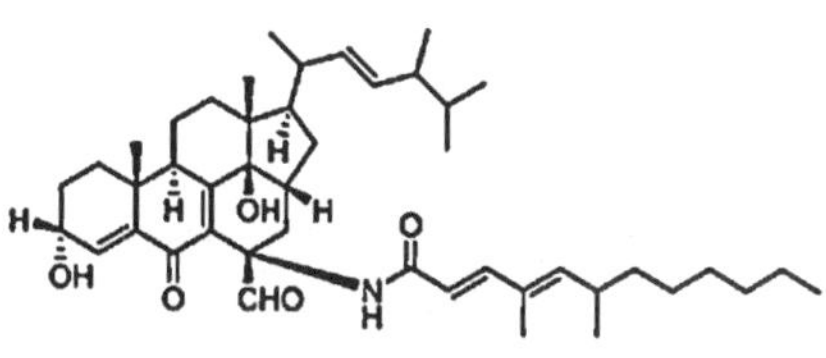

4

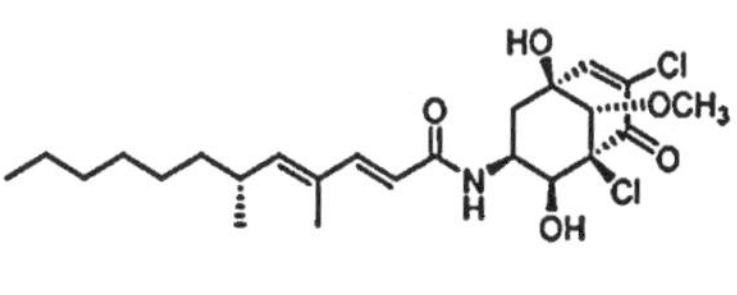

5

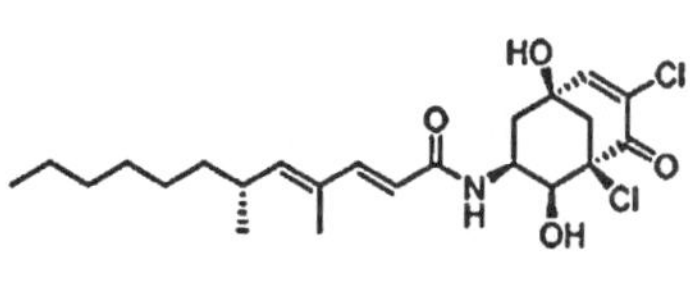

8

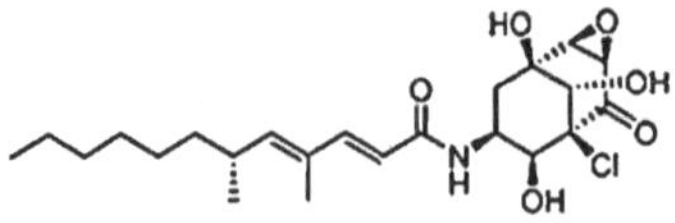

6

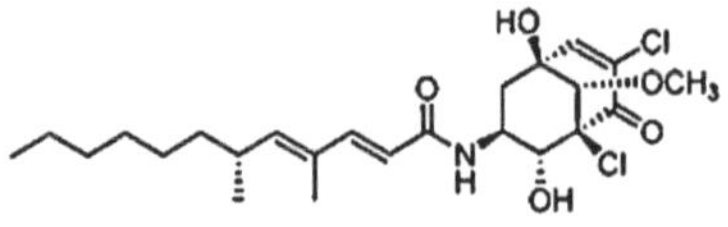

7

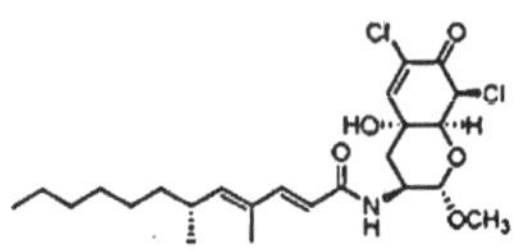

9

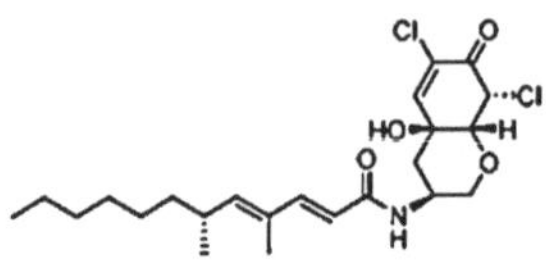

10

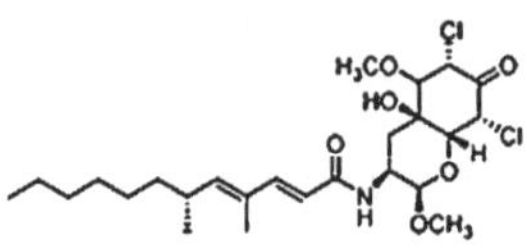

11

1-10: Cytotoxic alkaloids 11: Dankastatin C

Suberites domuncula

Associated fungi; and their pharmaceutical compounds and activities

Cytotoxicity: The ethyl acetate extract of the fungal species *Aspergillus ustus* derived from this marine sponge yielded new drimane-type sesquiter-penoids (1-3; 6-9) together with the related compounds viz. deoxyuvidin B (4), strobilactone B (5) and RES-1149-2 (10). Among these compounds, 6,7 and 10 displayed cytotoxic activity on a panel of human tumor cell lines, including L5178Y (mouse lymphoma), HeLa (human cervical cancer), and PC12 (rat pheochromocytoma) cells with IC50 values of 1.6, 15.8, and 19.3 μM, respectively (Elissawy et al. 2015).

Zhang et al.(2017) reported that three -pyrone derivatives (1-3) were characterized from *Petriella* sp. associated with this sponge species. While compound 1 exhibited pronounced cytotoxic activity against the L5178Y cell line, congeners 2 and 3 had moderate activity.

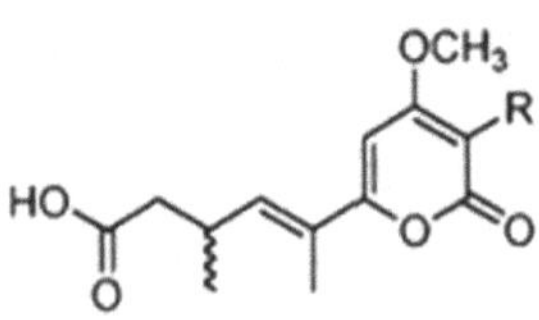

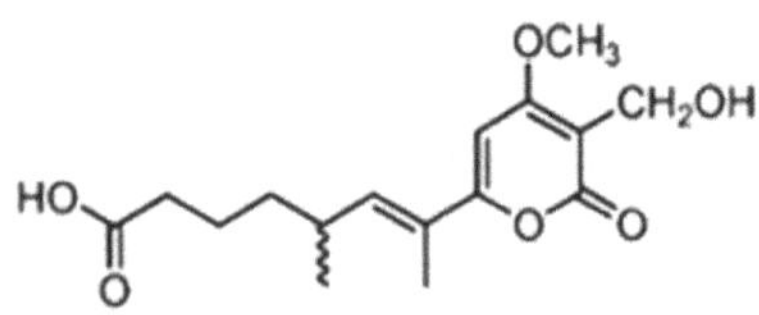

R=CH3; R=CH2OH

α-pyrone derivatives

Order: Tetractinellida Family: Ancorinidae

Jaspis johnstoni

Common name (s): Not reported

Global distribution: Western Central Pacific: Indonesia.

Habitat: Sessile.

Associated fungi; and their pharmaceutical compounds and activities

Anticancer activity:Tricyclic sesquiterpenes chloriolines A-C, coriolin B, and dihydrocoriolin C were derived from an unidentified fungus of the class Hyphomycetes associated with this sponge. Among these compounds, coriolin B displayed potent inhibitory activity on the CNS and human breast cell lines with IC50 values of 0.5 and 0.7 μg/mL, respectively (Thomas et al. 2010).

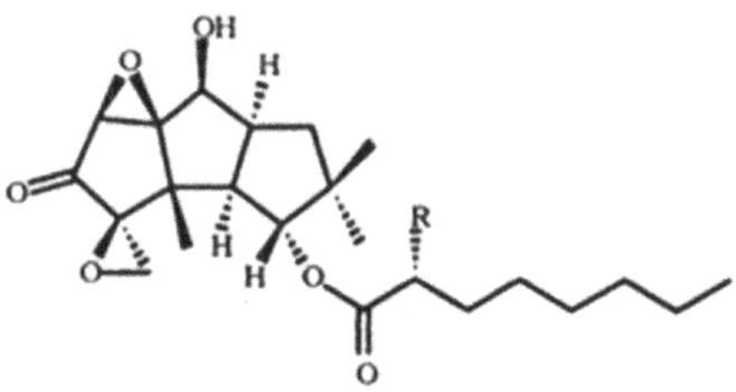

Coriolin B, R=H;

Dihydrocoriolin C, R=OH

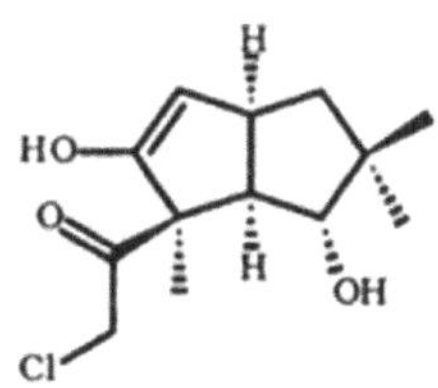

Chlorolin A

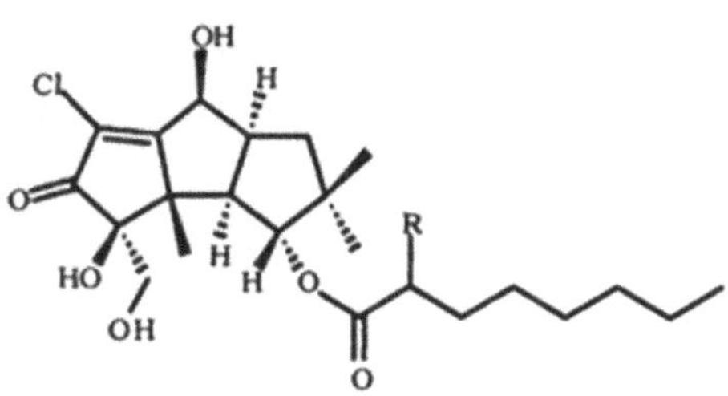

Chlorolin B,R=OH; Chlorolin C, R=H

Jaspis stellifera (= *Jaspis coriacea*)

Common name (s): Mound sponge

Global distribution: Indo-West Pacific.

Habitat: Sessile; depth range 2–25 m;; inhabits shallow lagoons where it encrusts dead corals, coral rubble and any other available hard substrates

Associated fungi: Therapeutic compounds and their bioactivities

The fungal strain *Aspergillus ochraceus* isolated form this sponge species yielded the compounds chlorocarolide A-B, R-mellein, penicillic acid, and hexylitaconic acid, the bioactivities of which are to be known (Holler 1999).

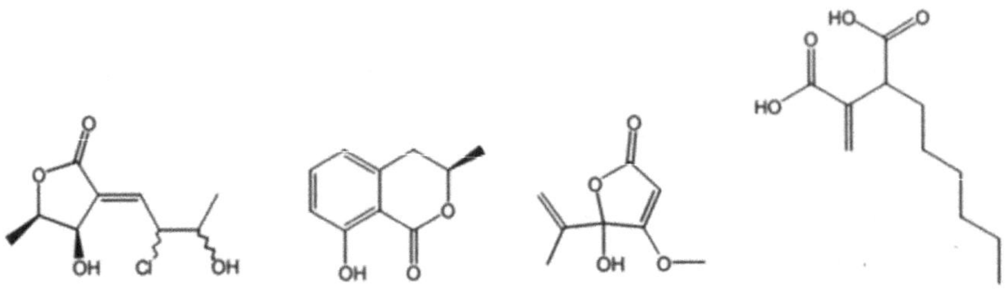

Chlorocarolide A, 7R*8S* R-Mellein Penicillic acid Hexalitaconic acid
Chlorocarolide B, 7S*8S*

Stelletta sp.

Common name (s): Grey wall sponge

Global distribution: Northen Cape to KwaZulu-Natal

Habitat: Rocky reefs

Associated fungi and their pharmaceutical compounds and activities

Anti-inflammatory activity: The fungus *Penicillium* sp. isolated from this sponge produced hexylitaconic acid derivatives and (3S)-methyl 9-hydroxy-3-methoxycarbonyl-2-methylenenonanoate. These compounds were found to be weak inhibitors of proinflammatory mediator interleukin-1β in murine macrophage (Xu et al. 2019).

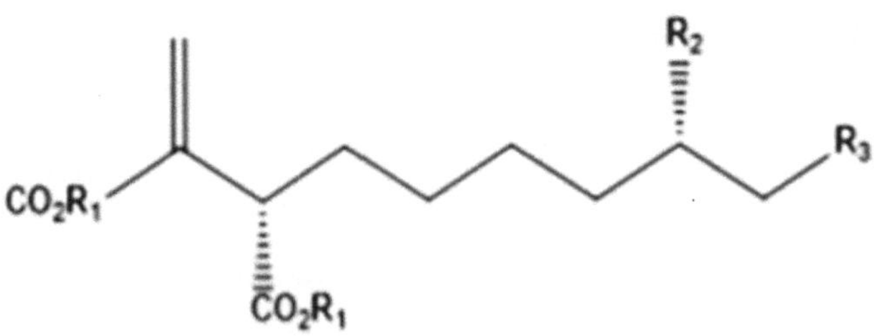

Hexylitaconic acid derivatives

R1= Me,R2=OH,R3=H; R1=R3=H, R2=OH; R1=R2=H, R3=OH

Antioxidative and antibacterial activity: The fungus *Aspergillus sydowii* derived from this sponge yielded compounds such as diorcinolic acid

(1), diphenylethers (2-7), -d-glucopyranosyl aspergillusene A (8), a diketopiperazine (9), a chromone (10) and a xanthone (11). Among these metabolites, compounds 1 and 8 displayed moderate cytotoxic activity on human nasopharyngeal carcinoma cells (KB), human colon cancer cells (HCT 116) and human liver cancer cells (HepG2). Further, compounds 2 and 4-7 exhibited moderate antibacterial activity on human pathogenic *Staphylococcus aureus*; and compounds 4 and 7 also showed anti-oxidative activity by weakly suppressing NO production (Liu et al. 2017a).

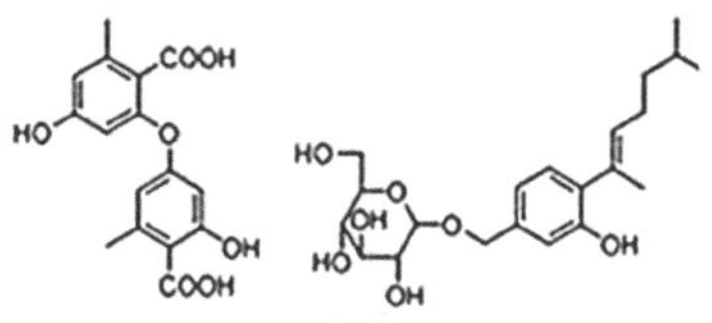

1 2

1: Diorcinolic acid 2: β-d-glucopyranosyl aspergillusene A

Order: Tetractinellida Family: Geodiidae

Geodia cydonium

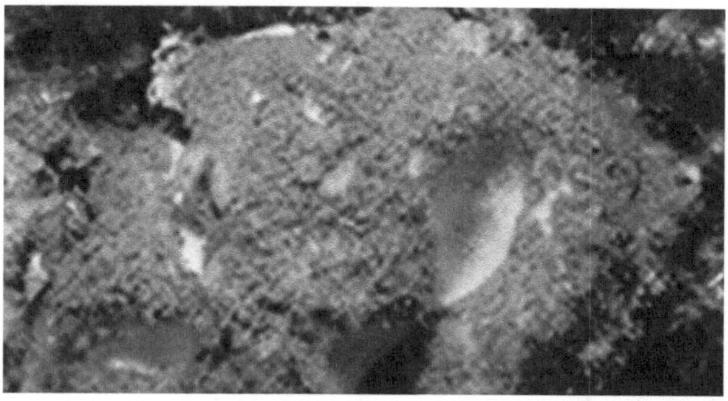

Image credit: Luis Sánchez Tocino, https://creativecommons.org/licenses/by-sa/4.0/

Common name (s): Not reported

Global distribution: Northeast Atlantic; British Isles, Roscoff, Norway; Mediterranean

Habitat: Pebble bottoms at 80-94 m

Associated fungi and their pharmaceutical compounds and activities

Antiproliferative activity: The methanol extract of the marine fungus *Arthrinium* sp. associated with this sponge yielded diterpenoids, arthrinins A,D and myrocins A, D; xanthone derivatives, norlichexanthone and anomalin A; and decarboxycitrinone and 2,5-dimethyl-7-hydroxychromone. Among these compounds, norlichexanthone and anomalin featured the strongest antiproliferative activities (IC50 values of 0.4074.0 μM). Further, norlichexanthone and anomalin A inhibited aurora-B, PIM-1, and VEGF-R2 kinases with IC50 values ranging from 0.3 to 11.7 μM. Furthermore, anomalin A and myrocins D and A were found to inhibit the endothelial cell sprouting with the C50 values of 1.8,2.6, and 3.7 μM, respectively (Elissawy et al. 2015).

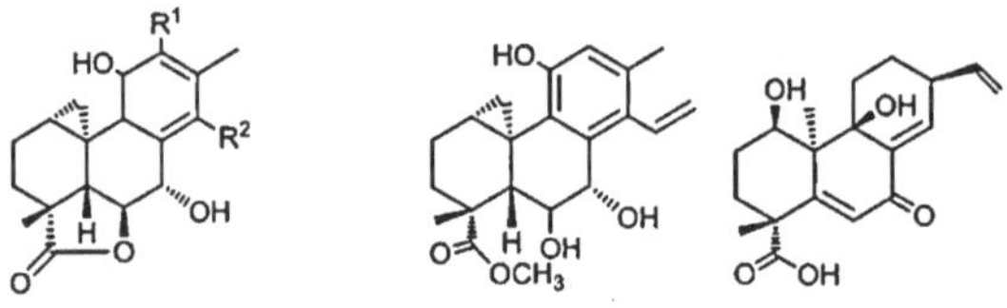

1 2 3

1: Arthrinin A (R1=H, R2= -C=CH2); Arthrinin B (R1=-C=CH2, R2=H); Arthrinin C (R1,R2=H); 2.Arthrinin D; 3: Myrocin D

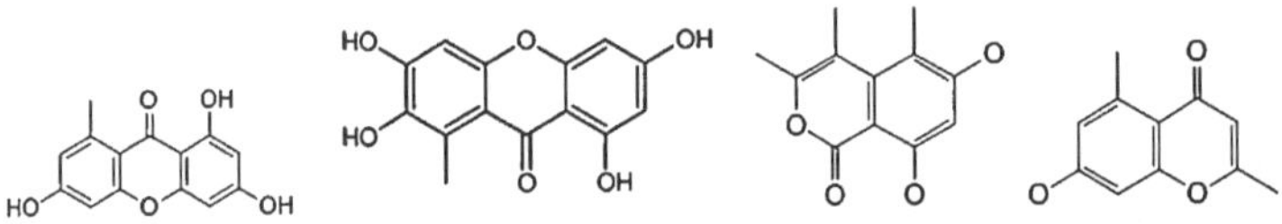

Norlichexanthone Anomalin A Decarboxycitrinon 2,5-dimethyl-7-hydroxychromone

Order: Tethyida Family: Tethyidae

Tethya aurantium

Image credit: Ed Bierman, https://creativecommons.org/licenses/by/2.0/

Common name (s): Golf ball sponge, orange puffball sponge,

Global distribution: From southern Namibia around the southern African coast to KwaZulu-Natal

Habitat: Usually on shallow reefs.

Associated fungi and their pharmaceutical compounds and activities

Antibacterial activity: Five meroterpenoid austalides M-Q have been purified from *Aspergillus* sp. which is associated with this sponge (Zhou et al. 2011). Further, the fungus *Penicillium chrysogenum* isolated from this sponge yielded a compound named cillifuranone which displayed weak antibacterial activity against *Xanthomonas campestris* and *Septoria tritici* (Wiese et al. 2011).

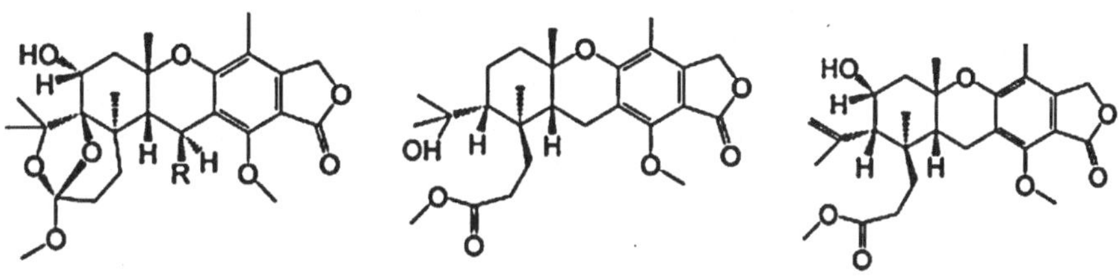

1 2 3

1: Austalide M (R=Ome); Austalide N (R=OAc); Austalide O (R=OH); 2. Austalide P; 3. Austalide Q

Antibacterial and anti-tumor activity: The fungus *Scopulariopsis brevicaulis* associated with this sponge yielded two cyclodepsipeptide compounds *viz.*, scopularides A and B, which showed weak activity against Gram-positive bacteria; and potent activity against a number of human tumor cell lines at a concentration of 10 μg/mL (Yu et al. 2008).

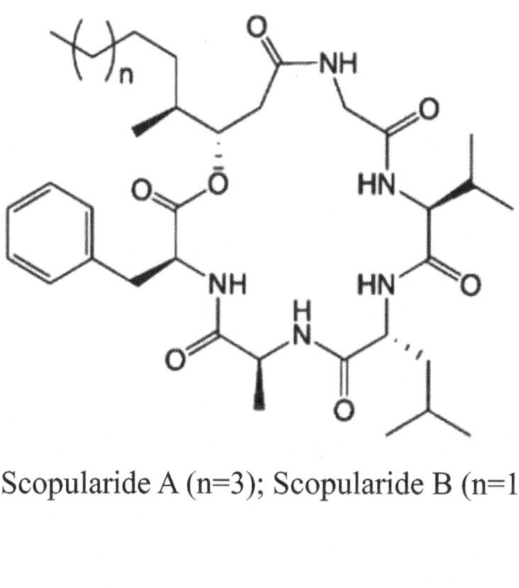

Scopularide A (n=3); Scopularide B (n=1)

Antimicrobial activity: The fungal species *Bartalinia robillardoides* strain LF550 associated with this sponge species produced compounds *viz*. helicusin A, deacetylsclerotiorin, helicusin E, isochromophilone X isochromophilone XI and bartanolide. All the compounds except the last one showed antimicrobial activity against pathogenic bacteria and fungi viz. *Bacillus subtilis, Staphylococcus lentus, Candida albicans, Septoria tritici* and *Trichophyton rubrum* (Jansen et al. 2013).

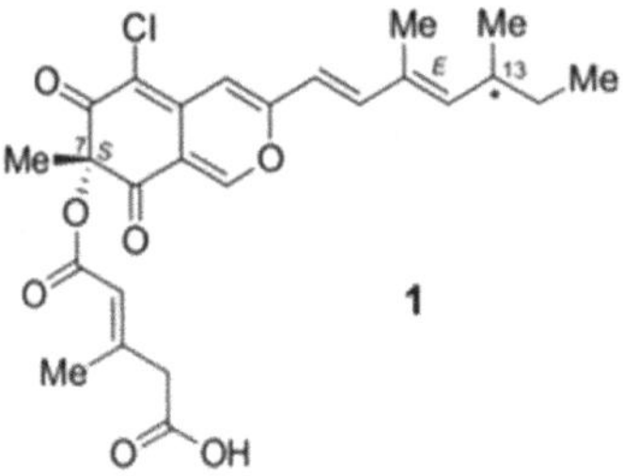

1

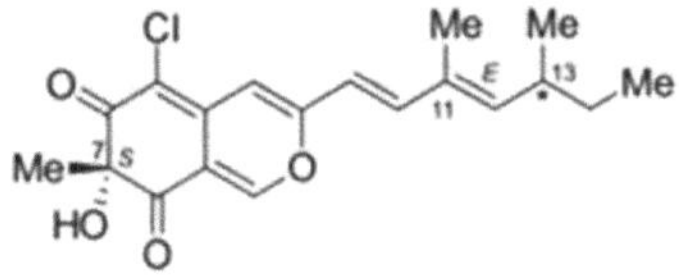

2

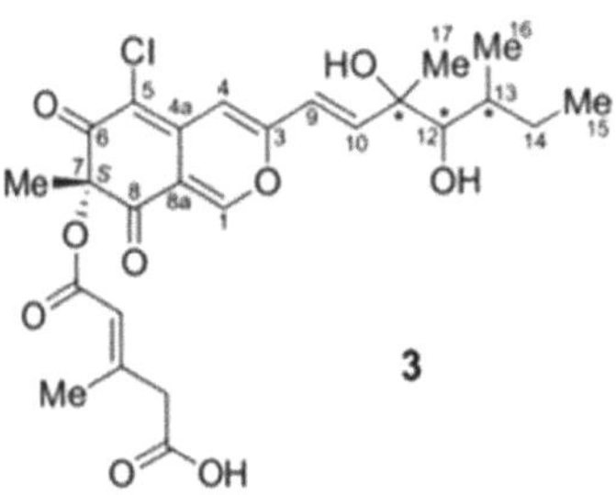

3

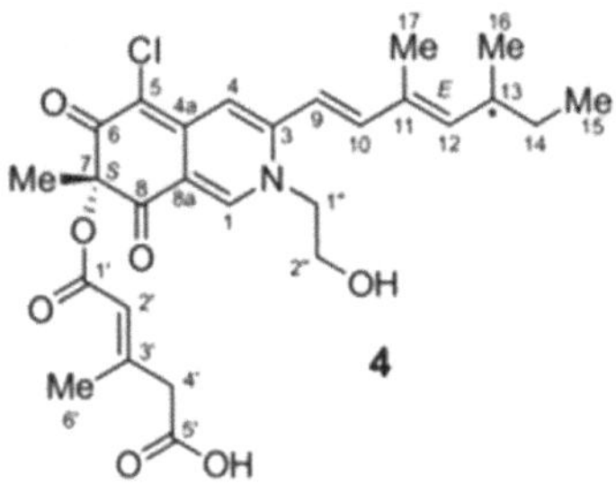

4

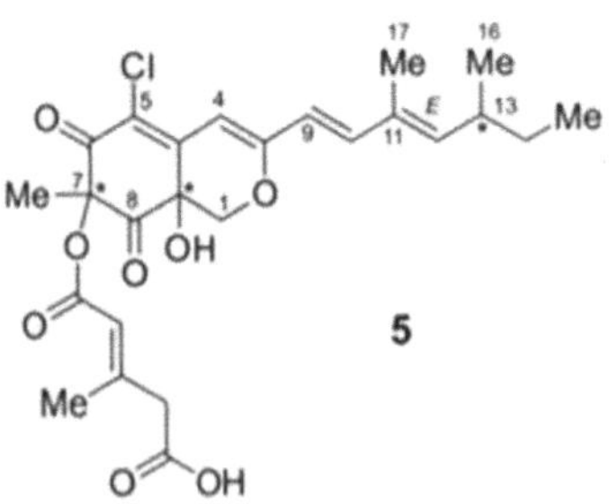

5

Helicusin A (1); deacetylsclerotiorin (2); helicusin E (3); isochromophilone X (4); isochromophilone XI (5)

Order: Tetractinellida Family: Tetillidae

Cinachyrella **sp.**

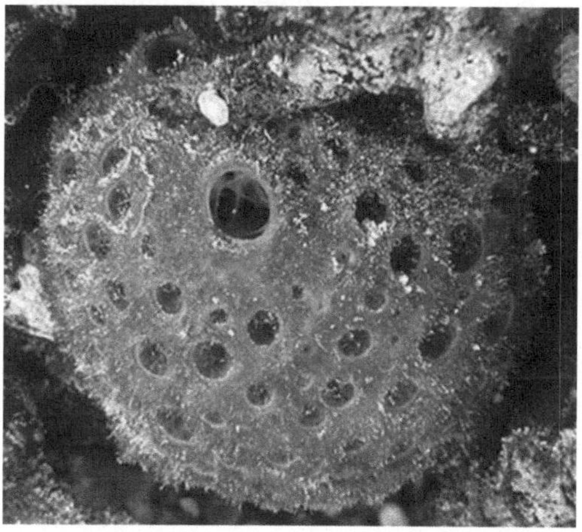

Associated fungi; and their pharmaceutical compounds and activities

Anticancer activity: Chemical analysis of the fungus *Emericella variecolor* XSA-07-2 isolated from this sponge afforded four lactones varioxiranols IL which displayed moderate cytotoxicity against human ovarian cancer (A2780) human colon carcinoma (HCT-116), gastric cancer (BGC-823), lung cancer stem cells (NCI-H1650), and liver hepatocellular carcinoma (HepG2) (Zhang et al. 2017).

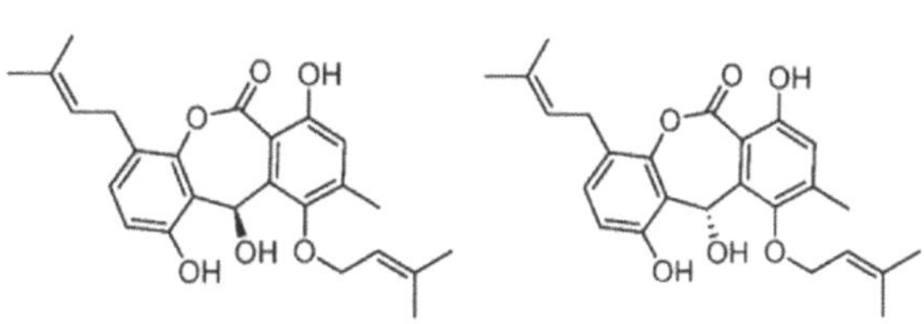

Varioxiranols I,J

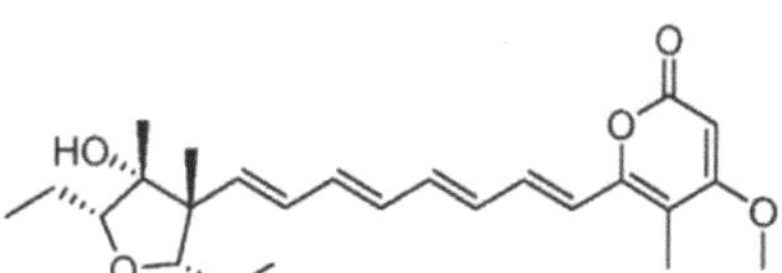

Varioxiranol K

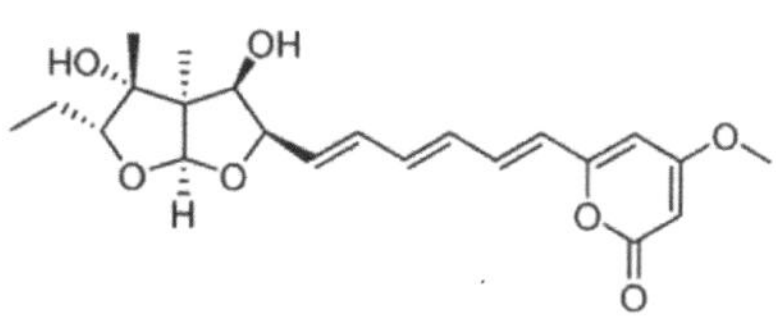

Varioxiranol L

Antibacterial activity: The extract of the fungus *Trichoderma reesei* PDSP 5.7 MG547722.1 isolated from this sponge species showed best antibacterial activity against *Salmonella enterica* ser. *typhi* (inhibition zone 14.72 mm), followed by ESBL *Escherichia coli,* (9.10 mm), *Staphylococcus haemolyticus* (9.04 mm) and methicillin-resistant *Staphylococcus aureus* (MRSA) (8.97mm) (Sibero et al., 2018).

Order: Verongiida Family: Aplysinidae

Aplysina aerophoba

Common name (s): Yellow Tube Sponge, golden sponge

Global distribution: Mediterranean, Portuguese, and NW Spain.

Habitat: This species prefers shallow water (0–20 m) habitats which are exposed to direct sunlight

Associated fungi and their pharmaceutical compounds and activities

Protein kinase C inhibitor: An unidentified fungal species of the genus microsphaeropsis produced a compound *viz.* 10-Hydroxy-18-methoxylbetaenone which is known to be a protein kinase C inhibitor. The latter is therapeutically very important since it plays an important role in carcinogenesis, tumour cell invasion and neoplastic transformation (Thomas et al., 2010).

Antioxidant properties: Jadulco et al. (2002) reported that a strain of *Cladosporium herbarum* which is associatd with this sponge species produced compounds herbarin A, herbarin B, citreoviridin A, phthalide and herbaric acid. Among these compounds, herbarins A,B showed antioxidant properties (Heimberger et al. 2015) and citreoviridin A showed antibacterial activity (Müller 2012).

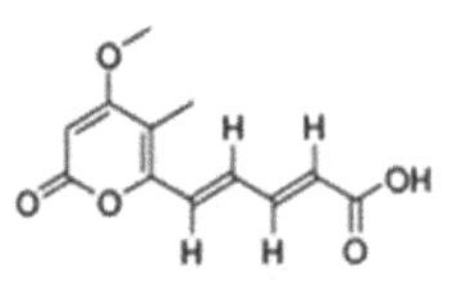
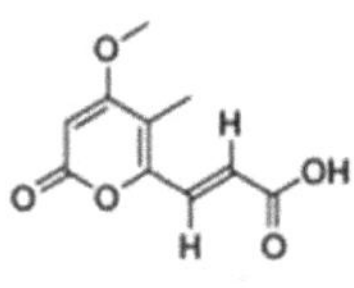
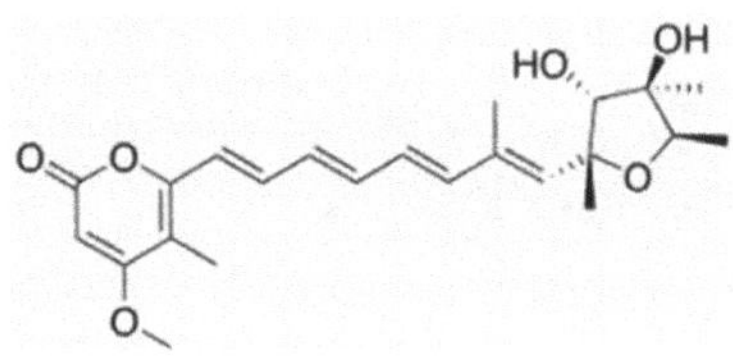

Herbarins A,B Citreoviridin A

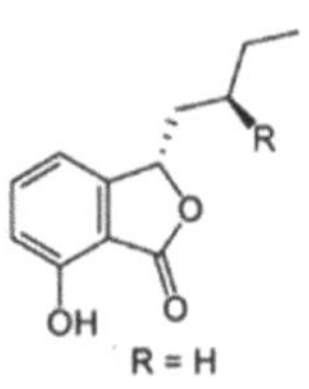
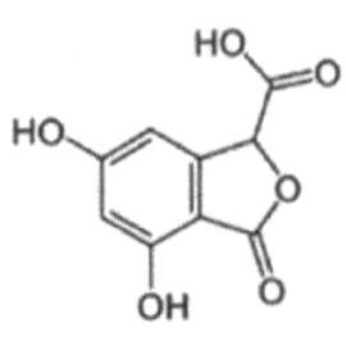

Phthalide Herbaric acid

Order: Verongiida Family: Pseudoceratinidae

Pseudoceratina purpurea

Associated fungi; and their pharmaceutical compounds and activities

Anti-tumor activity: The extract of the fungus, *Metarrhizium* sp. 001103 isolated from this sponge species yielded six N-methylated depsipeptides viz. desmethyl B, E and E2 chlorohydrin, and destruxins A, B, B2, Among these compounds, desmethyl B and E chlorohydrin and destruxins A, B2 exhibited inhibitory activity against certain human tumor cell lines. On the other hand, E2 chlorohydrin displayed cytotoxic activity on murine c38 cell line (Thomas et al. 2010).

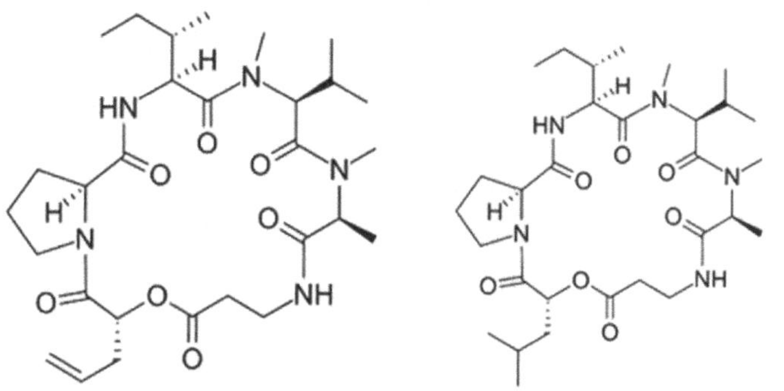

Destruxins A, B

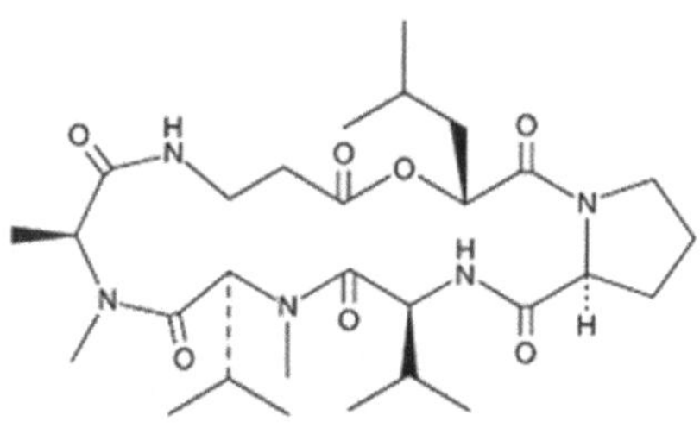

Destruxin B2

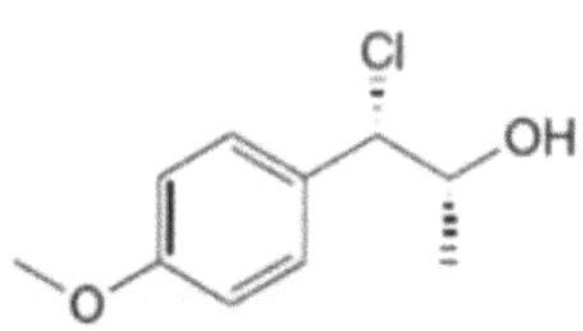

Chlorohydrin B

Antiproliferative activity: From a strain of the fungus *Emericella variecolor* derived from this marine sponge, the compound evariquinone was derived. This compound has been reported to exhibit antiproliferative effect on KB tumor cell line with 60% inhibition; and NCI-H460 cell line with 69% inhibition at 3.2 mg/mL (Zhang et al. 2017).

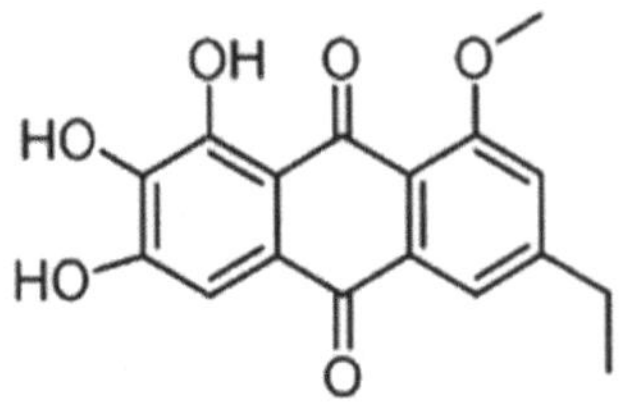

Evariquinone

Cytotoxic effects: The fungal strain *Tritirachium* sp., SpB081112MEf2, associated with this sponge produced three compounds viz. JBIR-97,98 and 99. All these compounds were shown to have cytotoxic effects on HeLa cells with IC50: values of 11, 17, and 17 μM, respectively; and on ACC-MESO-1 ce line with IC5 values of 31, 63, and 59 μM, respectively (Zhang et al. 2017).

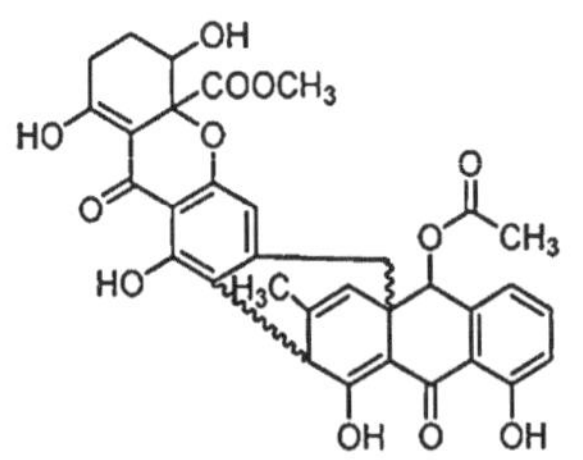

JBIR-97, 98

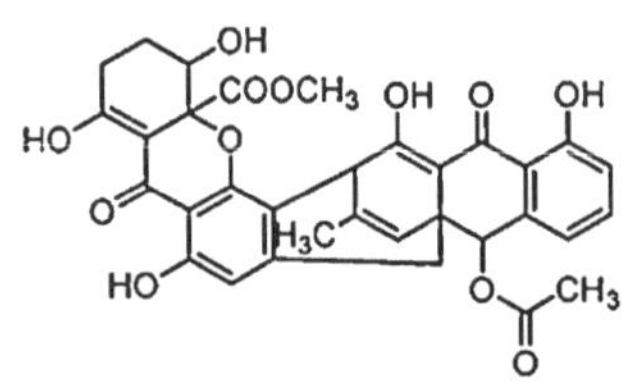

JBIR 99

Class: Calcarea Order: Leucosolenida Family: Grantiidae

Grantia compressa

Common name (s): Atlantic sponge

Global distribution: Northeast Atlantic and Northeast Pacific; British Isles; Arctic to Channel Islands.

Habitat: Reef-associated; sea shore where specimens are found attached to the underside of overhangs, often in abundance; they also occur commonly in the shallow sublittoral, e.g. on kelp stipes.

Associated fungi and their pharmaceutical compounds and activities

Antiviral activities: The crude organic extracts of the fungus *Eurotium chevalieri* MUT 2316 associated with this sponge yielded 10 compounds *viz.* echinulin, neoechinulin A, flavoglaucin, isodihydroauroglaucin, neoechinulin D, physcion, dihydroauroglaucin, asperflavin, cinnalutein, and cyclo-L-Trp-L-Ala. Among these compounds, neoechinulin D, physcion and dihydroauroglaucin displayed inhibitory activity on influenza A virus by 70,80 and 100% respectively. Further, the compounds isodihydroauroglaucin, physcion and neoechinulin D inhibited herpes simplex virus 1 totally. Furthermore, isodihydroauroglaucin exhibited potent bactericidal activity with an MIC value of 464 μg/mL (Bovio et al. 2019).

Antibacterial activity: Among all the aforesaid compounds, isodihydroauroglaucin displayed the most promising antibacterial activity against several Gram-positive bacteria including *Bacillus cereus* and *Salmonella typhimurium* and the MIC values recorded ranged between 4 and 64 μg/mL. The lowest MIC value was with the strain *S. pneumoniae* Monza-82. On the other hand, the compound dihydroauroglaucin showed activity against *E. faecalis*, *S. pneumoniae*, and *S. aureus* strains with MIC values of 64, 8 and 128 μg/mL, respectively. Further, asperflavin was active against all *S. aureus* strains, *S. pneumoniae* Monza-82, and *Mycobacterium tuberculosis* with MIC values of 64 μg/mL, 32 μg/mL and 25 μg/mL, respectively (Bovio et al. 2019).

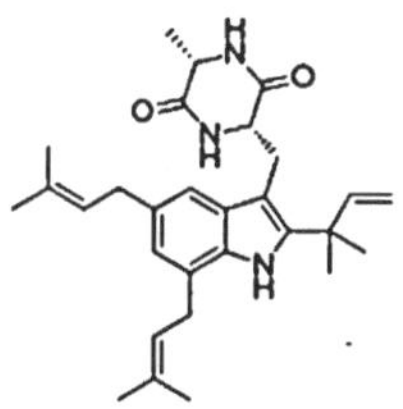

1

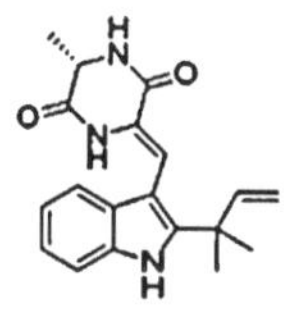

2

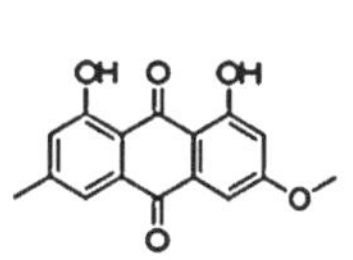

3

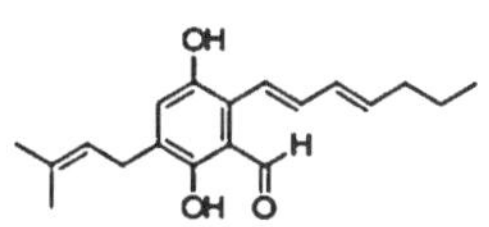

4

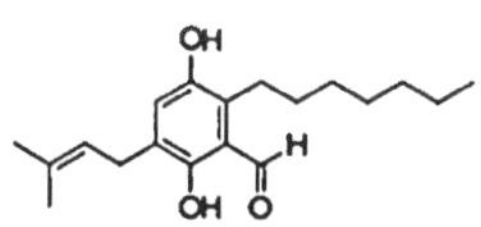

5

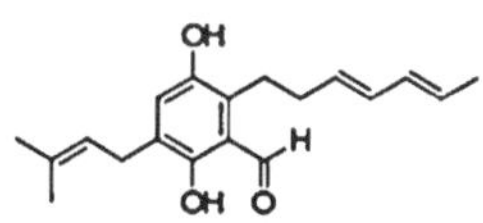

6

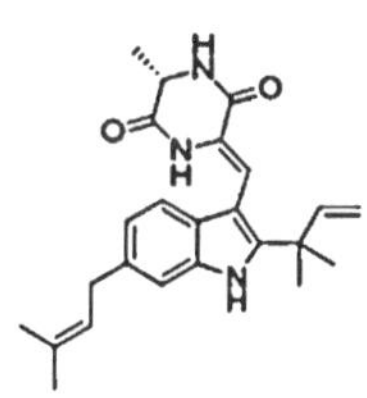

7

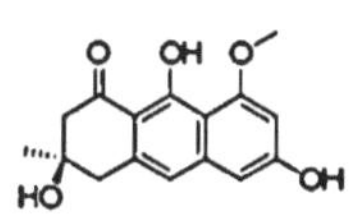

8

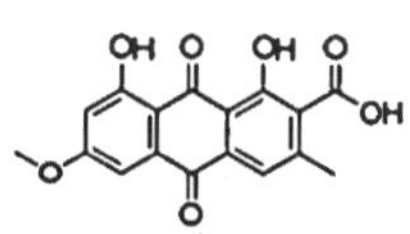

9

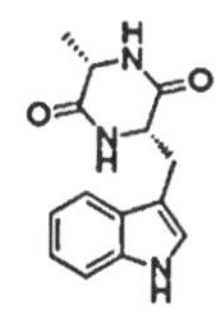

10

1: echinulin; 2: neoechinulin A; 3: physcion; 4: dihydroauroglaucin; 5: flavoglaucin, 6: isodihydroauroglaucin: 7:neoechinulin D, 8: asperflavin 9: cinnalutein; 10: cyclo-L-Trp-L-Ala

Unidentified sponge sp.1

Associated fungi and their pharmaceutical compounds and activities

Antimicrobial activity: The fungus *Aspergillus* sp. OPMF00272, isolated from this sponge yielded terretonin-type meroterpenes, terretonin G and terretonin. Compound terretonin G displayed very strong antibacterial effects on several Gram-positive bacterial species (Elissawy et al. 2015).

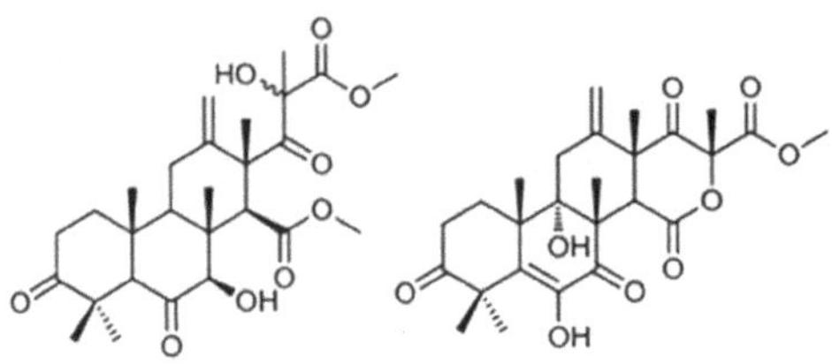

Terretonin G Terretonin

Unidentified sponge sp.2

Associated fungi; and their pharmaceutical compounds and activities

Antiproliferative activity: The fungus *Arthrinium sacchari* associated with this sponge yielded diterpene compounds libertellenone C, E and F; and myrocin A and.D. Among these compounds myrocin A and libertellenone C displayed antiprolifertive activity on human umbilical artery endothelial cells and human umbilical vein endothelial cells (Elissawy et al. 2015).

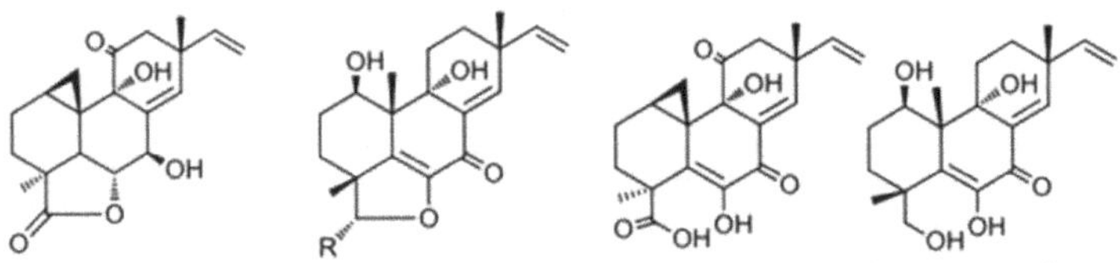

Myrocin D Libertellenone E,F Myrocin A Libertellenone C

Libertellenone E (R=OH); Libertellenone F (R=H)

Unidentified sponge sp.3

Associated fungi; and their pharmaceutical compounds and activities

Antibacterial and cytotoxic activities: The fungus associated with this sponge yielded compounds such as sagenomellamides A–C and and comazaphilone

E. Among these compounds, sagenomellamide A showed antibacterial activity against the strains of *Bacillus subtilis, Staphylococccus epidermidis, Salmonella typhimurium, Escherichia coli* and *Staphylococcus aureus* (MSSA) with MIC values of > 64 μg/mL. Further this compound had cytotoxicity against A-498 renal cancer, ACHN renal cnacer and MCF-7 breast cancer cell lines with IC50 values of 48.2, 47.6, and 51.6, respectively (Him 2015).

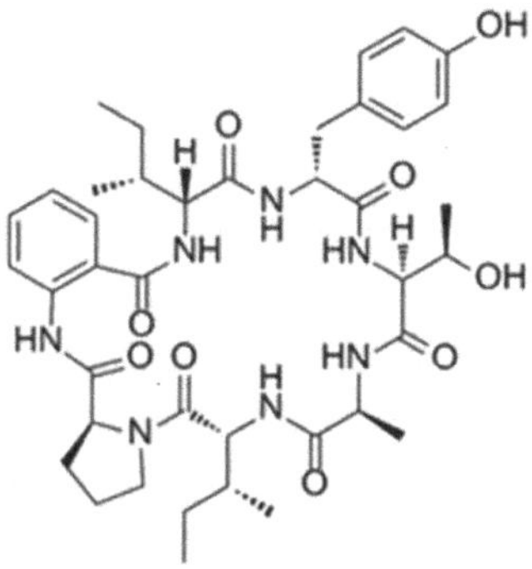

Sagenomellamide B

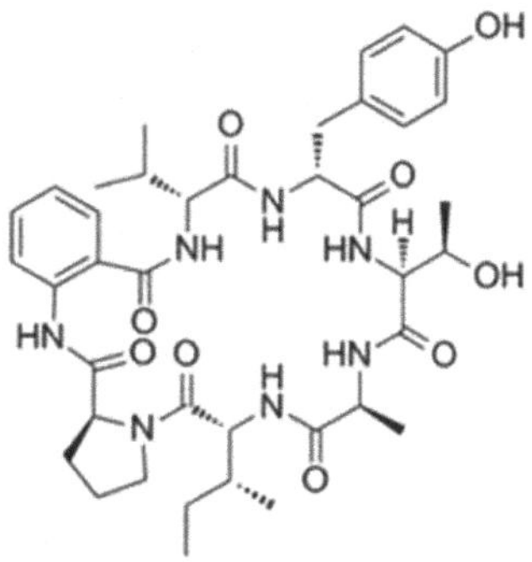

Sagenomellamide A

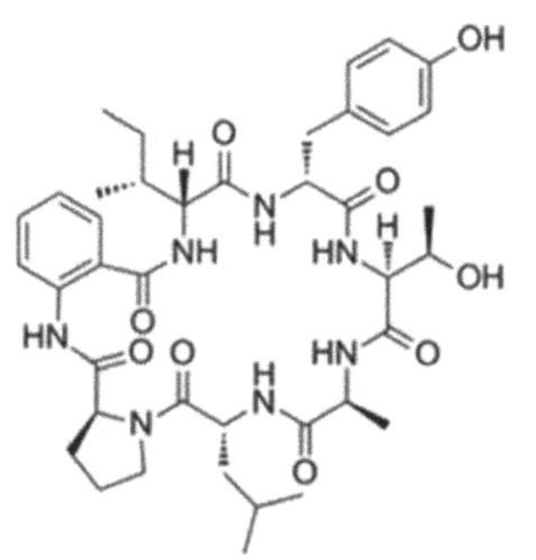

Sagenomellamide C

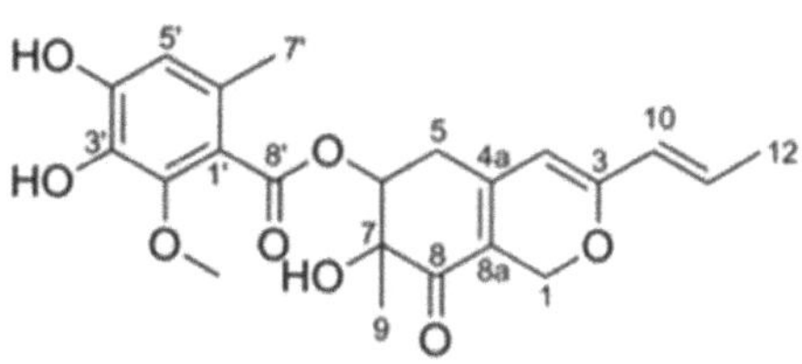

Comazaphilone E

***Unidentified sponge* sp.4**

Associated fungi and their pharmaceutical compounds and activities

From an unidentified Choristida sponge found in Korea, *Acremonium strictum* was isolated which yielded acremostrictin, a tricyclic lactone demonstrating strong antioxidant activity in the DPPH radical scavenging assay and inhibited death of human keratinocyte HaCat cells (Maharaj, 2015).

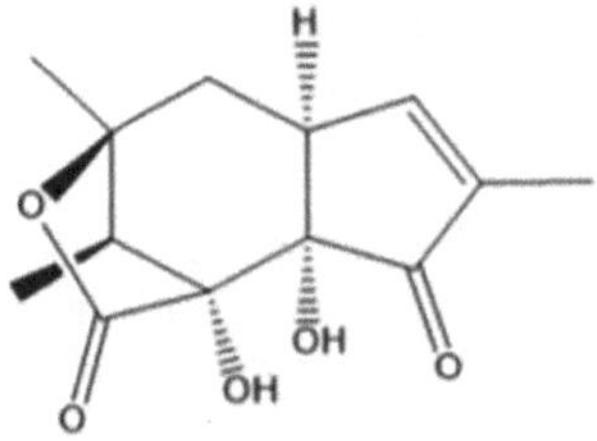

Acremostrictin

Unidentified sponge sp.5

Associated fungi; and their pharmaceutical compounds and activities

Antibiotic activity: The ethyl acetate extract of this sponge-derived fungal species viz. *Aspergillus similanensis* KUFA 0013 yielded the compounds viz. 6, 8-dihydroxy-3-methylisocoumarin, 6,8-dihydroxy-3,7-dimethylisocoumarin, 5-hydroxy-8-methyl-2H, 6H-pyrano[3,4-g]chromen-2,6-dione, reticulol (4), chevalone E (5), pyripyropene S (6),chevalone C, S14-95 (7), and pyripyropene E (8). Among them, chevalone E (5) was found to show activity against methicillin-resistant *Staphylococcus aureus* (MRSA) (Prompanya et al., 2014).

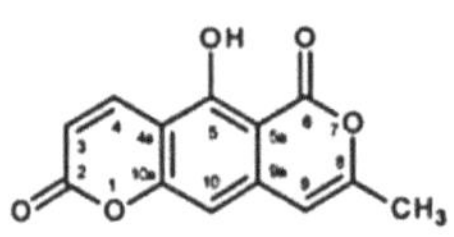

1

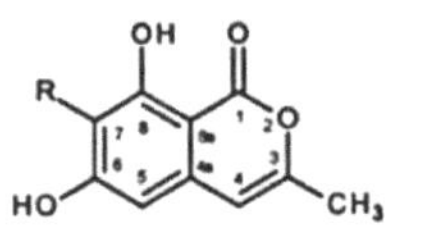

2,3,4

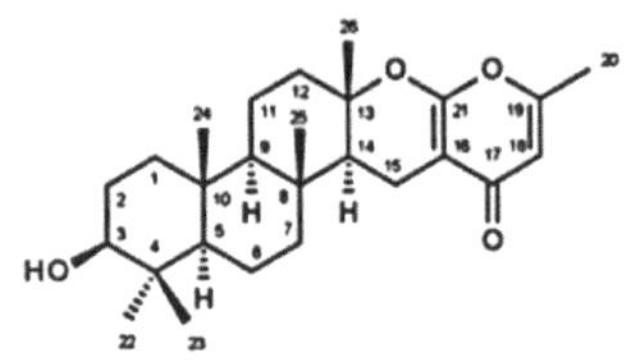

5

2: R=H; 3: R=Me; 4: R= Ome

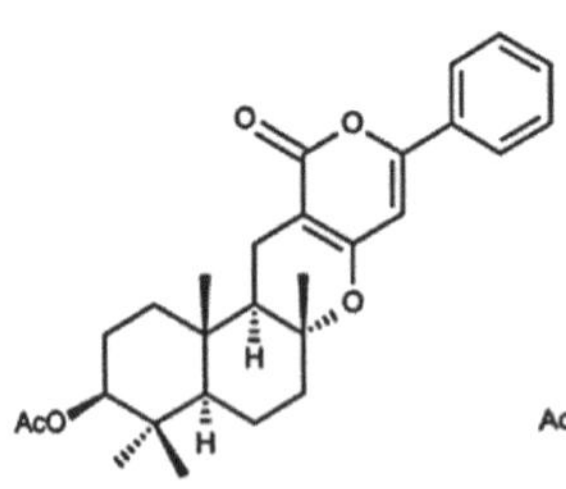

6

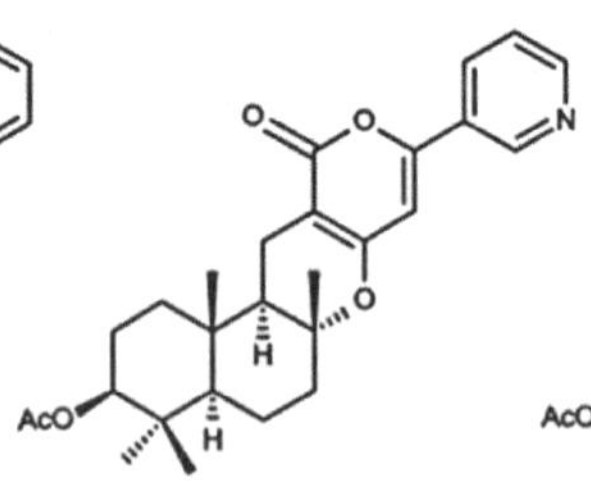

7

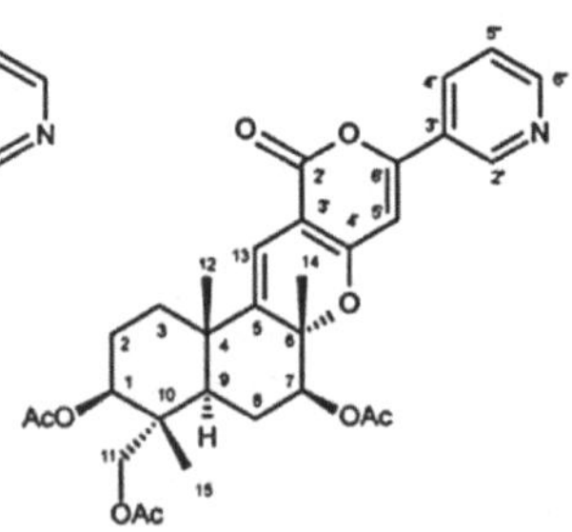

8

1: 5-hydroxy-8-methyl-2H, 6H-pyrano[3,4-g]chromen-2,6-dione, 2: 6, 8-dihydroxy-3-methylisocoumarin, 3: 6,8-dihydroxy-3,7-dimethylisocoumarin, 4: reticulol, 5: chevalone E, 6: pyripyropene S, 7:, chevalone C, S14-95, 8: pyripyropene E.

***Unidentified sponge* sp.6**

Associated fungi; and their pharmaceutical compounds and activities

Anticancer and antimicrobial activities: This Venezuelan sponge-associated fungus, *Emericella variecolor* yielded compounds *viz.* varixanthone and varitriol. While varixanthone showed antimicrobial activity, varixanthone anticancer activity on selected breast cancer cell lines, CNS and renal cell lines (Thomas et al. 2010).

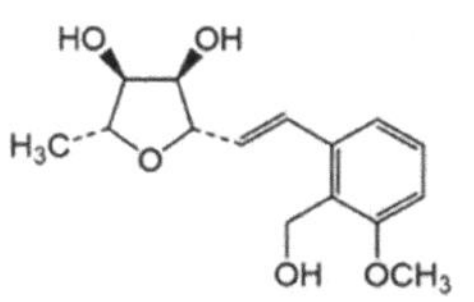

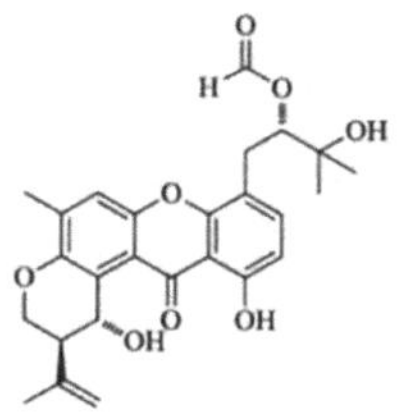

Varitriol Varixanthone

***Unidentified sponge* sp.7**

Associated fungi; and their pharmaceutical compounds and activities

Cystein protease inhibiting activity: The fungal strain *Microascus longirostris* SF-73 derived from this sponge yielded compounds such as cathestatin A, B, and C which exhibited potent inhibitory activity on ysteine proteases which are known to activate the pathogenic processes of several human diseases including AIDS and cancers. Hence these inhibitors may be of great use in clinical therapy as potentially powerful tools (Thomas et al. 2010).

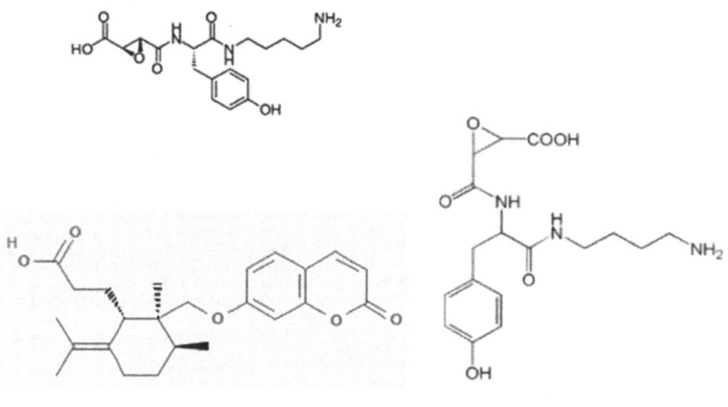

Cathestatins A, B

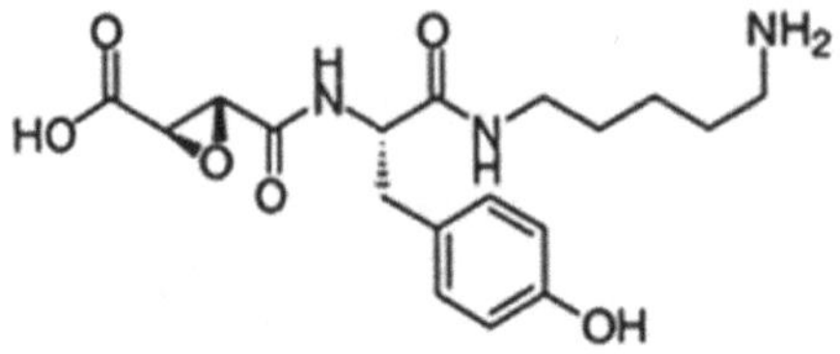

Cathestatin C

***Unidentified sponge* sp.8**

Associated fungi: Therapeutic compounds and their bioactivities

Cytoxic and antibacterial activities: The associated fungus *Aspergillus ostianus* 01F313 of this unidentified sponge yielded the compounds *viz.* 9-chloro-8-hydroxy-8,9-deoxyasperlactone, 9-chloro-8-hydroxy-8,9-deoxyaspyrone and 8-chloro-9-hydroxy-8,9-deoxyasperlactone which displayed antibacterial effects. Further, the above fungal strain also produced compounds *viz.* aspergillides A, B and C, aspinorene, and dihydroaspyrone which exhibited cytotoxic activity on lymphocytic leukemia cell line (Thomas et al. 2010).

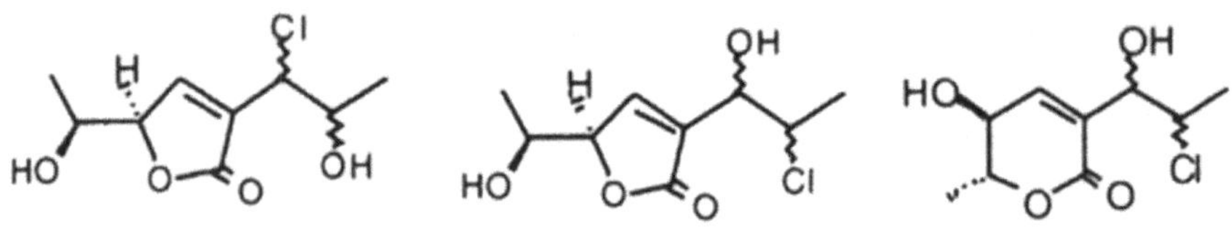

1 2 3

1: 8-chloro-9-hydroxy-8,9-deoxyasperlactone; 2: 9-chloro-8-hydroxy-8,9-deoxyasperlactone; 3: 9-chloro-8-hydroxy-8,9-deoxyaspyrone

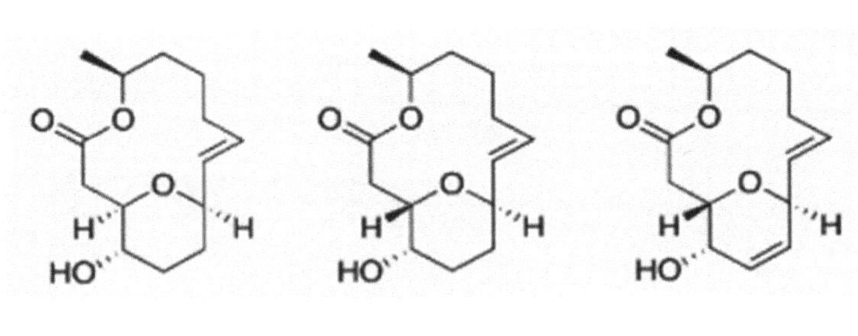

Aspergillides A- C

***Unidentified sponge* sp.9**

Associated fungi and their pharmaceutical compounds and activities

Antimycobacterial activity: The compounds diaporthein A and B derived from the fungus *Cryptosphaeria eunomia*, associated with this unidentified sponge antimycobacterial activity (Thomas et al. 2010).

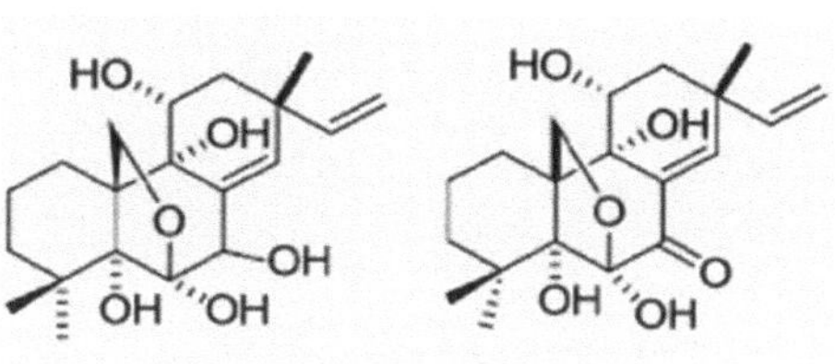

Diaportheins A, B

***Unidentified sponge* sp.10**

Associated fungi and their pharmaceutical compounds and activities

Cytotoxicity: The fungus *Chrysosporium articulatum* associated with this unidentified dictyoceratid sponge. yielded three benzolactones, *viz.* chrysoarticulins A-C which displayed poor cytotoxic activity on K562 and A549 tumor cell lines (Zhang et al. 2017).

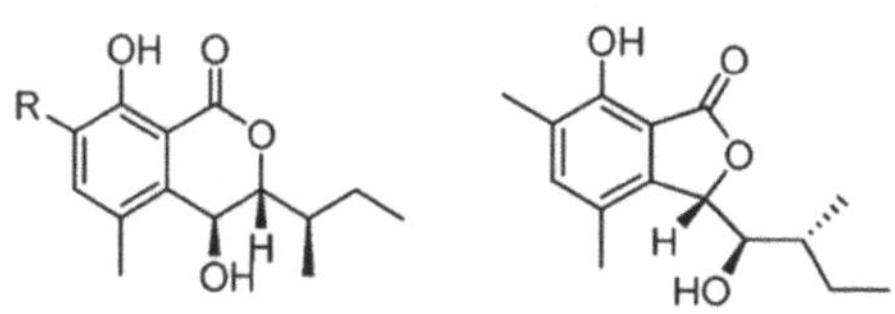

Chrysoarticulin A,B Chrysoarticulin C

(Chrysoarticulin A, R=CH3; Chrysoarticulin B, R=CH2OH)

***Unidentified sponge* sp.11**

Associated fungi and their pharmaceutical compounds and activities

Anti-tumor activity: The fungal strain *Penicillium citrinum* SpI080624G1f01, derived from this unidentified sponge yielded the compound JBIR-59 which possessed l-glutamate toxicity on N18-RE-105 tumor cell line with an EC50 value of 71 μM (Zhang et al. 2017).

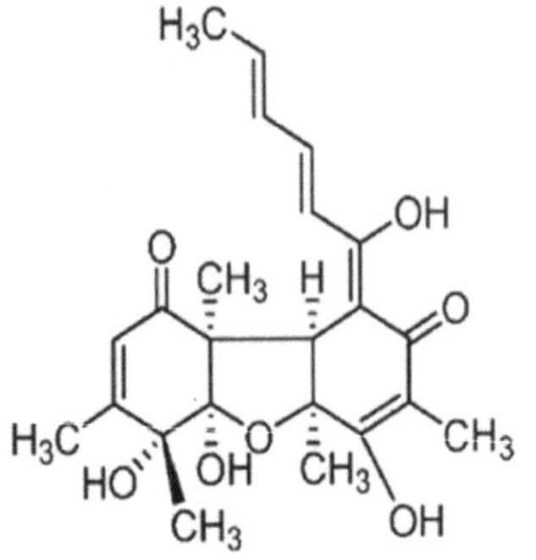

JBIR-59

Unidentified sponge sp.12

Associated fungi; and their pharmaceutical compounds and activities

Cytotoxicity and AchE inhibitory activity: Two 4-hydroxy-2-pyridone alkaloids, arthpyrones were isolated from the fungus *Arthrinium arundinis* ZSDS1-F3, which was obtained from an unidentified sponge Both the compounds displayed potent cytotoxic activity on MCF-7, U937, BGC823, K562, A549, Huh-7, MOLT-4, H1975, HL60, and Hela cell lines with IC50 values ranging from 0.24 to 45 M (Jin et al. 2016).

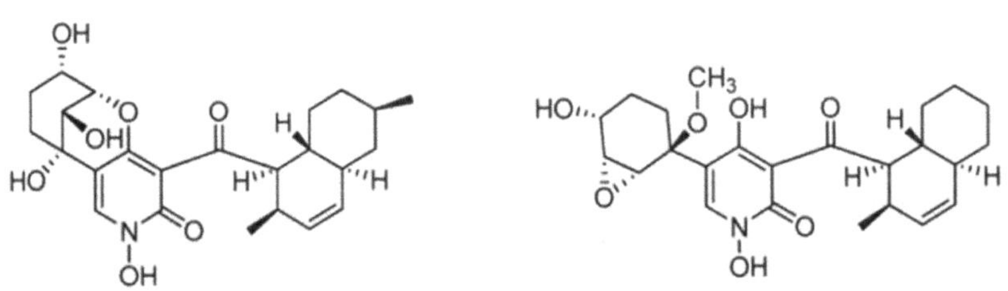

Arthpyrones

Unidentified sponge sp.13

Associated fungi: Therapeutic compounds and their bioactivities

Antibacterial activity: The fungal strain, *Diaporthaceae* sp. PSU-SP2/4 from an unidentified marine sponge generated a pentacyclic cytochalasin,diaporthalasin which displayed potent antibacterial activity against MRSA (methicillin-resistant *S. aureus*)and S.aureus with an MIC value of 2 μg/mL (Jin et al. 2016).

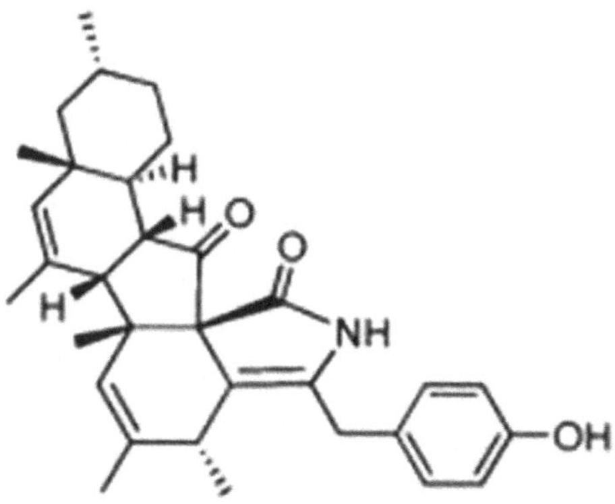

Diaporthalasin

Unidentified sponge **sp.14**

Associated fungi and their pharmaceutical compounds and activities

Antibacterial activity: A chevalone derivative, named chevalone E was isolated from the ethyl acetate extract of an undescribed marine sponge-associated fungus *Aspergillus similanensis* KUFA 0013. This compound was found to display synergism on MRSA (methicillin-resistant *S. aureus*) (Jin et al. 2016).

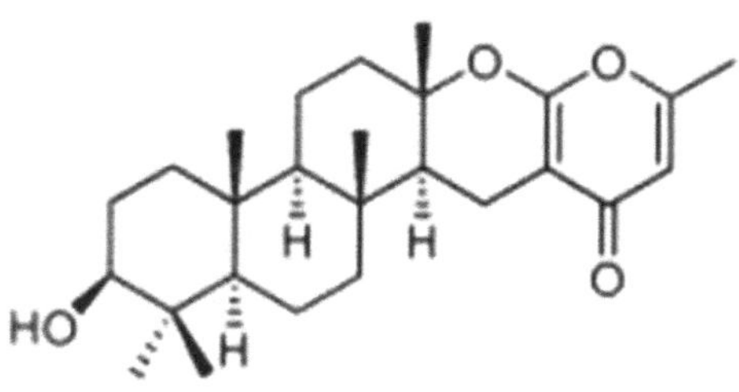

Chevalone E

Unidentified sponge **sp.15**

Associated fungi; and their pharmaceutical compounds and activities

Anticancer activity: The compound, xylarianaphthol-1, derived from an unidentified fungus of the order Xylariales displayed antitumor activity in osteosarcoma MG63 cell line (Jin et al. 2016).

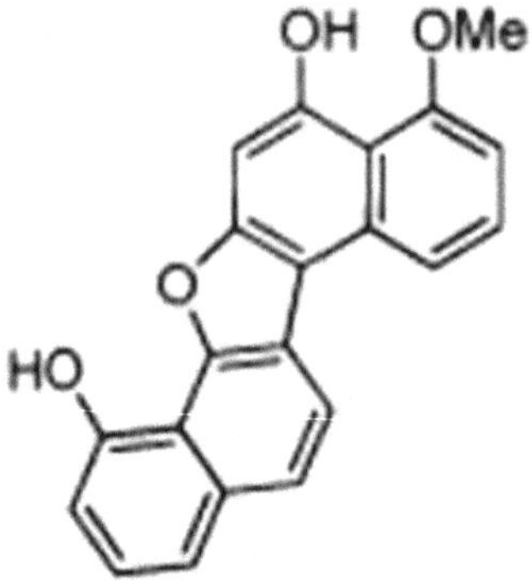

Xylarianaphthol-1

Unidentified sponge sp.16

Associated fungi and their pharmaceutical compounds and activities

Antibiotic activities: The polyketide, lindgomycin derived from the mycelia and culture broth of various Lindgomycetaceae strains associated with this unidentified sponge showed antibiotic activities with IC50 value of 5.1 M against MRSA (Jin et al. 2016).

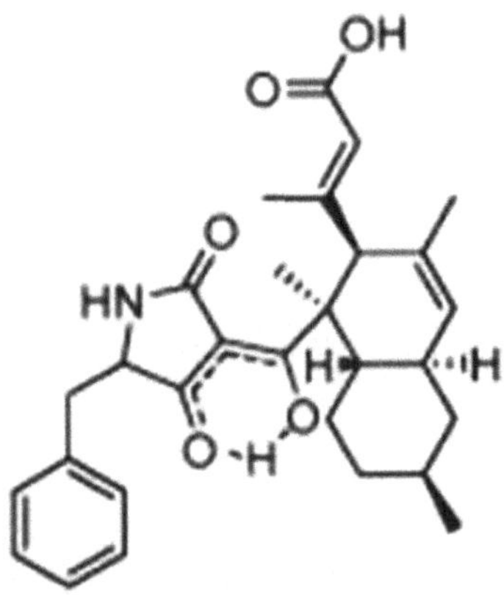

Lindgomycin

Unidentified sponge sp.17

Associated fungi and their pharmaceutical compounds and activities

Antitubercular activity: A polyketide (naphthoquinone dimer) compound, viomellein obtained from an unidentified sponge-associated fungus, *Aspergillus* sp. displayed significant antitubercular activity against *Mycobacterium bovis* BCG with an IC50 value of 1.56 µg/mL (Mayer et al. 2021).

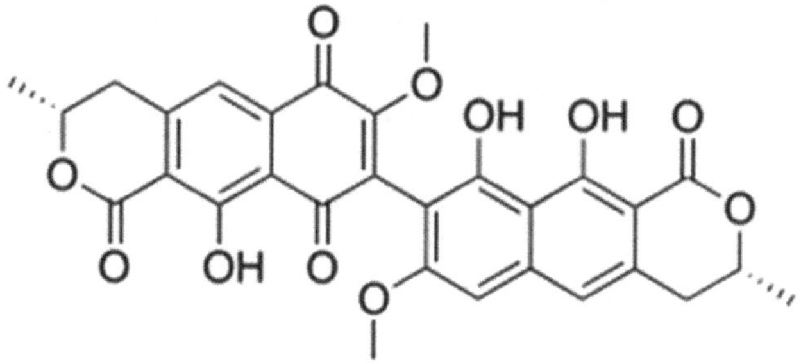

Viomellein

***Unidentified sponge* sp.18**

Anti-inflammatory activity: A polyketide compound, curvularin-type metabolite isolated from this sponge- associated fungus *Penicillium* sp. SF-5850 showed anti-inflammatory activity with IC50 values ranging from 1.9 to 2.7 µM. This compound has also been reported to significantly inhibit the release of the pro-inflammatory NO and PGE2 (Mayer et al. 2021).

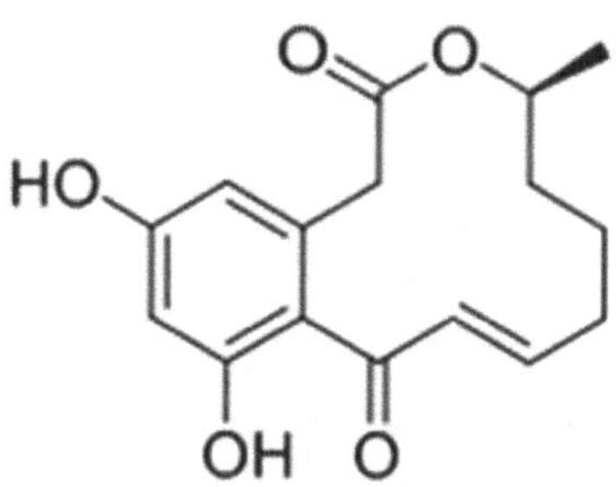

Curvularin-type compound

CNIDARIANS

Class: Scyphozoa Order: Rhizostomeae Family: Stomolophidae

Nemopilema nomurai

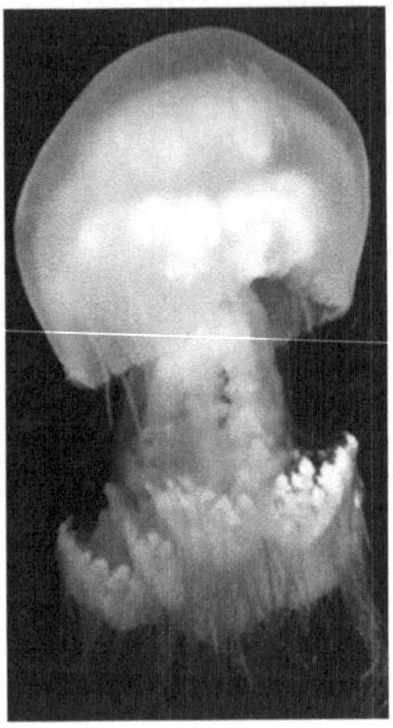

Image credit: Totti, https://creativecommons.org/licenses/by-sa/4.0/

Common name: Nomura's jellyfish

Global distribution: Northwest Pacific: endemic to the East Asian Marginal Seas.

Habitat: Pelagic

Associated fungi: Pharmaceutical compounds and activities

Antibacterial and antifungal activities: The symbiotic fungi from different tissues of this jellyfish include four genera *viz. Aspergillus, Cladosporium, Purpureocillium,* and *Tilletiopsis*. The antimicrobial results showed that all the 13 EtOAc extracts displayed different levels of antibacterial activity, three of which exhibited strong to significant antibacterial activity to the bacterial pathogens *Staphylococcus aureus* and *Salmonella entrica.* Anti-fungal activity indicated that the EtOAc extracts of *Aspergillus versicolor* and *Tilletiopsis* sp. showed maximal mycelial growth inhibition of 82.32% for *Rhizoctonia solani* and 48.41% for *Botrytis cinerea* at 200 μg/ml (Yue et al. 2015).

Antibacterial activity: The fungus *Paecilomyces variotii*, which was associated from this jellyfish produced showed polyketides which displayed antibacterial activity against multi-drug-resistant *Vibrio parahemolyticus* 7001 and methicillin-resistant *Staphylococcus aureus* 3089 with MIC vlaues ranging from 5 to 40 g/mL (Liu et al. 2011).

Class: Anthozoa Order: Alcyonacea Family: Alcyoniidae

Cladiella sp.

Image credit: Haplochromis, https://creativecommons.org/licenses/by-sa/3.0/

Common name: Finger leather coral

Associated fungi: Pharmaceutical compounds and activities

Antifungal activity: The fungus, *Aspergillus versicolor* strain LCJ-5-4 from this octocoral has been reported to produce radical-scavenging polyketides, cyclopentapeptides and alkaloid cottoquinazoline D. Among these compounds, cottoquinazoline D displayed antifungal activity against *Candida albicans* with an MIC value of 22.6 M (Raimundo et al. 2018).

Zhuang et al. (2011) reported that the fungus *Aspergillus versicolor* isolated from this coral produced three compounds, cottoquinazoline B–D from which only cottoquinazoline C showed modest antifungal activity against *C. albicans*.

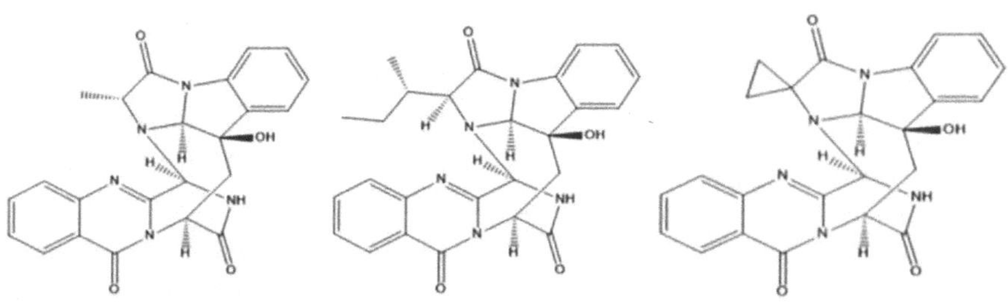

Cottoquinazoline B-D

Antioxidant activity: The fungus *Aspergillus versicolor* LCJ-5-4, retrieved from this finger leather coral produced the compound, tetraorcinol A which possessed antioxidant activity by scavenging DPPH radical (Raimundo et al. 2018).

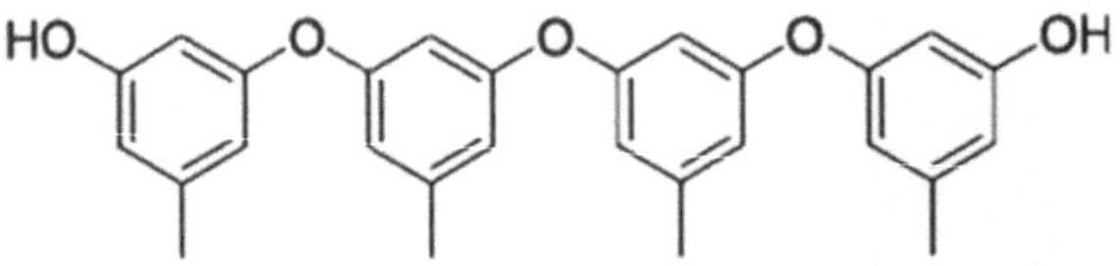

Tetraorcinol A

***Lobophytum* sp.**

Image credit: Public domain.

Common name: Finger Leather Coral, Lobed Leather Coral,

Global distribution: Not known.

Habitat: It occurs at shallow coral reef flats at depths of 6–18 m

Associated fungi: Pharmaceutical compounds and activities

Antimicrobial and antioxidant activity: The supernatant extracts of the fungus *Emericella unguis* 8429 isolated from this species were found to show significant activity against *Pseudomonas aeruginosa*, *Staphylococcus aureus* and *Candida albicans*. Further, an unidentified fungus FC3 isolated from this soft coral. had very low DPPH-free radical scavenging activity (El-Hady et al. 2014b).

Sarcophyton subviride

Common name: Not reported

Global distribution: Andaman and Nicobar Islands, India

Habitat: Shore reef flats

Associated fungi: Pharmaceutical compounds and activities

Anti-inflammatory effects: The fungal strain *Aspergillus terreus* associated with this species produced a lactone viz. versicolactone B which displayed anti-inflammatory activity by inhibiting the NO production in the macrophages of mouse (Raimundo et al. 2018).

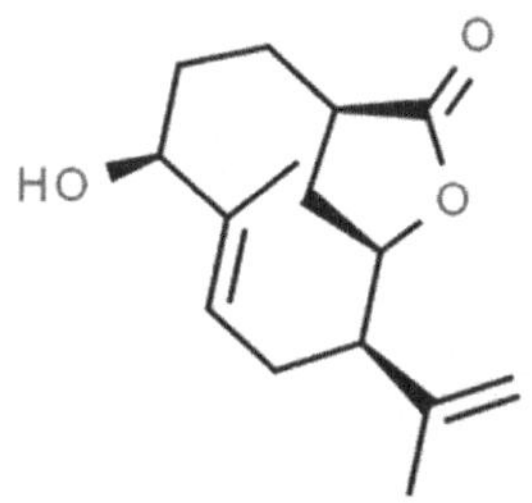

Versicolactone B

Sarcophyton tortuosum

Common name: Hainan softcoral

Global distribution: India, Fiji, New Caledonia, Taiwan

Habitat: Intertidal reef areas

Associated fungi: Pharmaceutical compounds and activities

Anticancer activity: The fungus *Chondrostereum* sp. associated with this species produced hirsutane sesquiterpenoids chondrosterins F–J. Among these compounds, chondrosterin J showed potent cytotoxicity on CNE-2 and 1 (human nasopharyngeal cancer cell lines) with IC50 values of 0.6 and 1.3 μM, respectively (lissawy et al. 2015; Blunt et al. 2015).

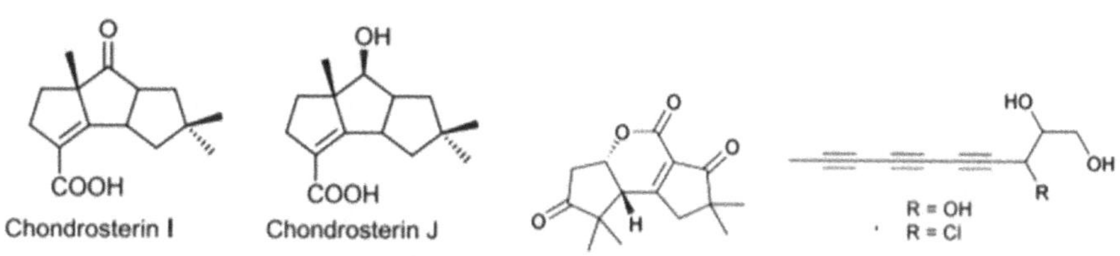

Chondrosterin F Chondrosterins G,H

Cytotoxic activities: The fungus *Chondrostereum* sp. collected from this soft coral yielded five triquinane-type sesquiterpenoids, chondrosterins AE and another sesquiterpenoid hirsutanol C. Of these metabolites, chondrosterin A displayed potent cytotoxicity against cancer lines *viz.* LoVo A549, and CNE2, with IC50 values of 5.5, 2.5, and 5 μM, respectively (Li et al. 2012a).

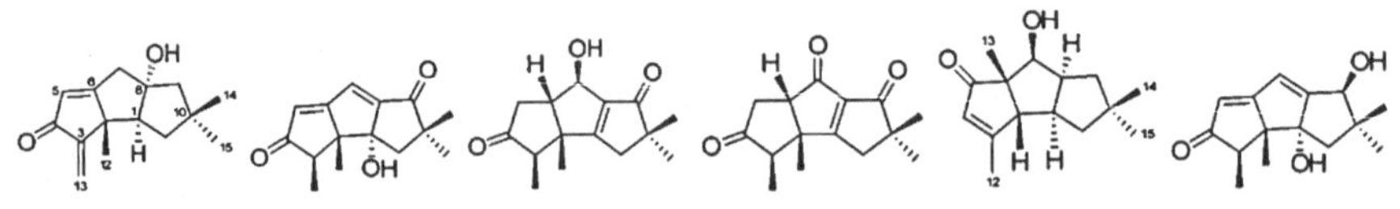

1 2 3 4 5 6

1-5: Chondrosterins A–E; 6: Hirsutanol C

Sarcophyton **sp.**

Image credit: Philippe Bourjon, https://creativecommons.org/licenses/by-sa/4.0/

Associated fungi: Pharmaceutical compounds and activities

Antibacterial activity: The fungus *Pestalotiopsis* sp. ZJ-2009-7-6, isolated from this species is the producer of the benzophenone derivative pestalachlorides C, D. Among these compounds, pestalachloride D was found to be active against several Gram-negative bacteria (Blunt et al. 2015; Raimundo et al. 2018).

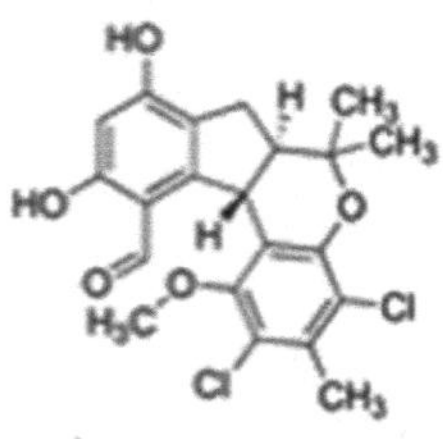

Pestalachloride C

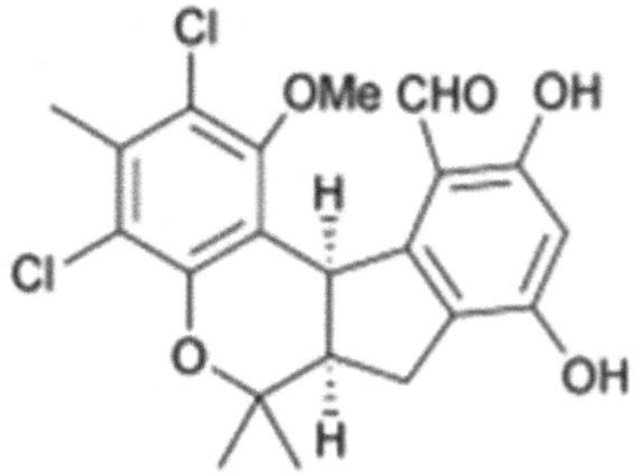

Pestalachloride D

Antibacterial activity: The fungal strain *Aspergillus elegans* ZJ-2008010 derived from this species produced the phenylalanine derivatives 4-O-methoxyasperphenamate and asperphenamate. These metabolites displayed antibacterial activity on *Staphylococcus epidermidis* which is believed to cause chronical infections in humans undertaking implanted medical devices (Raimundo et al. 2018).

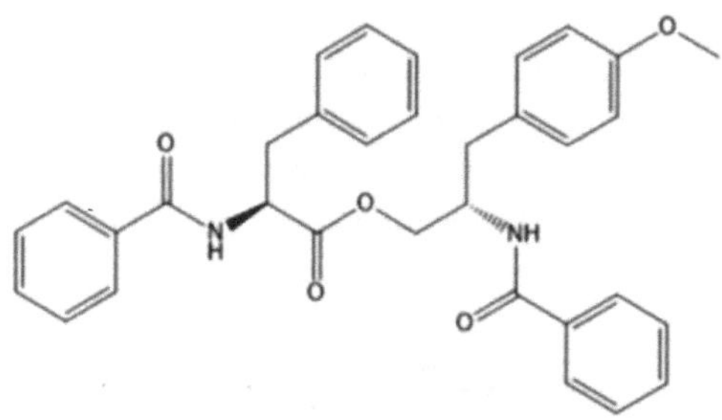

4′-O-methoxyasperphenamate

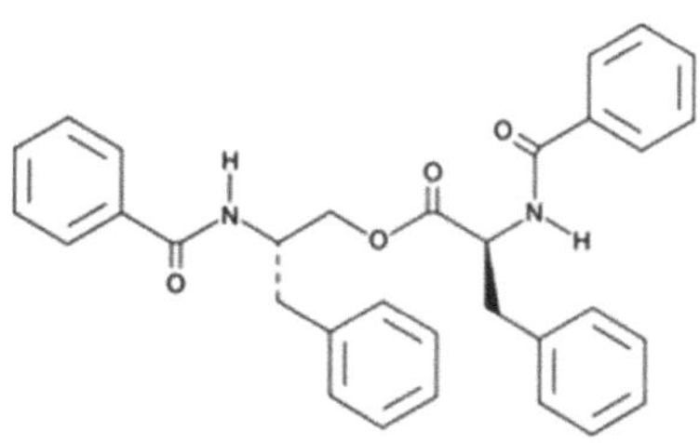

Asperphenamate

Antidiabetic activity: The fungal strain *Eurotium rubrum* SH-823 derived from this soft coral produced benzofuran derivatives, *viz.* eurothiocin A and B, aspernolide D, vermistatin, and methoxyvermistatin, zinniol, and butyrolactone I. Among these compounds eurothiocin A and B displayed significant inhibiton against -glucosidase activity (-glucosidase inhibitors can retard the liberation of d-glucose from dietary complex carbohydrates and delay glucose absorption, resulting in reduced postprandial plasma glucose levels and suppression of postprandial hyperglycemia). The IC50 (μM) values of -glucosidase inhibitory activity for all the above 1-7 compounds were 17.1, 42.6, 98.5, 110.8, 107.1, and 236, respectivley (Liu et al. 2014).

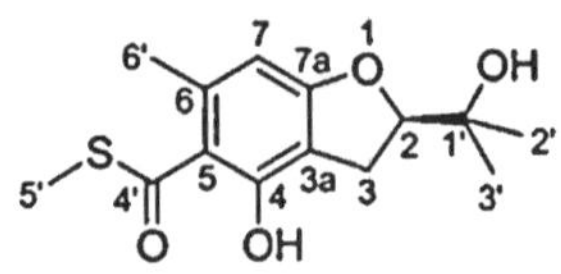

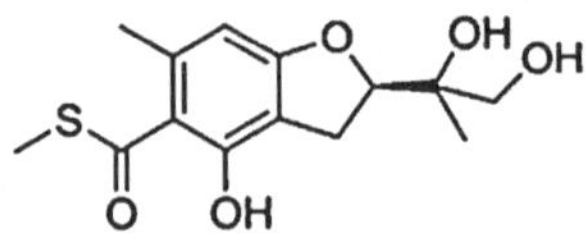

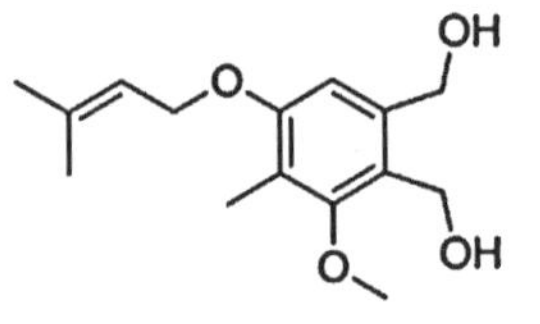

1 2 3

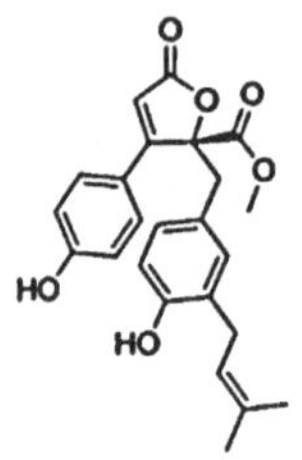

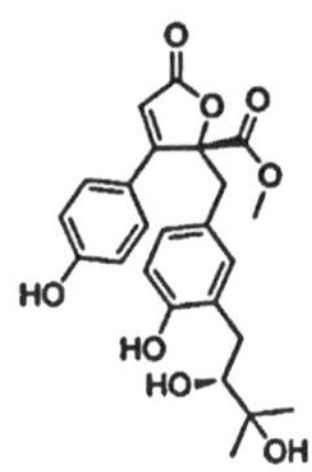

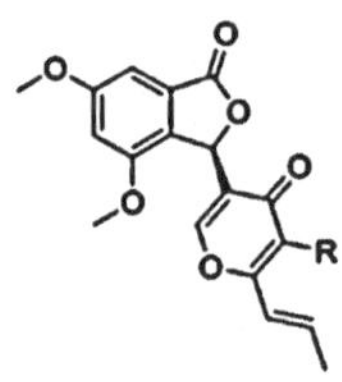

4 5 6,7 (6, R=H; 7, R= Ome)

1: Eurothiocin A; 2: Eurothiocin B; 3: Zinniol; 4: Butyrolactone I; 5: Aspernolide D; 6: Vermistatin; 7: Methoxyvermistatin

Antiviral activity: The fungus *Alternaria* sp. ZJ-2008003, retrieved from this species produced compounds alterporriol Q and tetrahydroaltersolanol C which displayed antiviral activity on the porcine respiratory and reproductive syndrome virus which is the common infectious disease affecting pigs (Raimundo et al. 2018).

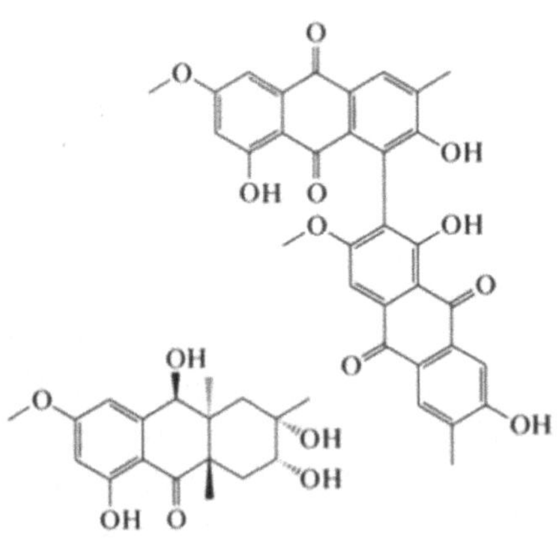

Tetrahydroaltersolanol C Alterporriol Q

The polyketidecyclopeptide (PKS-NRPS) metabolites () and (+)-pestaloxazine A produced by *Pestalotiopsis* sp., which is associated with this octocoral showed antiviral activity against EV71 (human Enterovirus 71 which is known to causeHFMD (hand, foot, and mouth disease) and associated neurological complications in infants (Raimundo et al. 2018).

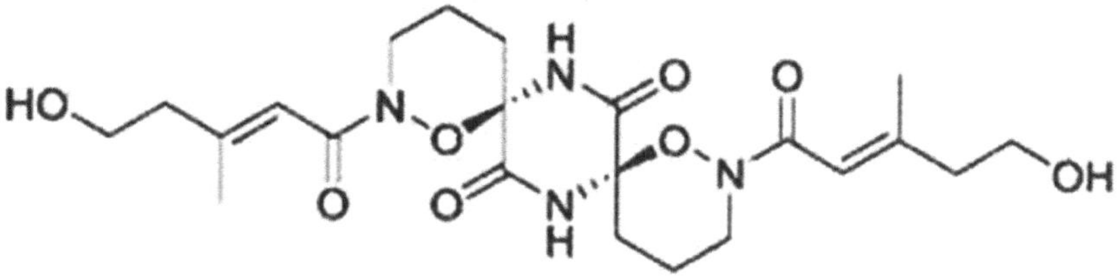

(+)-pestaloxazine A

***Sinularia* sp.**

Image credit: Paul Asman and Jill Lenoble, https://creativecommons.org/licenses/by/2.0/

Associated fungi: Pharmaceutical compounds and activities

Antibacterial and antifungal activities: A nor-dammarane triterpene viz. 6,16-Diacetoxy-25-hydroxy-3,7-dioxo-29-nordammara-1,17(20)-dien-21,24-lactone (174),,derived from the fungus *Aspergillus fumigatus* KMM4631 which is isolated from this soft coral displayed poor antibacterial and antifungal activities (Elissawy et al. 2015).

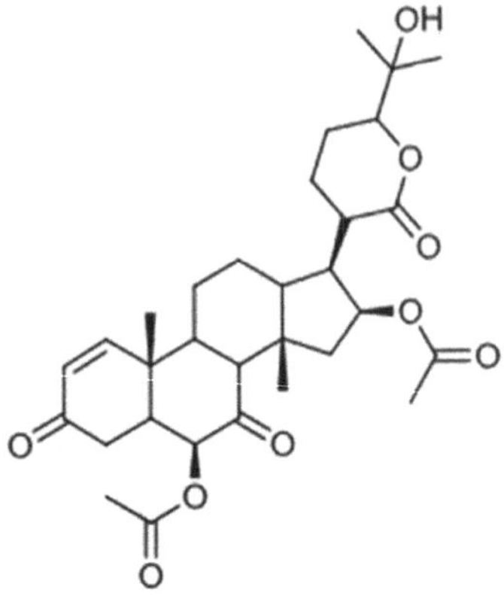

6β,16β-Diacetoxy-25-hydroxy-3,7-dioxo-29-nordammara-1,17(20)-dien-21,24-lactone

The supernatant and mycelial extracts from the static culture of the identified fungus *Emericella unguis* 8429 isolated from this soft coral was found to be highly effective against *Pseudomonas aeruginosa, Staphylococcus aureus* and *Candida albicans* (El-Hady et al. 2014).

Antioxidant (DPPH-free radical scavenging) activity: The supernatant and mycelial extracts from the static culture of the identified fungus FC2 (*Emericella unguis* 8429) and the unidentified fungus FC1 isolated from this soft coral had the highest free radical scavenging activity (FRSA) against DPPH radical (El-Hady et al. 2014).

AChE inhibitory activity: The mycelia of the fungus FC2 (*Emericella unguis* 8429) isolated from this soft coral showed very high Acetylcholinesterase (AChE) inhibitory activity (up to 80%) (El-Hady et al. 2014b).

Others: The cyclopeptide-alkaloid spirotryprostatin F has been extracted from the fungus *Aspergillus fumigatus,* retrieved from this soft coral (Raimundo et al., 2018).

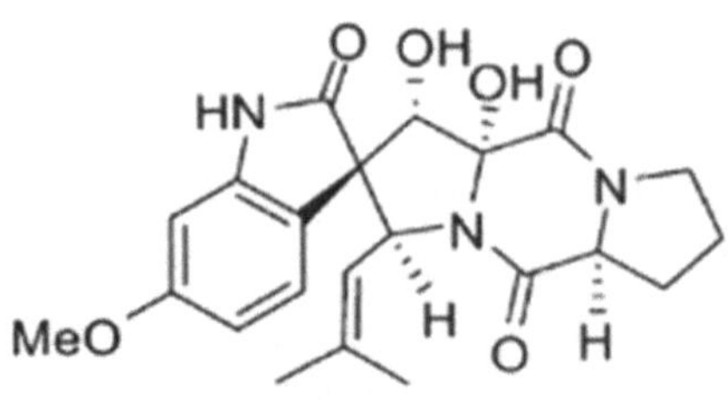

Spirotryprostatin F

Order: Alcyonacea Family: Acanthogorgiidae

Muricella abnormalis

Muricella sp.

Common name: Not reported

Global distribution: Japan

Habitat: Depth, 206 m

Associated fungi: Pharmaceutical compounds and activities

Antiviral activity: The ethyl acetate extract of *Aspergillus* sp. XS-20090B15 which was isolated from this octocoral yielded the hydroquinolone, 22-O-(NMe-l-valyl)-21-epi-aflaquinolone B. This compound showed strong antiviral activity on RSV (Respiratory syncytial virus) with an IC 50 value of 0.042 μM. This virus has been reported to cause bronchiolitis and pneumonia (Raimundo et al. 2018).

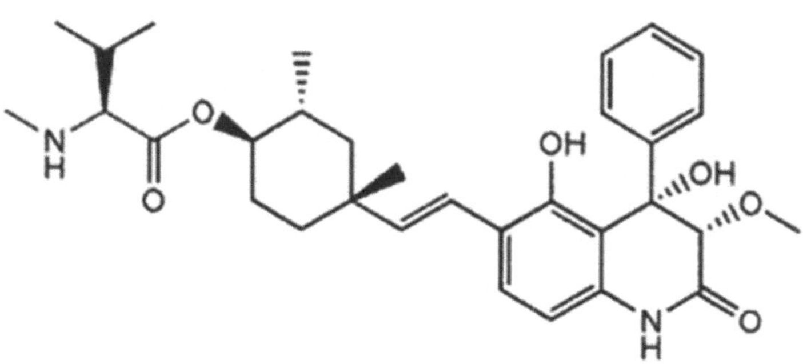

22-O-(NMe-l-valyl)-21-epi-aflaquinolone B

Antibacterial activity: The fungus, *Aspergillus* sp. XS-20090B15 associated with this octocoal produced the cyclopeptide asperpeptide A and derivative of the nucleoside aroyl uridine which was active against *Staphylococcus epidermidis* (Raimundo et al., 2018).

Wang et al.(2012) and Nicoletti and Trincone (2016) reported that the fungus *Penicillium commune* 518 obtained from this octocoral produced communols A–G, clavatol, 2,4-Dihydroxy-3-methylacetophenone, 2,4-Dihydroxy-3-methoxymethyl-5-methylacetophenone, and 2,4-Dihydroxy-5-methylacetophenone. Among these compounds, communols A, F,and G showed moderate antimicrobial activities against *Escherichia coli* and *Enterobacter aerogenes* with MIC values of 4.1/16.4, 6.4/25.8, and 23.8/23.8 $\mu mol \cdot L^{-1}$, respectively.

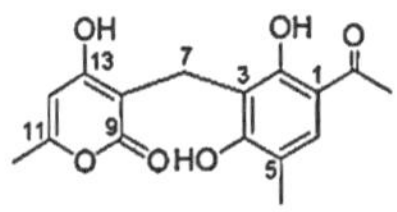

1

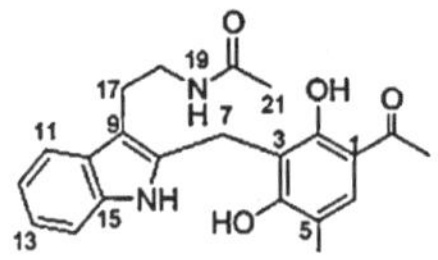

2

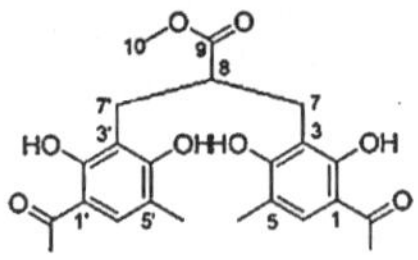

3

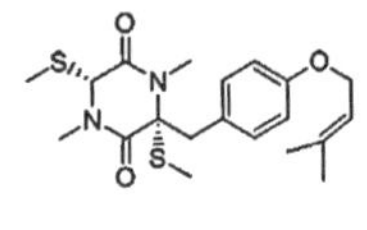

12

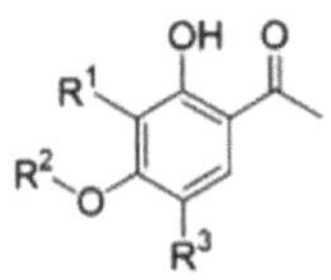

$R^1 = CH_2SOCH_3$, $R^2 = H$, $R^3 = CH_3$

$R^1 = CH_2CH_2COCH_3$, $R^2 = H$, $R^3 = CH_3$

$R^1 = CHO$, $R^2 = H$, $R^3 = CH_3$

$R^1 = CH_2OH$, $R^2 = R^3 = CH_3$

$R^1 = CH_2OCH_3$, $R^2 = H$, $R^3 = CH_3$

$R^1 = R^3 = CH_3$, $R^2 = H$

$R^1 = CH_3$, $R^2 = R^3 = H$

$R^1 = R^2 = H$, $R^3 = CH_3$

4-11

1-7: communols A—G; 8: 2,4-dihydroxy-3-methoxymethyl-5- methy lacetophenone; 9: clavatol; 10: 2,4-dihy-droxy-3-methylacetophenone; 11: 2,4-Dihydroxy-5-methylacetophenone.

Muricella flexuosa

Common name: Not designated

Global distribution: Tropical Indo-Pacific

Habitat: Reef slope.

Associated fungi: Pharmaceutical compounds and activities

Anticancer activity: The chromones oxalicumones A and B derived from the fungus by *Penicillium oxalicum* which is associated with this octocoral exhibited cytotoxicity against colon (SW-620) and skin (A375) cancer cells,. Further, oxalicumone A was reported to be more potent (Raimundo et al. 2018).

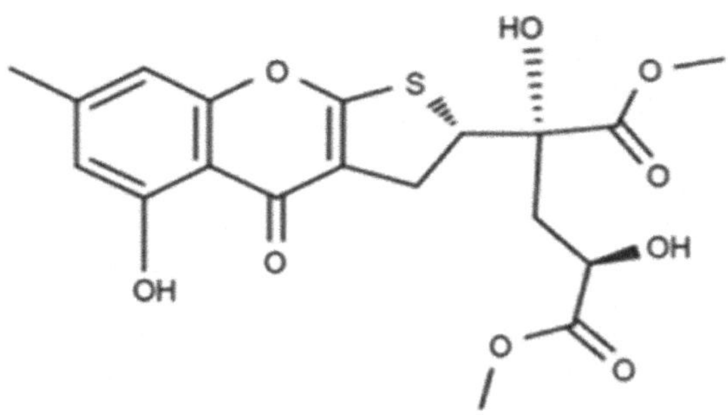

Oxalicumone A

Order: Alcyonacea Family: Clavulariidae

Carijoa **sp.**

Associated fungi: Pharmaceutical compounds and activities

Six dihydroquinolin-2-one-containing alkaloids, three monoterpenoids combined with a 4-phenyl-3,4- dihydroquinolin-2(1H)-one (13) and three 4-phenyl-3,4- dihydroquinolin-2(1H)-one alkaloids (46), were isolated from this gorgonian coral-derived fungus *Scopulariopsis* sp. The pharmaceutical activities of these compounds are to be known (Shao et al. 2015).

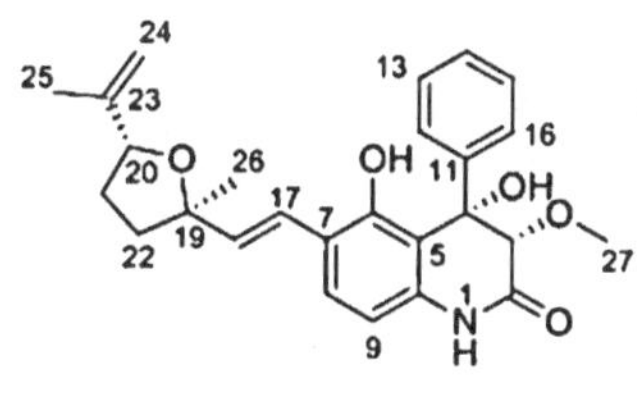

1

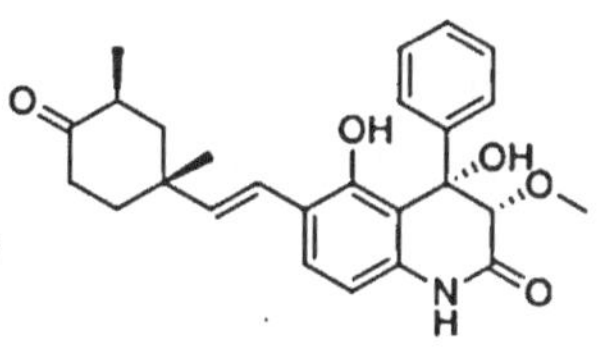

2

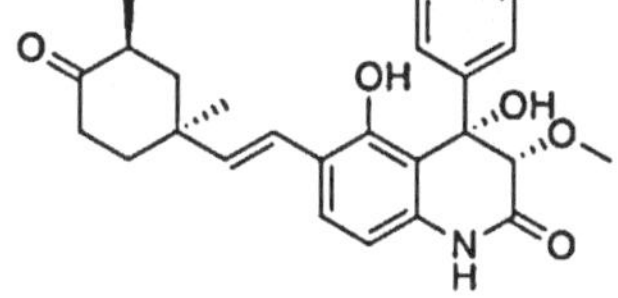

3

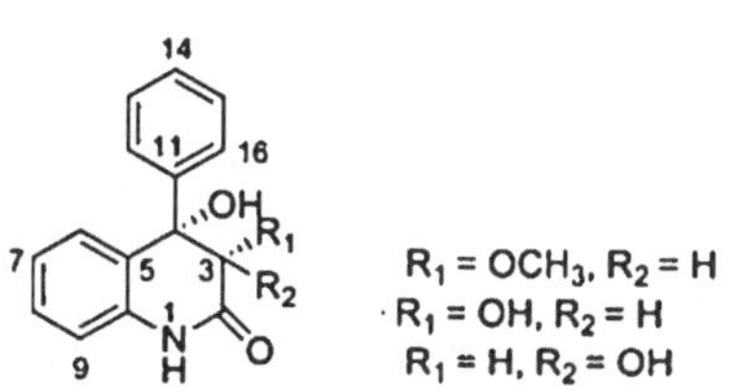

4,5,6

Shao et al. (2013) reported that the alkaloid, fumiquinazoline L extracted from the fungus *Scopulariopsis* sp. of this gorgonian showed no cytotoxic or antibacterial activities.

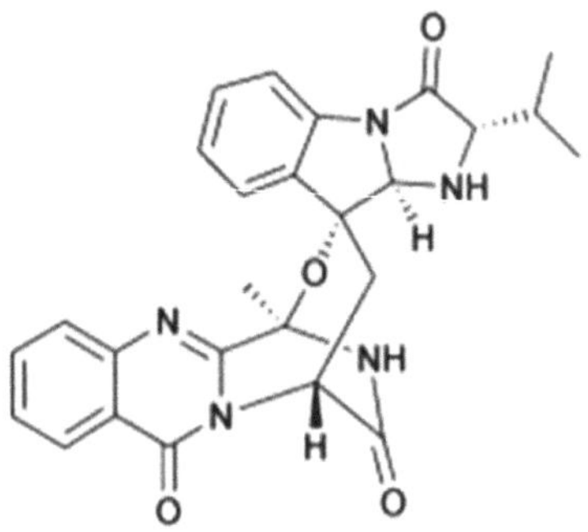

Fumiquinazoline L

Order: Alcyonacea Family: Ellisellidae

Dichotella gemmacea

Common name: Not reported

Global distribution: Indo-Pacific.

Habitat: Prefers deeper waters and inhabits coral reefs with hard and rocky substrates; depth range 15 - 2630 m

Associated fungi: Pharmaceutical compounds and activities

Antibacterial activity: Three new structurally related bisabolane sesquiterpenoids *viz.* methyl sydowate (1), 7-deoxy-7,14-didehydrosydonic acid (2), and 7-deoxy-7,8-didehydrosydonic acid (3) and other sesquiterpenes (+)-sydowic acid (4) and (+)-sydonic acid (5) have been derived i from the unidentified fungus, *Aspergillus* sp. associated with this gorgonian. Except compound 3, all other compounds displayed weak antibacterial activity on *Staphylococcus aureus* (Elissawy et al. 2015).

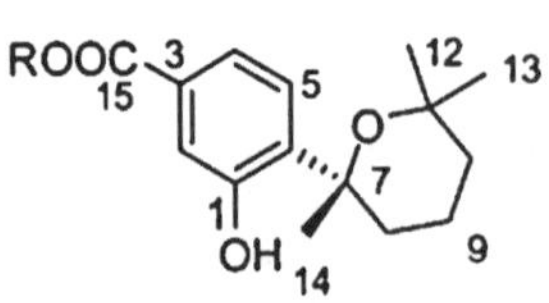

1: R=Me; 4: R=H

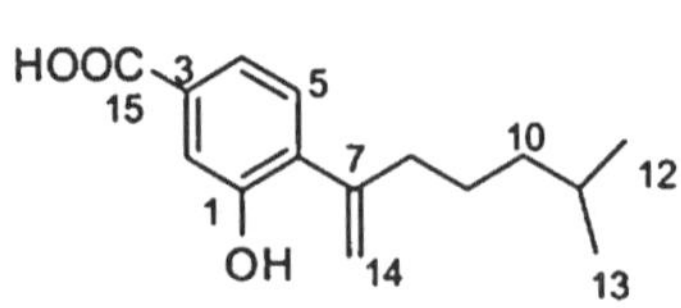

2

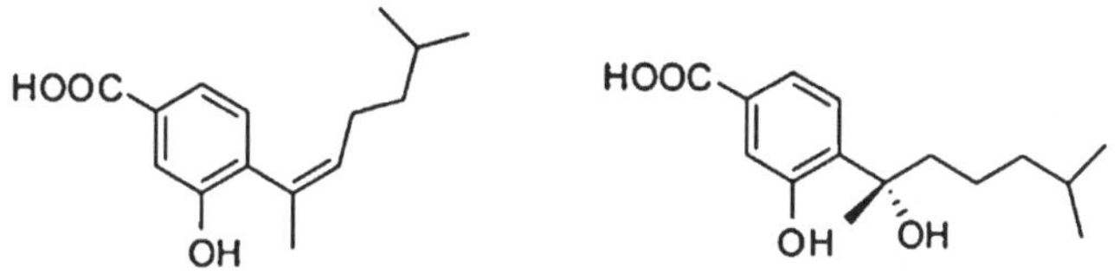

3 5

1: (+)-methyl sydowate; 2: 7-deoxy-7,14-didehydrosydonic acid 3: 7-deoxy-7,8-didehydrosydonic acid; 4: (+)-sydowic acid;; 5: (+)-sydonic acid)

Raimundo et al. (2018) reported that the cyclopeptide asperpeptide A and derivative of the nucleoside aroyl uridine extracted from the fungus *Aspergillus versicolor* associated with this gorgonian showed antibacterial activity against *Staphylococcus epidermidis*.

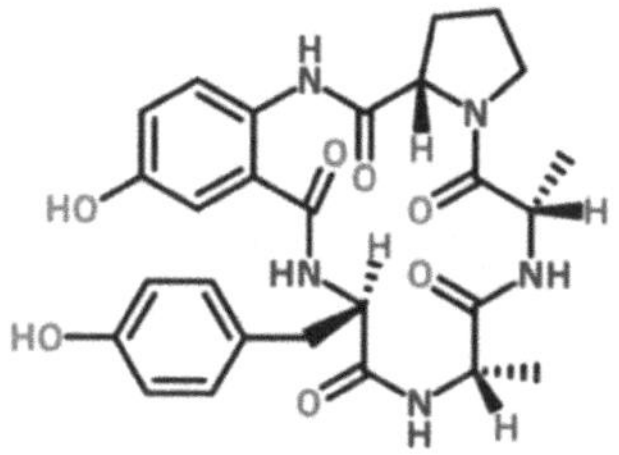

Asperpeptide A

Bao et al. (2013) reported on the isolation of compounds paecilin C, 6,8,5'6'-Tetrahydroxy-3'-methylflavone, emodin, citrorosein, isorhodoptilometrin, penicillixanthone A, and secalonic acid B–D from the fungal strain *Penicillium* sp. SCSGAF0023 associated with this species. Among these compounds, secalonic acid D, secalonic acid B and penicillixanthone A showed antimicrobial activity against *M.luteus P.nigrifaciens E.coli* and *B.subtilis* with MIC values ranging from 24.4 to 97.5 µg/mL; from 97.5 to 390.5 µg/mL; and from 24.4 to 97.5 µg/mL respectively (Bao et al. 2013a).

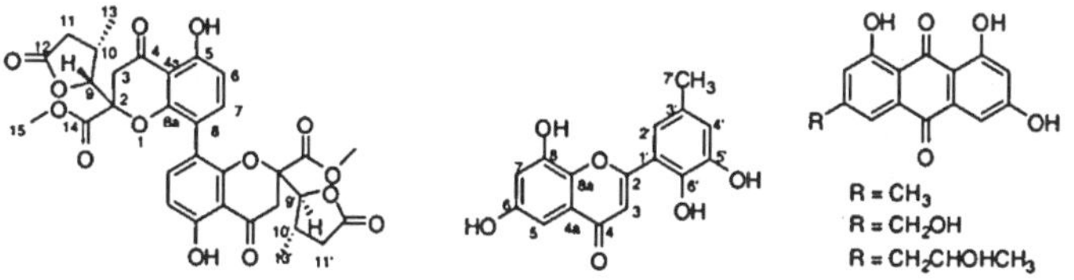

1 2 3-5

1: Paecilin C; 2: 6,8,5'6'-Tetrahydroxy-3'-methylflavone; 3: emodin, 4: citrorosein, 5: isorhodoptilometrin

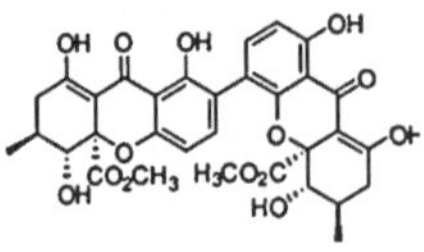

Penicillixanthone A

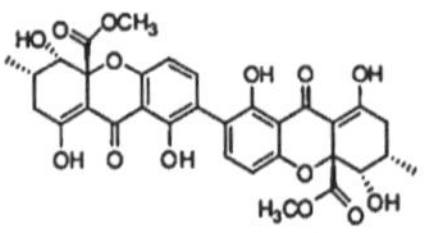

Secalonic acid D

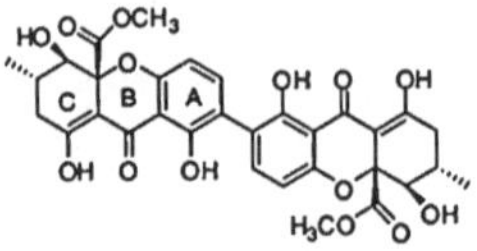

Secalonic acid B

Anticancer activity: The ethyl acetate extract of an *Aspergillus* sp. strain associated with this cnidarian species displayed potent cytotoxic activity against A-549 (human lung carcinoma) cell line. Further, the said extract also yielded an azaphilone derivative viz. aspergilone A which exhibited cytotoxicity on A-549 (human lung carcinoma), HL-60 (human promyelocytic leukemia), and MCF-7 (human breast adenocarcinoma) cell lines (Raimundo et al. 2018).

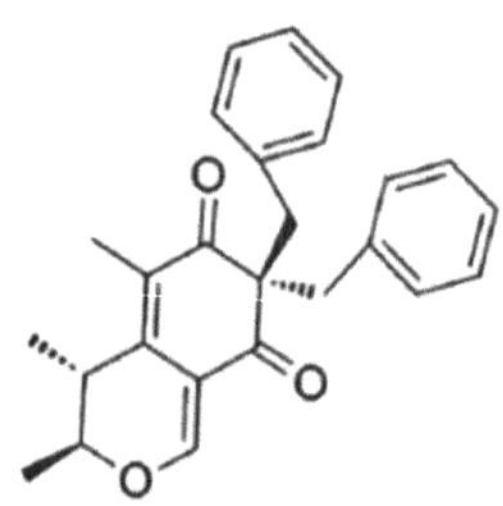

Aspergilone A

Cytotoxicity: The fungal strain *Penicillium oxalicum* SCSGAF0023 derived from this species produced compounds viz. oxalicumones A–E. Among these compounds, oxalicumones A,B showed moderate and mild cytotoxicity against SW-620 and A375 cell lines, respectively. Compound E showed cytotoxicity against several cell lines with IC50 values of less than 10μM (Li 2019).

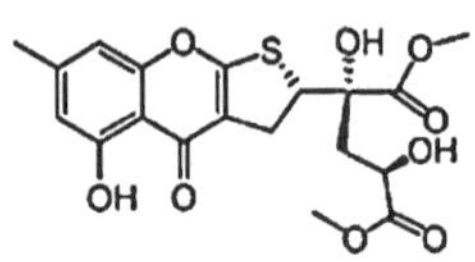

1

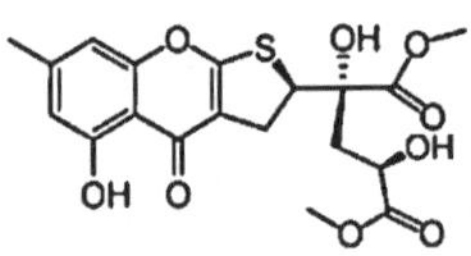

2

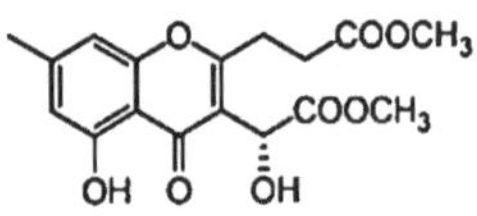

3

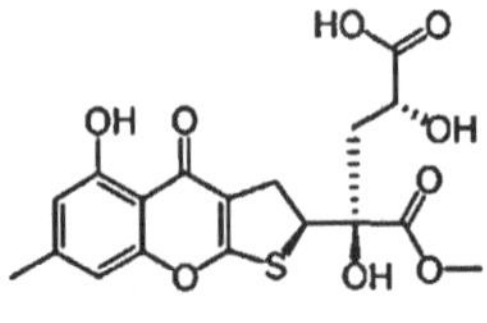

4

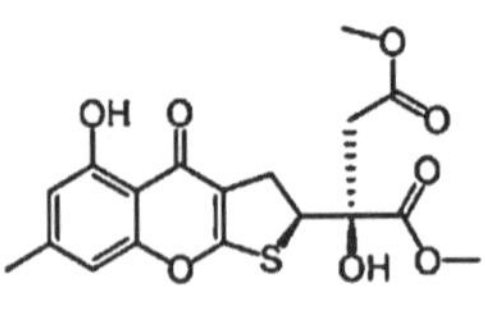

5

1-5:Oxalicumones A-E

Others: Nicoletti and Trincone (2016) reported on the isolation of compounds viz. methoxyroquefortine C, meleagrin, glandicoline B, neoxaline, (Z)-3-(1H-Imidazole-4-yimethylene)-6-(1H-indl-3-ylmethyl)-2,5piperazinediol, penilloid A, roquefortine C, and isoroquefortine C, from the fungal strain *Penicillium* sp. SCSIO00258 associated with this species. However, the bioactivities of these compounds have not been reported.

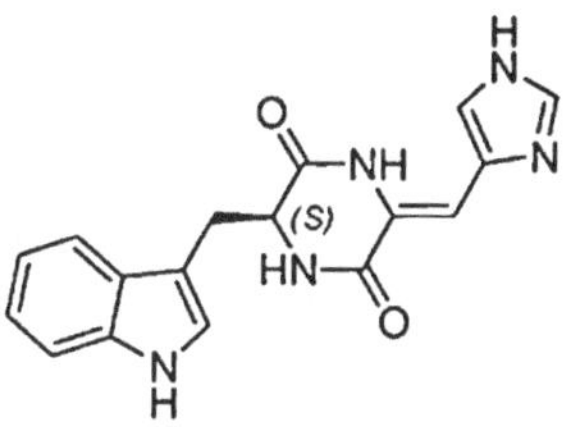

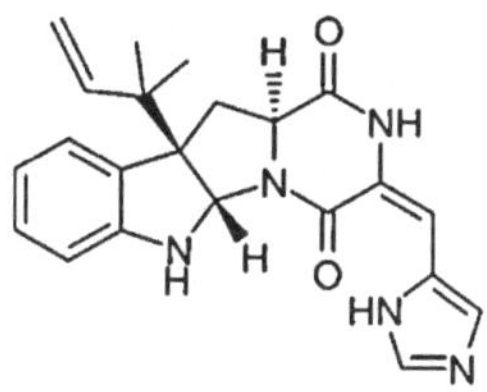

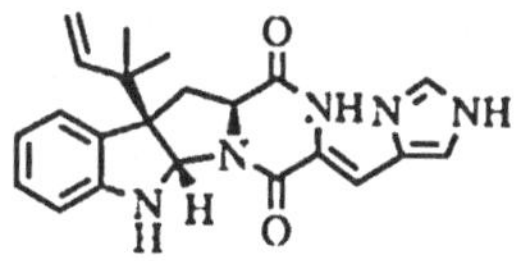

Penilloid A Roquefortine C Isoroquefortine C

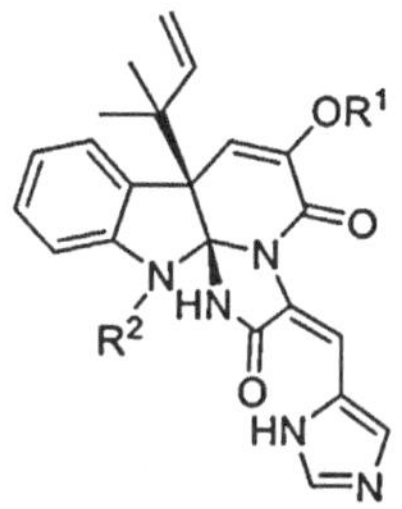

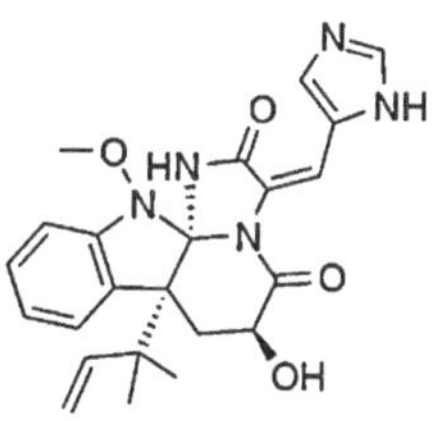

Meleagrin (R1=H, R2=OCH3; Glandicoline B (R1=H, R2=OH) Neoxaline

Meleagrin (R1=H, R2=OCH3; Glandicoline B (R1=H, R2=OH) Neoxaline

The fungus, *Aspergillus* sp. associated with this octocoral produced a mero-terpenoid compound *viz.* chrodrimanin B. The pharmaceutical activity of this compound is to be known (Raimundo et al. 2018).

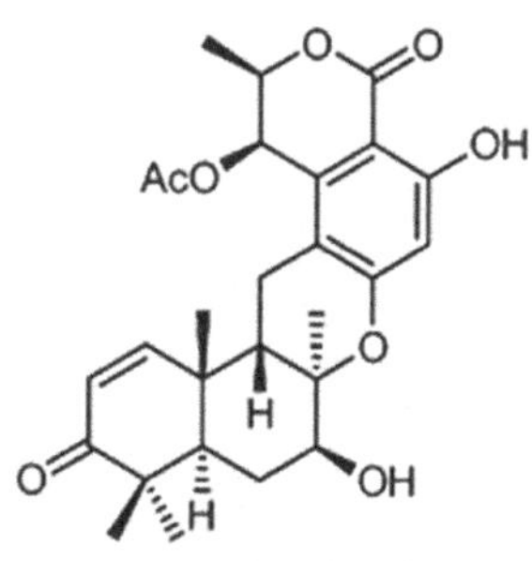

Chrodrimanin B

Verrucella umbraculum

Common name: Octocoral

Global distribution: Indian Ocean, Andaman Sea; China

Habitat: Depth 88 m

Associated fungi: Pharmaceutical compounds and activities

Antifungal activity: The fungal strain *Nigrospora oryzae* SCSGAF 0111 isolated from this gorgonian yielded two citrinins, nigrospins B and C which showed weak antifungal activity on the pathogenic fungus, *Aspergillus versicolor* with the inhibition zone dia. value of 8 cm at 50 g/paper disc (Jin et al. 2016).

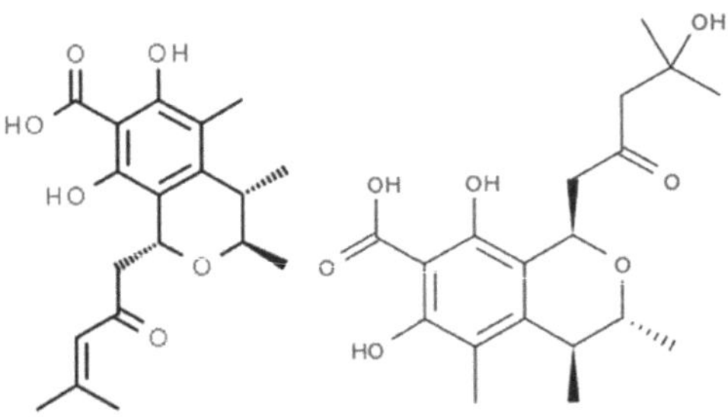

Nigrospins B, C

Cytotoxicity: The fungal strain *Aspergillus sydowii* SCSIO 00305 isolated from this species yielded compounds such as an indole diketopiperazine alkaloid, cyclotryprostatin E (1), and nine other compounds viz. [4-(2-methoxyphenyl)-1-piperazinyl][(1-methyl-1H-indol-3-yl)]-methanone (2), cyclotryprostatin B (3), fumiquinazoline D (4), fumitremorgin B (5), fumiquinazoline C (6), fumiquinazoline B (7), fumiquinazoline A (8), fumiquinazoline F (9), fumiquinazoline G (10). Among these compounds, compound 2 showed significant cytotoxicity against A375 cell lines with IC50 (half maximal inhibitory concentration) value of 5.7 mM (He et al. 2012; Blunt et al. 2015).

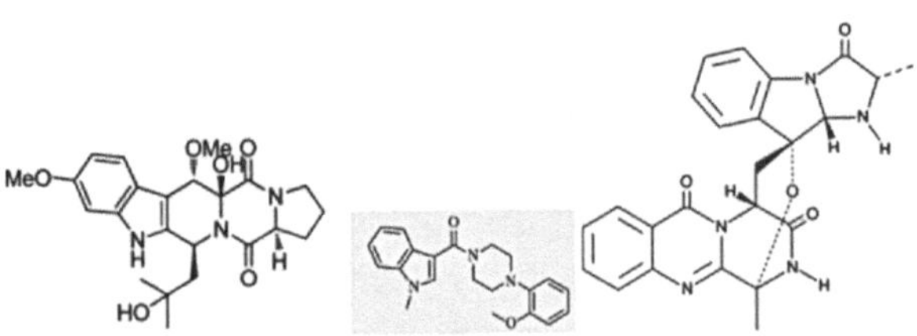

1 2 3

1: Cyclotryprostatin E 2: [4-(2-methoxyphenyl)-1-piperazinyl][(1-methyl-1H-indol-3-yl)]-methanone; 3: Fumiquinazoline D

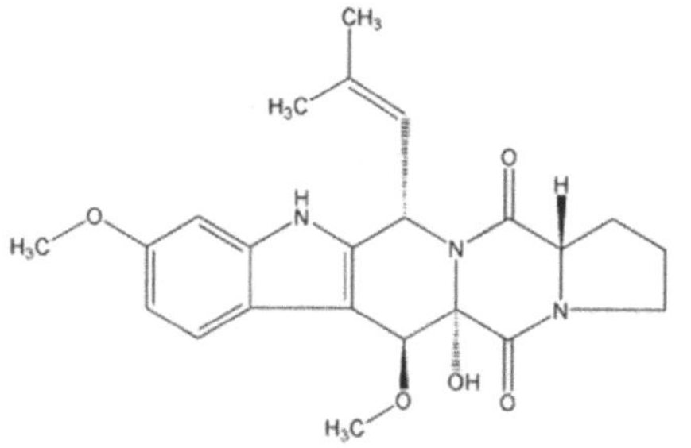

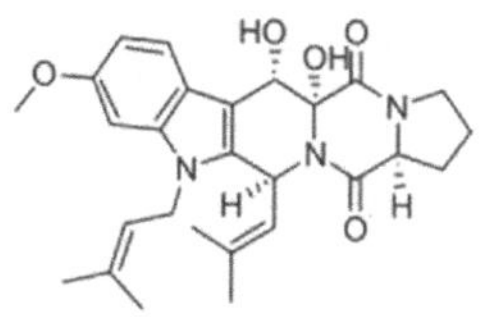

Cyclotryprostatin B Fumitremorgin B

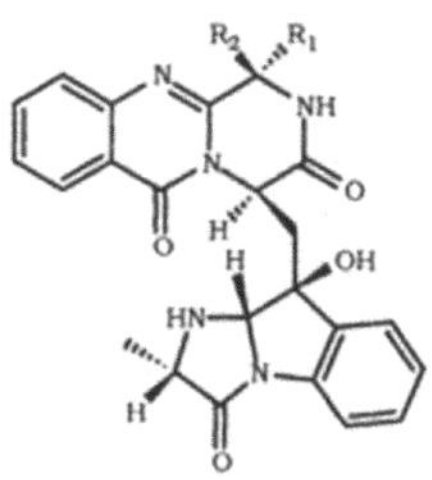

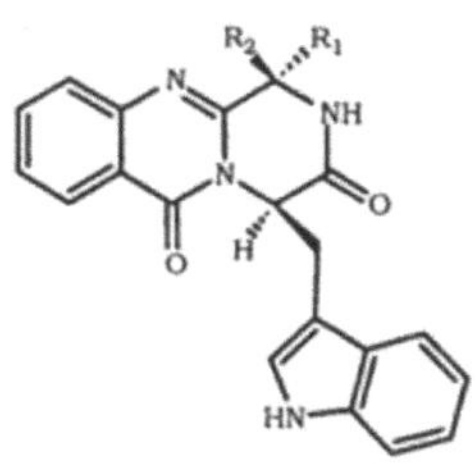

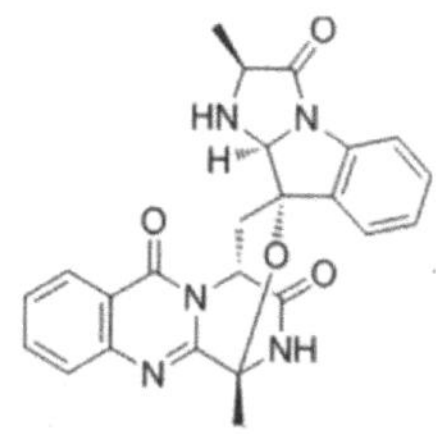

Fumiquinazoline A (R1=CH3, R2=H) Fumiquinazoline G (R1=CH3, R2=H) Fumiquinazoline C

Fumiquinazoline B (R1=H, R2=CH3) Fumiquinazoline H (R1=H, R2+ CH3)

Order: Alcyonacea Family: Gorgoniidae

***Rumphella* sp.**

Image credit: Haplochromis, https://creativecommons.org/licenses/by-sa/3.0/

Common name (s): Bushy Gorgonian

Global distribution: Indo-Pacific Ocean

Habitat: Found at various depths on reef slopes

Associated fungi; and their pharmaceutical compounds and activities

Antibacterial activity: The associated fungus, *Neosartorya siamensis* KUFA 0017 of this species has been reported to produce an indole alkaloid viz. neofiscalin A which displayed antibacterial activity against MRSA and *Escherichia faecalis* with an MiC value of 8 μg/mL (Nweze et al. 2020).

Order: Alcyonacea Family: Melithaeidae

Melitodes squamata

Common name: South China Sea gorgonian

Global distribution: Indian Ocean: Pulau Gomunu, south of Obi, Moluccas, Indonesia

Habitat: Not reported

Associated fungi: Pharmaceutical compounds and activities

Antiviral and cytotoxic activities: The fungal strain *Aspergillus* sp. SCSGAF 0076 isolated from this gorgonian yielded three polyketones, aspergillide D (1), (R)- semivioxanthin (2), (R)-semixanthomegnin (3) [3b], and four alkaloids (4-7), 5-(1H-indol-3-ylmethyl)imidazolidine-2,4-dione (4), 9,14-dihydroxy-6-p-nitrobenzoylcinnamolide(5), 7,14-dihydroxy-6-p-nitrobenzoylconfertifolin (6) and azonazine (7). Among these products, compound 5 exhibited an obvious inhibitory effect on H1N1 and H3N2 influenza virus strains, with IC50 values of 7.4 and 4.3 M, respectively. Compound 6 produced IC50 values of 36.0 and 12.0 M, respectively, while compound 7 had no activity. It is also reported that the compound 5 had strong cytotoxicity (Bao et al. 2013).

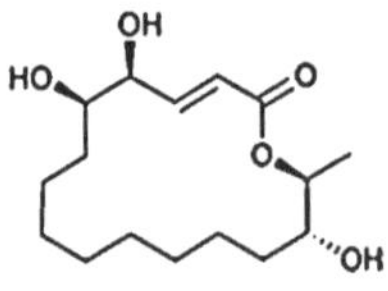

1

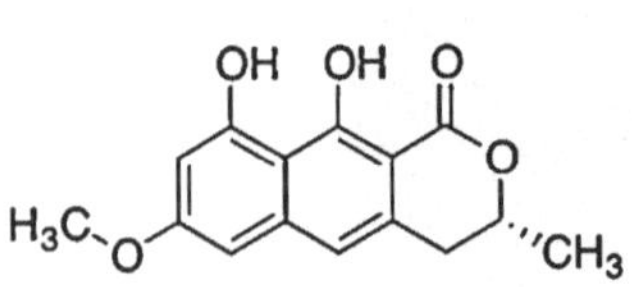

2

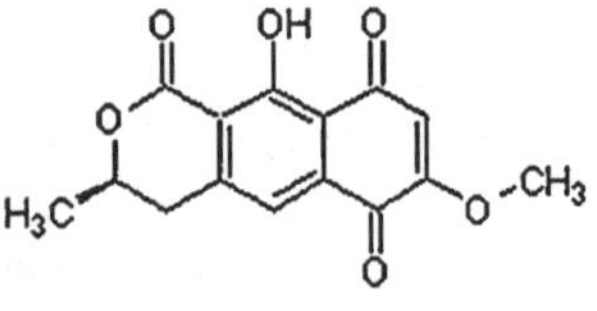

3

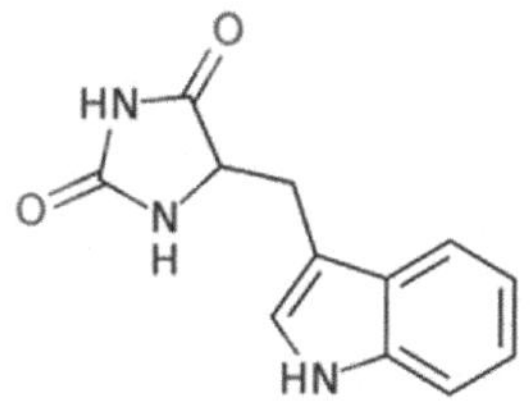

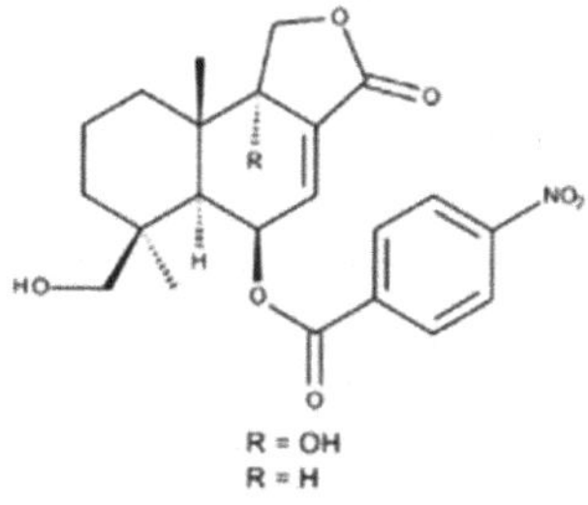

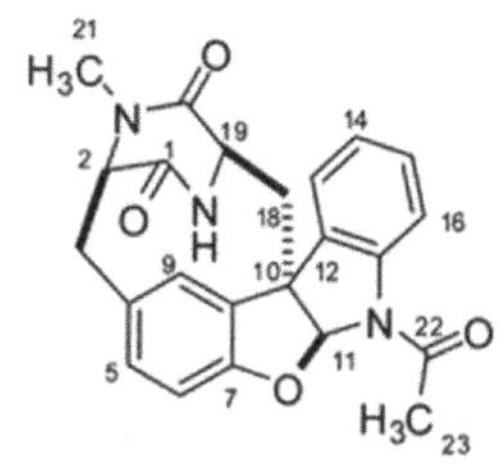

4 5,6 7

Aspergillide D (1), (R)- semivioxanthin (2), (R)-semixanthomegnin (3), 5-(1H-indol-3-ylmethyl)imidazolidine-2,4-dione (4), 9α,14-dihydroxy-6β-p-nitrobenzoylcinnamolide(5), 7α,14- dihydroxy-6β-p-nitrobenzoylconfertifolin (6), azonazine (7).

Antiviral activity: A pentapeptide compound, aspergillipeptide D obtained from this gorgonian - associated fungus *Aspergillus* sp. SCSIO 41501 showed moderate antiviral activity against HSV-1 (herpes virus simplex type 1) with an IC50 value of 9.5 μM (Mayer et al. 2021).

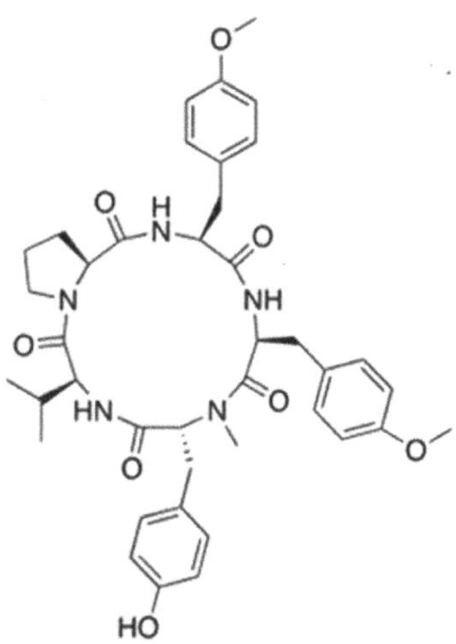

Aspergillipeptide D

Others: Blunt et al.(2015) reported on the isolation of the cyclic tetrapeptides aspergillipeptide A–C and asteltoxin B from *Aspergillus* sp. associated with this species.

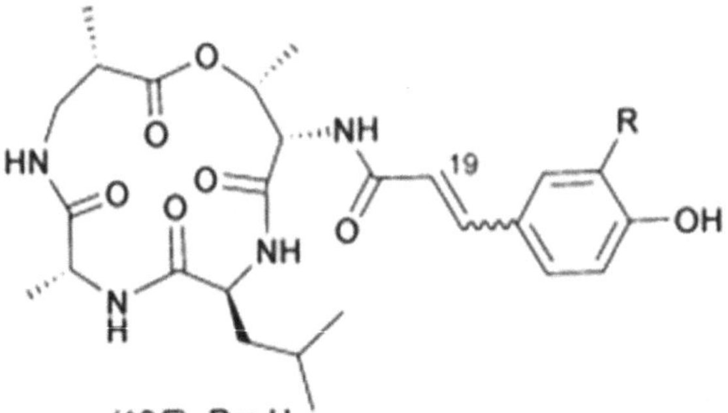

Aspergillipeptides A–C

Order: Alcyonacea Family: Nephtheidae

***Dendronephythya* sp.**

Image credit: Nhobgood Nick Hobgood, https://creativecommons.org/licenses/by-sa/3.0/

Common name: Tree soft coral, ball flowery soft coral.

Associated fungi: Pharmaceutical compounds and activities

Cytotoxic activity: The fungus *Penicillium janthinellum* isolated from this species yielded griseofulvin, dechlorogriseofulvin, janthinolide A–B, and deoxymycelianamide.Among these compounds, deoxymycelianamide showed strong cytotoxic activity against U937 cell line with IC50 values of 0.785 microg/ml (Yao et al. 2009; Nicoletti and Trincone 2016).

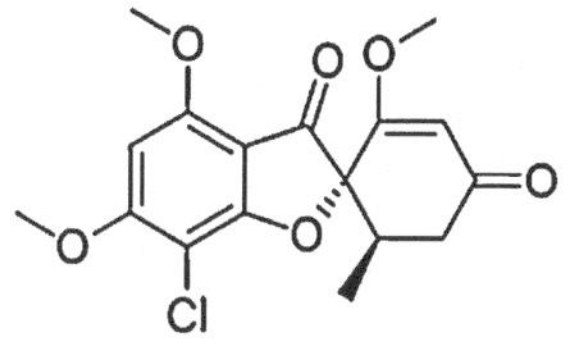

Griseofulvin

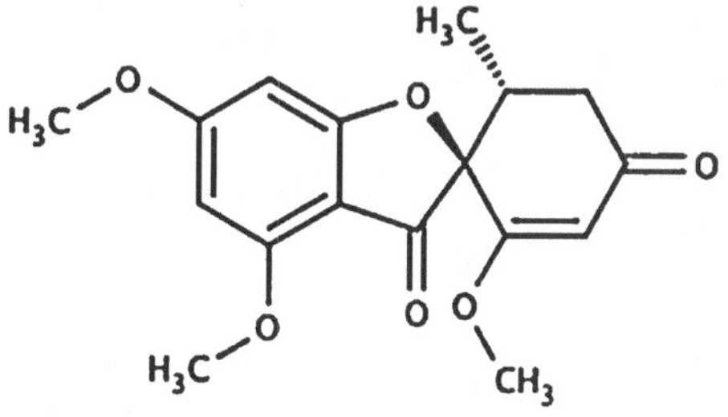

Dechlorogriseofulvin

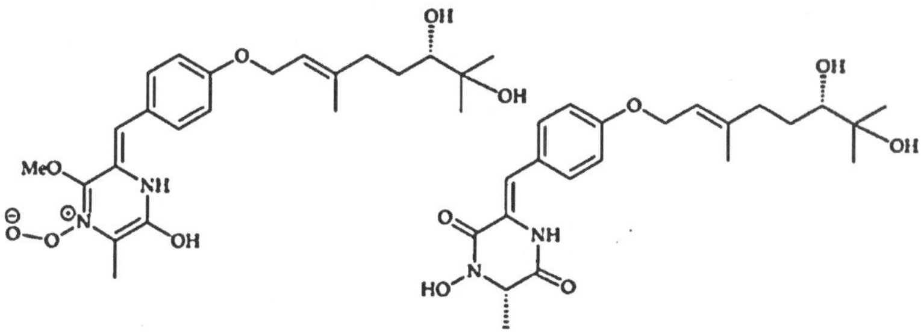

Janthinolides A,B

Order: Alcyonacea   Family: Plexauridae

Echinogorgia aurantiaca

Image credit: Chaloklum Diving. https://creativecommons.org/licenses/by-sa/4.0/

Echinogorgias sp.

Common name: Not reported

Global distribution: South China Sea

Habitat: Benthic

Associated fungi: Pharmaceutical compounds and activities

Cytotoxic and antiviral activities: The fungus *Aspergillus terreus* SCSGAFO162 isolated from this gorgonian yielded compounds such as a terrapeptide, asperterrestide A, an alkalide, terremide C and a butenolide, aspernolide E. Among these compounds asperterrestide A showed cytotoxic activity against human cancer cell lines U937a and MOLT 4. Further, this compound also displayed antiviral effects on the nfluenza virus strains, H1N1 and H3N2 (Atta-ur-Rahman, 2016).

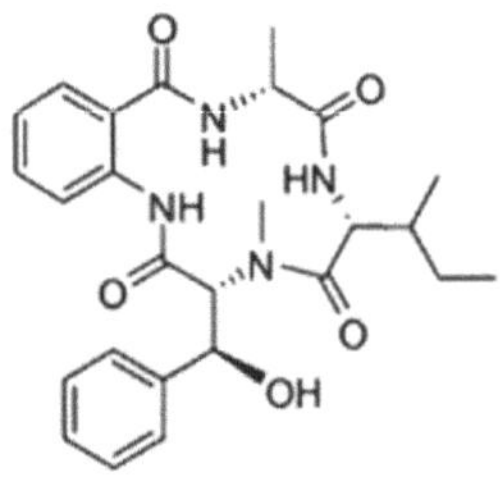
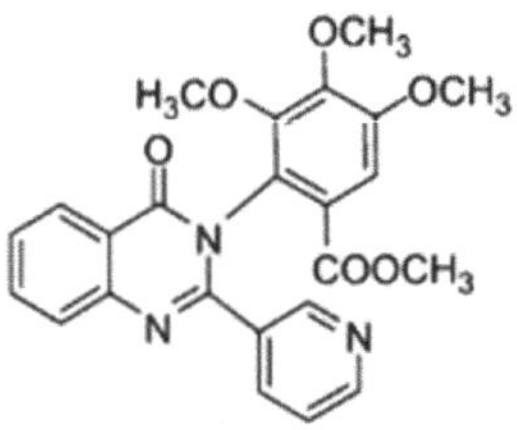
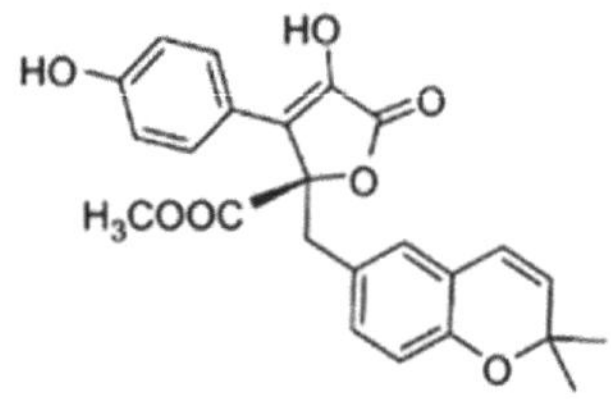

Asperterrestide A Terremide C Aspernolide E

Antiviral activity: The marine fungus, *Aspergillus terreus* SCSGAF0162 isolated from this species yielded four compounds including 11a-dehydroxyisoterreulactone A, arisugacin A, isobutyrolactone II and aspernolide A which displayed antiviral activity against HSV-1 with IC50 values of 33.38, 12.76, 62.08, and 68.16 μM, respectively (Moghadamtousi et al., 2015).

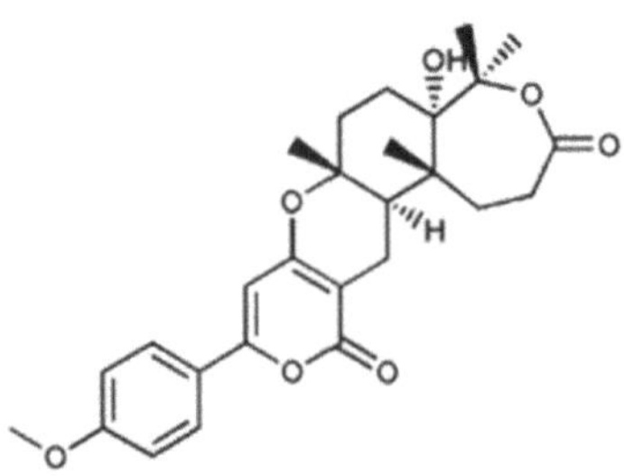
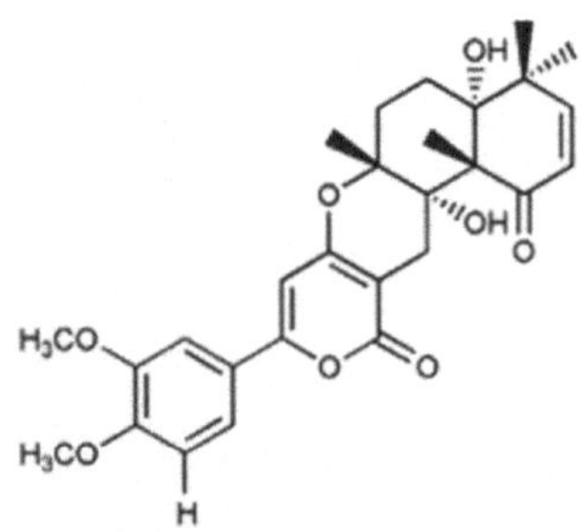

11a-dehydroxyisoterreulactone A Arisugacin A

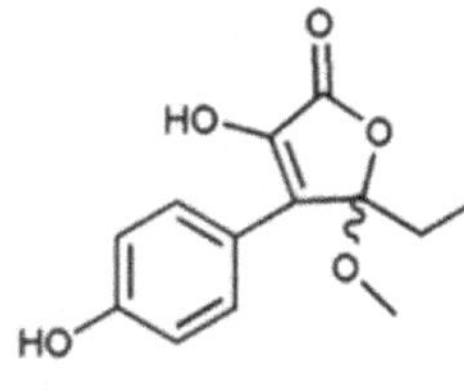

Isobutyrolactone II

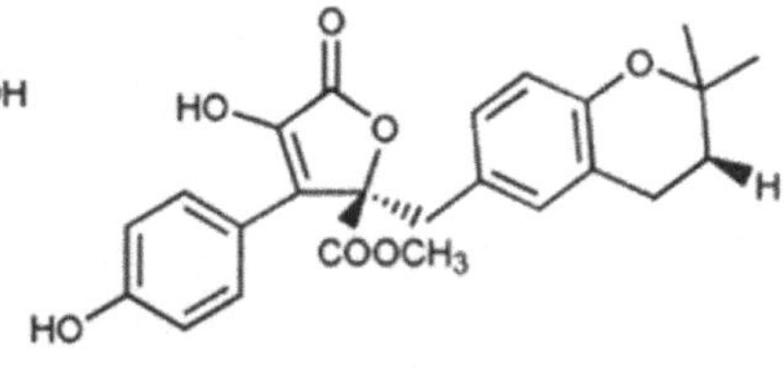

Aspernolide A

Moghadamtousi et al.(2015) also reported that the compound asperterrestide A a cyclic tetrapeptide isolated from the fungal strain of *A. terreus* SCSGAF0162 of this gorgaonian species displayed suppressive impacts on the influenza virus strains A/WSN/33 (H1N1) and A/Hong Kong/8/68 (H3N2) with IC50 values of 15 and 8.1 μM, respectively.

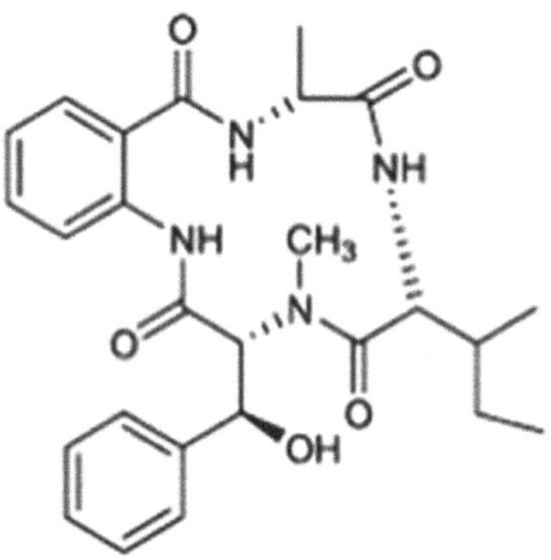

Asperterrestide A

AChE inhibitory activity: The territrem derivatives territrem D and E obtained from *Aspergillus terreus* strain SCSGAF0162 of this gorgaonian species showed potent (AChE) inhibitory activity with the IC50 values of 4.2 and 4.5 μM respectively (Raimundo et al., 2018). These AChE inhibitors have been reported to be of great value in as drug candidates in the treatment of Alzheimer's disease, a neurodegenerative disorder.

Cathepsin B inhibitor: The fungus *Penicillium citrinum* SCSGAF167 isolated from this gorgonian species yielded penicitrinol G–H, 2,11-Dihydroxy-1-methoxycarbonyl-9-carboxylxanthone, and chrysophanol. Among these compounds, chrysophanol exhibited inhibitory activity against cathepsin B with IC50 value of 1.7 μM (Sun et al. 2014; Nicoletti and Trincone 2016). Cathepsin B has been implicated in a variety of diseases including inflammation, parasite infection, and tumor metastasis.

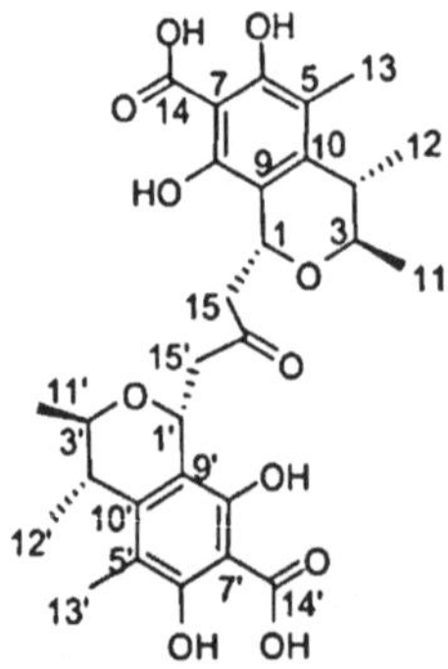

1

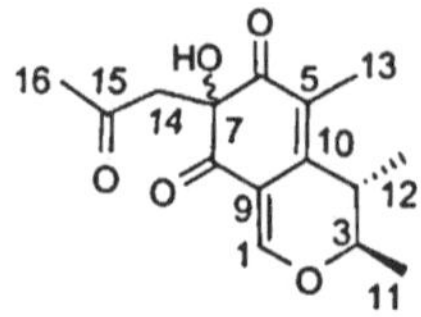

2 (7R:7S=1:1)

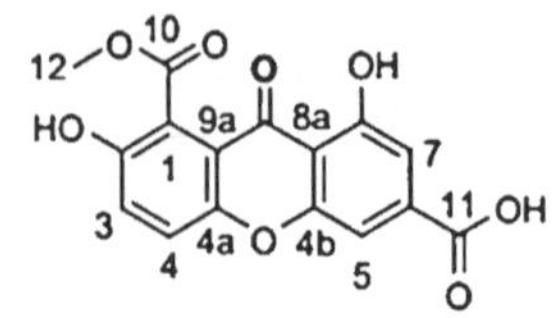

3

1,2: Penicitrinol G–H;; 3: 2,11-Dihydroxy-1-methoxycarbonyl-9-carboxylxanthone,

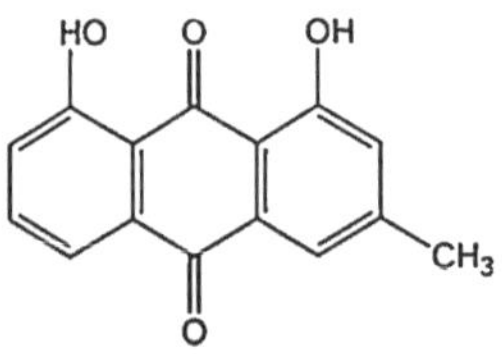

Chrysophanol

Order: Alcyonacea Family: Subergorgiidae

***Annella* sp.**

Associated fungi: Pharmaceutical compounds and activities

Antibacterial activity: The anthraquinone penicillanthranin A obtained from the fungal strain *Penicillium citrinum* PSU-F51 associated with this sea fan showed activity against MRSA strains with 16 μg/mL as MIC value (Raimundo et al. 2018).

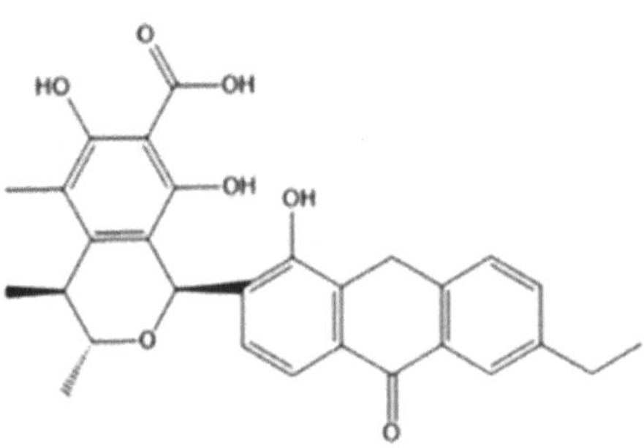

Penicillanthranin A

Antifungal sactivity: The fungal strain *Penicillium* sp. PSUF44 isolated from this species yielded metabolites, penicipyrone and penicilactone, and three macrolides (+)brefeldin A, (+) brefeldin C and 7oxobrefeldin A Among these metabolites, 7oxobrefeldin A displayed potent antifungal activity against *Microsporum gypseum* whereas the remaining compounds were inactive (Debbab et al. 2010).

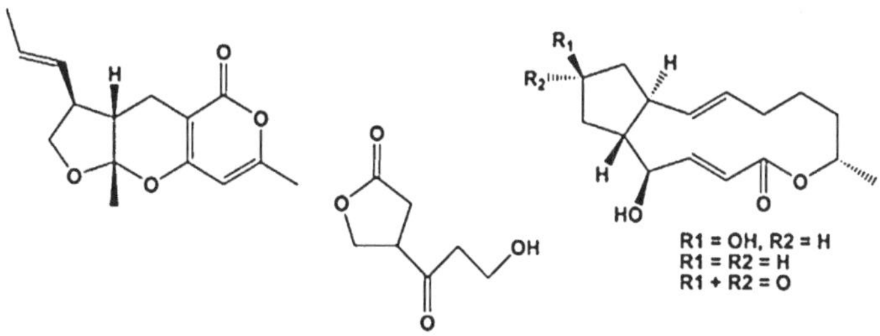

1 2 3-5

1: Penicipyrone; 2: penicilactone; 3: (+)-brefeldin A; 4: (+)-brefeldin C; 5: 7-oxobrefeldin A

Preedanon et al. (2016) reported that the compound giseofulvin extracted from the fungal strain *Nigrospora* sp. PSU-F13 associated with this species showed potent antifungal activity against *Microsporum gypseum* with an MIC valiue of 2 μg/mL.

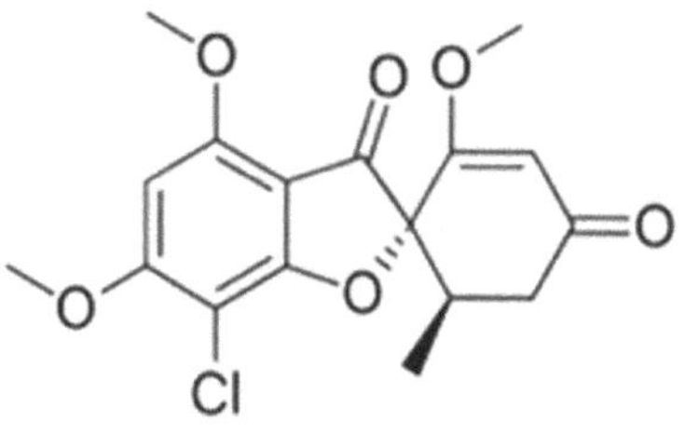

Giseofulvin

Antibacterial and cytotoxic activities: Li (2019) reported that the fungal strains *Nigrospora* sp.PSU-F5 and *Nigrospora* sp.PSU-F18 associated with this species displayed antibacterial and cytotoxicity; and antibacterial activity, respectively.

Others: The fungal strain *Penicillium* sp. (PSU-F40) isolated from this species produced compounds such as Penicipyrone, Penicipyranone, Penicisochroman A–E, Penicisoquinoline, Peniciphenol, TMC-120B, TMC-120C, 2-(2-Methoxybenzoyl)pyrrole, 1-(2,4-Dihydroxy-6-methylphenyl)-3-methyl-1-butanone (aspidinol) and Nicotinic acid (Nicoletti and Trincone 2016).

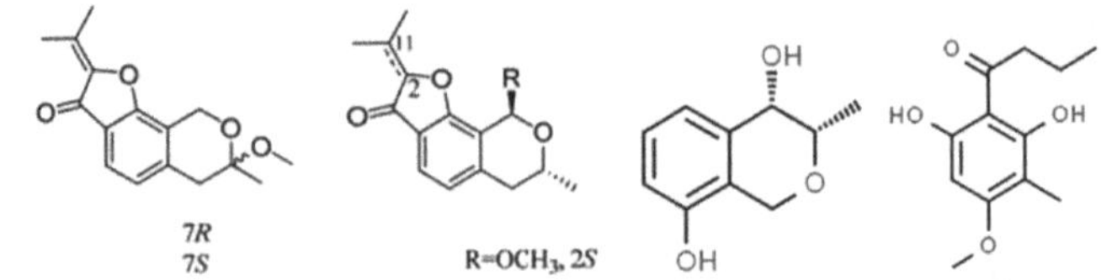

(-),(+) penicisochroman A Penicisochroman B Penicisochroman E Aspidinol

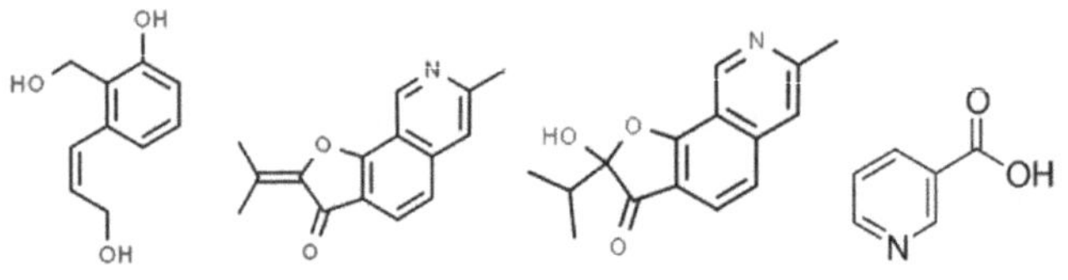

Peniciphenol TMC-120B TMC-120C Nicotinic acid

Raimundo et al. (2018) reported that the fungi, *Aspergillus sydowii* and *Curvularia* sp. were isolated from this species. The former fungi produced sesquiterpenes aspergillusene A and B, (+)-(7S)-7-O-methylsydonic acid and xanthones aspergillusone A and B; and the latter produced compounds curvulapyrone, curvulalide, and curvulalic acid.

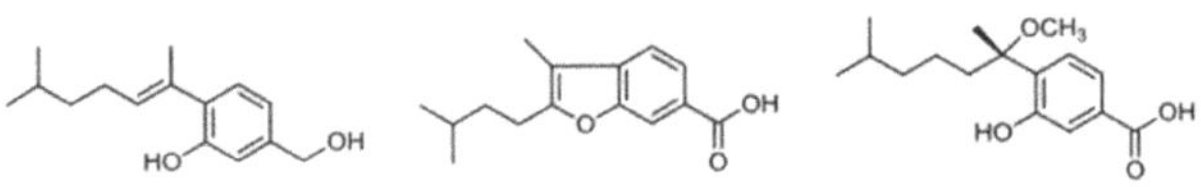

Aspergillusene A Aspergillusene B (+)-(7S)-7-O-Methylsydonic acid

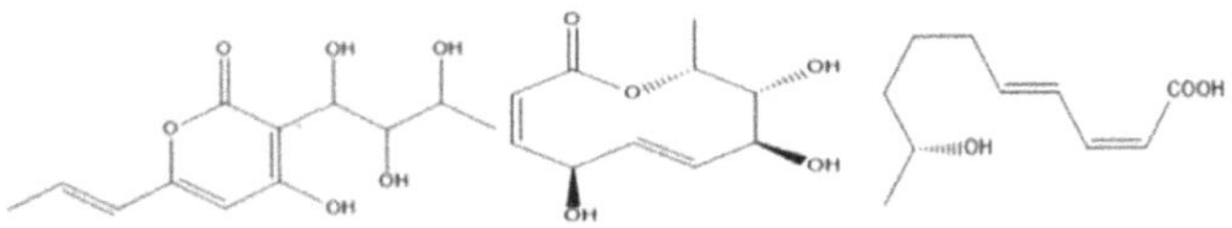

Curvulapyrone Curvulalide Curvulalic acid

Subergorgia suberosa

Subergorgia sp.

Image credit: Adapted from J Johnson. Copyright: © In-Depth Images Kwajalein

Common name: Red sea fan

Global distribution: Relatively common in tropical water around Australia and the Indo-Pacific.

Habitat: Sessile; found in turbid environments and clear water.

Associated fungi: Pharmaceutical compounds and activities

Enzyme inhibitory activity: The fungus *Talaromyces* sp. SBE-14 isolated from this coral yielded compounds, talaromycin A-C, penicillide, Δ1',3',-1'-dehydroxypenicillide, Purpactin A, C, and Tenellic acid A methyl esther. Among thse compounds, talaromycin A showed monoamine oxidase, acetylcholinesterase and PI3K inhibitory activity which was however, weak (Nicoletti and Trincone 2016; Zhang et al. 2019).

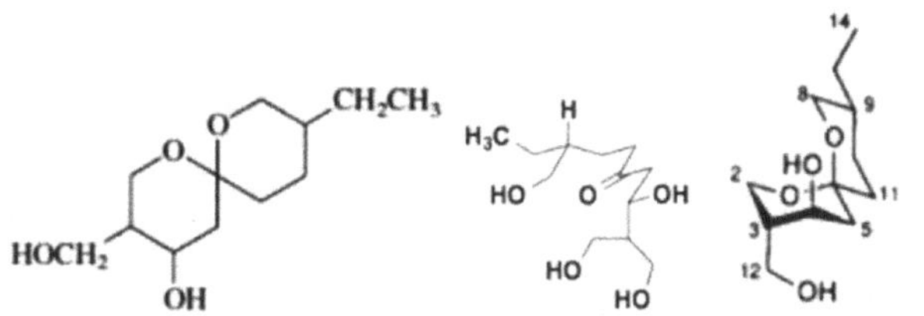

Talaromycin A-C

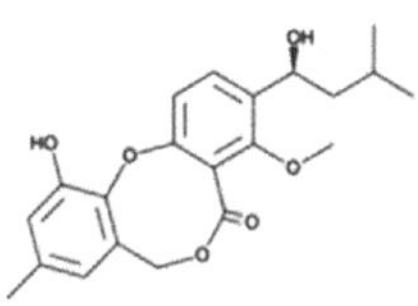

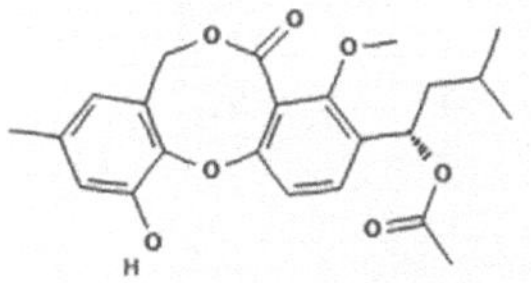

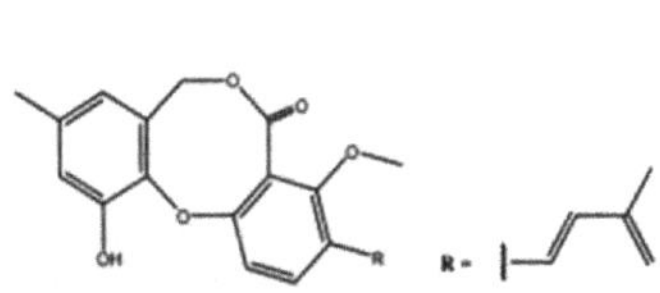

Penicillide Purpactin A Δ1',3',-1'-dehydroxypenicillide

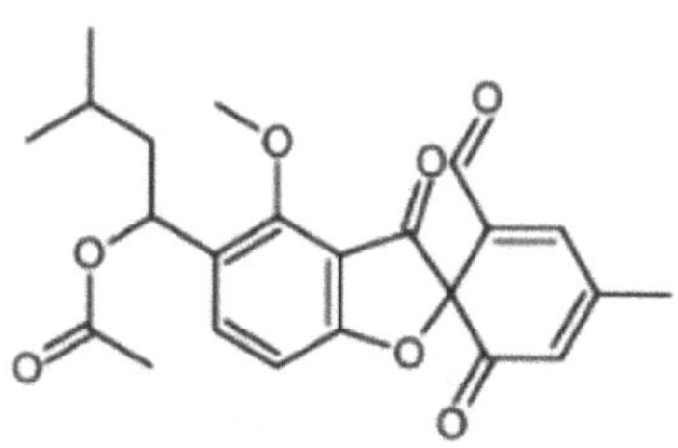

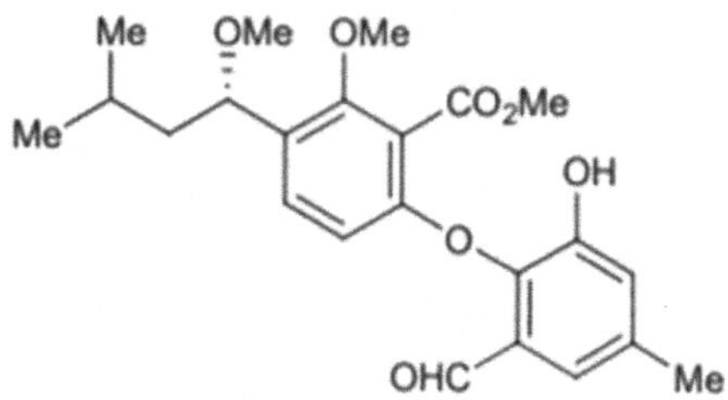

Purpactin C Tenellic acid A methyl ester

Order: Antipatharia Family: Antipathidae

Antipathes dichotoma

Associated fungi: Pharmaceutical compounds and activities

Antifungal activity: Four fungal orders: Eurotiales (30%), Hypocreales (6.6%), Pleosporales (1.7%) and Botryosphaeriales (1.7%) were found associated with this species. Among them,the most abundant and diverse fungi were observed in the genera *Penicillium* (3 species and 10 strains) and *Aspergillus* (two species and 7 strains). Among the species of these genera, *Aspergillus ochraceopetaliformis* SCSAAF0001 showed antifungal activity against *Aspergillus versicolor* and *Aspergillus sydowii* with inhibition zone values of 17.15 (strong activity) and 11.62 mm (moderate activity) respectively. Similarly, the fungus *Fusarium proliferatum* SCSAAF0009 of this species showed strong activity against *Aspergillus versicolor* with 15.45 mm (Zhang et al. 2012).

Order: Scleractinia Family:Euphyllidae

Galaxea fascicularis

Image credit: David Witherall and Jacob Crabtree, https://creativecommons.org/licenses/by/3.0/

Common name (s): Octopus coral, fluorescence grass coral

Global distribution: Indo-Pacific

Habitat: It is commonly found on coral reef slopes, rly where weak wave action exists and its depth range is 2-15 m

Associated fungi: Therapeutic compounds and their bioactivities

Antibacterial activity: The fungus, *Aspergillus tritici* SP2-8-1 derived from this coral species produced compounds such as 4-methyl-candidusin A, aspetritone A, and aspetritone B which exhibited moderate antibacterial activity against two strains of MRSA ATCC 43300 and CGMCC 1.12409 (Nweze et al. 2020).

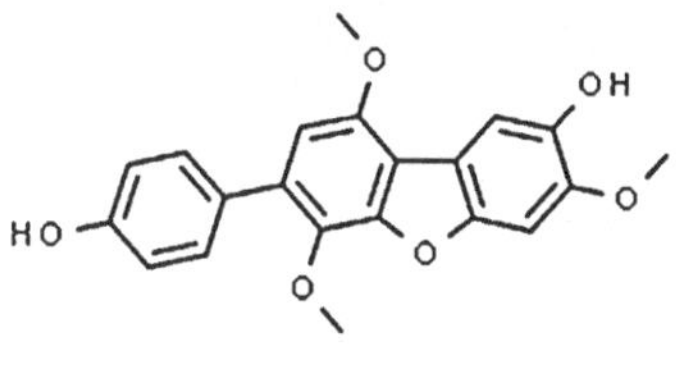

4-methyl-candidusin A

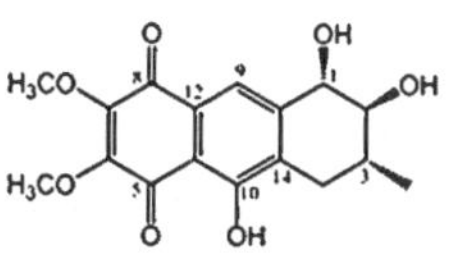

Aspetritone A

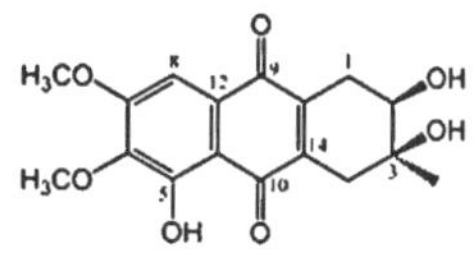

Aspetritone B

Order: Scleractinia Family: Faviidae

Favia **sp.**

Image credit: Emőke Dénes, https://creativecommons.org/licenses/by-sa/4.0/

Global distribution: Atlantic Ocean and the Indo-Pacific

Associated fungi: Pharmaceutical compounds and activities

Antibacterial activity: The extracts of the fungal species *Trichoderma longibrachiatum* associated with this coral displayed antibacterial activity against the pathogenic MDR *Staphylococcus haemolyticus* (Cristianawati et al. 2017).

Unidentified coral **sp.1**

Associated fungi: Pharmaceutical compounds and activities

Anti-yeast and cytotoxic activity: The fungus *Penicillium citrinum* isolated from this coral produced JBIR-27, petasol, sporogen AO-1 and dihydrosporogen

AO-1. Among these compounds, sporogen AO-1 exhibited anti-yeast activity against *Candida albicans* (MIC = 4.0 mM). Further, sporogen AO-1 and dihydrosporogen AO-1 showed toxicity against Ehrlich carcinoma cells with ED50 of 0.9 and 0.4 mM, respectively. JBIR-27 and petasol however, did not exhibit cytotoxic activity up to concentrations of 1.0 and 1.1 mM, respectively (Yurchenko et al. 2013a).

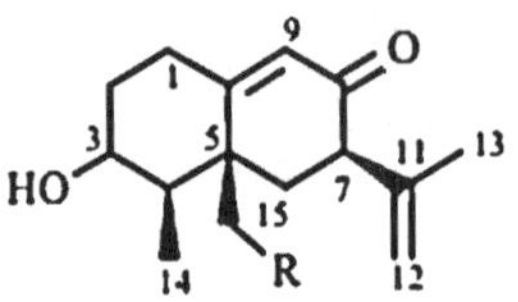

1,2

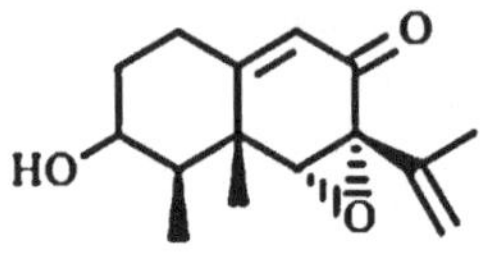

3

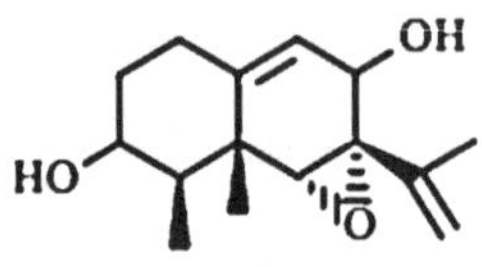

4

(1: R=OH; 2: R=H)

1: JBIR-27; 2: petasol; 3: sporogen AO-1 4: dihydrosporogen AO-1

***Unidentified coral* sp.2**

Associated fungi: Pharmaceutical compounds and activities

Anti-inflammatory activity: The fungus *Penicillium* sp. SF-5995 isolated from this coral yielded methylpenicinoline which inhibited lipopolysaccharide (LPS)-stimulated nitric oxide (NO) production in BV2 microglia. and RAW264.7 macrophages. It also attenuated prostaglandin E2 (PGE2) production by suppressing cyclooxygenase-2 (COX-2) expression in a concentration-dependent manner (from 10 M to 80 M) without affecting cell viability. In addition, this compound reduced the production of the pro-inflammatory cytokine interleukin-1 (IL-1) (Nicoletti and Trincone 2016).

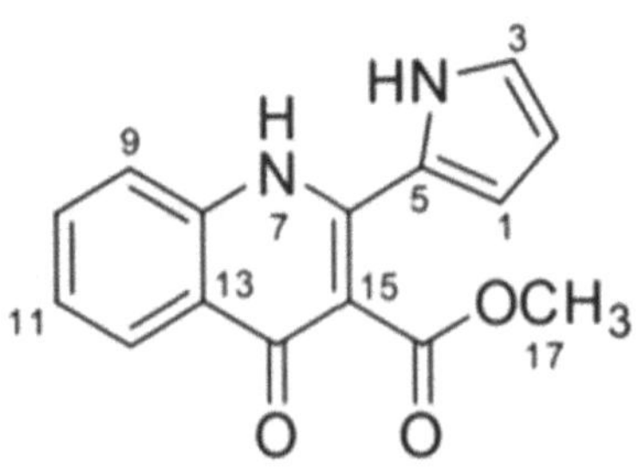

Methylpenicinoline

Cytotoxicity: Three azaphilone derivatives, pinophilins D–F (1–3) and another two unidentified compounds (4,5) were isolated from this gorgonian-derived fungus *Penicillium pinophilum* XS-20090E18. Among these compounds, compounds 4 and 5 showed cytotoxicity against Hep-2, RD, and HeLa cell lines with IC50 values ranging from 6.1 to 7.8 μM (Zhao et al. 2015).

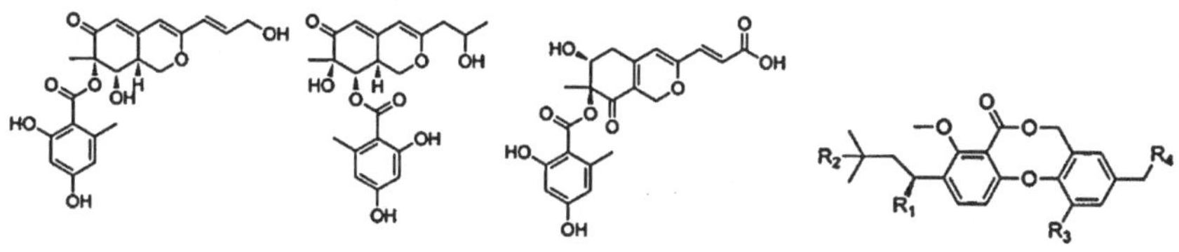

1 2 3 4,5

(4: R1=OH, R2=H, R3= OH, R4=OH; 5: R1=OH, R2=H, R3= OH, R4=H)

Pinophilins D–F (1–3);

Unidentified coral **sp.3**

Antibacterial activity: Asperteramide A, an unusual Nphenylcarbamic acid methyl ester trimer, was isolated from this coral-derived fungus *Aspergillus terreus*. This compound showed significant antibacterial activity against seven microbial pathogens viz. *Escherichia coli*, MRSA, *Klebsiella pneumoniae*, *Acinetobacter baumannii*, *Pseudomonas aeruginosa*, *Enterococcus faecalis*, and *Candida albicans*] with MIC values ranging from 2 to 64 μg/mL (Liu et al. 2019).

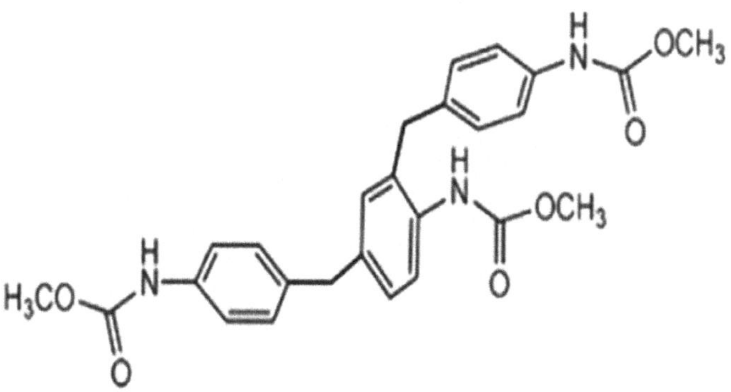

Asperteramide A

***Unidentified coral* sp.4**

Associated fungi: Pharmaceutical compounds and activities

Antimalarial activity: The tetronic acid nodulisporacid A produced by this Thailand octocoral-associated ascomycete *Nodulisporium* sp. CRIF1 exhibited potent antimalarial activity on the chloroquine-resistant strain *viz. Plasmodium falciparum* 94, with an IC50 value of 110 μM (Mander and Liu 2010).

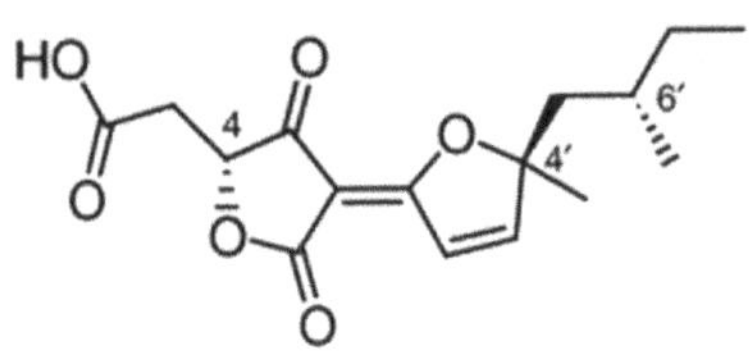

Nodulisporacid A

***Unidentified coral* sp.5**

Associated fungi: Pharmaceutical compounds and activities

Anti-inflammatory activity: The fungus *Aspergillus terreus* associated with this coral yielded the prenylated tryptophan derivative, luteoride E (1), a butenolide derivative, versicolactone G (2) and an aliphatic alcohol, (3E,7E)-4,8-dimethyl-undecane-3,7-diene-1,11-diol (3). Among these compounds, compound 2 displayed potent -glucosidase inhibitory activity with an IC50 value of 105 M. However, all the three compounds showed anti-inflammatory activity against NO production, with IC50 values ranging from 5.5 to 29 M (Liu et al. 2018).

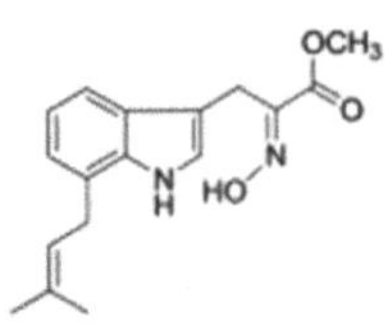

Luteoride E

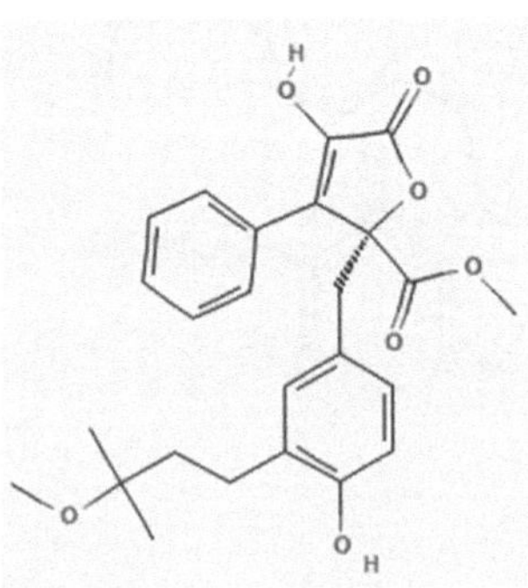

Versicolactone G

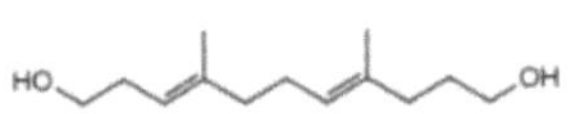

(3E,7E)-4,8-dimethyl-undecane-3,7-diene-1,11-diol

***Unidentified coral* sp.6**

Associated fungi: Pharmaceutical compounds and activities

Antibacterial activity: A dechlorinated benzophenone polyketide compound, pestalone derived from the soft coral-associated fungus *Pestalotiopsis* sp. showed activity against MR *Staphylococcus aureus* with an IC50 value of 6.25 μg/mL (Mayer et al. 2021).

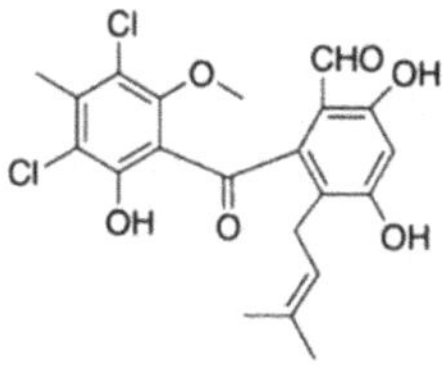

Pestalone

***Unidentified gorgonian* sp.1**

Associated fungi: Pharmaceutical compounds and activities

Others: The fungus *Penicillin pinophilum* (= *Talaromyces pinophilus*) XS-20090E18 isolated from this unidentified gorgonian yielded compounds such as purpactin A, penicillide, isopenicillide, hydroxypenicillide, azaphilone, Sch1385568, azaphilone Sch725680, pinophilins B,D–F, mitorubrin, mitorubrinol, and mitorubrinic acid (Nicoletti and Trincone 2016).

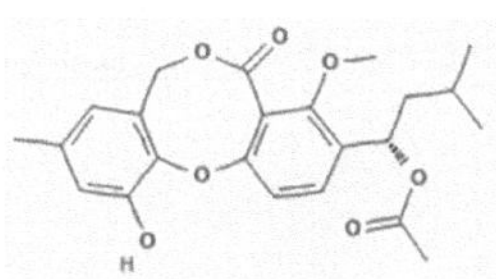

Purpactin A

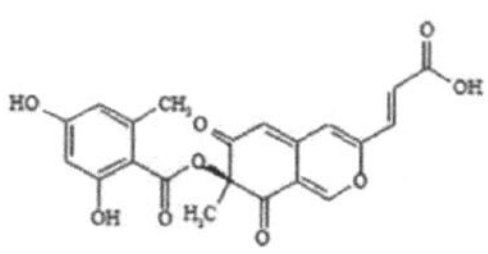

Mitorubrinic acid

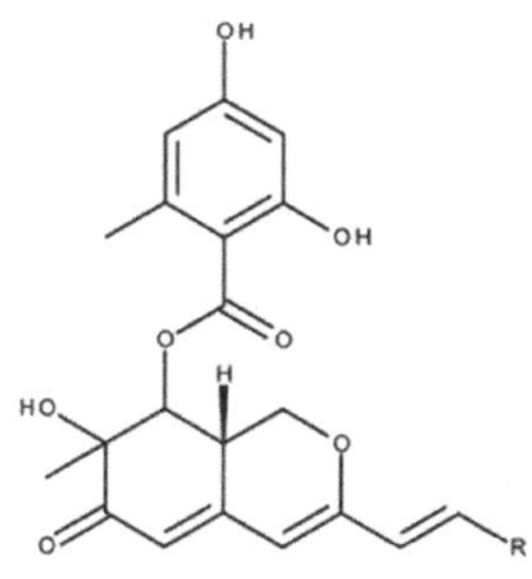

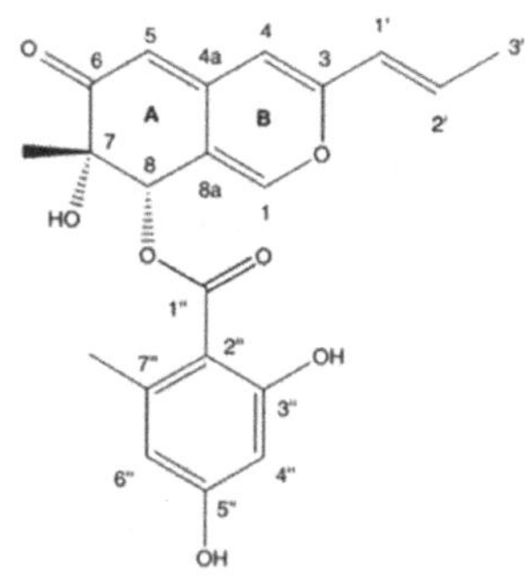

Pinophilin B (R=CH2OH); Sch725680 (R=Me) Sch1385568

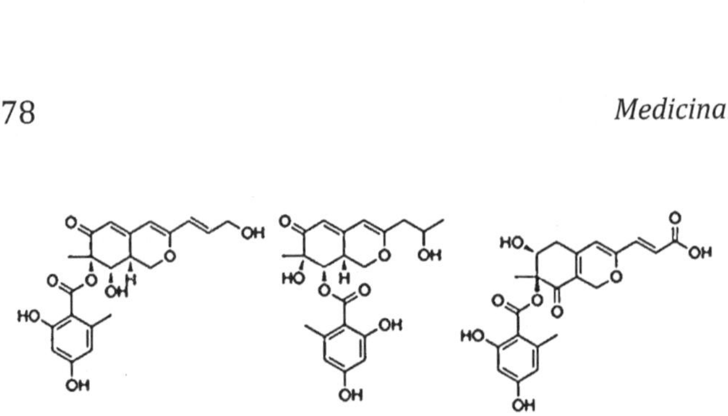

Pinophilins D-F

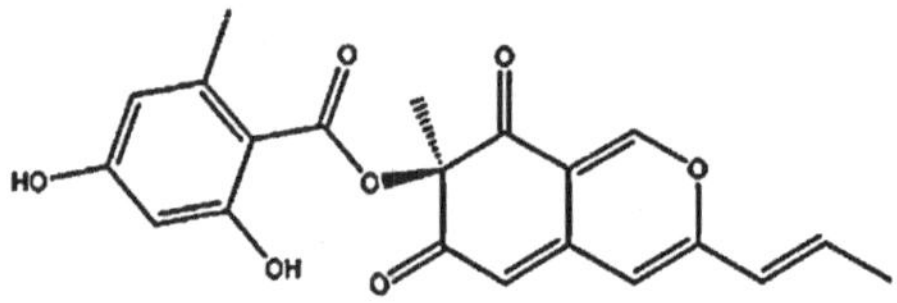

Mitorubrin

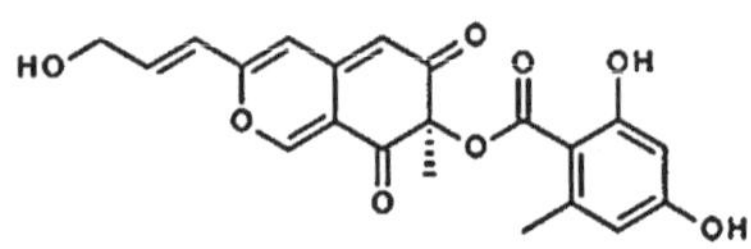

Mitorubrinol

***Unidentified gorgonian* sp.2**

Associated fungi: Pharmaceutical compounds and activities

Antibacterial activity: Two polyketide compounds (diphenyl ether derivatives) *viz.* phomaethers A, C isolated from the gorgonian-associated fungus *Phoma* sp. showed activity against *Escherichia coli* and *Staphylococcus aureus* with IC50 values ranging from 0.15 to 1.25 μM (Mayer et al. 2021).

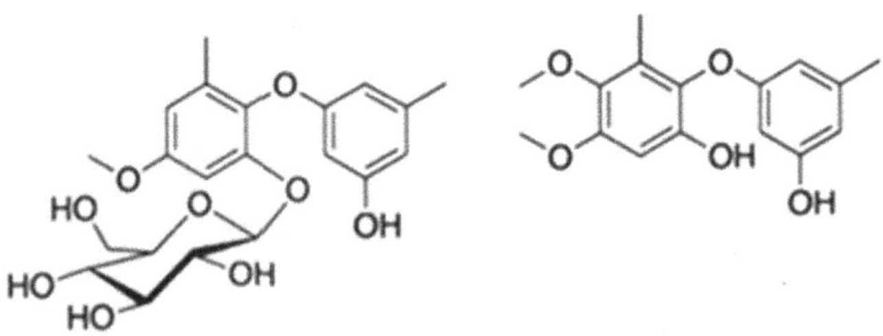

Phomaethers A, C

***Unidentified gorgonian* sp.3**

Associated fungi: Pharmaceutical compounds and activities

Antibacterial activity: The fungal species *Rhinocladiella similis* 35 and *Streptomyces rochei* MB037 associated with an unidentified staghorn gorgonian produced a chromone derivative which displayed antibacterial activity against MRSA with an MIC value of 0.195 μg/mL (Nweze et al. 2020).

BRYOZOANS

UNIDENTIFIED BRYOZOAN (1)

Associated bacteria; and their pharmaceutical compounds and activities

Antibacterial activity: Three cyclohexadepsipeptides, isaridin G (1), desmethylisaridin C1(2) and desmethylisaridin G (3) were isolated from the fungal strain, *Beauveria felina* EN-135.which is associated with this bryozoan. All these compounds showed antibacterial activity on the pathogenic *Escherichia coli* with MIC values of 64,8 and 64 g/mL, respectively (Jin et al. 2016).

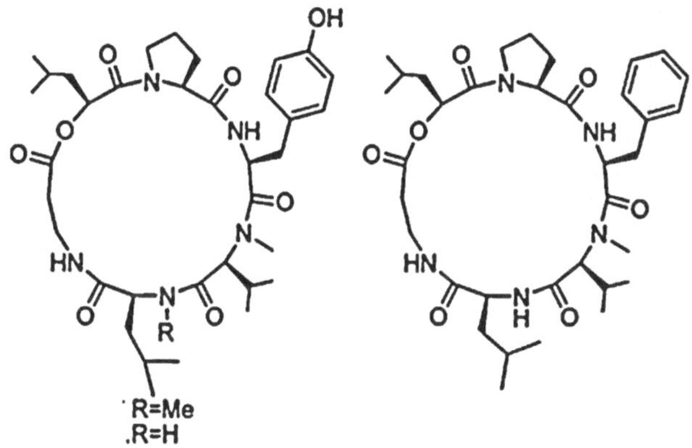

1,2 3

1: Isaridin G; 2: Desmethylisaridin G; 3: Desmethylisaridin C1

UNIDENTIFIED BRYOZOAN (2)

Associated bacteria; and their pharmaceutical compounds and activities

Antibacterial activity: The culture extract of the fungus *Beauveria felina* EN-135 associated with this bryozoan yielded compounds *viz.* O-containing heterocyclic compounds, felinones A and B (1,2) and a cyclodepsipeptide, iso-isariin D (3). Of these metabolites, only compound 2 displayed antibacterial activity against the pathogenic *Pseudomonas aeruginosa* with an MIC value of 32 g/mL (Jin et al. 2016).

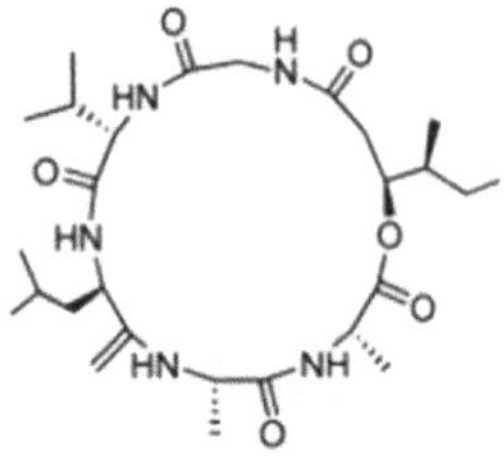

Iso-isariin D

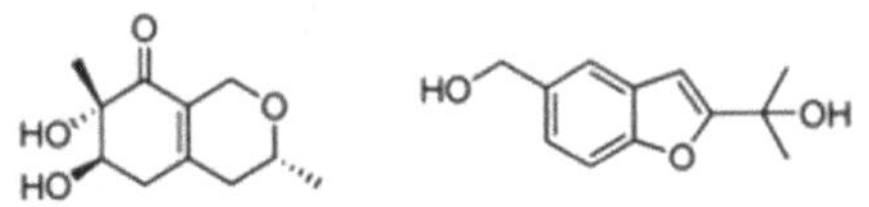

Felinones A, B

UNIDENTIFIED BRYOZOANS (3)

Associated bacteria;; and their pharmaceutical compounds and activities

Anti-inflammatory activity: The organic extract of the fungus *Penicillium* sp. SF-5292 isolated from an unidentified bryozoan yielded the compounds, penicillinolide A and cycloexpansamine AB. Among these compounds, penicillinolide A displayed anti-inflammatory activity by inhibiting the production of NO and PGE2 due to inhibition of the expression of iNOS and COX-2. Penicillinolide A also reduced TNF-, IL-1, and IL-6 production (Lee et al. 2013; Nicoletti and Trincone 2016).

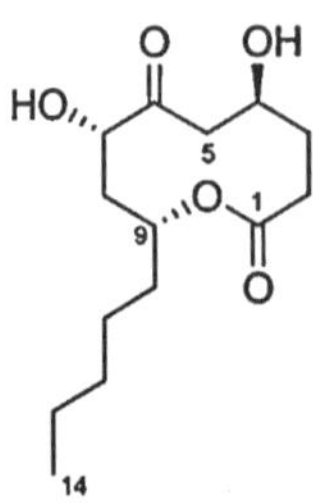

Penicillinolide A

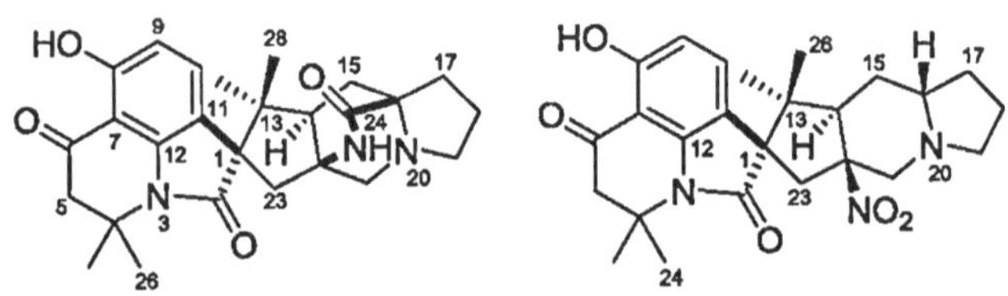

Cycloexpansamines A,B

CRUSTACEANS

ISOPODS

Phylum: Arthropoda Subphylum: Crustacea Class: Malacostraca

Order: Isopoda Family: Ligiidae

Ligia oceanica

Image credit: me'nthedogs. https://creativecommons.org/licenses/by/2.0/

Common names: Sea roach or sea slater

Global distribution: This temperate species is found distributed from Norway to the Mediterranean Sea waters; and from Cape Cod north to Maine

Habitat: It occurs commonly in rock pools, crevices and under stones

Associated fungi: Therapeutic compounds and their bioactivities

Cytotoxic activity: An aspochalasin, 20-methylthio-aspochalsin Q

(n= aspochalasin V, 44) was derived from the culture broth of *Aspergillus* sp., which was obtained in the gut of this marine isopod. This compound showed moderate cytotoxic activity against the PC3 (prostate cancer PC3) and (HCT116) cell lines with IC50 values of 30 and 39 μM, respectively (Jin et al. 2016).

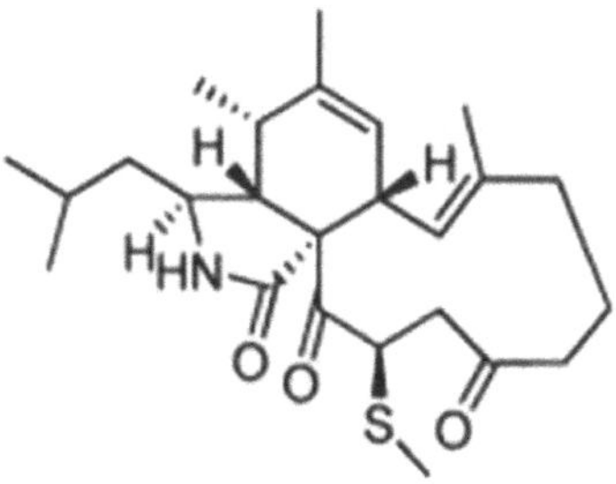

Aspochalasin V

SHRIMPS

Subphylum: Crustacea Class: Malacostraca Order: Decapoda
Family: Penaeidae

Penaeus vannamei

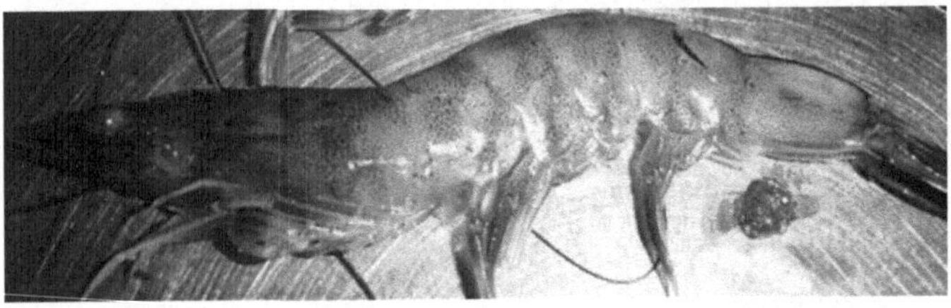

Image credit: Herman Gunawan, https://creativecommons.org/licenses/by-sa/3.0/

Common names: King prawn, whiteleg shrimp or Pacific white shrimp

Global distribution: It is native to the E.Pacific Ocean i.e from fMexican state of Sonora to N. Peru.

Habitat: Adults live and spawn in the open ocean, while their to euryhaline areas like coastal estuaries and mangroves. This species is also known to live in areas with above 20°C.

Associated fungi: Therapeutic compounds and their bioactivities

Cytotoxic and antibacterial activities: The fungal strain, *Aspergillus flavus* OUCMDZ-2205, obtained from this prawn yielded two indole-diterpenoids (1, 2). The first compound has been reported to display antibacterial activity on *Staphylococcus aureus* with an MIC value of 21 μM. This compound also exhibited PKC-beta inhibitory activity with IC50 value of 16 μM. On the other hand, at a concentration of 10 μM,compounds 1,2 have been reported to arrest the A549 cell cycle in S phase (Jin et al. 2016).

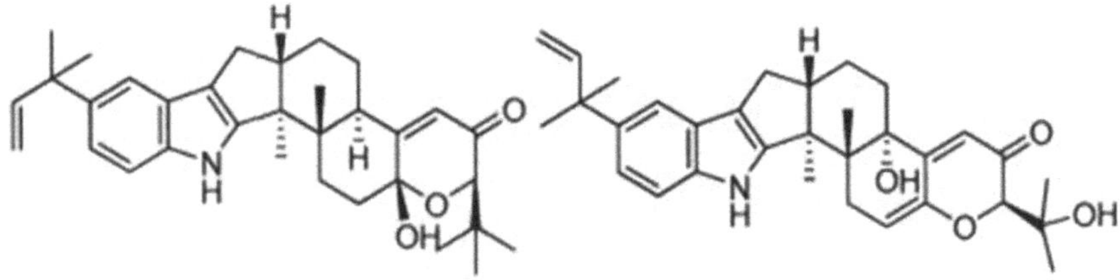

Indole-diterpenoids (1 and 2)

CRABS

Order: Decapoda Infraorder: Brachyura Family: Grapsidae

Pachygrapsus crassipes

Image credit: Peter D. Tillman, https://creativecommons.org/licenses/by-sa/3.0/

Common name (s): Striped shore crab; lined shore crab

Global distribution: It has wide distribution in Gulf of California, Japan, Korea

Habitat: This high and mid intertidal species is often seen in crevices, tide-pools, clay burrows, and mussel beds; and under rocks.

Associated fungi and their pharmaceutical compounds and activities

Antibacterial activity: The fungal strain, *Penicillium* sp. ZZ380 associated with this wild crab species produced the compounds penicipyrroether A and pyrrospirone J which displayed antibacterial activity against MRSA ATCC 43300 with MiC values of 5.0 μg/mL and 1.7 μg/mL, respectively (Nweze et al. 2020).

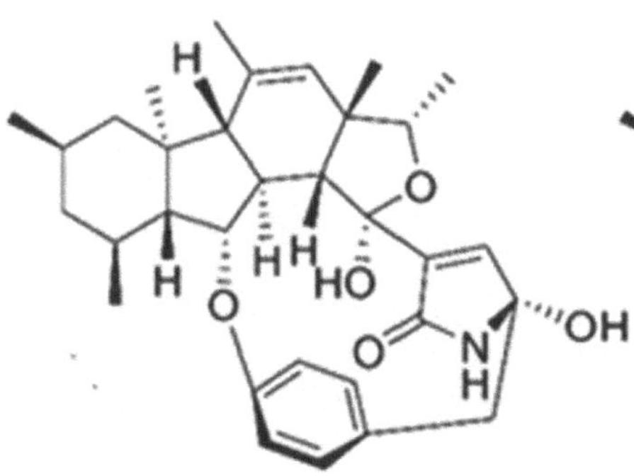

Penicipyrroether A

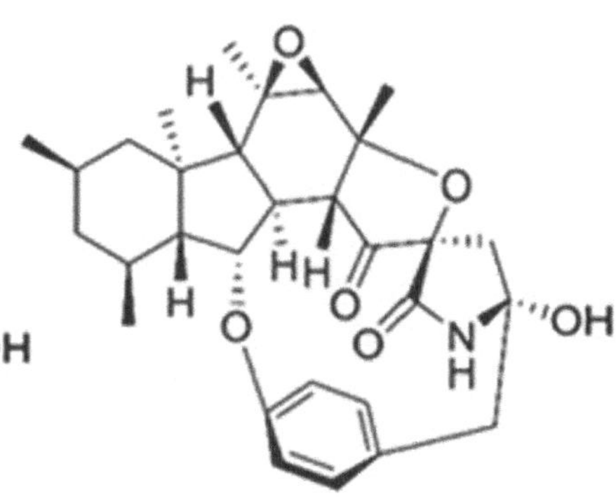

Penicipyrroether J

Infraorder: Brachyura   Family: Xenograpsidae

Xenograpsus testudinatus

Image credit: SSR2000, https://creativecommons.org/licenses/by-sa/3.0/

Common name: Hydrothermal Vent Crab

Global distribution: Northwest Pacific.

Habitat: Demersal; depth above 15 m

Associated fungi: Pharmaceutical compounds and activities

Antiproliferative activity: The cyclodepsipeptides clavatustides A and B and a D-phenyllactic acid residue, isolated from the fungus *Aspergillus clavatus* associated with this crab showed antiprolifertive activity by suppressing the proliferation of hepatocellular carcinoma (HCC) cell lines (HepG2, SMMC-7721, and Bel-7402) in a dose-dependent manner, and induced an accumulation of HepG2 cells in G1 phase and reduction of cells in S phase (Jiang et al. 2013; Blunt et al. 2015).

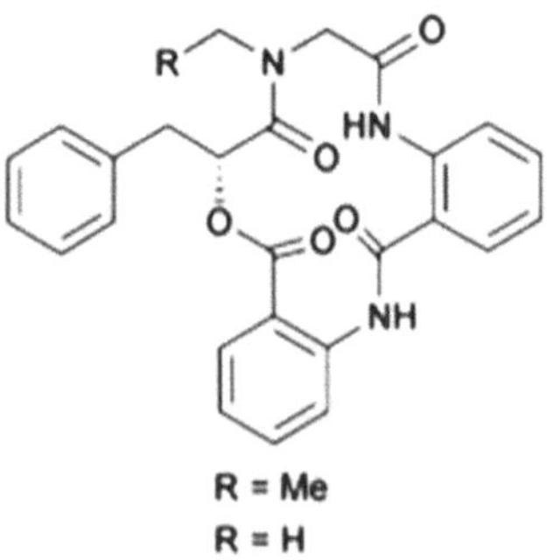

Clavatustides A,B

Phylum: Mollusca Class: Bivalvia Order: Mytilida Family: Mytilidae

MOLLUSKS

Mytilus edulis

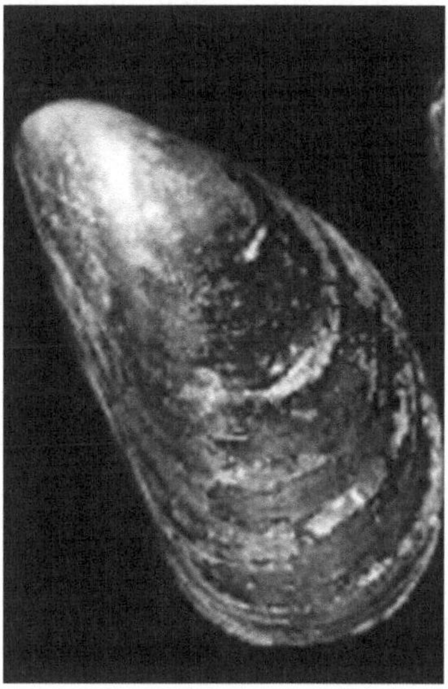

Image credit: Pallbo, Public domain.

Common name: Blue mussel, common mussel

Global distribution: Commonly found on beaches around the world.

Habitat: Intertidal areas attached to rocks and other hard substrates by strong (and somewhat elastic) thread-like structures called byssal threads.

Associated fungi: Pharmaceutical compounds and activities

Cytotoxicity: The fungi *Penicillium marinum* and *P. restrictum* strains were isolated from this mollusk. Among them the extracts of *P. restrictum* yielded the mycotoxin patulin which displayed cytotoxicity (Geiger et al. 2013a).

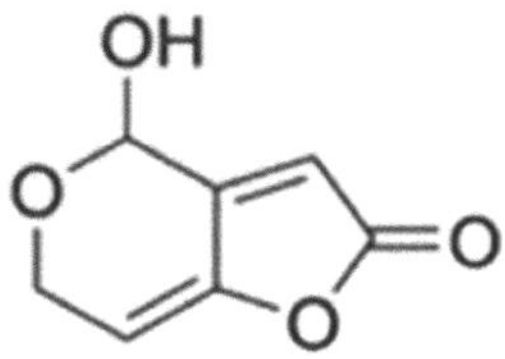

Patulin

Antimicrobial and cytotoxic activity: The fungus *Trichoderma longibrachiatum* MMS151 isolated from this from blue mussel produced six peptaibols (linear peptides). Among them, two peptides have been reported to display antifungal activity on the pathogenic fungus viz. *Aspergillus fumigatus,* and cytotoxicity against KB cell line (Mohamed-Benkada et al. 2016).

Others: The fungus *Penicillium ubiquetum* MMS330 isolated from this blue mussel yielded nine meroterpenoids *viz* 22-deoxyminiolutelide A (1), d 4-hydroxy-22-deoxyminiolutelide B (2), 22-deoxyminiolutelide B (3), miniolutelide A (4), miniolutelide B ((5), berkeleyacetal A(6), f miniolutelide C(7), 22-deoxy-10-oxominiolutelide B (8),and eupeniacetal A (9) (Hoang et al. 2018).

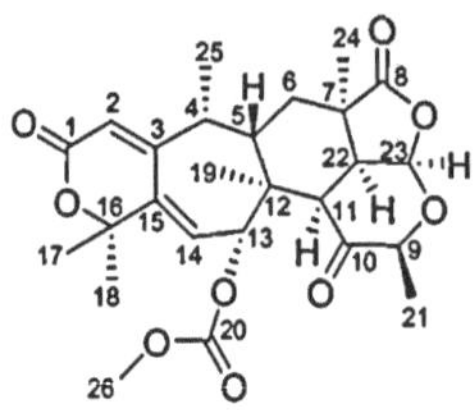

1

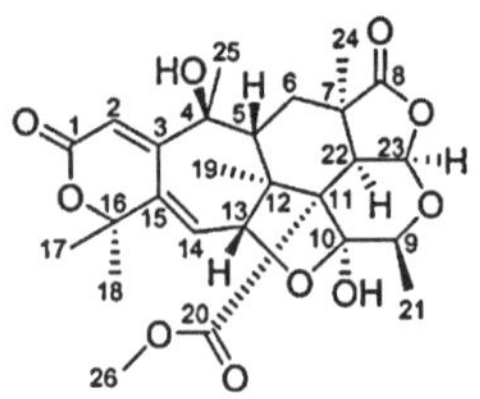

2

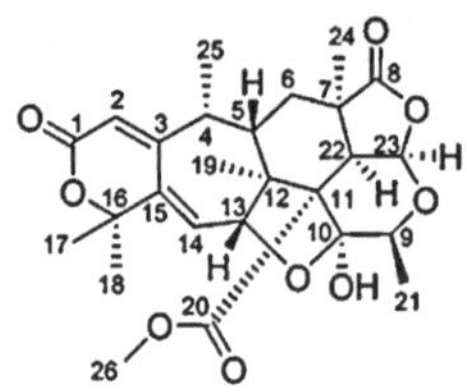

3

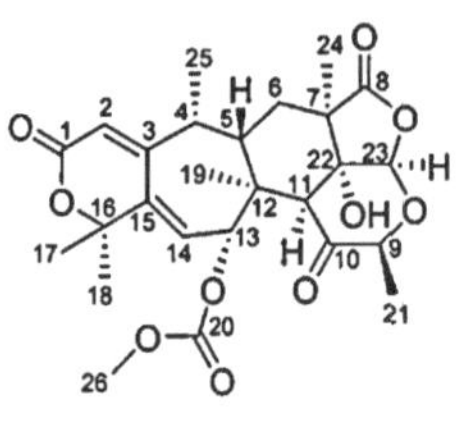

4

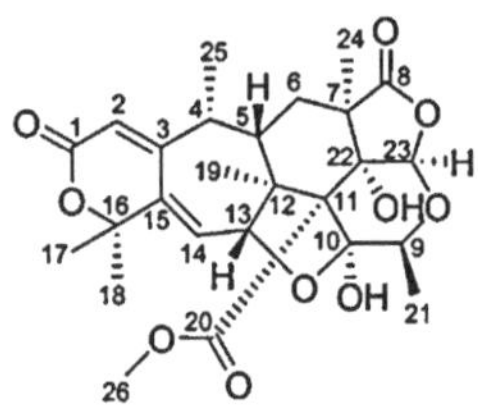

5

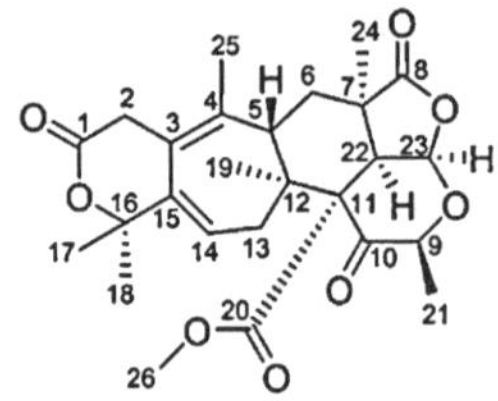

6

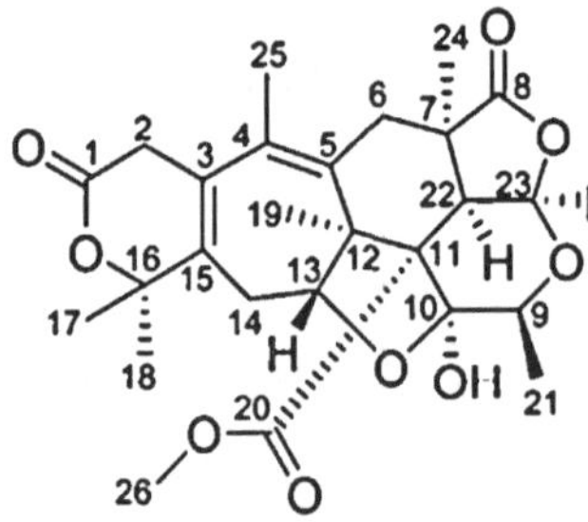

7

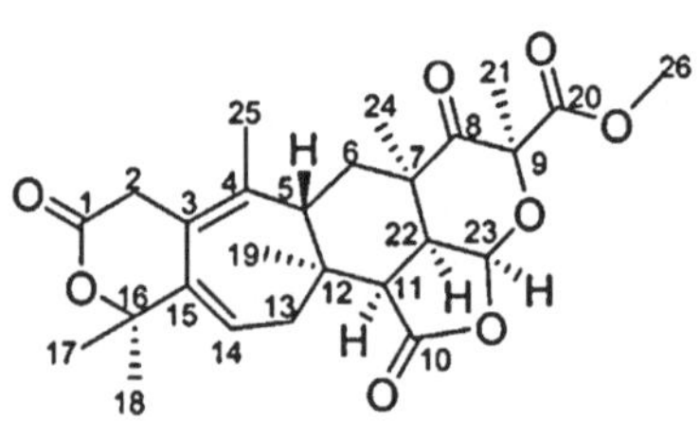

9

22-deoxyminiolutelide A (1), d 4-hydroxy-22-deoxyminiolutelide B (2), 22-deoxyminiolutelide B (3), miniolutelide A (4), miniolutelide B ((5), berkeleyacetal A(6), f miniolutelide C(7), eupeniacetal A (9)

Cytotoxicity: The prenylated indole alkaloids (notoamides A-D, and I) have been isolated from a culture strain of *Aspergillus* sp. obtained from this mussel. Among them, notoamides A–C and I showed moderate cytotoxicity against HeLa and L1210 cells with IC50 values in the range of 21–52 μg/mL. However, the IC50 of notoamide D was higher than 100 μg/mL (Lee et al. 2013).

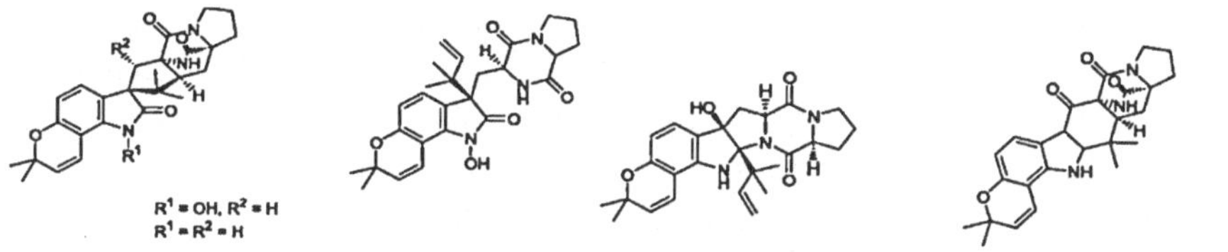

Notoamides A,B   Notoamide C   Notoamide D   Notoamide I

The fungal strain of *Aspergillus* sp. isolated from this mollusk produced himeic acids E–G, the pharmaceutical activities of which are to be known (Blunt et al. 2015).

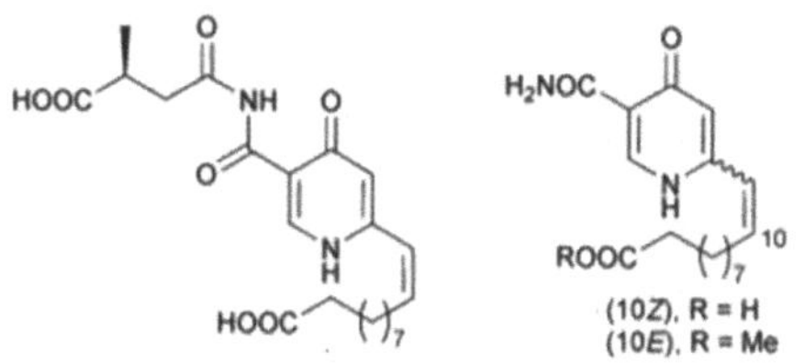

Himeic acid E   Himeic acids F,G

Order:Mytilida   Family: Mytilidae

Mytilus unguiculatus (= *Mytilus coruscus*)

Image credit: Jin-Long Yang et al., https://creativecommons.org/licenses/by-sa/4.0/

Common name: Korean mussel or hard-shelled mussel,

Global distribution: This subtropical species is found distributed in the western Pacific Ocean; and East Sea and Yellow Sea.

Habitat: Lower intertidal; shallow subtidal; depth range, 5–20 m

Associated fungi: Pharmaceutical compounds and activities

The fungal strain, *Penicillium* sp. K029 isolated from this species showed the presence of compounds *viz.* coruscol A and herquline A (Nicoletti and Trincone, 2016; Kagata, et al., 2000). Komatsu et al. (2000) reported that the fungal strain *Penicillium* sp. K036 of this bivalve yielded compounds viz. sculezonone A–B. Among these compounds, herquline A has been reported to inhibit platelet aggregation at 100 µg/ml (https://www.kitasato-u.ac.jp/lisci/Splendid_Page/Herquline.pdf).

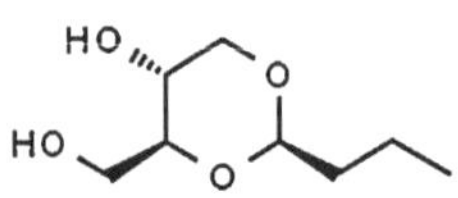

Coruscol A

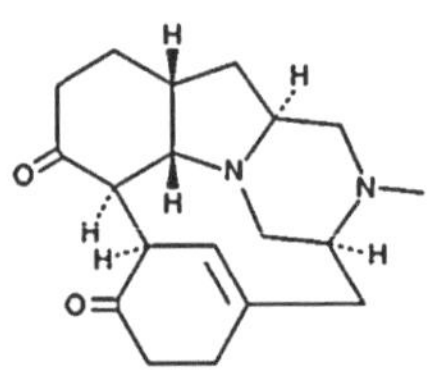

Herquline A

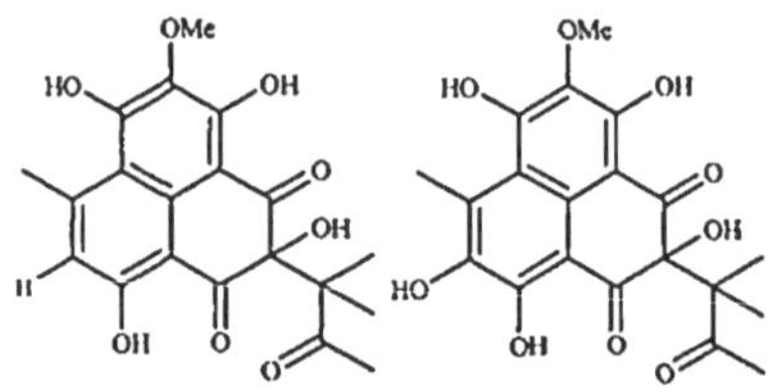

Sculezonone A–B

Unidentified cockle

Associated fungi: Pharmaceutical compounds and activities

Bioactivities: Nicoletti and trincone (2016) reported that the fungal strain *Penicillium antarcticum* MMS14 isolated from an unidentified cockle yielded chrysogine, cladosporin (=asperentin), 5-Hydroxyasperentin, antarone A, violaceic acid, patulin, terrestric acid. Among these compounds, the compound, patulin showed cytotoxicity, and cladosporin has been reported to possess antifungal, antimicrobial and antimalarial activities (Geiger et al., 2013a). Further, another strain, *Penicillium antarcticum* MMS15 isolated from an unidentified cockle produced compounds such as cladosporin, chrysogine, 5-Hydroxyasperentin, aurantioclavine, antarone A, patulin and terrestric acid (Geiger et al., 2013a). Furthermore, the strain *Penicillium restrictum* MMS417 of a cockle species yielded compounds viz. pestalotin, hydroxypestalotin and 5,6-Dihydro-4-methoxy-6-(1-oxopentyl)-2H-pyran-2-one. Among these compounds, pestalotin derivatives have been described to exhibit only weak cytotoxicity against the human tumor cell line U-251.

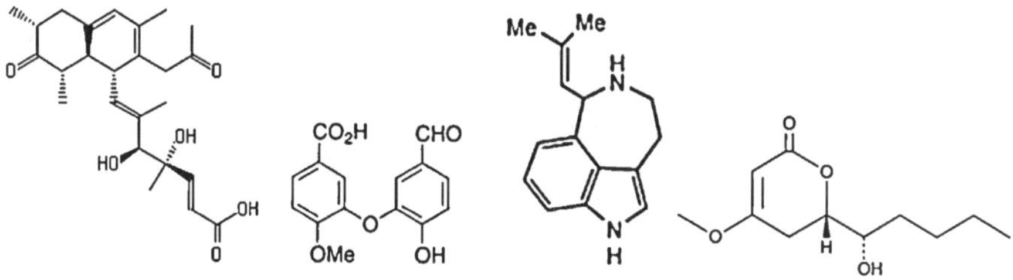

Antarone A Violaceic acid Aurantioclavine Pestalotin

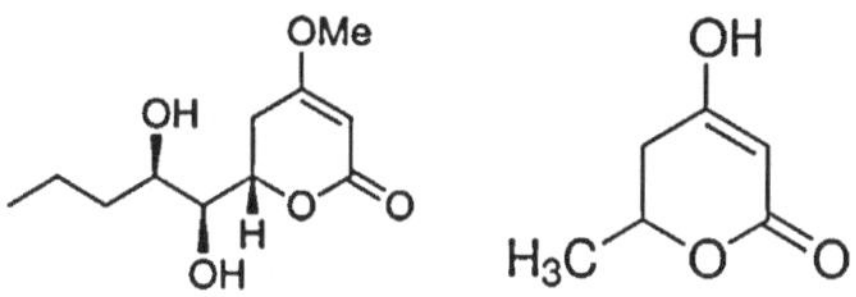

Hydroxypestalotin 5,6-Dihydro-4-methoxy-6-(1-oxopentyl)-2H-pyran-2-one

Unidentified mussel

Associated fungi: Pharmaceutical compounds and activities

Cytotoxicity: The fungal strain *Penicillium antarcticum* MMS163 isolated from this unidentified mussel species yielded bioactive compounds such as patulin, chrysogine, cladosporin, 5-Hydroxyasperentin and terrestric acid. Similarly, its *Penicillium marinum* MMS266 showed the presence of the compound, fusoxysporone (Geiger et al. 2013a; Nicoletti and Trincone 2016).

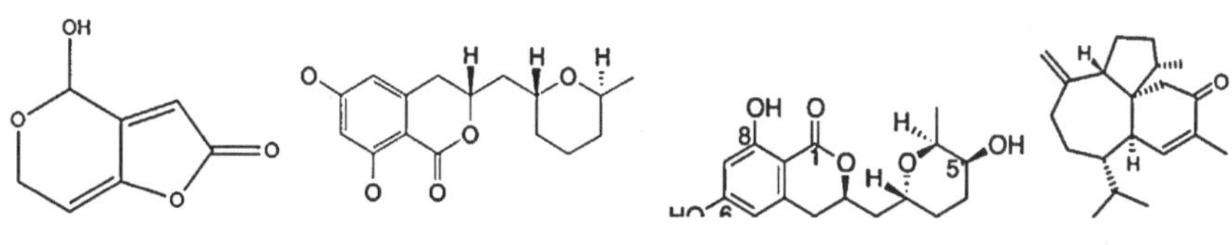

Patulin Cladosporin 5-Hydroxyasperentin Fusoxysporone

Ten strains of the fungus *Penicillium* viz. MMS 50, 163, 231, 266, 270, 330, 393, 399, 404 and 417 associated with this mussel showed cytotoxicity with IC50 vlaues ranging from 5 to >30, µg ml-1 (Geiger et al. 2013a).

Unidentified shellfish

Associated fungi: Pharmaceutical compounds and activities

Nicoletti and Trincone (2016) reported on the presence of compounds meleagrin, roquefortine C–D, chrysogine, aurantioclavine, maculosin, glandicolin A–B, terrestric acid and verruculotoxin from the fungal strain *Penicillium chrysogenum* MMS5 isolated from this unidentified shellfish.

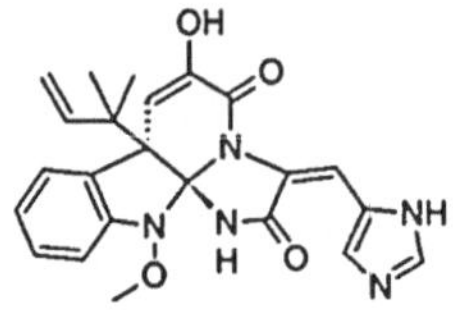

Meleagrin

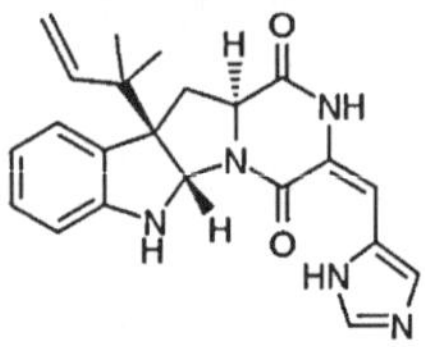

Roquefortine C

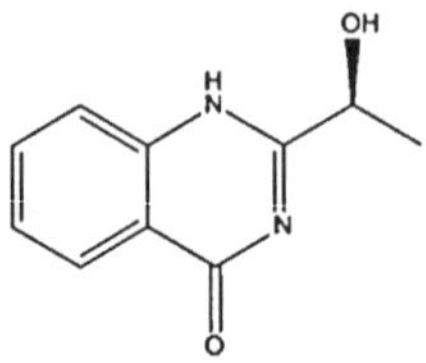

Chrysogine

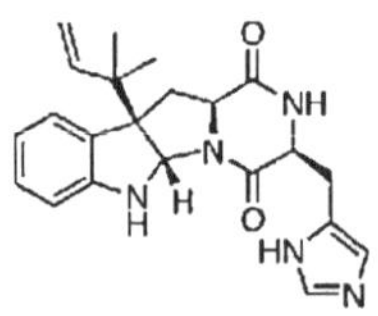

Roquefortine D

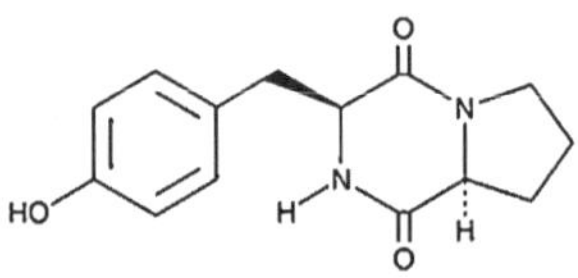

Maculosin

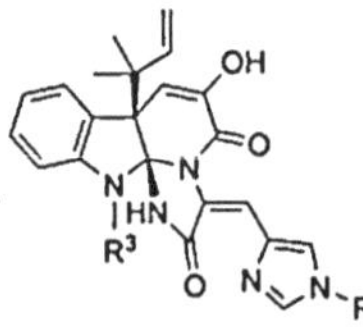

Glandicolin A,B

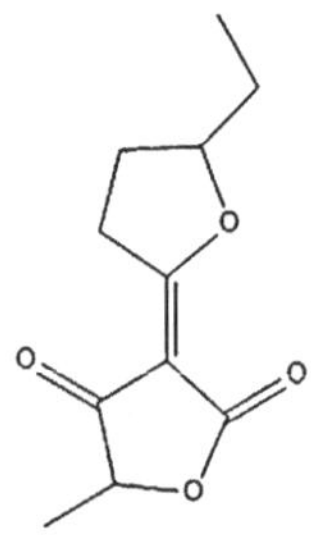

Terrestric acid

(Glandicolin A, R3=H, R4=H; Glandicolin B, R3=OH,R4:H)

ECHINODERMS

Phylum: Echinodermata Class: Asteroidea

Order: Valvatida Family: Acanthasteridae

Acanthaster planci

Associated fungi: Pharmaceutical compounds and activities

Antidiabetic activity: The crude extract (CLCE) of the fungus, *Ceriporia lacerata* derived from this star fish produced the compounds, lanostanes (3β-acetoxy-15α-hydroxylanosta-8,24-dien-21-oic acid) which have the ability to specifically protect insulinoma cell line -1 (INS-1) cells from dexamethasone -induced cytotoxicity through the modulation of the PI3K/PKB pathway. It is suggested that the crude extract of this fungus may be applied for the prevention and/or treatment of steroid diabetes (Kim et al. 2013; Blunt et al. 2015).

Antibacterial and cytotoxic activity: The compound, gliotoxin was produced from the fungus, *Neosartorya pseudofischeri* derived from this star fish. This compound displayed potent antibacterial activity against 3 multidrug-resistant bacterial strains viz. MRSA (R3708) and *Escherichia coli* (ATCC25922), and *Staphylococcus aureus* (ATCC29213). Further, gliotoxin also displayed cytotoxic activity against human cancer cell lines like HEK 293 (human embryonic kidney), HCT-116 (human colon cancer) and RKO (a poorly categorized colon carcinoma) (Jin et al. 2016).

Cytotoxic activity: The compound, pseudaboydins A, an isobenzofuranone derivative was isolated from the marine fungus, *Pseudallescheria boydii*, derived from this starfish. This compound showed moderate cytotoxic activity against GLC82,HONE1, SUNE1 and with IC50 values of 87.2, 37.1, and 46.5 μM, respectively (Jin et al. 2016).

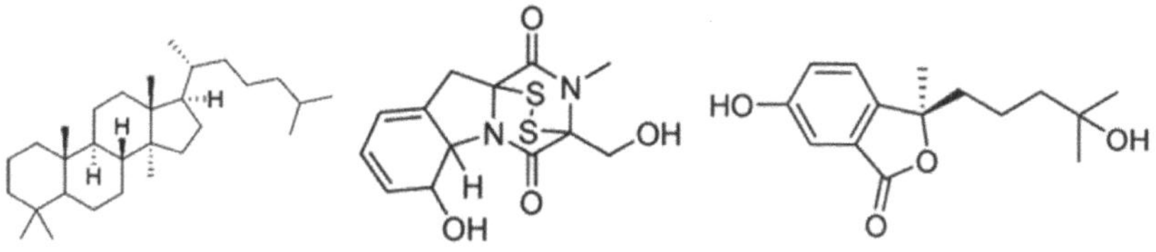

Lanostane Gliotoxin Pseudaboydin A

Class: Echinoidea Order: Camarodonta Family: Echinometridae

Heliocidaris crassispina *(= Anthocidaris crassispana)*

Image credit: Marine species (https://www.marinespecies.org/aphia.php?p=image&tid=513782&pic=144110 https://creativecommons.org/licenses/by-sa/4.0/)

Associated fungi: Pharmaceutical compounds and activities

Cytotoxicity: The fungus *Aspergillus versicolor*, the symbiont of this species produced the tetrahydropyran diterpenes, anthcolorins A–F. Among these compounds, anthcolorins B–D showed moderate growth inhibition of P388 cell line (Blunt et al. 2015).

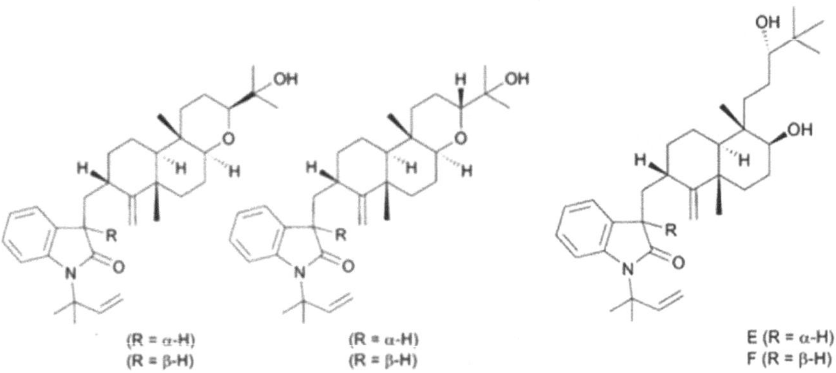

Anthcolorin A,B Anthcolorin C,D Anthcolorin E,F

Order: Dendrochirotida Family: Cucumariidae

Cucumaria frondosa japonica 1868 *(= Cucumaria japonica)*

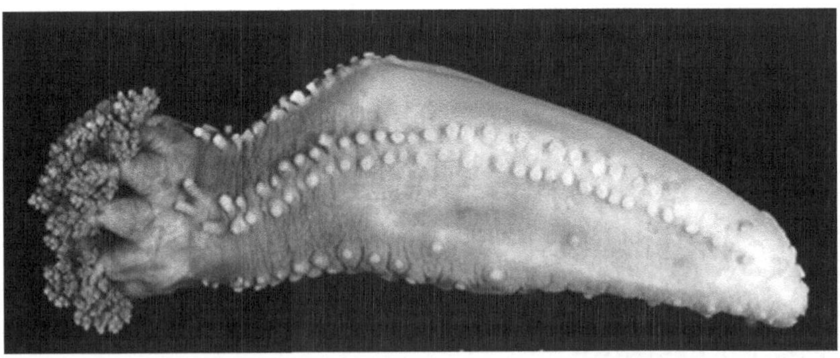

Image credit: Eric A. Lazo-Wasem, Public domain.

Common name: Japanese cucumaria

Global distribution: Northwest Pacific.

Habitat: Benthic; depth range 0 - 300 m

Associated fungi: Pharmaceutical compounds and activities

Proteolytic activity: The fungal species *Cladosporium sphaerospermum* and *Cladosporium brevicompactum* isolated from the coelom of this species showed proteolytic activity with the values of 1.06 cm and 1.26–8.06 cm, respectively (Pivkin 2000))

Order: Spatangoida Family: Schizasteridae

Brisaster latifrons

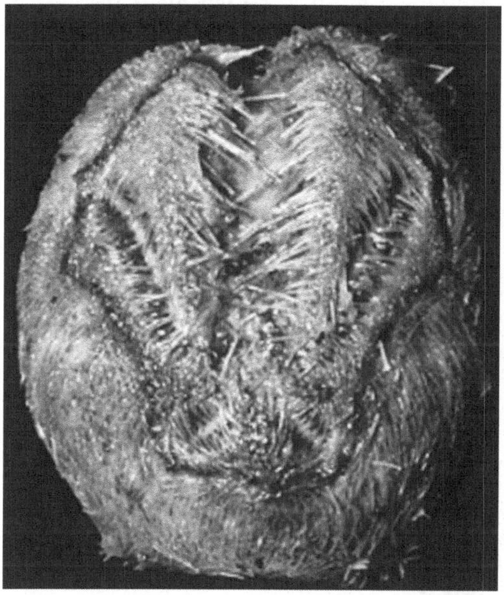

Image credit: Jimi N, Kimura T, Ogawa A, Kajihara H, https://creativecommons.org/licenses/by/4.0/

Common name: Northern heart urchin

Global distribution: Eastern Pacific: Alaska to USA.

Habitat: Benthic; depth range 20 - 1800 m

Associated fungi: Pharmaceutical compounds and activities

Antibiotic activity: The fungus *Penicillium* sp. SF-6013 isolated from this species produced tanzawaic acids A–B and D–E. Among these compounds, tanzawaic acids A and B showed reduced toxicity in bacterial, fungal or human cells through bacterial conjugation which is the main mechanism for the dissemination of multiple antibiotic resistance in human pathogens (Getino et al. 2016; Nicoletti and Trincone 2016).

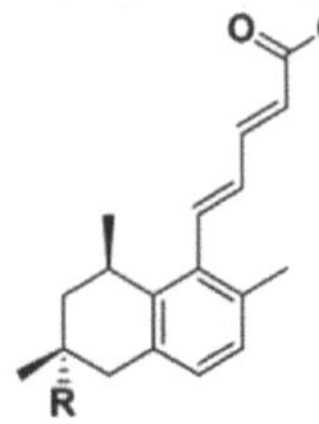

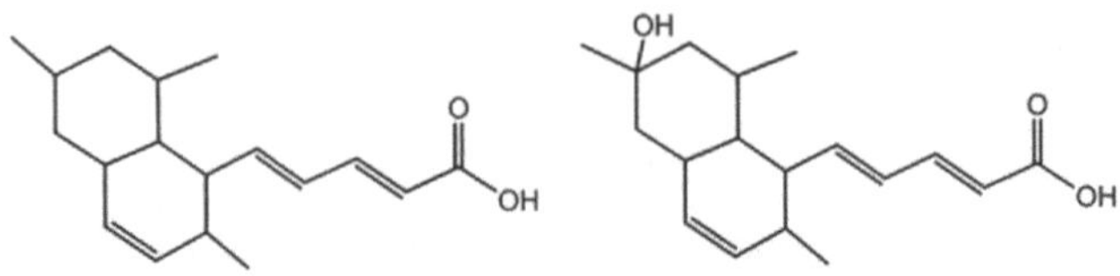

Tanzawaic acid A (R=H); B (R=OH)   Tanzawaic acid B,E

Class: Crinodea   Order: Comatulida   Family: Himerometridae

Himerometra robustipinna *(= Himerometra magnipinna)*

Image credit: Holobionics, https://creativecommons.org/licenses/by-sa/4.0/

Common name (s): Not designated

Global distribution: Indian Ocean and western Pacific

Habitat: Coastal waters with coral reefs

Associated fungi; and their pharmaceutical compounds and activities

Antiviral activity: A meroterpenoid compound, stachybonoid A derived from this crinoid-associated fungus, *Stachybotrys chartarum* 952, displayed moderate antiviral activity against the replication of dengue virus with an IC50 value of 25 μM (Mayer et al. 2021).

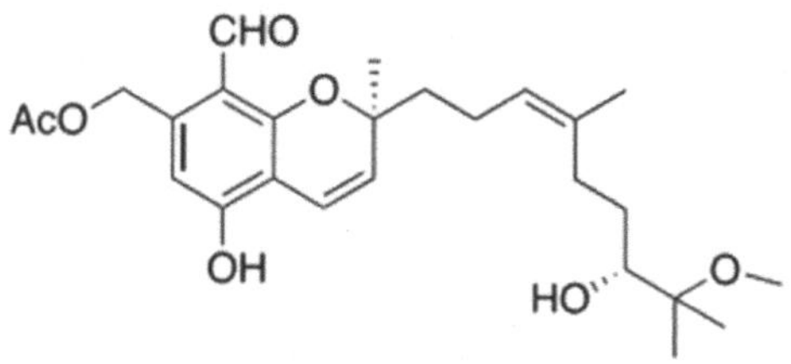

Stachybonoid A

Class: Holothuroidea Order: Dendrochirotida Family: Sclerodactylidae

Eupentacta fraudatrix

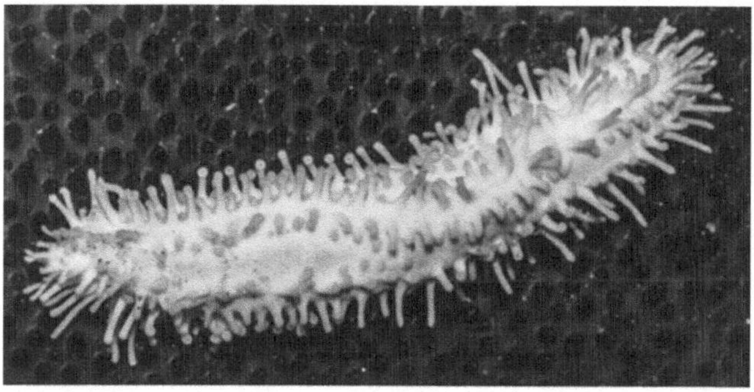

Eupentacta sp

Image credit: Jerry Kirkhart, https://creativecommons.org/licenses/by/2.0/

No other information

Associated fungi: Pharmaceutical compounds and activities

Cytotoxic activity: Three diterpene glycosides, virescenosides O, P, and Q, have been isolated from a marine strain of *Acremonium striatisporum* KMM 4401 associated with this holothurian. All these compounds exhibited cytotoxic action against tumor cells of Ehrlich carcinoma (IC50)) 20–100 µM) *in vitro* (Afiyatullov et al. 2002).

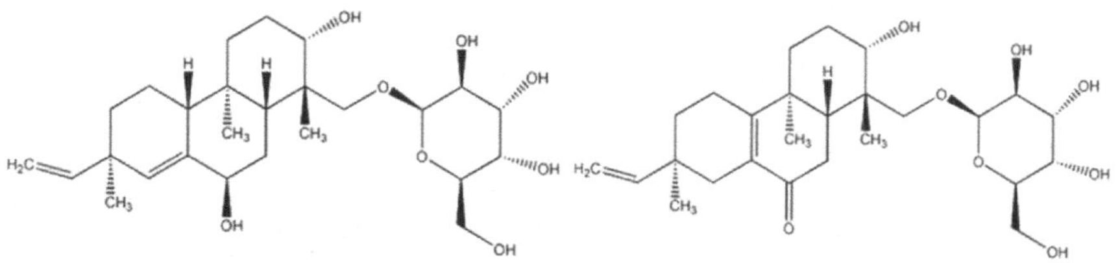

Virescenoside O,P

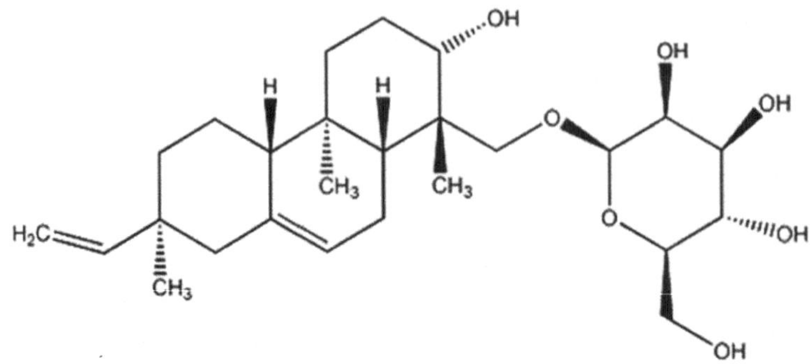

Virescenoside Q

Others: The fungus *Acremonium strictum* isolated from this holothurian contributed four diterpene glycosides virescenosides Z4-Z7 (Afiyatullov et al. 2011).

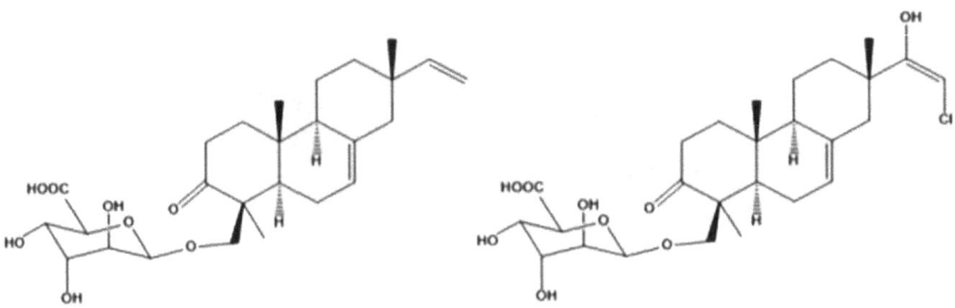

Virescenosides Z4,Z5

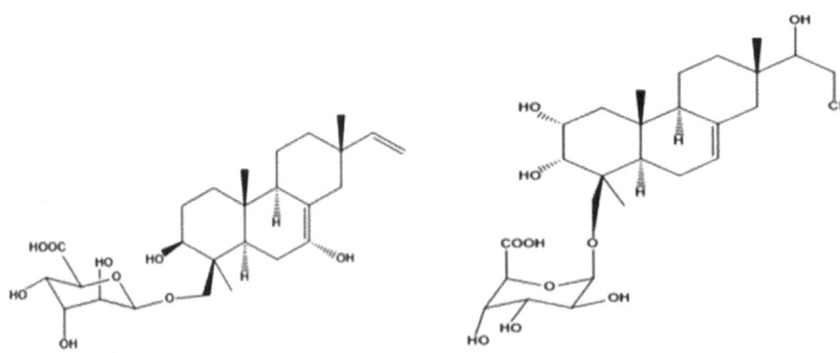

Virescenosides Z6,Z7

Order: Holothuriida Family: Holothuriidae

Holothuria (Microthele) nobilis

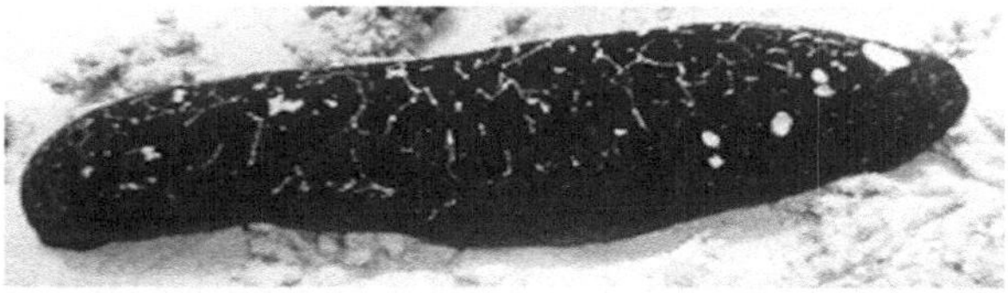

Image credit: Fernando Herranz Martín, https://www.gnu.org/licenses/gpl-3.0.html

Common name: Black sea cucumber, black teatfish

Global distribution: Indo-Pacific. Tropical and subtropical.

Habitat: Reef-associated; depth range 0 - 45 m

Associated fungi: Pharmaceutical compounds and activities

Cytotoxicity: The fungus *Dendrodochium* sp. associated with this species produced macrolides dendrodolide A–M. Among these compounds, dendrodolides A–E, G–I, K and L had modest inhibitory activity against two human tumor cell lines viz. SMMC-7721 and HCT116 (Blunt et al. 2015).

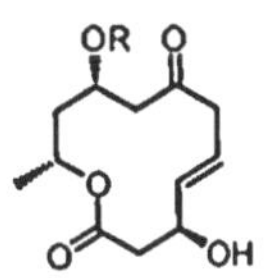

A,C

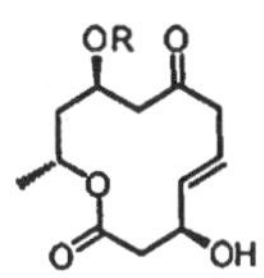

B,D

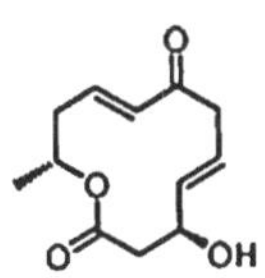

E

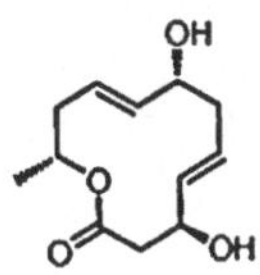

F

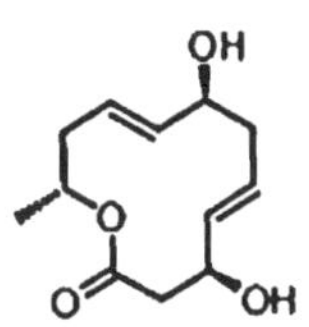

G

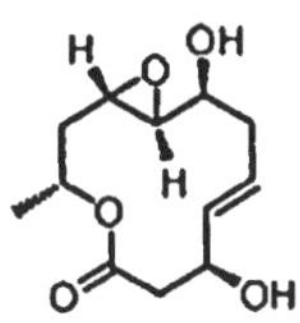

H

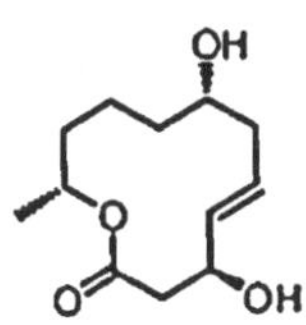

I

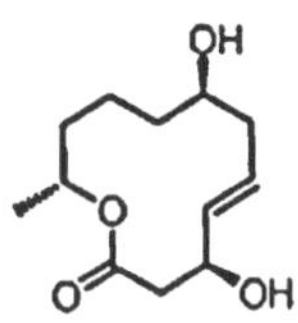

J

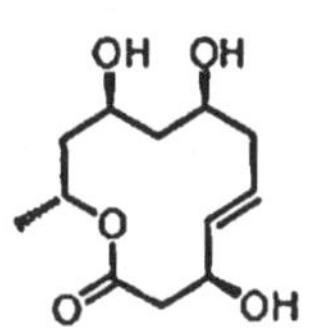

K

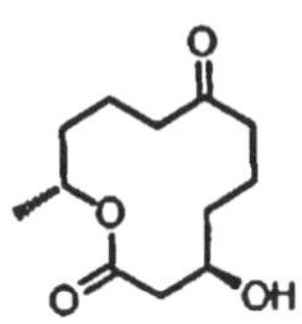

L

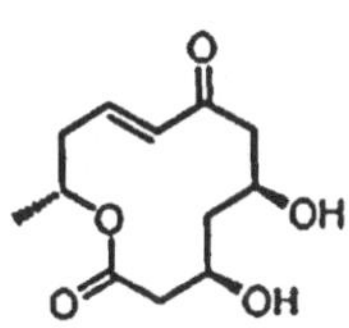

M

Dendrodolide A-M (Dendrodolide A, R=Me; Dendrodolide C, R=H; Dendrodolide B, R=Me; Dendrodolide D, R=H)

Order: Synallactida Family: Stichopodidae

Apostichopus japonicus *(= Stichopus japonicus)*

Image credit: OpenCage, https://creativecommons.org/licenses/by-sa/2.5/

Common name: Japanese spiky sea cucumber

Global distribution: Northern regions of the Pacific coastal waters of Japan, northeastern China, the Korean peninsula, Sakhalin, Kuril Islands and Alaska.

Habitat: Shallow coastal sea bottoms from the intertidal zone to 80 m depth

Associated fungi: Pharmaceutical compounds and activities

Antidiabetic activity: The fungus *Epicoccum* sp. isolated from this species yielded isopimarane diterpene (1), together with two known compounds, 11-deoxydiaporthein A (2) and iso-pimara-8(14),15-diene (3) which showed in vitro inhibitory activities against α-glucosidase. Compounds 1 and 2 exhibited more effective inhibitory activities, with IC50 values of 4.6 and 11.9 μM, respectively, while compound 3 did not exhibit inhibitory activities (IC50 > 200 μM) (Xia et al. 2015).

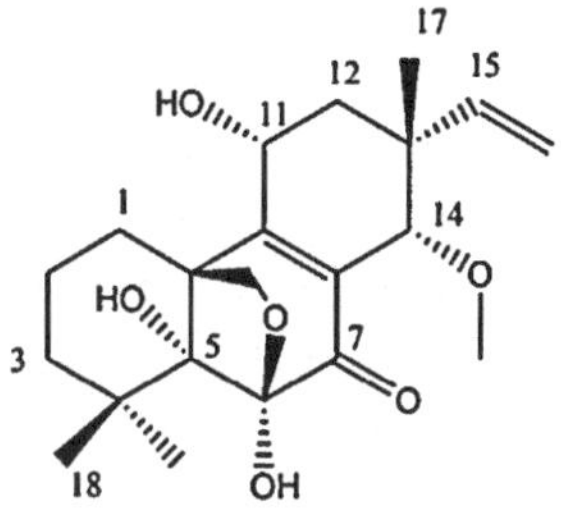

1

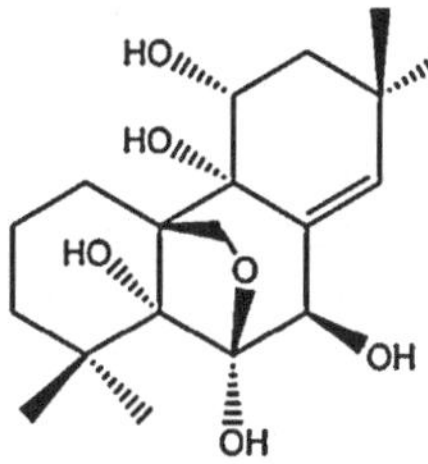

2

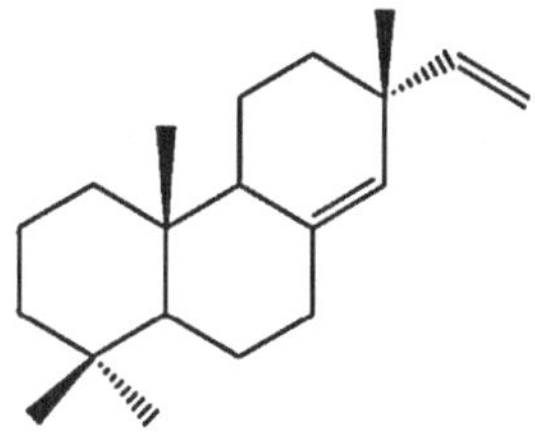

3

1: Isopimarane diterpene; 2: 11-deoxydiaporthein A; 3: iso-pimara-8 (14),15-diene

Cytotoxic activity: Pimarane diterpenoids (1-3) together with diaporthin B (4) were isolated from the fungal strain *Epicoccum* sp. HS-1, derived from this sea cucumber. All isolated compounds showed cytotoxic activities against KBv200 and KB cell lines. While the metabolite 1 showed potent activity on these cell lines with IC50 values of 6.8 and 10.1 μM, respectively, the 2nd compound had IC50 values of 17.9 and 10.6 μM respectively. Metabolite 2 was only weakly active (IC50 > 45 μM), whereas compound 3 showed no activity (Elissawy et al. 2015).

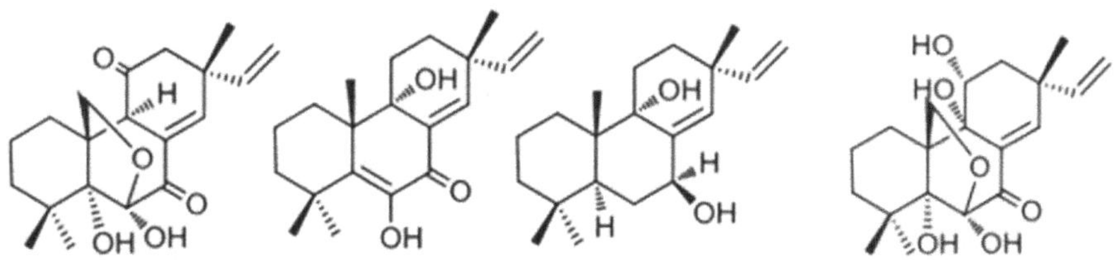

Pimarane diterpenoids Diaporthin B

Compounds viz. an unidentified compound (1), spirotryprostatins C–E (2–4), 2 derivatives of fumitremorgin B (5 and 6), 13-oxoverruculogen (7), spirotryprostatin A (8),10 13-oxofumitremorgin B (9),18–20 fumitremorgin B (10) 3,4 verruculogen (11),21 3-b hydroxy cyclo-L-tryptophyl-L-proline (12),22,23 cyclo-L-tryptophyl-L-proline (13),24 tryprostatin B (14),7,8 tryprostatin A (15),7,8 N-prenyl-cyclo-L-tryptophyl-L-proline (16),25 fumitremorgin C (17),8 12,13-dihydroxyfumitremorgin C (18), 8 and cyclo-tryprostatins A (19) were derived from the symbiotic fungus *Aspergillus fumigatus* of this species. Among these compounds, compounds 1-7 showed cytotoxicity. While compound 1 acted against A549, all the other compounds showed cytotoxicity with all cell lines. viz. MOLT-4, A549, HL-60, and BEL-7420. Compound 1 showed an IC50 value of 125 μM against A549; compound 2 had toxicity against all the four cell lines with IC50 values ranging from 26 to 69 μM; compound 3 with values ranging from 18 to 45 μM; compounds 4, 5, 6, and 7 showed ranges as 2–98 μM; 3–11 μM; 5–12 μM; and 2- 26 μM, respectively (Wang et al. 2008).

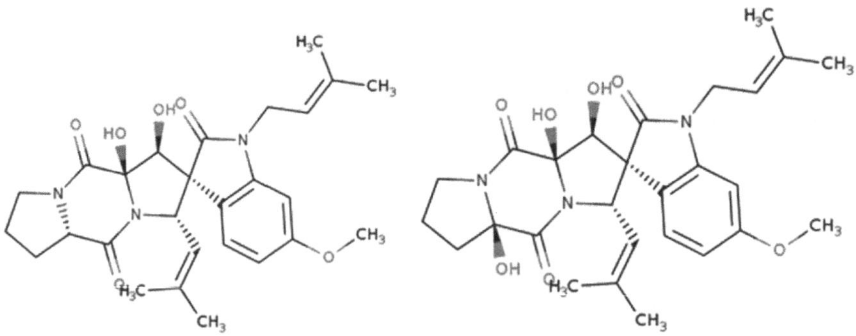

Spirotryprostatin C, D

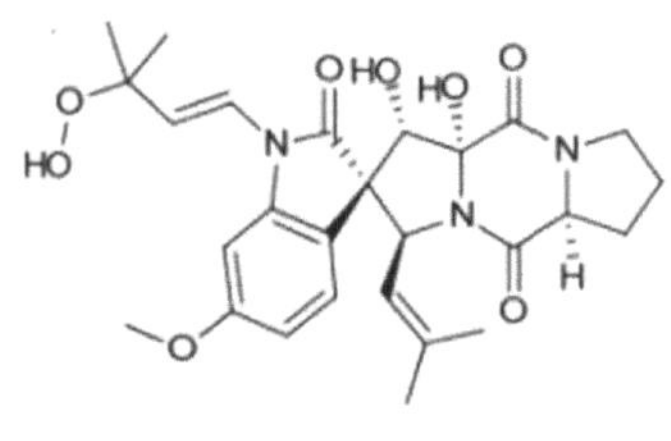

Spirotryprostatin E

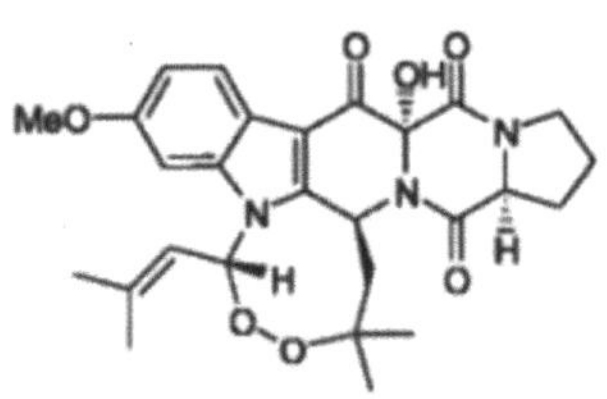

13-oxoverruculogen

Proteolytic activity: The fungal species *Cladosporium sphaerospermum* and *Cladosporium brevicompactum* isolated from this species showed proteolytic activity with the values of 0.56–5.06 cm and 0.86–6.06 cm, respectively (Pivkin 2000).

TUNICATES

Phylum: Chordata Subphylum: Tunicata Class: Ascidiacea

Order: Aplousobranchia Family: Clavelinidae

Nephtheis fascicularis *(= Oxycorynia fascicularis)*

Image credit: Nhobgood Nick Hobgood, https://creativecommons.org/licenses/by-sa/3.0/

Common names: Blue palm coral or lollipop coral

Global distribution: Indonesia

Habitat: This shallow water species is commonly seen on reef drop-offs

Associated fungi: Pharmaceutical compounds and activities

The fungus, *Pithomyces* sp. isolated from this tunicate generated the compounds, pitholides A–D and (R)-5-methylmellein the bioactivities of which are to be known (Chen et al. 2018).

Pitholide A,B

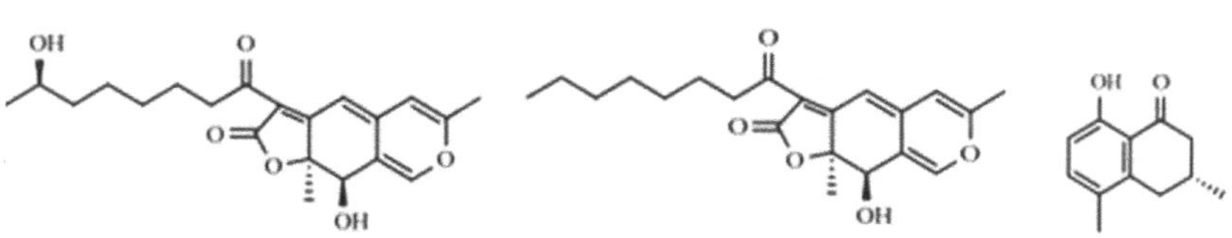

Pitholide C,D (R)-5-methylmellein

Order: Aplousobranchia Family: Didemnidae

Didemnum granulatum

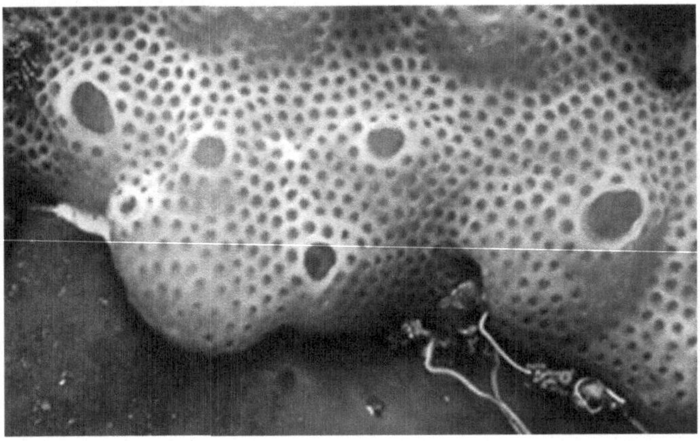

Didemnum sp.

Image credit: Parent Géry, Public domain.

Common name: Not reported

Global distribution: Indo-Pacific and Atlantic Ocean: East South America.

Habitat: Reef-associated; depth range 6–18 m

Associated fungi: Pharmaceutical compounds and activities

Antimicrobial and/or cytotoxic activity: The *Penicillium* sp. strain DG(M3)6'C derived from this tunicate yielded 13-desoxyphomenone which showed antimicrobial and/or cytotoxic activity (Kossuga et al. 2012).

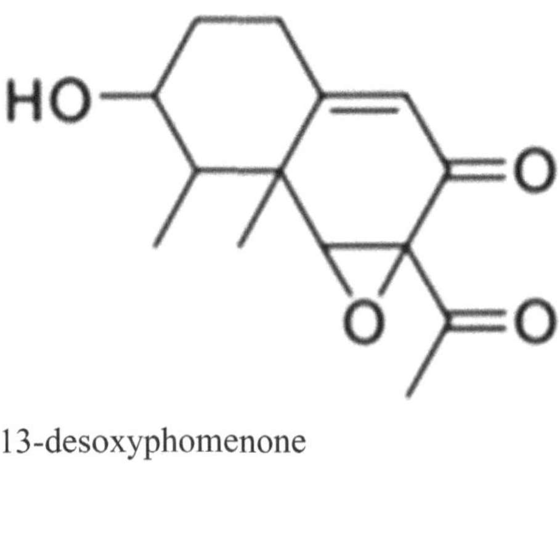

13-desoxyphomenone

Didemnum molle

Image credit: Univ of Guelph students, https://creativecommons.org/licenses/by/3.0/

Common name: Urn ascidian

Global distribution: Native to the western Indo-Pacific region

Habitat: Grows on rock or dead coral on shallow reefs.

Associated fungi: Pharmaceutical compounds and activities

Cytotoxicity: The fungal species *Penicillium* sp. strain SS080624SCf1 derived from this tunicate - produced sesquiterpenoids JBIR-27 and JBIR-28, together with two other compounds sporogen-AO1 and phomenone which showed cytotoxicity against HeLa expect for JBIR-27 (Chen et al. 2018).

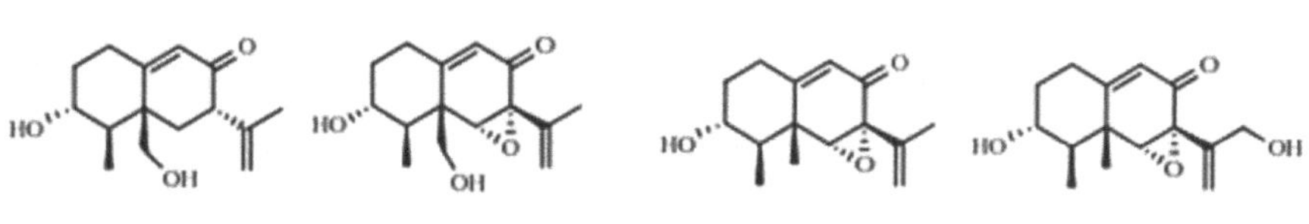

JBIR-27,28 Sporogen-AO1 Phomenone

Antimicrobial and cytotoxic activity: The fungus, *Trichoderma virens* associated with this species produced trichodermamides, A and B). Among these compounds, trichodermamide B exhibited antimicrobial activity against methicillin-resistant *Staphylococcus aureus*, vancomycin-resistant *Enterococcus faecium* and amphotericin-resistant *Candida albicans* (Chen et al. 2018).

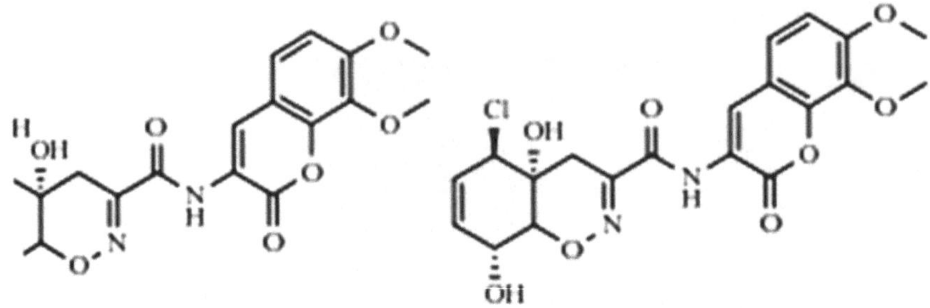

Trichodermamides A,B

Didemnum **sp.**

Associated fungi: Pharmaceutical compounds and activities

Antimicrobial and cytotoxic activity: The marine fungus, *Penicillium* sp. isolated from this tunicate, produced cerebrosides, penicillosides A and B. Among these compounds, penicilloside B illustrated antibacterial activities against *Escherichia coli* and *Staphylococcus aureus;* penicilloside A displayed antifungal activity on *Candida albicans*; and both the metabolites displayed poor cytotoxic activity on HeLa cells (Murshida et al. 2016).

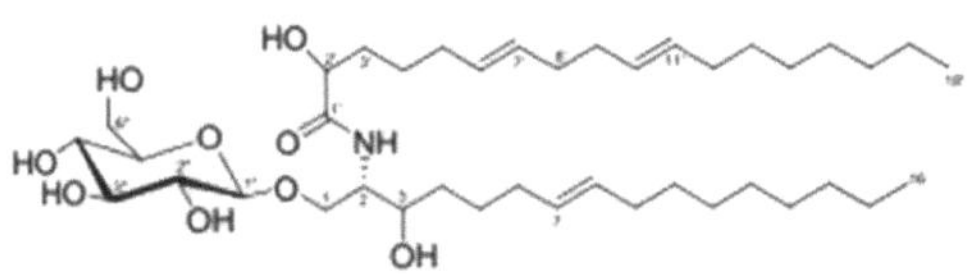

Penicilloside A

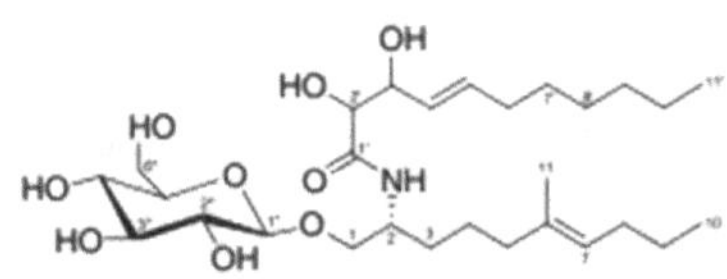

Penicilloside B

Lissoclinum patella

Associated fungi: Pharmaceutical compounds and activities

Antimycobacterial and cytotoxic activity: The filamentous fungus, Eurotiomycetes strain 110162 isolated from this tunicate produced oxazinin A which displayed antimycobacterial activity against *Mycobacterium tuberculosis;* cytotoxicity on CEM-TART T-cell leukemia cells; and moderate activity of TRP(transient receptor potential) channels (Lin et al. 2014).

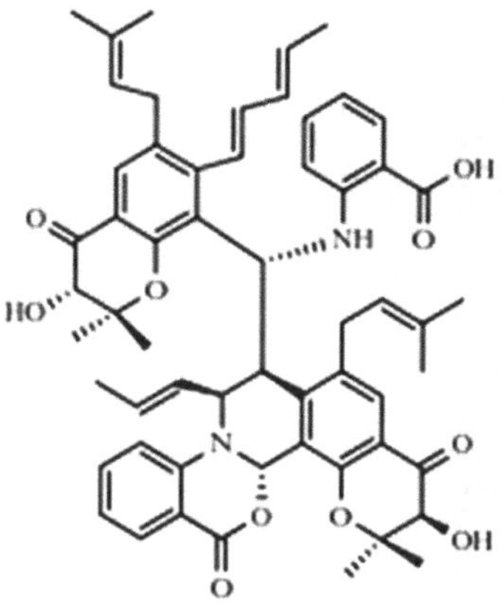

Oxazinin A

Order: Aplousobranchia   Family: Polycitoridae

Eudistoma vannamei

Common name: Not reported

Global distribution: Southwest Atlantic: Brazil.

Habitat: Sessile

Associated fungi: Pharmaceutical compounds and activities

Cytotoxicity: The fungus, *Aspergillus* sp. isolated from this tunicate yielded compounds, mellein, cis-4-hydroxymellein, trans-4-hydroxymellein, and penicillic acid. Among these compounds, only penicillic aciddisplayed cytotoxic activity on HCT-8 and MDA-MB 435 cell lines with IC50 values of 8.8 and 4.4 and μg/ml, respectively (Montenegro et al. 2012).

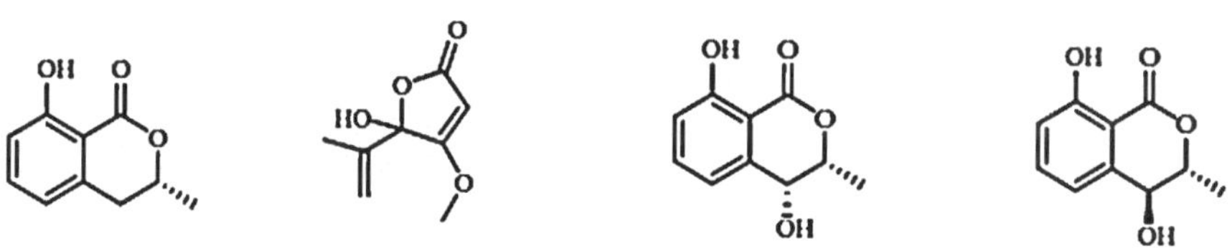

1 2 3 4

1: Isocoumarins mullein; 2: cis-4-hydroxymellein; 3: trans-4-hydroxymellein; 4: penicillic acid

Order: Aplousobranchia   Family: Polyclinidae

Aplidium sp.

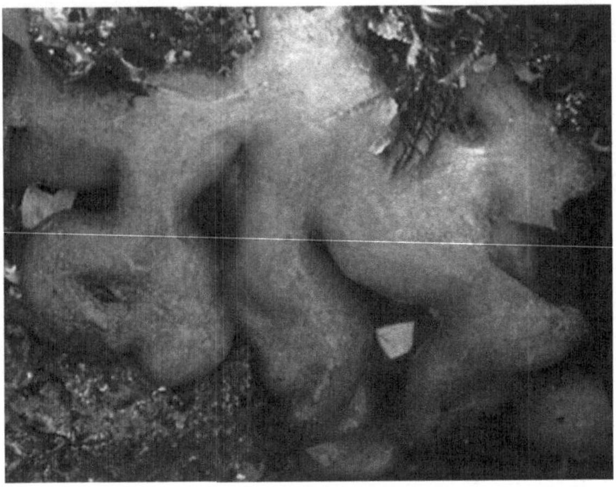

Image credit: jkirkhart35, https://creativecommons.org/licenses/by/2.0/

Associated fungi: Pharmaceutical compounds and activities

Antimicrobial activities: The fungus *Aspergillus niger* (F97S11 associated with this tunicate yielded yanuthones AE together with 1-hydroxyyanuthone A, 1-hydroxyyanuthone C, and 22-deacetylyanuthone A. All these compounds displayed poor antimicrobial activity on vancomycin-resistant *Enterococcus* sp. and MRSA (Chen et al. 2018).

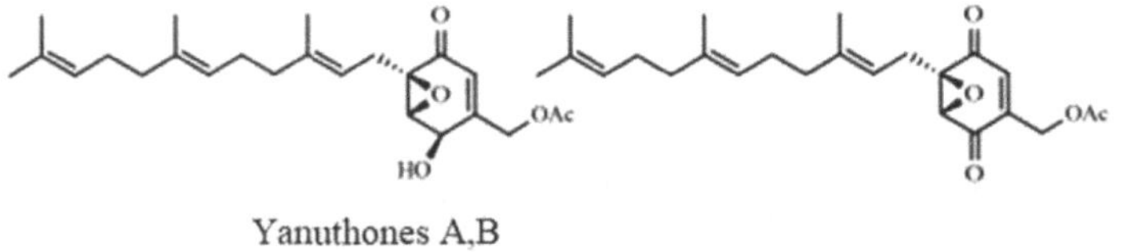

Yanuthones A,B

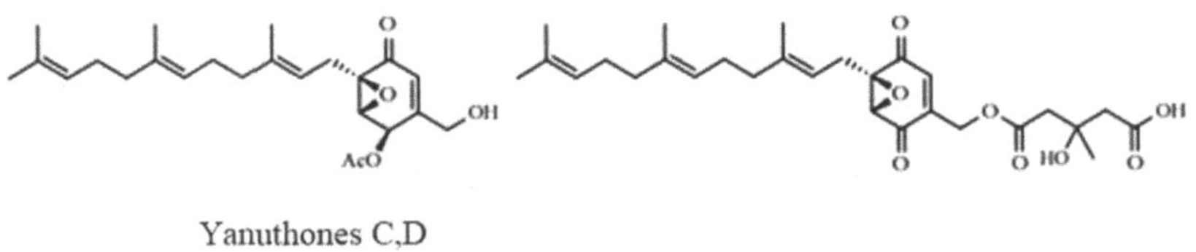

Yanuthones C,D

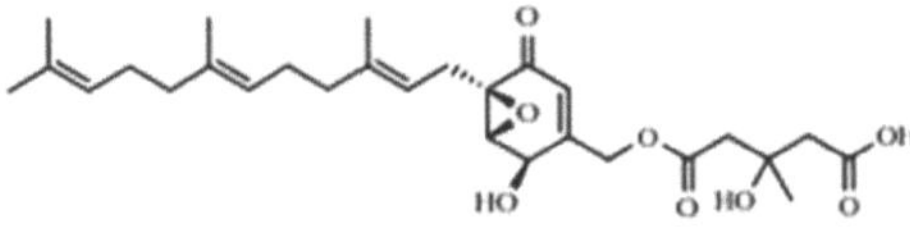

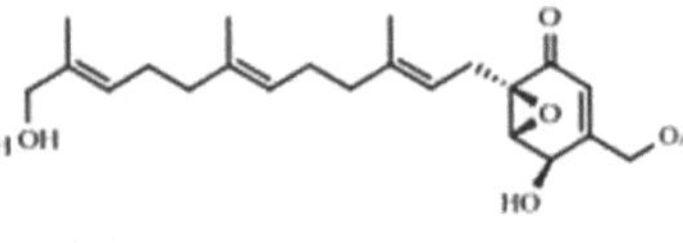

Yanuthone E 1-hydroxyyanuthone A

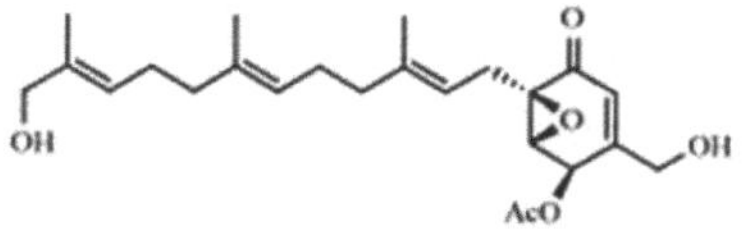

1-hydroxyyanuthone C

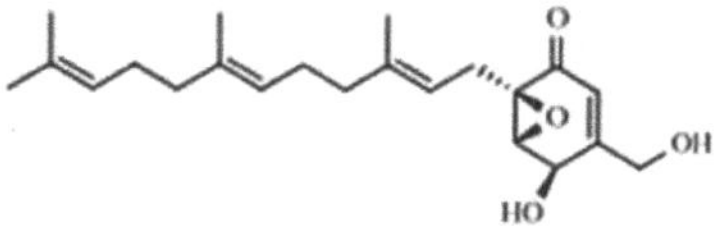

22-deacetylyanuthone A

Order: Phlebobranchia Family: Cionidae

Ciona intestinalis

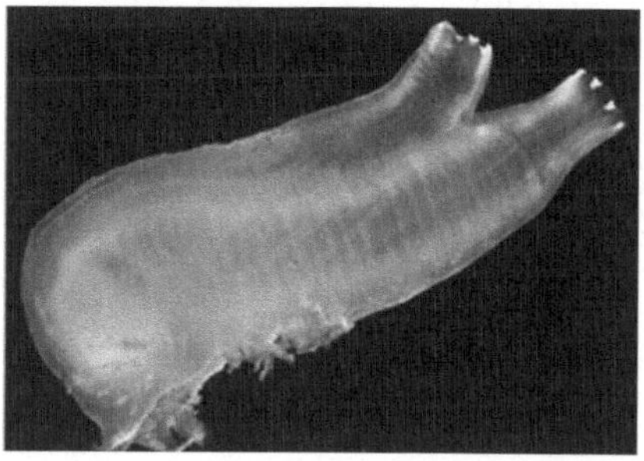

Image credit: Eric A. Lazo-Wasem, https://creativecommons.org/publicdomain/zero/1.0/

Common name: Vase tunicate, sea vase

Global distribution: This cosmopolitan species inhabits tropical, temperate and sub-arctic waters

Habitat: It is known to grow on submerged habitats like eelgrass, kelp and rock; and on anthropogenic materials like wood, metal, concrete docks, etc.

Associated fungi: Pharmaceutical compounds and activities

The fungus *Meyerozyma* sp. associated with this species yielded the depsipeptide JBIR-113, the pharmaceutical activity of which is to be known (Chen et al. 2018).

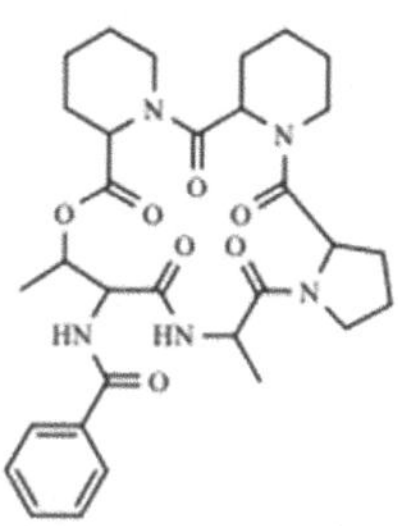

JBIR-113

Order: Phlebobranchia Family: Perophoridae

Ecteinascidia turbinata

Associated fungi: Pharmaceutical compounds and activities

Anifungal and anti-inflammatory activities: Three new oxepincontaining natural products viz. oxepinamides A,C (13) and two fumiquinazoline metabolites, fumiquinazolines H and I have been isolated from the organic extracts of the culture broth and mycelia of the fungus, *Acremonium* sp. derived from this tunicate. Among these compounds, oxepinamide A displayed potent anti-inflammatory activity; and fumiquinazolines H and I exhibited weak anifungal activity against *Candida albicans* (Belofsky et al. 2000).

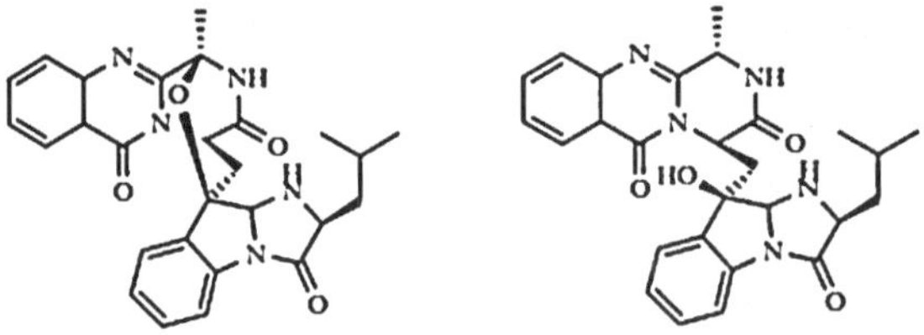

Fumiquinazolines H, I

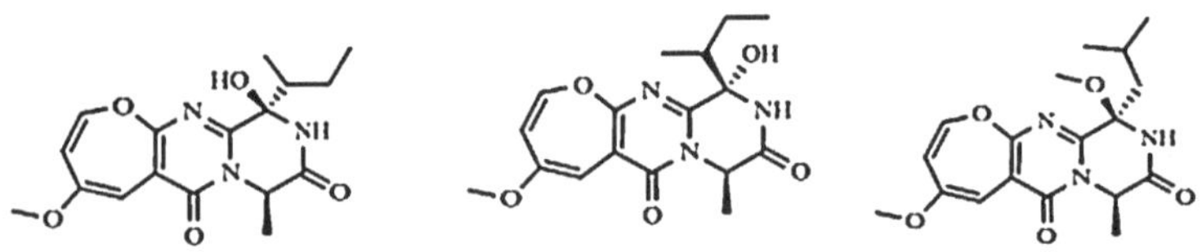

Oxepinamides A–C

Order: Stolidobranchia Family: Styelidae

Polycarpa aurata

Image credit: Mudasir Zainuddin. https://creativecommons.org/licenses/by-sa/4.0/

Common names: Ox heart ascidian, gold-mouth sea squirt or ink-spot sea squirt

Global distribution: Tropical W. Pacific and E. Indian Ocean

Habitat: This shallow water species occurs at depths of 5–50 m.

Associated fungi: Therapeutic compounds and their bioactivities

Antidiabetic effects: The fungus, *Talaromyces verruculosus* isolated from this tunicated yielded compounds viz. chrodrimanins A (1), B (2) and H (3); and verruculides A and B (4,5). Among these compounds, compounds 2,3 and 5 showed inhibitory activity on protein tyrosine phosphatase 1B which is known to serve as an 'insulin sensitizer' in the inhibition of type 2 diabetes and the cardiovascular/metabolic syndrome (Goldstein et al. 2001).

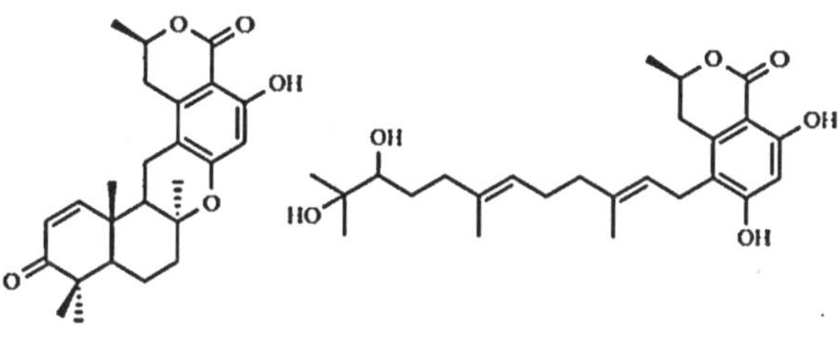

Verruculides A, B

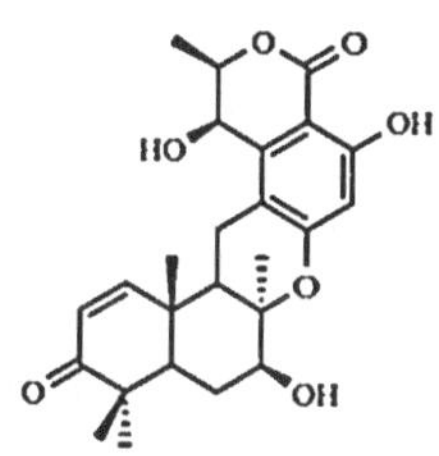

Chrodrimanin A

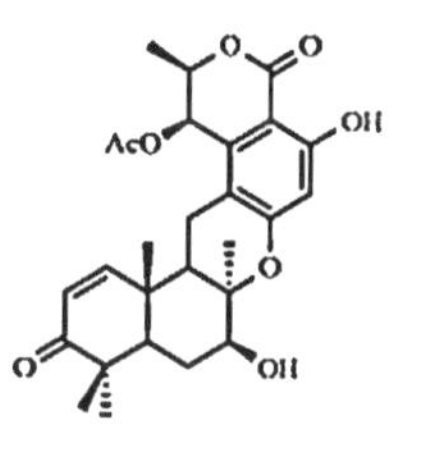

Chrodrimanin B

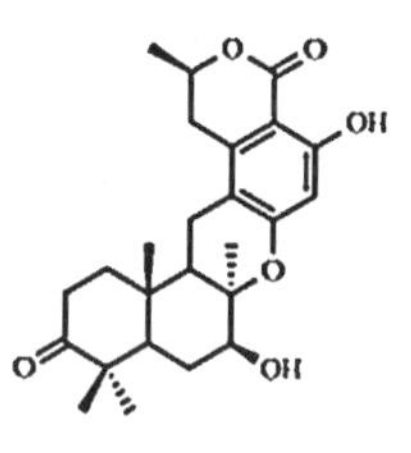

Chrodrimanin H

Unidentified ascidian 1

Associated fungi: Pharmaceutical compounds and activities

The fungus *Talaromyces* sp. (CMB-TU011) isolated from this tunicate yielded the compound, talarolide A, which showed no antifungal ability when tested (Dewapriya et al. 2017)

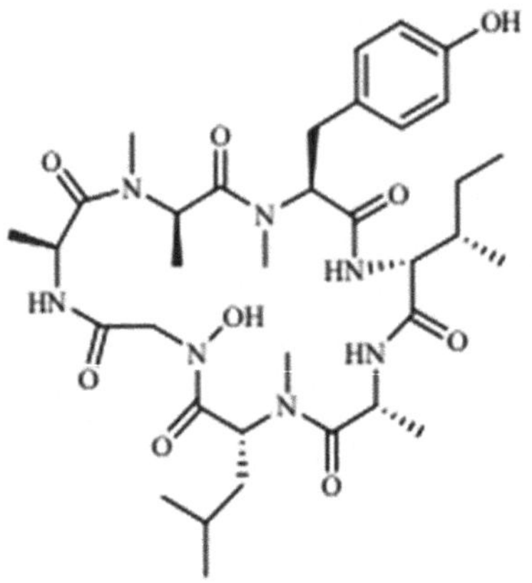

Talarolide A

Unidentified ascidian 2

Associated fungi: Pharmaceutical compounds and activities

Antidiabetic effects: The strain of the fungus viz. *Penicillium albobiverticillium* TPU1432 derived from this tunicate produced monodictyphenone which showed moderate inhibitory activities against CD45 tyrosine phosphatase (CD45), vaccinia H-1-related phosphatase (VHR), tyrosine phosphatase (PTP) 1B, and T cell PTP (TCPTP). Among these enzymes, The PTP1B inhibitor may serve as an 'insulin sensitizer' in the control of type 2 diabetes and the metabolic syndrome (Sumilat et al. 2017).

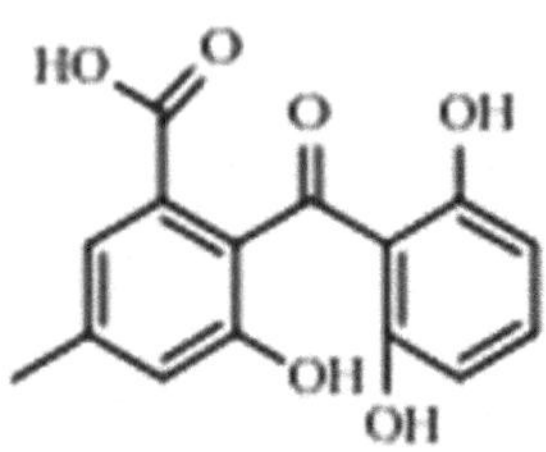

Monodictyphenone

Unidentified ascidian 3

Associated fungi: Pharmaceutical compounds and activities

The fungus, *Penicillium steckii* isolated from this tunicate yielded two carboxylic acids, tanzawaic acid E and F; and 3,7-dimethyl-1,8-dihydroxy-

6-methoxy-isochroman and 3,7-dimethyl-8-hydroxy-6-methoxyisochroman. The pharmaceutical activities of these compounds are to be known (Malmstrøm et al. 2000).

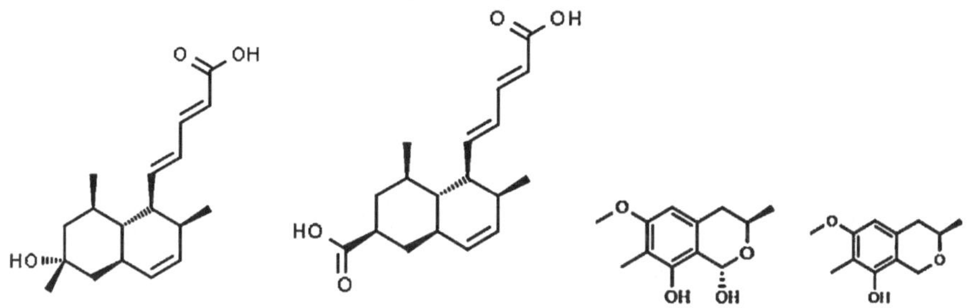

1 2 3 4

1: Tanzawaic acid E; 2:Tanzawaic acid,F; 3: 3,7-dimethyl-1,8-dihydroxy-6-methoxyisochroman; 4: 3,7-dimethyl-8-hydroxy-6-methoxyisochroman

Unidentified ascidian 4

Associated fungi: Pharmaceutical compounds and activities

Cytotoxicity: The fungus, *Penicillium stoloniferum* QY2-10 isolated from this tunicate yielded stoloniferols A (1) and B (2), together with a sterol, 5α,8α-epidioxy-23-methyl-(22E,24R)-ergosta-6,22-dien-3β-ol (3). Among these compounds, compound 3 showed in vitro cytotoxicity to the P388 cell line with an IC50 value of 4.07 μM (Xin et al. 2007).

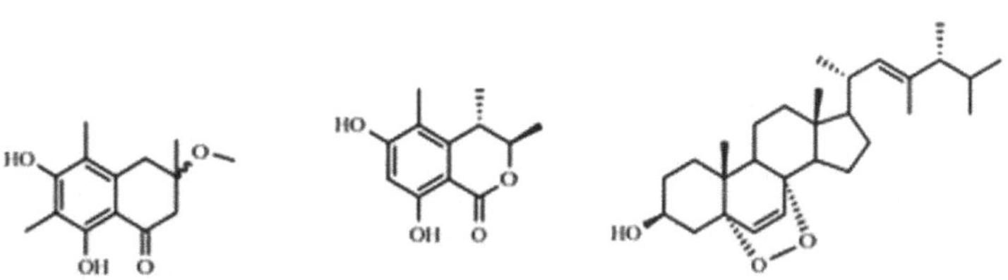

Stoloniferols A,B 5α,8α-epidioxy-23-methyl-(22E,24R)-ergosta-6,22-dien-3β-ol

Unidentified ascidian 5

Associated fungi: Pharmaceutical compounds and activities

Antidiabetic activity: The fungus *Talaromyces albobiverticillius* TPU1432 isolated from this tunicate produced 2-hydroxy-6-(20-hydroxy-30 -hydroxy-methyl-5-methylphenoxy)-benzoic acid which exhibited moderate inhibitory activity on T cell PTP (TCPTP), CD45 tyrosine phosphatase (CD45) and protein tyrosine phosphatase (PTP) 1B (Chen et al. 2018).

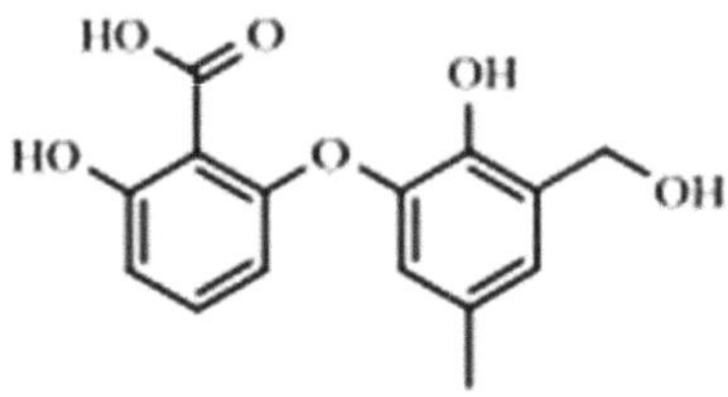

2-hydroxy-6-(20 -hydroxy-30 -hydroxymethyl-5-methylphenoxy)-benzoic acid

Unidentified ascidian 6

Associated fungi: Pharmaceutical compounds and activities

Antibacterial and cytotoxic activities: The fungal strain, *Humicola fuscoatra* KMM 4629 derived from this tunicate yielded fuscoatrol A (1),11-epiter-pestacin (2) and and -nitro-propionic acid (3). Compound 1 possesses an antimicrobial activity against *Staphylococcus aureus* and *Bacillus subtilis* (MIC = 12.5 μg mL1) and exhibits a cytotoxic action (MIC50 = 40 μg mL1). Compound 2 has an antimicrobial activity against *S. aureus* and *B. subtilis* (MIC = 100 μg mL1), while 3 is active against *S. aureus* (Smetanina et al. 2004).

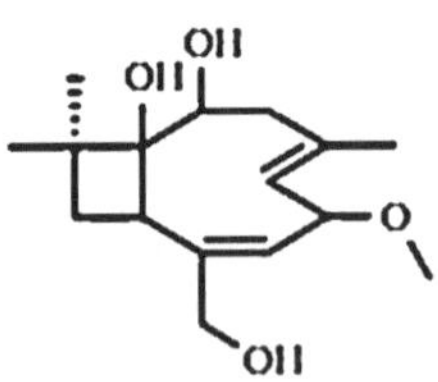
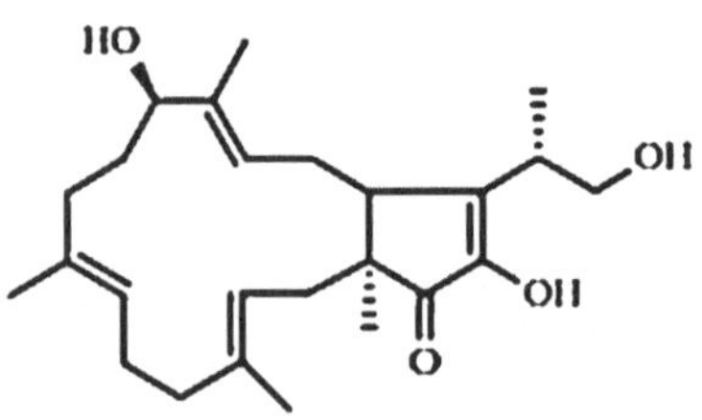
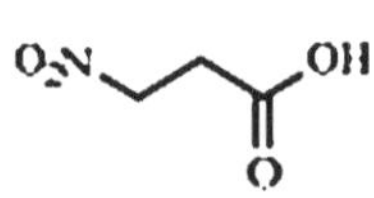

Fuscoatrol A 11-epiter-pestacin β-nitro-propionic acid

Unidentified ascidian 7

Associated fungi: Pharmaceutical compounds and activities

Cytotoxicity: The fungal strain *Aspergillus candidus* KMM 4676 associated with this tunicate produced terphenyllin (1), 4-dehydroxy-30 -hydroxyterphenyllin (2), 30 -hydroxyterphenyllin (3), candidusin A (4), 4-dehydroxycandidusin A (5) and chlorflavonin (6). Among these metabolites, compounds 4 and 5 showed cytotoxicity on LNCaP (hormone-sensitive prostate cancer cell line) (Yurchenko et al. 2017).

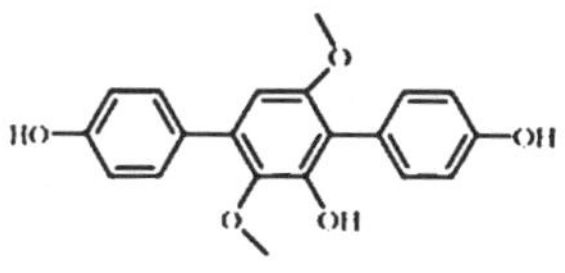
1

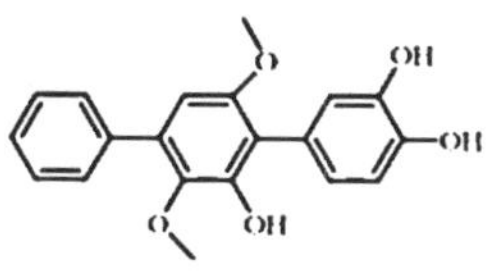
2

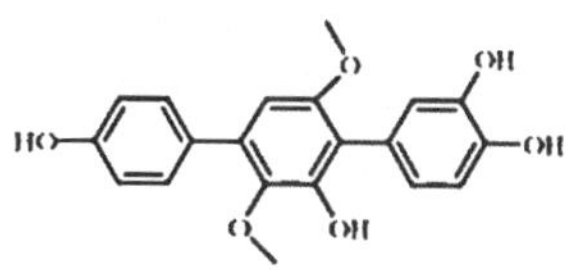
3

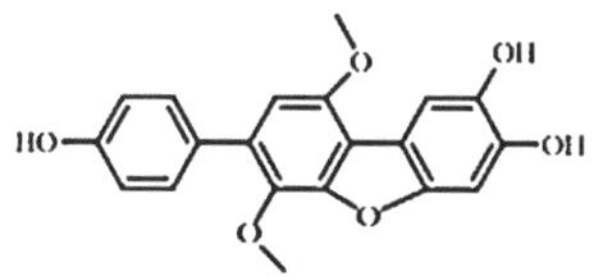
4

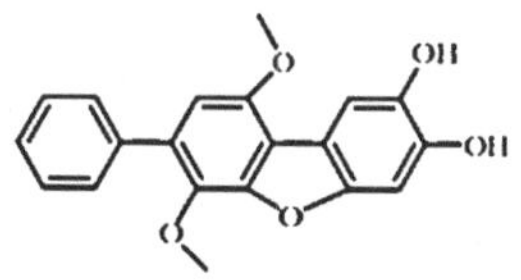
5

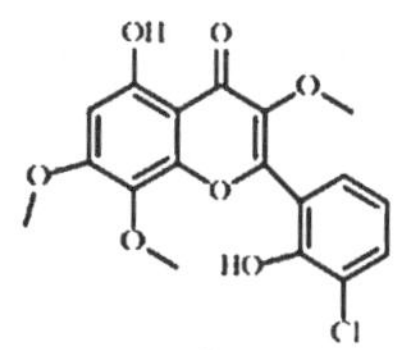
6

1: Terphenyllin; 2: 4”-dehydroxy-30–hydroxyterphenyllin; 3: 30–hydroxyterphenyllin; 4: candidusin A; 5: 4”-dehydroxycandidusin A; 6: chlorflavonin

Unidentified Ascidian 8

Associated fungi: Pharmaceutical compounds and activities

Antifungal properties: This tunicate-associated fungal strain, *Talaromyces* sp. (CMB TU011 yielded talarolide A (1) and a diterpene glycoside sordarin (2). Among these compounds, sordarin showed antifungal properties. On the other hand, talarolide A (1) did not inhibit growth of human embryonic kidney (HEK-293) or colorectal (SW-620) adenocarcinoma cells, or the fungus *Candida albicans* (ATCC 90028), or the Gram-negative bacteria viz. *Pseudomonas aeruginosa* (ATCC 10145) and *Escherichia coli* (ATCC 11775) or the Gram-positive bacteria viz. *Bacillus subtilis* (ATCC 6633 and ATCC 6051) and *Staphylococcus aureus* (ATCC 9144 and ATCC 25923) with IC50 vlaues of above 30 μM (Dewapriya et al. 2017).

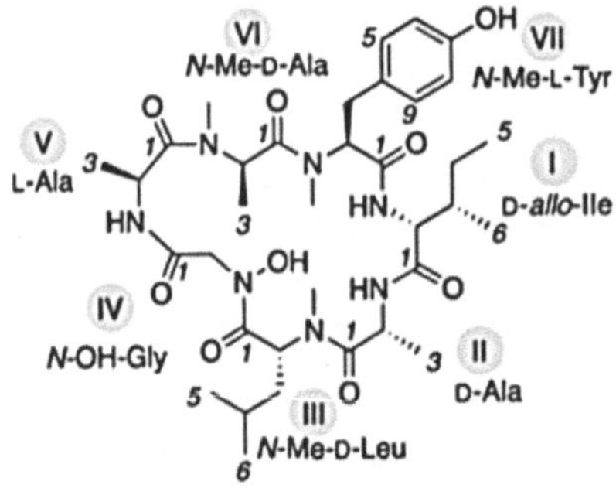

Talarolide A

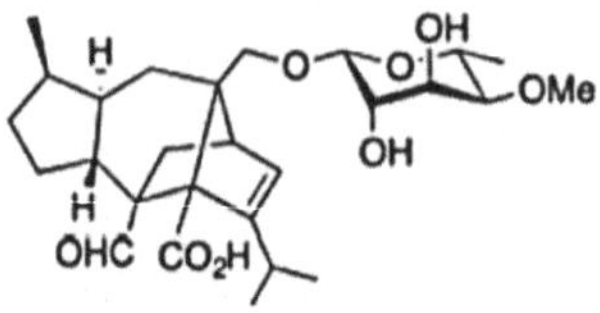

Sordarin

Unidentified ascidian 9

Associated fungi: Pharmaceutical compounds and activities

Anticancer and antifungal activities: This tunicate-derived fungus, *Penicillium* sp. CYE-87 yielded compounds, 1,4-diazepane skeleton, terretrione D, methyl-2-([2-(1H-indol-3-yl)ethyl]carbamoyl)acetate, tryptamine, indole-3-carbaldehyde, 3,6-diisobutylpyrazin-2(1H)-one and terretrione C. Among these compounds, terretrione C and D showed significant antimigratory activity against MDA-MB-231, as well as antifungal activity against *C. albicans* (Shaala and Youssef 2015).

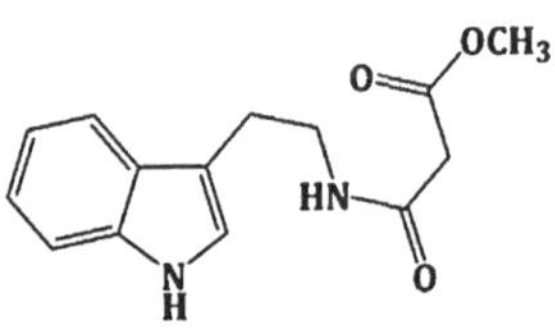

1

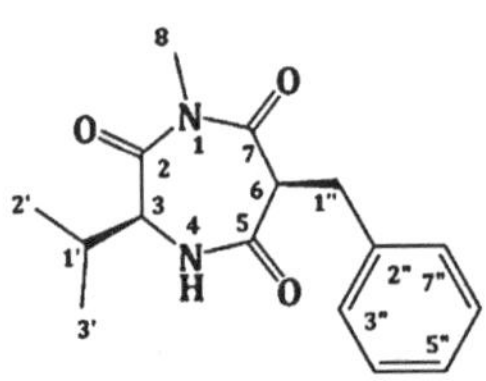

2

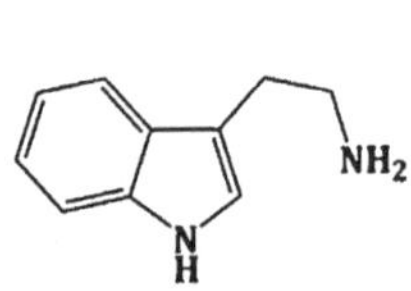

3

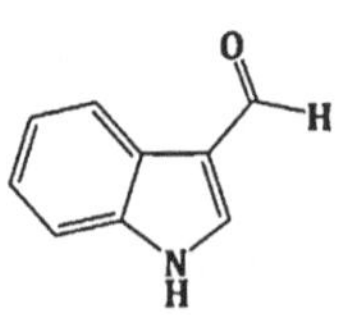

4

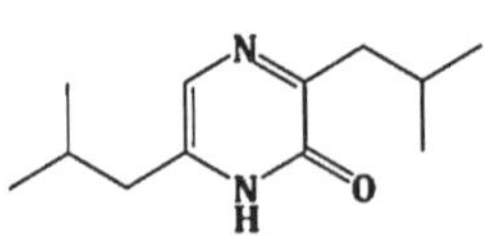

5

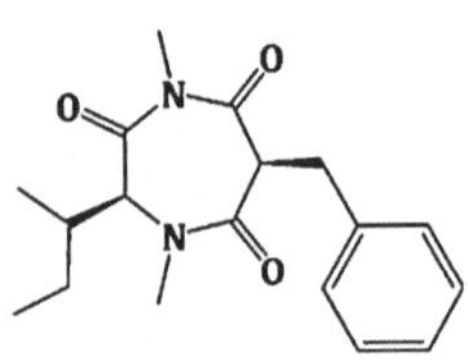

6

1,4-diazepane skeleton, terretrione D (2), methyl-2-([2-(1H-indol-3-yl)ethyl] carbamoyl)acetate (1), tryptamine (3), indole-3-carbaldehyde (4), 3,6-diisobutylpyrazin-2(1H)-one (5), terretrione C (6)

Unidentified ascidian 10

Associated fungi: Pharmaceutical compounds and activities

Cytotoxicity: Sesquiterpenoids named JBIR27 (1) and -28 (2), together with sporogen-AO1 (3) and phomenone (4) were isolated from the culture broth of *Penicillium* sp. SS080624SCf1 derived from this tunicate. Among these compounds, sporogen-AO1, phomenone, and JBIR 28 showed cytotoxicity against human cervical carcinoma cell line HeLa at IC50 values of 8.3, 19 and 92 mM, respectively, whereas compound 1 was inactive at a concentration of 80 mM (Motohashi et al. 2009).

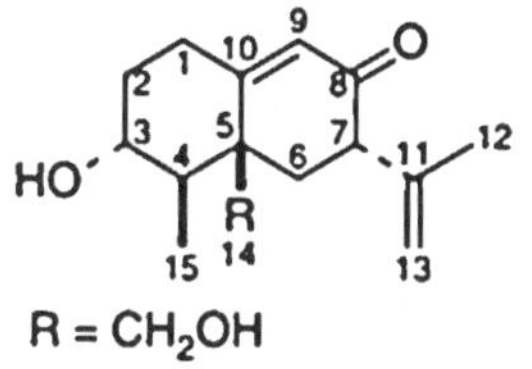

1

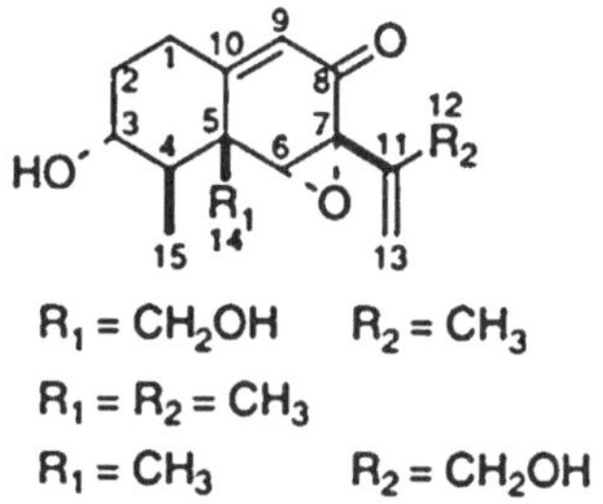

2,3,4,

1: JBIR27; 2: JBIR28; 3: Sporogen-AO1; 4: Phomenone

FUNGI ASSOCIATED WITH MARINE FISHES

Order: Perciformes Family: Apogonidae

Ostorhinchus endekataenia (= *Apogon endekataenia*)

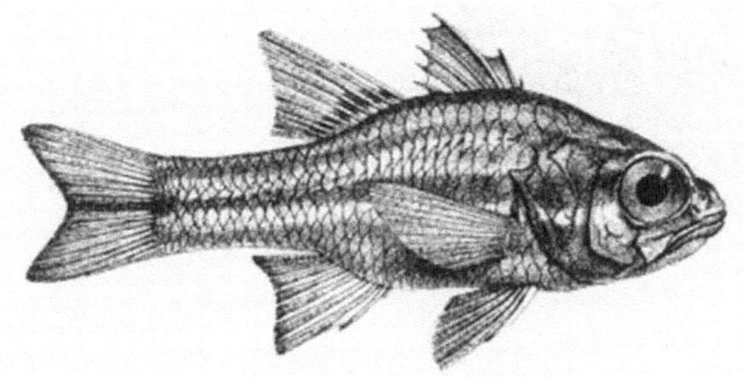

Ostorhinchus sp.

Image credit: George Henry Ford - Day F. Public domain.

Common name: Candystripe cardinalfish

Global distribution: Western Pacific: Tokyo Bay to Taiwan and southward along the western Pacific.

Habitat: Marine; reef-associated; depth range 25-30 m

Associated fungi: Pharmaceutical compounds and activities

Proteasome Inhibitory activity: The fungus *Penicillium fellutanum* (= *P. dierckxii*) derived from this fish produced the compounds fellutamides AB. Among these metabolites, fellutamide B potently induced NGF (nerve growth factor) release from glial-derived cells and fibroblasts; and inhibited significantly the proteasome catalytic activity. This fungus and its compounds are therefore considered as potential candidates in the treatment of neurodegenerative diseases and neuronal injury (Hines et al. 2008; Nicoletti and Trincone, 2016).

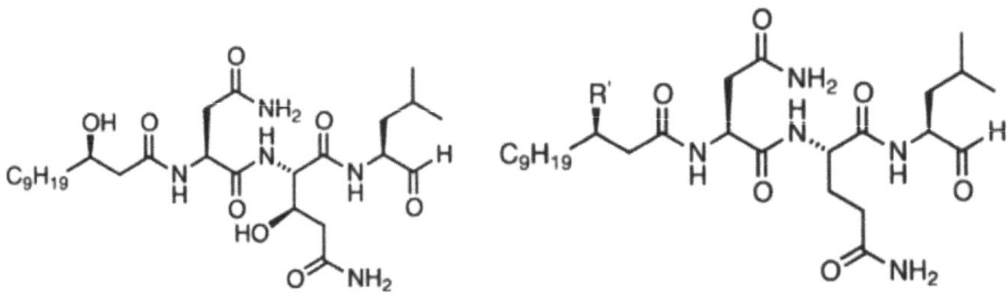

Fellutamides A, B

Order: Perciformes Family: Scaridae

Scarus ovifrons

Image credit: Keiga Kawahara, Public domain.

Common name: Knobsnout parrotfish

Global distribution: Northwest Pacific: Japan

Habitat: Reef-associated

Associated fungi: Pharmaceutical compounds and activities

Antiproliferative activity: The fungus, *Penicillium citrinum* N055 isolated from this species produced perinadine A and scalusamides A–C. Among these compounds, scalusamide A showed significant antiproliferative activity against the human triple negative breast cancer cell line MDA-MB-231 with an IC50 value of 14.8 μM (Nicoletti and Trincone, 2016; Mady et al. 2017).

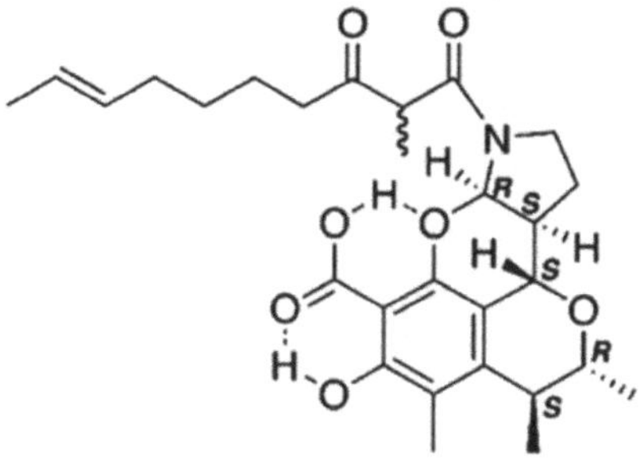

Perinadin A

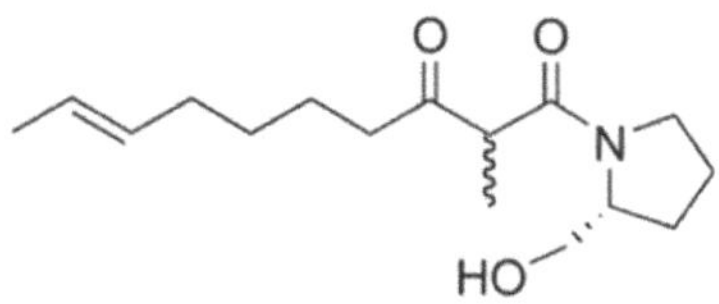

Scalusamide A

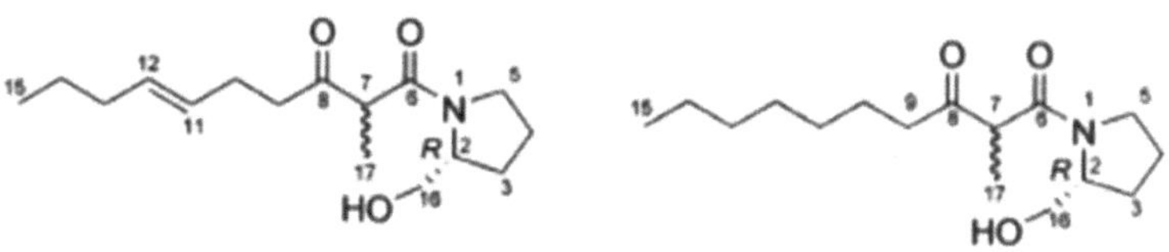

Scalusamides B,C

Order: Mugiliformes Family: Mugilidae

Mugil cephalus

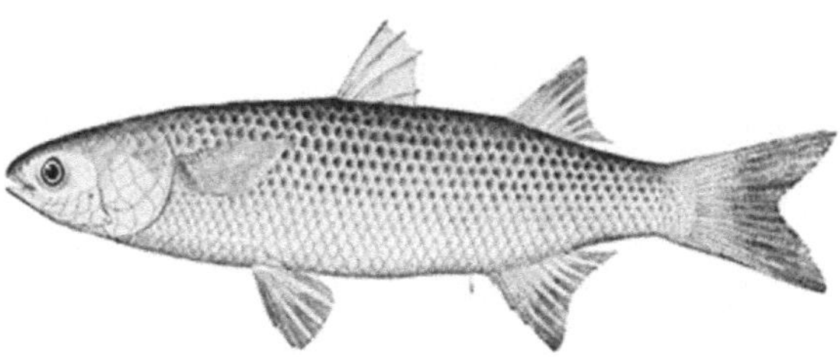

Image credit: The Fisheries and Fisheries Industries of the United States, by George Brown Goode, Public domain.

Common name: Flathead grey mullet

Global distribution: Eastern Pacific, Western Pacific, Western Indian Ocean, Western Atlantic, and Eastern Atlantic.

Habitat: Marine; freshwater; brackish; benthopelagic; catadromous; depth range 0–120 m

Associated fungi: Pharmaceutical compounds and activities

Anticancer activity: From this fish species, the marine fungus *Chaetomium globosum* was isolated. This fungus produced three compounds, Chaetomugilins P-R demonstrating strong inhibitory activity to several cancer lines (Maharaj 2015).

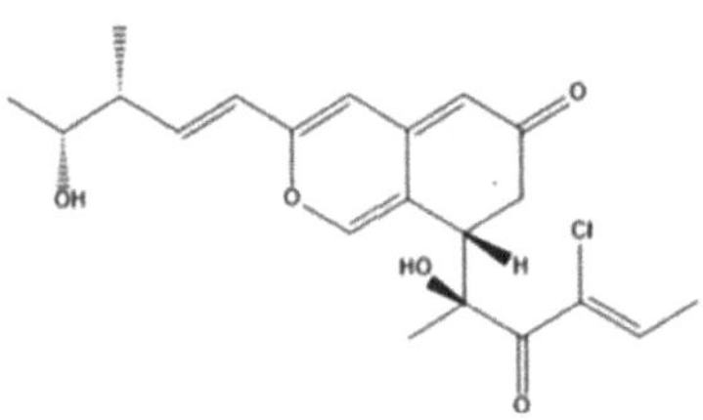

Chaetomugilins P

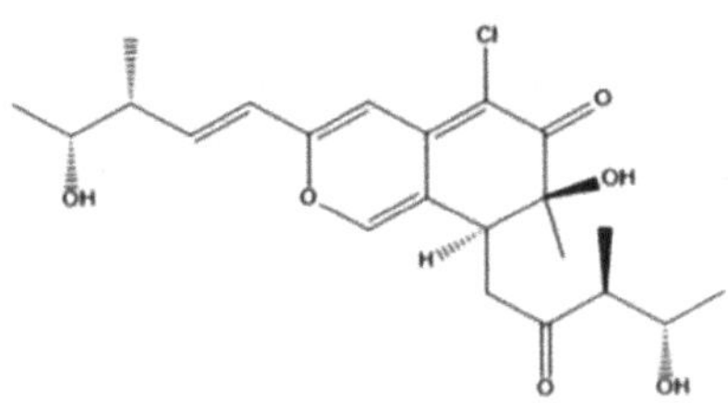

Chaetomugilins Q

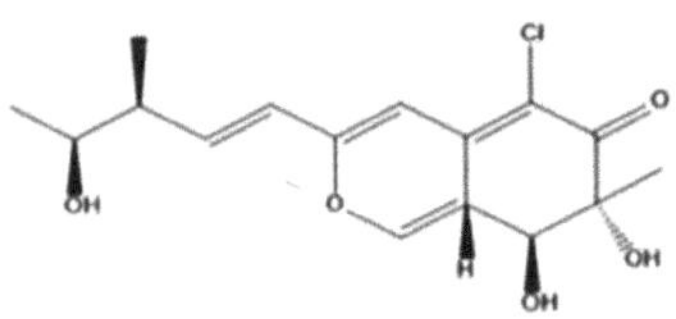

Chaetomugilins R

Yasuhide et al. (2008) reported on the isolation of chaetomugilins A-F from a fungal strain of *Chaetomium globosum*. These metabolites displayed potent growth inhibitory activity on HL-60 and P388 cancer cell lines. Further, chaetomugilins A, C, and F have also been reported to possess selective cytotoxicity on 39 human tumor cell lines.

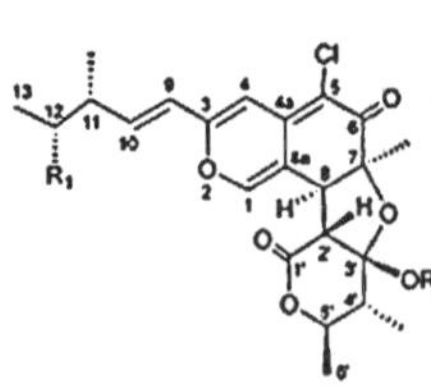

1,2,4,5

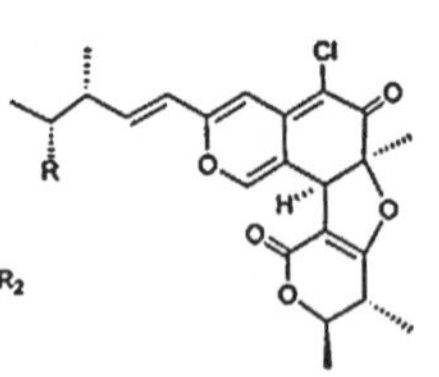

3,6

(1: R1=OH, R2=H; 2: R1=OH, R2=CH3); 4: R1=H, R2=H; 5: R1=H, R2=CH3; 3: R=OH; 6: R=H)

Chaetomugilins A-F (1-6)

Yamada et al. (2012) reported on the isolation of three compounds viz. dechloro-chaetomugilin A, dechloro-chaetomugilin D, and chaetomugilin S from a fungal strain, *Chaetomium globosum* originally obtained from this marine fish. All these metabolites have been reported to possess moderate inhibitory activity on the growth of cultured L1210, KB, P388, and HL-60 cancer cell lines.

Order: Mugiliformes Family: Mugilidae

Planiliza haematocheila *(= Chelon haematocheilus)*

Image credit: OpenCage. https://creativecommons.org/licenses/by-sa/2.5/

Common name: So-iuy mullet

Global distribution: Northwest Pacific, Vietnam, Black Sea and the Mediterranean

Habitat: Marine; freshwater; brackish; pelagic-neritic; catadromous (Ref. 46888); depth range 5 - 10 m

Associated fungi: Pharmaceutical compounds and activities

Cytotoxicity and antiviral and antioxidant activities: Two ubrolides, rubrolides R (1) and S (2), were isolated from the fermentation broth of the fungal strain P388, HL-60, which was found associated with the viscera of this estuarine species. The metabolite rubrolide R showed antioxidation against ABTS radicals with an IC50 value of 1.33 M. On the other hand, compound rubrolide S showed anti-influenza A (H1N1) virus with an IC50 value of 87.1 M. Further, both the compounds showed weak cytotoxicity against the K562 cell line with IC50 values of 12.8 and 10.9 M, respectively, (Jin et al. 2016).

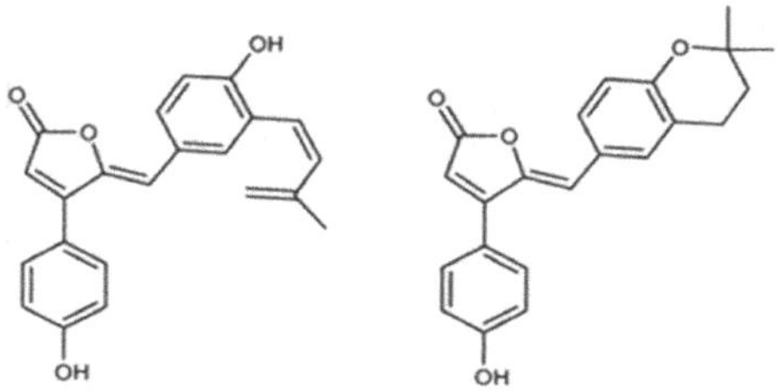

Rubrolides R,S

Shark gills

Associated fungi: Pharmaceutical compounds and activities

Antibacterial activity: Nine chiral compounds *viz.* an acetylenic aromatic ether, aurantiomide C, fructigenine A, cyclopenin, cyclopenol, (R)-penipratynolene, (3S,4S)-3,4-dihydro-3,4,8-trihydroxyl-naphthalenone, verrucosidin, and norverrucosidin were isolated from the culture of a shark gill-derived fungus *Penicillium polonicum* AP2T1. In antimicrobial tests, compounds 1–4, 6, and 8–9 showed weak antibacterial activity against *Staphylococcus aureus, Bacillus subtilis,* and/or *Escherichia coli*. Compounds 3, 8, and 9 also exhibited cytotoxicity against human colon cancer cell line HCT116 (Feng et al. 2017).

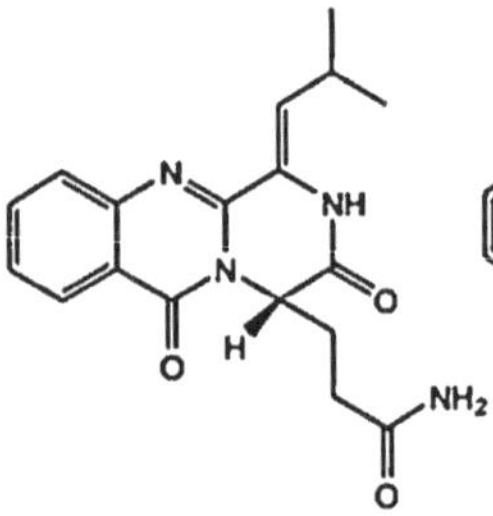

1

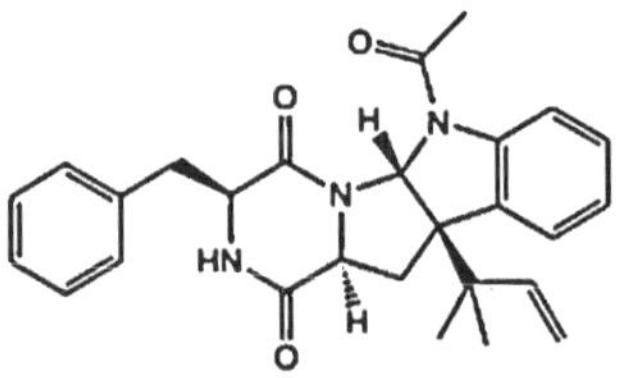

2

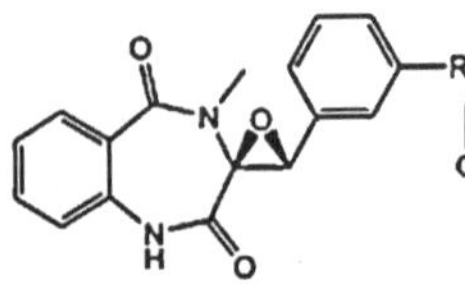

3,4

(3: R=H,; 4: R=OH)

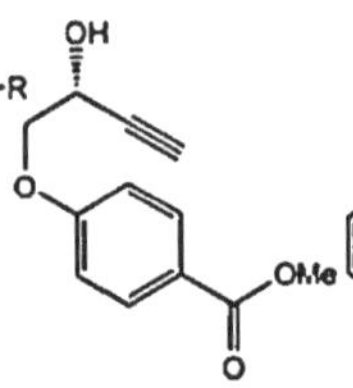

5

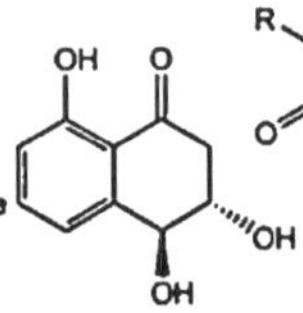

6

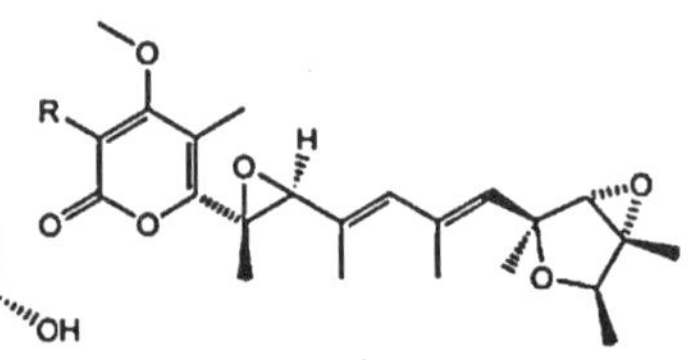

7,8

(7: R=Me; 8: R=H)

1: Aurantiomide C; 2: fructigenine A; 3: cyclopenin; 4: cyclopenol; 5:(R)-penipratynolene; 6: (3S,4S)-3,4-dihydro-3,4,8-trihydroxyl-naphthalenone; 7: verrucosidin; 8: norverrucosidin

Index

A

Acetic acid, 192
Acetylcholine (ACh), 37, 131, 132
Acetylcholinesterase (AChE), 87, 184, 187, 202, 216, 217, 238, 240, 372, 452, 467, 471
Achnanthes, 130
Acidimicrobium, 215
Acinetobacter, 269, 317, 332, 475
  baumannii, 475
Acremonium, 12, 348, 379, 381, 435, 495, 508
Actin, 53, 107, 110, 114
Actinobacteria, 11, 189–192, 207, 212, 242, 248, 262, 290, 330
Actinobacterial genera, 299
Actinobacterium, 214, 232, 258
Actinocyclus, 130
Actinokinespora sp, 191
Actinomyces, 215
Actinomycete, 11, 165, 172, 203, 208, 220, 224, 235, 242, 249, 256, 259, 299, 304–306, 314
Adenocarcinoma, 41, 55, 57, 66, 68, 80, 94, 100, 114, 136, 139, 235, 402, 413, 458, 513
Adhesion, 137, 325
Adriatic sea, 202, 215, 321
Aegean sea, 81, 202, 215
Aerobactor aerogens, 266, 267
Aeromonas, 175, 195, 205, 228, 290, 295, 300, 330–332
Agonist, 37, 297
Agrobacterium, 12, 199, 201, 205, 209, 234, 315, 327
Ala-Tyr-Ala-Pro-Glu (AYAPE), 88, 89
Alcaligenes faecalis, 196, 205, 239
Aldehydes, 92, 103, 105
Algae, 1–3, 17, 156, 171, 174, 209, 226, 347, 350, 371, 387
Algal
  bloom, 9, 17, 106
  extracts, 41
Aliphatic alcohol, 476
Alismatales, 173–177, 359
Alkaloids, 5, 9, 12, 15, 18, 189, 219, 228, 229, 440, 455, 462, 487
Alphaproteobacteria, 252, 262, 290, 330
Alteramide A, 231
Alternaria, 12, 40, 180, 199, 201, 205, 209, 450
Alteromonas, 11, 12, 184, 190, 228, 231, 274, 275, 278, 279, 283, 287, 302, 317, 333
Alzheimer's disease, 34, 125, 129, 149, 155, 202
Amphidinol 18 (AM18), 136, 138
Amphidinolide (AMP), 9, 18, 21, 141, 143
  N (AMP-N), 142
Amphora, 20, 130
Amphoteric iminium compounds, 158
Amyloid levels, 149
Analgesic, 9, 18, 137, 146, 158, 272, 299
Andrimid, 199
Androgen receptor, 326
Androgenic hormones, 326
Antarctic, 92, 211, 213, 218, 226, 234, 314, 358
Antarctica, 3, 213, 222, 250, 358
Antibacterial activity, 29, 32–34, 45, 52, 69, 70, 73, 81, 82, 88, 91, 93, 94, 99, 100, 103, 105, 125, 136, 164, 166, 168–178, 182, 184, 186, 188, 191, 195, 196, 199, 202–208, 210, 211, 213, 215–220, 222, 224, 226, 229, 230, 236, 239, 244, 248–250, 256–258, 261, 263, 264, 266–268, 275, 277–284, 291, 292, 294, 306–308, 310, 314, 316, 317, 321, 322, 324, 326, 327, 330–333, 335, 336, 339, 341, 350, 356, 359, 360, 364, 368, 370, 378, 379, 385, 390, 393–395, 400, 408, 415, 427, 430, 433, 440, 441, 444, 448, 449, 454, 456, 462, 468, 472, 473, 475, 477–479, 483, 520

Antibiotic, 1, 9, 11, 12, 52, 69, 166, 188, 199, 232, 248, 256, 257, 259, 316, 388, 393, 418, 420, 442, 493
Antibodies, 5, 6
Anticancer, 5, 8, 9, 18, 37–39, 45, 47, 50, 52, 55, 60, 64, 72, 73, 75–78, 80–82, 86, 87, 90, 92, 94, 95, 97, 99, 100, 102–105, 114, 115, 136, 139, 141, 151, 161, 166, 168, 170, 197, 204, 228, 235, 257, 271, 275, 277, 298, 299, 309, 312, 313, 315, 325, 336, 353, 357, 364, 372, 413, 423, 429, 437, 441, 447, 454, 458, 514, 518
Anticarcinogenic, 9
Anticoagulant, 99
Antidepressant activity, 297
Antidiabetic, 79, 90, 165, 366, 367, 369, 449, 490, 498, 509, 510, 512
AntiESBL, 172
Antifungal, 9, 16, 18, 33, 41, 53, 103, 119, 129, 130, 136–138, 141, 142, 146, 155, 156, 179, 180, 189, 197, 199, 201, 205, 209, 233, 234, 242, 243, 245, 256, 261, 265, 287, 296, 305, 338, 342, 347, 349, 361, 373, 377, 378, 383, 388, 389, 396, 404, 412, 414, 418, 444, 445, 451, 460, 469, 472, 486, 488, 504, 509, 513, 514
  activity, 33, 41, 103, 119, 129, 130, 136, 138, 141, 142, 155, 179, 180, 189, 197, 199, 201, 205, 209, 233, 242, 243, 245, 261, 265, 287, 296, 305, 342, 347, 349, 361, 373, 378, 388, 389, 396, 404, 412, 414, 445, 460, 469, 472, 486, 504, 514
Antigens, 6, 372
Anti-HIV, 5, 11, 12, 198, 241
Antiinflammatory, 5, 30, 31, 33, 41, 49, 74, 83, 88–90, 94, 101, 102, 147, 151, 158, 192, 259, 260, 325, 443, 447, 476, 480, 508
  activity, 31, 74, 89, 147, 158, 259, 260, 325, 443, 447, 476, 480, 508
Anti-inflammatory properties, 90, 158
Antileishmanial activity, 52, 55
Antileukemic activity, 158
Antimalarial, 47, 52, 55, 61, 214, 241, 251, 476
  activity, 52, 55, 60, 61, 69, 214, 251, 476
Antimicrobial, 5, 9, 33, 40, 49, 72, 91, 103, 118, 120, 123–125, 140, 154, 161, 165–167, 180, 188, 190, 192, 193, 197–201, 204, 205, 207, 209, 210, 214, 215, 221–225, 228, 231, 233, 234, 236, 239–242, 245, 246, 253–255, 261, 269, 270, 274, 288, 290, 302, 316, 326, 329, 332, 336, 338, 349, 352, 355, 359–361, 363, 373, 375, 384, 388, 392, 396, 398, 404, 407, 409, 410, 418, 420, 428, 434, 437, 444, 446, 454, 457, 486, 488, 502, 503, 506, 512, 520
  compounds, 242
Antineoplastic potential, 158
Antiosteoclastic, 137, 146
Antioxidant, 9, 16, 31–34, 38, 45, 47, 49, 50, 62, 63, 71, 74, 76, 78, 79, 83, 85, 87, 88, 90, 93, 98, 151, 158, 162, 192, 222, 230, 252, 316, 334, 336, 348, 349, 352, 371, 374, 398, 430, 435, 446, 452, 519
  activity, 31, 33, 34, 38, 45, 47, 49, 62, 63, 76, 78, 83, 85, 87, 93, 222, 252, 316, 336, 348, 349, 371, 435, 446
Antioxidative, 41, 93
Antiparasitic activity, 191
Antiproliferating effects, 51
Antiproliferative activity, 30, 35, 152, 154, 405, 406, 419, 425, 431, 434, 484, 517
Antiprotozoal, 227
Antiresorptive drug, 158
Antitubercular activity, 87, 103, 345, 443
Antituberculosis, 87
Antitumor, 12, 45, 48, 58, 68, 87, 137, 141, 142, 146, 161, 214, 226, 232, 239, 242, 265, 325, 327, 340, 346, 384, 392, 406, 407, 410, 412, 441
Anti-tyrosinase
  activity, 354
  effects, 354
Antiviral, 5, 9, 16, 44, 45, 49, 50, 52, 96, 99, 106, 120, 123, 145, 151, 154, 161, 189, 338, 397, 432, 450, 451, 453, 462, 463, 466, 494, 519
  activity, 189
Aplousobranchia, 307, 308, 314, 317, 500
Apoptosis, 37–39, 42, 56, 79, 81, 89, 97, 102, 109, 147, 153, 214, 235, 407
Apoptosis, 79, 81
Aqabamycin, 261
Aquatic, 71

Arctic, 72, 92, 138, 234, 236, 318, 363, 412, 432
Arthrinium sp., 425
Arthrobacter, 179, 211, 213, 218, 222, 262, 308, 330
Arthrospira, 2, 48
Ascidian, 513
Aspergillus, 12, 40, 41, 129, 136, 138, 140, 142, 156, 200, 201, 233, 247, 340, 344, 347, 349, 350, 358–360, 365, 376, 388–390, 392–397, 400, 402–406, 413–415, 418, 420, 422, 424, 427, 434, 436, 438, 441, 443–447, 449, 451–454, 456–460, 462, 463, 466, 467, 470, 472, 475, 476, 481, 482, 484, 486, 487, 492, 499, 505, 506, 513
Astaxanthin, 2, 31, 33, 371
Asthma, 24, 50, 147, 342
Astronauts, 50
Ateramide A, 265
Atherosclerosis, 83, 89, 94, 101, 214
Atlantic, 75, 76, 80, 86, 91, 92, 100, 105, 107, 113, 114, 116, 118, 120, 124, 126, 131, 136, 138, 144, 152, 161, 163, 182, 185, 196, 198, 209, 210, 215, 217, 218, 220, 221, 224, 225, 229, 232, 237, 245, 246, 251, 254, 259, 267, 273, 274, 277, 279, 280, 282, 283, 286, 306, 310, 312, 317, 321, 324, 329, 330, 332, 339, 341, 343, 349, 351, 356, 363, 375, 376, 385, 392, 393, 396, 398, 408, 412, 425, 432, 473, 502, 505, 518
  Ocean, 92, 113, 126, 131, 136, 152, 163, 218, 220, 221, 229, 237, 259, 267, 280, 283, 317, 321, 339, 349, 356, 473, 502
Autophagy, 41, 79
Azaspiracid (AZA), 17, 21, 115–117
Azores, 182, 225

B

Bacillus, 11, 12, 33, 40, 41, 93, 103, 140, 169, 171, 173–175, 180, 184–186, 188, 190, 194–205, 207–210, 215, 219, 220, 222, 224, 228, 230, 233, 234, 236, 242, 244–246, 249, 250, 255, 261, 263, 264, 266–268, 273–275, 277–285, 290, 292, 294, 302, 306, 307, 309, 318, 321, 326, 329, 331–333, 335, 336, 356, 373, 379, 382, 383, 388, 395, 400, 404, 415, 418, 420, 428, 433, 435, 512, 513, 520
  cereus, 40, 41, 199, 209, 230, 285, 329, 332, 336, 418, 433
  subtilis, 40, 93, 103, 140, 169, 173–175, 186, 188, 198–202, 204, 205, 209, 215, 222, 224, 228, 234, 236, 244, 245, 261, 264, 275, 277–285, 309, 326, 356, 382, 383, 388, 395, 400, 404, 415, 420, 428, 435, 457, 512, 513, 520
Bacircines (BI), 198
Bactericidal activity, 286, 433
Bacteriodetes, 189, 248, 256
Bacterium, 209
Baltic sea, 35, 38, 39, 85, 121, 126, 161
Barramundi, 332
Bartanolide, 428
Basic fibroblast growth factor (bFGF), 119
Bays, 136, 226, 259, 287, 340, 363
Bedrock, 281, 284
Belize, 130, 224
Benthic, 264, 275, 466, 492, 493
  communities, 158
Benthos, 60
Benzoic acid, 207, 261, 512
Bermuda, 305, 315
Betaine, 83, 101
Betaproteobacteria, 219
Bioactive
  compounds, 3, 5, 9, 10, 12, 15, 16, 196, 240, 277, 290, 489
  molecules, 10
Bioactivity, 16, 89, 141, 163, 222, 323, 337, 392, 488
Biodiesel, 2
Bioimaging, 6
Biomarkers, 6
Biomedical applications, 9, 18
Bioprocessing, 10
Biosensors, 6
Biotechnological applications, 5, 6, 318
Biotechnology, 12
Biotoxins, 11
Bivalves, 290
Black sea, 96, 279, 519
Blood, 20, 42, 87, 94, 95, 101, 104, 105, 123, 132, 372
  cell, 158
  vessels, 132

Blooms, 9, 18, 69, 146
Blue-green algae, 1, 3, 17
Bone, 8, 141
  marrow, 141
Botrytis cinerea, 199, 201, 205, 209, 444
Brackish, 33, 36, 39, 40, 73, 76, 82, 84, 85, 90, 97, 103–105, 114, 119, 144, 161, 165, 195, 196, 217, 224, 287, 323, 324, 330–332, 335, 368, 392, 396, 398, 408, 518, 519
  water, 36, 39, 40, 73, 76, 82, 84, 85, 90, 97, 103, 144, 161, 323, 331, 332, 335
Brain, 20, 31, 101, 128, 147, 151
Breast cancer, 42, 52, 64, 99, 146, 161, 235, 271, 416, 435, 437, 517
Brevetoxin (BTX), 9, 11, 17, 18, 20, 24, 147
Brown algae, 156
Bryostatins, 276, 277
Bryozoa, 10, 275, 283
Butyrylcholinesterase (BChE), 313

C

Callyspongiidae, 204
Calothrixins, 47
Cancer, 20
  cell, 7, 37, 39, 46, 47, 58, 61, 64, 70, 87–89, 94, 95, 99, 100, 102, 104, 114, 141, 142, 144, 146, 152, 153, 168, 170, 227, 309, 313, 336, 351, 376, 380, 383, 409, 416, 417, 425, 455
    lines, 20, 21, 64, 72, 73, 75, 78, 81, 82, 86, 99, 107, 110, 117, 121, 122, 125, 140, 148, 156, 161, 214, 224, 232, 304, 309, 380, 389, 405, 406, 435, 437, 447, 466, 491, 518, 519
  treatments, 276
Candida albicans, 24, 41, 72, 136, 140, 154–156, 200, 201, 204, 209, 221, 233, 245, 247, 254, 256, 261, 296, 305, 336, 347, 359, 361, 388, 398, 404, 418, 428, 445, 446, 452, 474, 475, 503, 504, 508, 513
Carbon, 7, 153
Caribbean, 35, 55, 130, 132, 154, 157, 196, 200, 217, 226, 247, 259, 292, 305, 312, 315, 340, 341, 375, 384, 390, 398
  sea, 35, 55, 130, 132, 154, 157, 259, 305, 315, 341
Carmabin A, 52, 55
Carnobacterium, 207
Carotene, 2, 49, 87, 372
Carotenoid, 2–4, 31, 49, 87, 93, 152, 153, 231, 371
Caves, 193, 215
Cell, 7, 8, 11, 30, 33, 34, 37, 39, 41, 42, 46, 47, 49, 52, 55–58, 60, 61, 64, 66, 70, 72, 79–81, 84, 85, 87–90, 92, 94–97, 99, 100, 102–105, 109, 114, 115, 123, 125–127, 133, 141, 142, 144–148, 150, 152–154, 158, 168, 170, 214, 227, 229, 232, 235, 236, 251, 256, 271, 298, 307–309, 313, 316, 325, 326, 336, 344, 351, 352, 364, 376, 380, 381, 383, 389, 390, 403, 409, 414, 416, 417, 419, 422, 425, 429, 432, 434, 435, 455, 474, 484, 487, 490, 493, 495, 504, 513, 516
  lines, 20, 51, 53, 72, 75–78, 80, 81, 99, 117, 126, 136, 147, 154, 161, 185, 212, 231, 239, 243, 254, 261, 307, 346, 352, 403–405, 411, 413, 415, 419, 421–423, 427, 431, 437, 439, 440, 458, 460, 475, 481, 484, 497, 499, 500, 505, 518
  migration, 119
  walls, 6
Cercospora theae, 199, 201, 205, 209
Cervical cancer, 61, 87, 100, 142, 422
Chaetomugilins, 518
Chelonodon, 336
Chemical
  analysis, 93
  classes, 8, 12, 15, 18
  structure, 17
  synthesis, 26
Chemotherapeutic
  agents, 109
  drug, 146
Chile, 106, 118
Chinese hamster ovary (CHO), 19, 56, 344
  cells, 344
Chitinase, 189
Chlorellales, 34
Chlorocarolide, 424
Chlorophyll, 49, 72, 152
Cholera, 200, 205, 209
Cholesterol, 30, 35, 50, 88, 102, 115, 146, 162, 371, 372
Chondramide D, 237
Chondromyces crocatus, 237

Chondrostereum sp, 447, 448
Chordata, 500
Chromohalobacter sp, 211, 216, 228, 229, 257, 263, 266, 267
Chronic
obstructive pulmonary disease (COPD), 147
pain, 128
Ciguatoxin (CTX), 9, 11, 17, 18, 20, 127, 128, 130, 131
Ciliates, 158
Cirripedes, 284
Citricoccus, 222
Cladosporium, 342, 364, 388, 395, 430, 444, 493, 500
Clostridium difficile, 211
Cnidarians, 130, 158
Coastal, 34, 35, 75, 179, 184, 206, 289, 292, 322, 370, 494
areas, 130, 308
waters, 69, 85, 87, 93, 100, 103, 104, 107, 114, 120, 154, 157, 161, 168, 182, 188, 282, 293, 315, 319, 330, 339, 498
Coasts, 103, 120, 185, 209, 245, 253, 255, 273, 274, 282, 412
Coccolithophores, 8
Colorants, 2
Comorbidity, 71, 73, 80, 81
Compound
actinosporin A, 191
fuscol, 158
Cone snails, 299
Conioscleroderolide, 345
Consumption, 17, 372
Copepods, 285
Coral, 5, 8, 51, 57, 58, 69, 73, 128, 130, 132, 133, 145, 154, 156–158, 174, 175, 185, 188, 216, 248, 250, 260, 261, 267, 268, 270, 271, 289, 304, 341, 384, 392, 401, 423, 445, 446, 448, 449, 451, 452, 455, 456, 464, 471–477, 494, 501, 503
reefs, 57, 58, 69, 145, 156, 188, 270, 304, 456, 494
rubble, 130, 154, 157, 175, 216, 250, 289, 341, 384, 423
Coronary heart disease, 146
Corynebacterium sp, 206, 216, 257, 266, 268
Cosmopolitan, 57, 58, 76, 92, 95, 96, 101, 103, 105, 110, 343, 507
Crustacea, 284–286, 480
Crustaceans, 284, 480
Cryptocodinium, 2
Cryptomonadales, 35
Cryptophycin, 52
Cryptophyta, 35
Cucumber, 497–499
Culture, 2, 10, 26, 85, 88, 102, 392, 393, 398, 420, 442, 452, 479, 481, 487, 515
broth, 442, 481, 508, 515
Curvulalic acid, 470
Curvulalide, 470
Curvulapyrone, 470
Cyanobacteria, 1, 3, 5, 8, 17, 19
Cyanotherapeutics, 5
Cyclic
adenosine monophosphate (cAMP), 344
decapeptide, 12, 64
imines, 9, 18
peptide, 68, 301
Cyclodepsipeptide carriebowmide, 55
Cyclooxygenase-2 (COX-2), 30, 31, 89, 158, 474, 480
Cyclopenin, 363, 520
Cyclopeptine, 363
Cymodoceaceae, 173, 175, 176, 359
Cyprus, 131
Cystic fibrosis (CF), 147, 213, 218
Cytokines, 82, 161
Cytophagas, 228
Cytoplasm, 3, 326
Cytotoxic, 9, 11, 12, 18, 33, 45, 49, 51–53, 58, 66, 99, 107, 114, 136, 139, 140, 149, 155, 195, 207, 224, 227, 229, 231, 232, 236, 237, 241, 243, 254, 260, 261, 298, 301, 304, 306, 308, 309, 315, 316, 325, 326, 338, 340, 342, 353, 360, 364, 375, 376, 380–382, 389, 390, 392, 403–405, 407, 409, 410, 415, 416, 421, 422, 425, 431, 432, 434, 438–440, 448, 456, 458, 462, 464, 466, 469, 473, 474, 481, 482, 486, 491, 495, 499, 502–505, 512
Cytotoxicity, 16, 35, 48, 52, 55, 57, 58, 60, 61, 63, 64, 66, 68, 70, 80, 96, 97, 99, 104, 107, 110, 111, 114, 115, 117, 119, 121, 122, 129, 133, 139–143, 145, 146, 148, 150, 152, 154, 156, 161, 202, 207, 212, 227, 228, 231, 232, 235–237, 241, 254,

256, 260, 261, 265, 271, 304, 312, 314, 315, 325, 326, 336, 340, 351, 352, 361, 379, 381, 383, 384, 388, 390, 393, 402, 403, 405, 414–417, 419, 422, 429, 435, 439, 440, 447, 448, 455, 458, 460, 462, 469, 475, 485–490, 492, 497, 500, 503, 504, 511, 513, 515, 518–520

D

Dead organic matter, 12
Deep sea, 12, 390
Deinococcus sp, 266
Deltaproteobacterium, 237
Demosponges, 158
Dendroceratida, 191
Dendrodolide, 498
Deoxymycelianamide, 464
Deoxyribonucleic acid (DNA), 7, 35, 97, 256, 312, 326
Depsipeptide, 52, 57, 66, 312, 507
DGCC (1,2-diacylglyceryl3-O-carboxyhydroxymethylcholine), 83, 94
Diabetes, 24, 49, 50, 79, 101, 128, 155, 490, 509, 510
Diadinoxanthin, 49, 152
Diameter, 6, 174, 175, 177, 186, 195, 204, 211, 216, 220, 224, 250, 388
Diapolycopenedioic acid, 230
Diatoms, 1, 5–8, 17, 19, 20, 91, 130
Diatoxanthin, 49
Dictyoceratida, 196, 198, 201
Dictyodendrillidae, 191
Dietary supplement, 49, 50
Dihydromaltophilin, 296
Dihydrosporogen, 473, 474
Dihydrotestosterone, 326
Diketopiperazine (DKP), 393, 425, 460
Dinoflagellates, 1, 8, 17, 19, 130, 151
Dinophysistoxins (DTX), 9, 17, 18, 110, 111, 113–115, 155, 156
Diorcinolic acid, 424
Diseases, 1, 9, 18, 20, 24, 25, 83, 107, 128, 135, 146, 147, 155, 158, 334, 342, 363, 402, 437, 467, 516
Disk abalone, 294
Ditches, 30
Diterpenoids, 324, 425, 482, 499
Docosahexaenoic acid (DHA), 2, 32, 83, 88, 101, 151, 371, 372
Domoic acid (DA), 17, 20, 346
DPPH (2,2-diphenyl-1-picrylhydrazyl), 45, 78, 83, 87, 93, 98, 349, 352, 371, 374, 398, 435, 446, 452
Dragonamide A, 52, 55
Drug, 1, 6–9, 11, 12, 26, 38, 46, 72, 104, 107, 251
  carriers, 105
  delivery, 6, 92
  discovery, 16
Dunaliella salina, 2, 29

E

Ebola virus disease (EVD), 135
Echinoderms, 11
Echinogorgia reticulata, 266
Ecology, 29–37, 39, 40, 44–48, 51, 55, 57, 58, 60, 62–64, 67, 69, 71–76, 78, 80–85, 87, 88, 90–97, 99–101, 103–108, 110–128, 130–133, 135–139, 144–146, 148–158, 161
Ecteinascidin, 16
Efrapeptins, 380
Eicosapentaenoic acid (EPA), 2, 32, 34, 83, 88, 94, 100, 101, 161, 372
Emeniveol, 350
Endosymbionts, 158
Enterococci, 418
Enterococcus, 12, 33, 171, 188, 190, 191, 200, 201, 207, 215, 224, 228, 309, 314, 316, 336, 379, 475, 503, 506
Enteromorpha, 165, 343–345
Enzymatic activity, 37
Epicoccum sp., 498, 499
Epidermal carcinoma, 231
Epiphytes, 128
Epithelial cells, 39, 104, 147, 352
Epithelium, 41
Ergosterol, 30, 31, 371, 372
Escherichia coli, 29, 33, 40, 41, 72, 88, 93, 140, 154, 175, 186, 188, 190, 198–201, 203, 205, 207–209, 215, 219–222, 224, 226, 227, 234, 249, 261, 264, 266, 268, 269, 279, 281, 285, 292, 306, 308, 321, 327, 373, 388, 400, 404, 430, 435, 454, 457, 475, 478, 479, 491, 504, 513, 520

Esters, 33, 108, 230, 255, 340
Estuaries, 85, 92, 113, 114, 120, 136, 144, 287, 293, 303, 360, 363, 482
Estuarine
environments, 51, 125
waters, 146, 156
Ethanol, 2, 33, 40, 88, 94, 103, 152, 192
Ethers, 195, 414
European Medicines Evaluation Agency (EMEA), 52, 64
Eurothiocin, 450
Eurotium cristatum (ECE), 411
Eustigmatales, 33
Eustigmatophyceae, 33
Exopolysaccharides (EPS), 144, 332, 336
Extended spectrum beta lactamase, 172

F

Fatty acids, 1, 2, 35, 49, 72
Fenders, 277
Fermentation, 234
Ferrimicrobium, 215
Ferrithrix, 215
Fibrosis, 147, 213, 218
Fiji, 286, 298, 447
Filaments, 63
Firmicutes, 190, 198, 207, 210, 242, 248, 290
Flatworms, 158
Flavobacteriaceae, 226
Flavonoids, 41, 233, 372
Florida, 103, 224, 226, 237, 247, 279, 305, 312, 315, 340, 384, 398
Fluorescent markers, 2
Food
FDA (Food and Drug Administration), 52, 64
production, 31
supplement, 2
Foraminiferans, 158
Free
radical scavenging activity (FRSA), 93, 374, 398, 446, 452
radicals, 162, 349
French Polynesia, 200, 394
Fucoxanthin, 87, 90, 93, 94, 100, 101
Fumiquinazolines H, 508
Fungal metabolites, 402
Fungerin, 197
Fungi, 1, 5, 9–12, 15, 16, 141, 161, 167, 179, 180, 226, 337–361, 363–378, 381, 382, 384, 385, 387–390, 392–406, 408–416, 419–425, 427–432, 434–442, 444–448, 451, 453–456, 460, 462, 464, 466, 468, 470–474, 476–478, 481–485, 488–490, 492–495, 497, 498, 501–520
Fungus, 15, 129, 136, 234, 287, 340, 342–346, 348–353, 355–360, 362–367, 369–372, 374–376, 378, 379, 381, 382, 384, 386, 388, 389, 392–398, 401–406, 410, 412, 414–416, 420, 423–425, 427, 429–432, 434, 437–441, 443–448, 450–452, 454–457, 459, 460, 462–464, 466, 467, 471–480, 484, 486, 489–494, 497–499, 501, 503–514, 516–518, 520
Fusarium, 12, 33, 103, 180, 199, 201, 205, 209, 361, 362, 400, 472

G

Galicia, 193, 197, 198, 225
Gambieric acids (GA), 128, 129
Gambierol, 24, 128
Gammaproteobacteria, 242, 290, 331
Gammaproteobacterial strains Xp, 216
Gastric cancer, 256, 407, 429
Gastroenteritis, 331
Gastropods, 293
Genus, 11, 62, 73, 76, 83, 115, 130, 158, 240, 269, 270, 274, 338, 430
Geodia cydonium, 425
Germicidin, 323, 420
Giant clams, 158
Gifhornenolone A, 326
Glembatumumab vedotin (GV), 52, 64
Glucosidase, 165, 198, 322, 366, 367, 369, 449, 476, 498
Glycerol, 30, 185, 217, 232, 353, 374
Glycosides, 61, 495
Glycosphingolipid-1 (GSL-1), 252
Gobiidae, 336
Gonyautoxin (GTX), 9, 17, 18, 24, 25, 123, 142, 143
Gorgonian, 265, 455–457, 460, 462, 463, 466, 467, 475, 477, 478
Gram negative
proteobacterium, 239
bacterium, 230, 364

Granaticin, 308, 322, 326
Grassypeptolide D, 55
Green algae, 1, 3, 17
Greenland, 209, 329
Growth inhibition (GI50), 55, 80, 95, 103, 223, 444, 492
Guanidine, 219
Gymnodinium breve, 146

H

Hadromerida, 204
Haematococcaceae, 31
Haematococcus pluvialis, 2
Halichondria
  japonica, 230
  okadai, 230
Halomonas, 190, 207, 228
Hamsters, 161
Hand, foot, and mouth disease (HFMD), 451
Haplosclerida, 204, 214
Harbors, 275, 409
Harmful algal blooms (HABs), 17, 106
Health benefits, 3
Hemolysis, 326
Hepatocellular carcinoma (HCC), 97, 120, 123, 127, 154, 157, 336, 352, 429, 484
Hepatoprotective, 88
Herpes simplex virus type 1 and 2 (HSV-1 and HSV-2), 96, 433
Heterococcoliths, 84, 85
Hexylitaconic acid, 351, 424
Hirsutane sesquiterpenoids chondrosterins, 447
Histamine, 49, 132
Hormones, 326
Hot springs, 3
Human
  acute myelogenous leukemia, 229
  cancer cell lines, 232
  colon tumor cell line, 140, 265
  embryonic kidney (HEK-293), 37, 39, 344, 491, 513
  enterovirus, 451
  gingival fibroblasts (HGF), 161
  health, 9, 18, 49
  hepatic carcinoma cell, 224
  immunodeficiency virus (HIV), 5, 9, 11, 12, 44, 49, 96, 135, 198, 241, 389
  leukocyte elastase or human neutrophil elastase (HLE), 345
  lung carcinoma cell, 224
  mammalian cancer cells, 409
  nasopharyngeal cancer cell lines, 447
  peripheral blood
    mononuclear cells (HPBMC), 94, 95, 104, 105, 123
    promyeloblast cell, 224
  serotonin, 297
  T-lymphoid cell line, 41, 42
Hydrocharitaceae, 174, 176, 177, 359
Hydrogen, 2, 83
Hydroxmethyl, 83, 100
Hydroxysclerosporin, 345
Hyper-polymerization, 53
Hyphomycetes, 423
Hypocholesterolemic, 87, 102

I

Iamia, 215
Ibuprofen, 179
Ice cover, 12
Iceland, 329
Immune
  response, 94, 95, 252, 330
  system, 50, 128, 141, 155
Immunomodulatory, 31, 32, 82, 94, 95, 104, 105, 123, 161, 286, 372
  role, 286
Implants, 90
Indian Ocean, 111, 132, 144, 154, 206, 218, 229, 238, 268, 300, 356, 460, 462, 494, 509, 518
Indole, 47, 233, 234, 317, 346, 460, 462, 482, 487, 514
  oligomer compound, 239
Industrial applications, 1
Infectious disease, 450
Inflammation, 47, 50, 89, 137, 158, 260, 345, 467
Influenza A virus, 96, 433
Inhibitory activity, 46, 60, 94, 96, 139, 158, 178, 180, 183, 185, 198, 220, 225, 235, 246, 256, 257, 313, 325, 326, 331, 343, 345, 363, 366, 413, 414, 416, 418, 421, 423, 431, 433, 437, 440, 449, 452, 467, 471, 476, 482, 497, 498, 509, 510, 512, 518, 519

Insulin, 79, 509, 510
Insulinoma cell line-1 (INS-1), 490
Interleukin (IL), 6, 30, 32, 42, 49, 94, 95, 104, 105, 123, 161, 259, 260, 424, 429, 474, 480
  1 (IL-1), 42, 424, 474, 480
  5 (IL-5), 260
  6 (IL-6), 30, 32, 42, 49, 94, 95, 161, 480
Intestinal tracts, 286
Invertebrates, 10–12
Irciniidae, 196
Isolation, 10, 16, 205, 215, 236, 320, 362, 374, 389, 407, 457, 459, 463, 518, 519
Isopoda, 480

J

Jamaica, 224
Janthinolide A–B, 464
Japan, 93, 103, 106, 111, 118–120, 123, 139, 146, 152, 230, 286, 287, 290, 292, 294, 298, 319, 369, 416, 419, 453, 483, 498, 516
  ecology, 148
  sea, 354
  waters, 93
Japanese
  abalone, 294
  oyster, 293
  spiky sea cucumber, 498
  wireweed global distribution, 167
Jaspis
  coriacea, 423
  sp., 237
  stellifera, 423
Java habitat, 239
Jellyfish, 158, 259, 444
Junceella juncea, 263
Junk food, 1
Jurkat, 261
  cells, 41

K

Kalkitoxin, 19
Karenia brevis, 20, 24, 146
Karlotoxin (KmTx), 24, 145, 146
Kinases, 426
Klebsiella, 29, 33, 88, 168, 172, 199–201, 205, 209, 228, 244, 285, 302, 335, 475
Knolls, 239
Kocuria, 166, 200, 201, 207, 208, 256
Kuruma shrimp, 286
Kytococcus, 200, 201

L

Lactam alteramide A, 231
Lactic acid bacteria (LAB), 286
Lactobacillus fermentum, 201
Lactococcus lactis, 286
Lactone, 5, 366, 389, 407, 416, 429, 435, 447, 451
Lagocephalus, 336
Lagoon, 113, 120, 177, 190, 196, 226, 259, 261, 268, 271, 297, 340, 363, 398, 423
LDL cholesterol, 102
Leather, 260, 261, 445, 446
Leishmania donovani, 52, 55
Leotiomycetes, 234
Leptogorgia alba, 264
Leptolyngbyaceae, 73
Leukemia, 21, 38, 42, 100, 132, 136, 139, 146, 148, 158, 161, 212, 214, 228, 229, 231, 232, 260, 315, 392, 393, 409, 414, 415, 417–419, 421, 438, 458, 504
Limestone, 67, 387
Linear polyketides, 9, 18
Lipid, 5, 33, 49, 72, 77, 83, 86, 87, 94, 101, 102, 126, 335, 371, 401
Lipopeptides, 58
Lipophilic extract, 55
Lipopolysaccharide (LPS), 364, 474
Livestock feed, 2
Lungs, 20
Lutein, 33
Luteoride E, 476
Lyngbya, 5, 19, 38, 50, 51, 55, 57

M

Macrocyclic trichothecenes, 378
Macrolactin V, 180, 264, 325
Macrolide, 5, 9, 18, 60, 61, 119, 148, 180, 314, 356, 362, 469, 497
Macrophages, 102, 144, 364, 447, 474
Macrophomina phaseolina, 199, 201, 205, 209
Magnoliopsida, 173, 359, 365, 373
Main protease (Mpro), 36, 52

Maitotoxin (MTX), 9, 17, 18, 114, 127, 128
Malacostraca, 286, 480
Malay Archipelago, 286
Malformin, 360
Malyngamide C, 51
Mangrove, 15
  plant, 366–372
  roots, 188, 224, 284
  shrub, 367
Manoalide, 196
Marennine, 96
Marine, 1, 2, 4, 6, 10, 11, 13, 15–17, 19, 20, 24, 29, 64, 67, 74, 163, 330, 332, 335, 337, 516, 518, 519
  animals, 12
  bacteria, 10–12, 234, 316
  cyanobacteria, 3, 5, 19
  derived fungi, 15
  environments, 12, 72, 80, 94, 130
  fungi, 15
  habitats, 44, 58, 69
  invertebrates, 10, 12
  microalgae, 1–3, 8, 10, 11, 17, 18, 101
  microorganisms, 10, 16
  phycotoxins, 9, 18
  sponges, 81
Marinobacter, 209, 228
Marinomonas, 228, 300
Matrix metalloproteinase, 30, 89
Medicinal compounds, 29–40, 44–48, 51, 55, 58, 60–65, 67–85, 87–101, 103–108, 110–128, 130–133, 135–158, 161
Mediterranean, 113, 115, 117, 120, 131, 133, 138, 149, 150, 182, 185, 188, 193, 197, 198, 200, 209, 215, 218, 221, 225, 232, 234, 236, 245, 246, 251, 253, 273, 275, 277–279, 281–283, 286, 321, 324, 339, 349, 356, 376, 377, 386, 388, 393, 412, 425, 430, 481, 519
  sea, 113, 115, 117, 120, 131, 133, 138, 149, 150, 182, 188, 200, 215, 232, 275, 277–279, 282, 324, 339, 481
Melanoma, 11, 52, 64, 99, 104, 136, 152, 314, 411, 413, 415
Meleagrin, 459
Mellein, 412, 424, 505
Metabolic
  disorders, 313
  syndrome, 509, 510
Metabolism, 126
Metabolites, 5, 9, 10, 196, 315, 361, 411, 419, 425, 448, 449, 451, 469, 479, 504, 508, 513, 516, 518, 519
Metacycloprodigiosin, 224
Methane, 2, 87
Methanol, 40, 45, 94, 103, 125, 136, 370, 425
Methanolic extracts, 49
Methicillin
  resistant Staphylococcus aureus (MRSA), 12, 33, 74, 91, 136, 167, 171, 172, 174, 175, 177, 233, 236, 244, 245, 256–258, 291, 309–311, 315, 324, 326, 361, 368, 370, 378, 393, 394, 403, 407, 418, 420, 430, 436, 440–442, 444, 462, 468, 473, 475, 478, 483, 491, 503, 506
  sensitive *Staphylococcus aureus* (MSSA), 136, 311, 326, 435
Methoxyvermistatin, 450
Microalgae, 1–5, 8–11, 17, 18, 29, 101, 371
Microbacterium sp, 232, 379
Microbes, 1, 10, 11, 176, 219, 316
Microbial
  metabolites, 11
  symbionts, 16, 251
Microbiota, 1, 17, 29, 163, 337
Micrococcus, 11, 173, 191, 197, 207, 215, 217, 220–222, 226, 246, 261, 267, 268, 336
Microcoleaceae, 48
Microorganisms, 1, 9–11, 16, 165
Microthrix, 215
Mid intertidal zone, 347
Minimal effective concentration (MEC), 138
Mitochondria, 42, 63
Miyagi oyster, 293
Mollusk, 158, 290, 292–295, 299, 300, 302, 485, 487
Monodictyphenone, 510
Monodopsidaceae, 33
Mouse
  erythrocytes, 326
  lymphoma cell, 224
Mud, 286, 288, 300, 412
Multidrug-resistant (MDR), 60, 165, 167, 172, 175, 205, 224, 233, 263, 269, 327, 418, 473
  bacteria, 165, 224, 491

Murine
leukemia, 158
macrophage, 41, 424
melanoma cell line, 265
preadipocytes, 49, 72
Mycale cecilia, 409
Mycelia, 361, 363, 442, 452, 508
Myeloma, 207
Myeloperoxidase (MPO), 83
Myida, 290
Mytilus edulis, 485

N

Nakata cell lines, 158
Nanoparticles, 6, 8
Natural product, 9, 10, 315, 397, 508
Necrosis, 30
Nemertea, 271, 272
Neosaxitoxin (NeoSTX), 9, 18, 117
Neosymbioimine, 158
Neritic
waters, 55, 60, 95, 110, 114, 138, 155
zone, 103
Nerve
cell depolarization, 298
growth factor (NGF), 127, 516
Nervous system, 132, 372, 415
Neuro2a, 114
Neuroactivity, 232
Neuroblastoma, 20, 52, 55, 57, 66, 326
Neurological
complications, 451
toxins, 299
Neuroprotective activity, 149, 158, 376
Neuroprotector, 149
Neurotoxin, 17, 107, 123, 155, 272, 273, 289, 293, 295, 300, 302
Neutral lipid accumulation, 49, 72
Nicotinic acetylcholine receptors (nAChRs), 37
Nigericin, 16
Nitric oxide (NO), 41, 78, 85, 87, 89, 93, 94, 364, 399, 425, 443, 447, 474, 476, 480
Nitrogen, 3, 103
Nitzschia, 2, 90, 130
Nocapyrones, 298
Nodularin, 19
Noscomin, 40
Nostocaceae, 37
Nostocales, 37, 46
Nuclear erythroid 2-related factor 2 -antioxidant response element (Nrf2-ARE), 74
Nutraceuticals, 35

O

Obesity, 50, 71–73, 77, 78, 80, 81, 94, 101
Ocean, 92, 100, 105, 118, 144, 341
waters, 12
Ochrophyta, 33, 89
Octocoral, 11, 445, 451, 453, 454, 459, 476
Okadaic acid (OA), 9, 17, 18, 20, 21, 24, 26, 107, 108, 110–112, 114, 115, 130, 155–157, 231
Okadaxanthine, 231
Open sea, 349
Organic
compounds, 211
crude extract, 49, 72
Organisms, 3, 5, 10–12, 67, 158, 342
Oscillatoria, 5, 62
Oscillatoriales, 47, 48
Osteoclasts, 56
Osteoporosis, 24, 158
Osteosarcoma, 55, 441
Ovarian cancer, 99, 429
Ovatoxins (OVTX), 133
Oxalic acid, 192
Oxidative stress, 34, 83, 88
Oxidized low-density lipoprotein (Ox-LDL), 89
Oxygen, 3, 49, 69, 231

P

Pacific
abalone, 294
Ocean, 48, 92, 100, 105, 118, 119, 132, 149, 164, 190, 229, 261, 268, 271, 273, 274, 300, 333, 341–343, 356, 462, 482, 488
oyster, 293
Pacifigorgia sp, 265
Paenibacillus, 242, 290, 294
Palaemonidae, 286
Papain-like protease (PLpro), 36, 46
Parasites, 9, 183, 246

Parasitic diseases, 135
Pathogenic
bacteria, 261
microorganisms, 165
Pathogens, 88, 172, 173, 175, 177–179, 195, 196, 199, 204, 211, 218, 256, 279, 284, 326, 444, 475, 493
Pearl harbors, 409
Pectenotoxin (PTX), 9, 17, 18, 107–109
2 (PTX-2), 20, 107, 108, 113, 114
Pelagic species, 444
Penicillic acid, 424, 505
Penicillium, 12, 156, 341, 342, 353, 357–359, 361, 363, 364, 371, 372, 377, 379, 386, 389, 390, 402, 404, 409, 410, 424, 427, 439, 443, 454, 457–459, 464, 467–469, 472–475, 480, 483, 485, 486, 488–490, 493, 502–504, 510, 511, 514–517, 520
Penicipyrone, 469
Penicisochroman, 469
Peptide, 3, 15, 33, 35, 52, 64, 65, 68, 70, 79, 83, 88, 242, 256, 299, 372, 402, 486
Perciformes, 329, 331, 332, 336, 515, 516
Peripheral blood mononuclear cells (PBMC), 94, 95, 104, 105, 123
Petasol, 473, 474
Pharmaceutical
applications, 3
compounds, 163–180, 272–275, 277–281, 283, 284, 286–289, 338–348, 350–361, 363, 366–373, 375–378, 381, 382, 384, 385, 387–390, 392–403, 405, 406, 408–414, 416, 419, 420, 422–425, 427, 429–432, 434–437, 439–442, 462, 479, 480, 483, 494
industry, 33
Pharmacophores, 9, 18
Phenalenone, 345
Phenazine, 222, 248, 326
Phenolic compounds, 49, 71, 186
Phoma, 12, 364, 384, 418, 478
Phomopsis thae, 199, 201, 205, 209
Phosphorylation, 37, 149
Photobacterium, 207, 333
Photoluminescence, 6, 90
Phthalide, 363, 397, 430
Phycobiliproteins, 49, 71
Phycocyanin, 2
Phycocyanobilin, 49
Phycotoxins, 9, 17, 18, 26, 111–115, 117, 132, 150
Phylum, 3, 29, 33–36, 82, 89, 92, 105, 106, 160, 163, 173, 178, 181, 271, 272, 275, 283, 284, 337, 349, 359, 365, 374, 480, 490, 500
Phytoalexin, 79
Phytoplankton, 18
toxins, 18, 26
Phytosterols, 30, 101, 102
Pigment, 2, 3, 31, 33, 87, 93, 96, 99–101, 152, 153, 202, 371
Pilings, 284, 323
Pinkfish, 305
Pinnacles, 196, 392
Pinnatoxins (PnTX), 9, 18, 150
Plankton, 2, 85
Plantae, 359
Plasma malondialdehyde, 49
Plasmodium, 47, 60, 61, 69, 135, 214, 228, 241, 358, 397, 476
Platelet-activating factor (PAF), 83, 94, 101
Pleurosigma, 130
Plexauridae, 266
Pneumonia, 172, 199, 200, 205, 209, 211, 453
Polar
ice, 12
regions, 84
Polycyclic tetramic macrolactam compound, 296
Polyether palytoxin (PLTX), 131, 132, 135
Polyketide, 5, 9, 12, 15, 18, 45, 80, 141, 143, 145, 165, 169, 188, 242, 243, 251, 259, 265, 271, 276, 298, 316, 341, 351, 353, 364, 366, 367, 369, 371, 388, 393, 397, 403, 415, 442–445, 477, 478
pyrones, 298, 299
Polyol-polyene compounds, 137, 146
Polysaccharides, 3, 5, 41, 42, 48, 49, 71, 82, 98, 145, 161, 162, 372
Polyunsaturated fatty acids (PUFAs), 1–3, 49, 72, 93
Pontibacillus sp, 216, 268
Porcine respiratory, 450
Poria hypolateritia, 199, 201, 205, 209

Porifera, 10, 181, 374
Porphyridium, 2, 160
Postmenopausal women, 158
Potent
cytotoxic activity, 232, 380, 390, 440, 458
singlet oxygen quencher, 231
Prodigiosins analogs, 224
Propionibacterium, 135, 215
Prostaglandin E2 (PGE2), 443, 474, 480
Prostate, 146, 481, 513
cancer, 481, 513
Proteases, 36, 437
Protein, 5, 8, 37, 42, 49, 72, 89, 101, 335
phosphatase type 1 (PP1), 37, 39
tyrosine phosphatase (PTP), 509, 512
Proteobacteria, 184, 189, 193, 198, 199, 202, 207, 212, 236, 246, 248, 253
Proteobacterium, 236, 239, 243
Proteus vulgais, 261
Protists, 5
Provitamin A, 2
Pseudoalteromonas, 99, 188, 190, 202, 207, 211, 215, 218, 223, 234, 240, 242, 246, 250, 253, 260, 265, 268, 269, 274, 277, 278, 280, 282, 284, 290, 305, 327, 333
Pseudogymnoascins, 234
Pseudomonas, 12, 29, 33, 40, 52, 167, 172, 174, 175, 178, 188, 190, 199–201, 203–205, 207, 209, 211, 212, 215, 222, 224, 231, 233, 236, 241, 242, 261, 273, 274, 285, 290, 291, 294, 295, 300, 302, 306, 307, 309, 321, 326, 330, 332, 335, 336, 359, 360, 388, 446, 452, 475, 479, 513
aeruginosa, 29, 33, 52, 167, 172, 174, 175, 190, 205, 207, 209, 211, 212, 222, 233, 285, 291, 302, 306, 307, 309, 321, 326, 335, 336, 359, 360, 388, 446, 452, 475, 479, 513
Pseudovibrio, 190, 200, 201, 210, 222, 262, 279, 314
Psychrobacter, 207, 211, 239, 267, 292, 330
Psychroserpens burtonensis, 232
Pufferfishes, 336
Pulicatins, 297
Purpactin, 471
Purpureocillium, 444
Pyrenocine J, 409
Pyrone, 304, 323, 343, 354, 422
Pyrrolo, 205

Q

Queensland, 103, 352
Quencher, 231
Quenching, 49
Quinolone, 241
Quorum sensing (QS), 52, 69, 70, 196

R

Radiation, 45, 71
Raspailiidae, 204
Rat glial cells, 326
Red
algae, 387
sea, 190, 206, 227, 261, 286, 300, 304
Reproductive syndrome virus, 450
Respiratory syncytial virus (RSV), 145, 453
Resveratrol, 79
Rhabdastrella globostellata, 238
Rheumatoid arthritis (RA), 24, 128
Rhizoctonia solani, 199, 201, 205, 209, 444
Rhodobacter, 232
Rhynchopetala, 368
Rivulariaceae, 46
Rockpools, 357
Rocky surfaces, 293
Roridin A, 390, 409
Rubritalea squalenifasciens, 230
Ruegeria, 209, 236, 274, 333

S

Saccharomyces cerevisiae, 40, 234, 261
Saccharopolyspora sp, 224, 325
Salinicoccus roseus, 201
Salinity, 94
Salinospora, 11, 248
Salt lakes, 29, 30
Sandy areas, 293, 297
Sargassum muticum, 167
SARS-CoV-2 virus, 36
Saxitoxin (STX), 9, 11, 17–19, 25, 38, 117, 118, 120, 123–125, 142–144
Schizasteridae, 493
Schizochytrium, 2, 371, 372
Scleroderolide, 345

Sclerodinol, 345
Sclerosporin, 345
Sclerotiniaceae, 354
Scytonemin (SCY), 41, 47, 71
Sea anemones, 158
Seagrasses, 295
Seamounts, 239
Seaside cleodendrum, 367
Seaweeds, 10, 15, 51, 128, 131–133, 135, 281
Secomanoalide, 196
Secondary metabolites, 3, 8, 10, 12, 16, 18, 179, 256
Sediment, 10–12, 15, 51, 85, 133, 150, 155, 339
Seepweeds, 180
Septoria tritici, 427, 428
Serine proteinase, 345
Sesquiterpene, 392
Sesquiterpenoids, 515
Sessile, 45, 175, 177, 184, 186, 190, 195–198, 201, 204, 206, 209, 214, 218, 220, 224, 228, 236, 240, 241, 246, 253, 260, 277–280, 282, 283, 306, 310, 312–314, 317, 318, 321, 324, 356, 358, 375–377, 382, 385, 386, 392, 395, 398–400, 404, 406, 410, 414, 423, 471, 505
  species, 293
Seychelles, 238, 261
Shallow subtidal habitats, 294
Shellfish poisoning, 17
Shells, 167, 209, 236, 281, 284, 323, 387, 412
Shewanella, 171, 174, 211, 213, 244, 274, 280, 281, 300, 330, 379
Shrimps, 286
Skin diseases, 135
Slopes, 179, 206, 261, 263, 270, 462, 472
Softcoral, 447
Solid substrates, 167
Solirubrobacter, 215
Soritids, 158
Southeast Asia, 239, 289, 366
Southern hemisphere, 34
Spain, 90, 146, 209, 234, 245, 246, 253, 255, 282, 349, 430
Spatangoida, 493
Spawn, 482
Spectroscopy, 6
Sphingomonas, 252, 275, 278, 279, 282, 283, 290, 294, 332
Spikeweed, 356
Spirolides (SPX), 9, 17, 18, 24, 121, 122, 149
Spitsbergen, 209
Spongiobacter, 190
Staphylococci, 316, 418
Staphylococcus, 12, 29, 40, 41, 45, 72, 74, 81, 82, 91, 93, 94, 103, 104, 125, 136, 140, 167, 171–173, 180, 184, 186, 188, 190, 191, 198–201, 205, 209, 211, 215, 222, 224, 230, 233, 234, 236, 239, 241, 245, 246, 254, 255, 262, 264, 267–269, 275, 277–285, 302, 309, 310, 314, 315, 321, 324, 326, 329, 335, 336, 345, 359–361, 379, 382, 383, 388, 395, 398, 400, 404, 415, 425, 428, 430, 435, 436, 444, 446, 449, 452, 454, 456, 457, 473, 477, 478, 482, 491, 503, 504, 512, 513, 520
  aureus, 12, 29, 40, 45, 72, 74, 81, 82, 91, 93, 94, 103, 104, 125, 136, 140, 171–173, 180, 184, 186, 188, 191, 198–201, 205, 209, 211, 215, 222, 224, 230, 233, 234, 239, 241, 245, 246, 254, 255, 264, 268, 269, 302, 309, 310, 314, 315, 321, 324, 326, 329, 335, 336, 345, 359–361, 382, 383, 388, 395, 398, 400, 415, 425, 430, 435, 436, 444, 446, 452, 456, 477, 478, 482, 491, 503, 504, 512, 513, 520
  pyogenes, 103, 167
Starfish, 303, 304, 491
Stelletta tenuis, 238
Stem cells, 429
Stomozoidae, 317
Streptomyces, 11, 164, 165, 172, 175, 179, 182, 183, 187, 189, 191, 212, 215, 220, 221, 227–229, 232, 233, 235, 242, 246, 247, 256, 257, 259, 265, 267, 271, 296, 297, 299, 304–306, 308, 313, 315, 322, 323, 325–327, 416, 478
  sp, 164, 165, 172, 179, 183, 187, 189, 191, 212, 227, 232, 233, 246, 247, 256, 259, 265, 271, 296, 297, 304–306, 308, 313, 315, 325–327
Suberitida, 235
Suberitidae, 235

Subtidal zone, 293, 349
Subtropical, 69, 72, 198, 236, 385
Sulfated polysaccharides, 82, 98, 145, 161
Sulfur, 2
Sundarbans, 370
Surface-enhanced Raman spectroscopy (SERS), 6
Symbiodinolide, 158
Symbioimine, 24, 158
Symbiotic
  bacteria, 12
  bacterium, 12, 222, 277
Synechococcales, 73, 76, 79, 80

T

T cell PTP (TCPTP), 510, 512
Takifugu, 336
Tanea lineata, 295
Tanzawaic acid, 510
Tasmania, 118
Temperate zone, 85
Tenachibaculum, 190
Teredinidae, 290
Terpenes, 5, 15
Terpenoids, 12
Terretrione C, 514
Terreusinone, 358
Testosterone, 326
Tetracenoquinocin, 212
Tetraene, 99
Tetraspyrole chromophore, 49
Tetrodotoxin (TTX), 11, 24, 123, 272, 273, 286, 289, 293, 295, 300, 302, 303, 305, 336
Tetronamycin, 16
Therapeutic, 66, 105
  activities, 5, 12, 125, 126, 143
  compounds, 219, 226, 232, 267, 278, 282, 295, 302, 304, 315, 319, 323, 331, 332, 349, 359, 365, 374, 390, 395, 414, 415, 421, 423, 438, 440, 472, 481, 482, 509
  functions, 9, 18, 26
  potential, 132, 133, 135
Thermomonospora, 215
Thiazoline, 297
Thiomarinol, 12
Thiopeptide, 230
Thrombosis, 101
Tilletiopsis, 444
Tissue degradation, 345
Titanium dioxide (TiO2), 7
TNF-related apoptosis-inducing ligand (TRAIL), 214
Toxicity, 57, 88, 97, 114, 150, 272, 407, 439, 474, 493, 500
Toxins, 9, 18, 19, 26, 38, 109, 111, 117, 127, 157, 252, 299
Tracheophyta, 173, 178, 359, 365
Trebouxiophyceae, 34
Tree species, 368, 370
Trichoderma, 12, 360, 361, 364, 367, 379, 385, 390, 395, 396, 409, 410, 415, 419, 430, 473, 486, 503
Trichophyton rubrum, 428
Trididemnum, 310, 312
Tropical, 80, 188, 197, 220, 224, 240, 241, 263, 288, 334, 371, 375, 390, 454, 497, 509
Trypanosoma brucei, 70, 183, 191, 227, 246, 256
Tryptamine, 330, 514
Tryptophan, 207, 476
Tuberculosis, 87, 95, 103, 165, 178, 263, 433, 504
Tumor
  cells, 109, 256
  necrosis factor α (TNF-α), 30, 49, 141, 161
Tunicate, 16, 306, 309, 310, 312–316, 319, 321–323, 326, 327, 501–515

U

Ubiquinone, 334
  Q9, 326
Ulocladium botrylis, 396
Ulocladol, 396
Ultraviolet B (UVB), 35, 89
Ulva yielded, 345
Ulvaceae, 164, 342
Ulvales, 164, 342
Undecylprodigiosin, 224
Unidentified
  bryozoan, 480
  compound, 191, 236, 499
Upper subtidal areas, 296

V

Vaccinia H-1-related phosphatase (VHR), 510
Vancomycin-resistant, 12, 171, 175, 315, 503, 506
  *Enterococcus faecalis (*VRE), 171, 385
Vascular cell adhesion molecule 1 (VCAM-1), 137
Vasoconstriction, 158
Vasoconstrictive activity, 158
Venezuela, 123, 247, 312
Versicolactone, 447, 476
Vibrio, 32, 34, 70, 88, 89, 98, 99, 103, 125, 190, 195, 199, 200, 203–205, 207, 209, 222–224, 233, 246, 250, 261, 262, 268–270, 272–274, 280, 281, 285, 289, 293, 295, 300–303, 317, 321, 322, 333, 336, 444
Vibrioanguillarum, 100
Vietnam, 298, 300, 369, 519
Virgibacillus, 205, 250, 266
Viruses, 9, 49, 96, 312
Vitamin, 33, 49, 83, 94, 101
  B12, 49, 84
Volatile organic compounds (VOCs), 211
Voltagegated Na+ channel, 38
Voltage-sensitive, 158

W

Weak
  anifungal activity, 508
  anticancer, 55
Weeds, 130
West
  Africa, 225, 251, 306, 329
  Indian species, 409
Western
  Australia, 298, 333
  Central Pacific, 184, 198, 201, 214, 220, 251, 295, 298, 356, 382, 396, 400, 404, 423
Wild jasmine, 367

X

X isochromophilone XI, 428
Xanthidae, 288
Xanthomonas sp, 294
Xanthone, 12, 191, 396, 403, 425, 426
  aspergillusone, 470
Xanthophylls, 49
Xiphosura, 289
Xylosyl esters, 230

Y

Yeast, 83, 211, 254, 473, 474
Yellow, 186, 238, 245, 247, 430, 488
  Sea, 488
Yessotoxins (YTX), 9, 17, 18, 20, 125, 126
Yield, 10, 26, 142, 233, 355, 358, 420
Yondelis, 16, 315
Yongeichthys, 336

Z

Z-coniosclerodinol, 345
Zeaxanthin, 33
Zinniol, 450
Zone, 45, 71, 73, 80, 82, 96, 103, 161, 168, 170, 171, 173–175, 177, 179, 180, 195, 201, 202, 204, 206, 209, 211, 215, 216, 220, 221, 224, 229, 232, 234, 250, 257, 263, 266, 267, 280, 281, 285, 292, 305, 342, 343, 347, 349, 350, 359, 360, 363, 388, 390, 396, 400, 404, 430, 460, 472, 498
  inhibition, 266
Zooxanthellactone, 158
Zooxanthellae, 158
Zooxanthellamide D, 158
Zooxanthellatoxins (ZTs), 158